S0-AKY-003

Interactions of Surfactants with Polymers and Proteins

Editors

E. D. Goddard

Former, Corporate Research Fellow
Union Carbide Corporation
Tarrytown Technical Center
Tarrytown, New York

K. P. Ananthapadmanabhan

Unilever Research U.S. — Edgewater Laboratory
Edgewater, New Jersey

CRC Press

Boca Raton Ann Arbor London Tokyo

Library of Congress Cataloging-in-Publication Data

Interactions of surfactants with polymers and proteins / editors E.D. Goddard, K.P. Ananthapadmanabhan.
p. cm.
Includes bibliographical references and index.
ISBN 0-8493-6784-0
1. Polymer solutions. 2. Surface active agents. I. Goddard, E. D. (Errol Desmond), 1926- . II. Ananthapadmanabhan, Kavssery P., 1952- .
QD381.9.S65P65 1992
547.70454--dc20 92-27127
CIP

This book represents information obtained from authentic and highly regarded sources. Reprinted material is quoted with permission, and sources are indicated. A wide variety of references are listed. Every reasonable effort has been made to give reliable data and information, but the author and the publisher cannot assume responsibility for the validity of all materials or for the consequences of their use.

Direct all inquiries to CRC Press, Inc., 2000 Corporate Blvd., N.W., Boca Raton, Florida, 33431.

International Standard Book Number 0-8493-6784-0

Printed in the United States 1 2 3 4 5 6 7 8 9 0

Library of Congress Card Number 92-27127

Printed on acid-free paper

THE EDITORS

Dr. E. D. Goddard received his B.S. and M.S. from Rhodes University, South Africa and his Ph.D. from Cambridge University, England, where he specialized in insoluble monolayer research. During tenure of a post-doctorate fellowship at the NRC in Ottawa he pioneered the field of direct measurements of the heat of micellization of surfactants and the notion that water structure effects provide the key explanation of the low heats observed. During his industrial career, split roughly equally between Unilever and Union Carbide, where he held the positions of Principal Scientist and Corporate Research Fellow, respectively. He continued work in surface chemistry, including studies of surfactant interaction effects in monolayers, solution and anhydrous bulk phase; surfactants in foaming, detergency, wetting and flotation; foam control; the unique properties of silicone surfactants; and silane interactions with solid surfaces. The work for which he is perhaps best known has, however, over the last 2 decades been that on polymer/surfactant systems and he is regarded as a pioneer in this field, in particular of polyion/charged surfactant systems as they pertain to phase and gelling behavior. He is chief contributor to this Monograph on Polymer/Surfactant Interactions.

Dr. Goddard has published over 125 scientific papers, edited three books, authored 25 U.S. patents and is a founding editor of the journal *Colloids and Surfaces*. He has been a frequent contributor at domestic and international meetings on surfactants and polymers; has held several positions, including that of chairman, in the Colloid and Surface Chemistry Division of the American Chemical Society; and he is a former Chairman of the Gordon Conference on Chemistry at Interfaces.

As a member of the American Oil Chemists' Society he was a co-winner in successive years, 1979 and 1980, of the Soap and Detergent Association best paper award in the category of soaps, detergents and cosmetics.

Dr. K. P. Ananthapadmanabhan (Ananth) received his B. Tech. (1974) degree from the Indian Institute of Technology, Bombay, India, and his M.S. (1976) and D. Eng. Sci. (1980) in Surface Chemistry and Mineral Processing from Columbia University, New York. From 1980 to 1982, he was a Research Associate and from 1982 to 1983 an Adjunct Assistant Professor at Columbia University. While at Columbia, Dr. Ananth was involved in a number research projects including solution chemistry of surfactants and its role in adsorption at interfaces, mineral-solution equilibria, polymer-surfactant interactions, flotation/flocculation separation of minerals and enhanced oil recovery by micellar flooding.

During 1983 to 1990, Dr. Ananthapadmanabhan was a member of the Surface Chemistry Skill Center at the Tarrytown Technical Center of Union Carbide Corporation. His research work at Carbide included biological and other novel separations using aqueous two phase systems, lubrication and polymer-surfactant interactions.

Dr. Ananthapadmanabhan is currently a group leader with Unilever Research U.S. in Edgewater, New Jersey. His current research activities include interaction of surfactants with polymers and proteins, surfactant-skin interactions, polymer adsorption and bacterial adhesion.

Dr. Ananthapadmanabhan has published over 50 papers and has 10 U.S. patents. He is a member of the ACS and the SME/AIME (Society for Mining, Metallurgy and Exploration). He was a co-recipient, in 1986, of the Taggart Award from SME for a technical paper on solution chemistry of mineral separation by flotation.

CONTRIBUTORS

K. P. Ananthapadmanabhan
Senior Group Leader, Polymer Science
Unilever Research U.S. — Edgewater Laboratory
Edgewater, New Jersey

Eric Dickinson
Professor of Food Colloids
Procter Department of Food Sciences
University of Leeds
Leeds, United Kingdom

E. D. Goddard
Former, Corporate Research Fellow
Union Carbide Corporation
Tarrytown Technical Center
Tarrytown, New York

Björn Lindman
Professor
Physical Chemistry 1, Chemical Center
Lund University
Lund, Sweden

Ulrich P. Strauss
Professor Emeritus
Department of Chemistry
Rutgers University
New Brunswick, New Jersey

Kyrre Thalberg
Doctor, Research Engineer
Physical Chemistry 1, Chemical Center
Lund University
Lund, Sweden

Matthew Tirrell
Professor
Department of Chemical Engineering and Materials Science
University of Minnesota
Minneapolis, Minnesota

Françoise M. Winnik
Xerox Research Centre of Canada
Mississauga Ontario
Canada

TABLE OF CONTENTS

Chapter 1

INTRODUCTION

E. D. Goddard

Two areas of active and continuing research are concerned with the physical chemistry of aqueous surfactant solutions and of aqueous polymer solutions. Both aforementioned solute species have unique properties and it is not surprising that their mixed solutions can reveal rather unusual interaction effects. Historically, interest in the properties of mixtures of polymers and surfactants in aqueous solution is quite old. The formation and existence of lipoprotein aggregates in biological fluids were, for example, well recognized in the early part of this century.[1] Likewise, in the foods, cosmetics, and other industrial sectors, e.g., mineral processing, it has long been appreciated that interesting and unusual effects can be obtained by employing mixtures of surfactants and polymers.

The foundations of today's activities on mixed polymer/surfactant systems were laid in work carried out in two separate areas. The first, in the 1940s and 1950s, involved protein (and, to a lesser extent, acidic polysaccharide)/synthetic ionic surfactant pairs. In these, the importance of electrical forces of attraction was easy to recognize, the interaction was generally referred to as "binding" of the charged surfactant by the macromolecule, and an awareness of changes in the conformation of protein molecules during the binding process was developed. The second, in the 1950s and 1960s, involved water-soluble synthetic polymers which were uncharged and surfactants which were again charged. Though the sites for binding the surfactant molecules on such polymers were less easy to identify, the notion of "binding" of the former persisted in this case also. It should be pointed out that interest in charged pairs has again developed in the 1970s, 1980s, and early 1990s, in systems in which the polyelectrolyte is either synthetic or natural, including various acidic and basic polypeptides.

Five years ago this author prepared a review[2,3] of the field of polymer/surfactant research covering the previous two decades of activity. In this compilation the investigative methods employed and the factors influencing the associative reactions were reviewed in detail, and an overview of extant theories of complex formation was offered. Two reasons have prompted the undertaking of the present larger review. The first is that there has been a significant increase in research activity in this field in the last five years and it has seemed desirable to broaden the previous compilation to include an account of these recent developments. The second reason concerns the decision, in the interest of limiting the length of the 1986 review, to omit the important field of protein/surfactant interactions. For the present compilation

0-8493-6784-0/93/$0.00 + $.50

this decision has been reversed in the hope of expanding the potential utility and interest of the work to embrace other fields, in particular the biosciences.

Furthermore, to set the stage for a better appreciation and understanding of the properties of polymer/surfactant pairs it was decided to add chapters on the physical chemistry of the separate components themselves, i.e., surfactants and polymers in solution. While each of these chapters covers its respective subject in some detail, it is hoped that the latter one, in particular, will help to inspire more thought and research work on polymer/surfactant systems as viewed from the vantage point of the polymer since, to date, it is true to say that most interpretations have emphasized the role and fate of the surfactant member of the pair. Proteins are recognized as being sufficiently important in their own right that separate chapters are devoted to them and to their mixtures with surfactants. In consequence of the foregoing, the undertaking has increased considerably in scope over the original reviews (which appear in their original form as Chapter 4, Parts 1 and 2) and a very practical decision was taken to expand authorship to include several other contributors who are recognized experts in their respective fields.

It is appropriate to comment briefly on some recent important developments, all of which will be elaborated upon in the text. First, in the category of interacting nonionic polymer/ionic surfactant pairs the notion of site binding of the surfactant, still useful conceptually as regards detailed binding behavior, has given way progressively to mechanisms based on a perturbation of the micellization of the surfactant by the polymer. If one examines the structure of a typical ionic surfactant micelle (spherical) one can easily appreciate that two factors unfavorable to the formation of the micelle are the electrostatic repulsion of the assembled headgroups and the residual contact of the first few peripheral carbon atoms of the hydrocarbon tails with water (see Figure 1).

The current view of polymer/surfactant interaction is that, in the complex, segments of the polymer are wrapped around the micelle to relieve both of these stresses. Refined models presented in terms of the various free energy contributions to the overall aggregation process continue to be developed, and a summary of the current status is given in Chapter 5. Another growing recognition has been that of the great importance of hydrophobic groupings in the polymer in promoting interaction with surfactants. These hydrophobic entities can be as small as methyl groups. With such hydrophobic substitution the reactivity of certain polymers can be increased to the point of promoting interaction even with (certain types of) nonionic surfactants, which have heretofore been considered relatively inert. In some polymers fluorescent probe groups, introduced covalently for mechanistic studies, themselves provide the hydrophobic centers which favor interaction with surfactants and, indeed, their site binding. A noteworthy point is the greatly expanded use of the technique of fluorescence spectroscopy, in general, to study these and related systems. In view of this development an appendant section, which describes the underlying principles and use of fluorescent dyes for studying

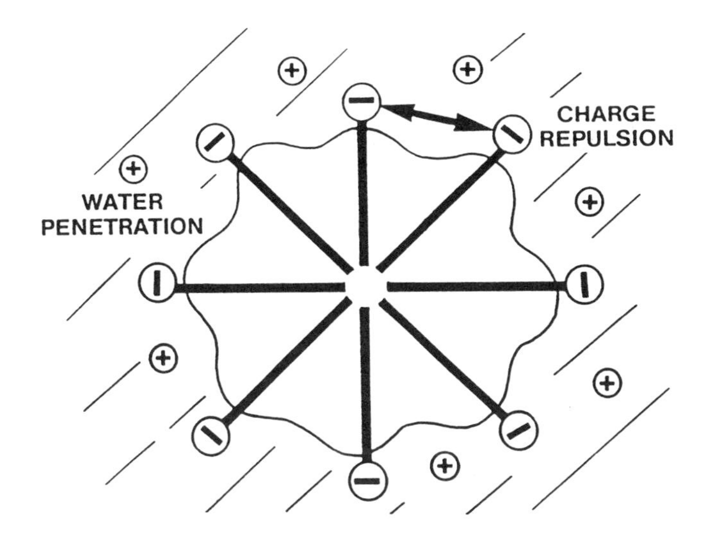

FIGURE 1. Schematic diagram of an anionic surfactant micelle.

aggregation in solution, and polymer/surfactant interaction in particular, is included as the next to last chapter of the book. It augments and expands the many references to the use of the fluorescent dye technique found in several of the other chapters.

A great deal of renewed interest is being shown in what can be termed ''polymeric surfactants'', i.e., soluble polymers in which the hydrophobic groups are long ($>C_{10}$) alkyl groups and which are akin to preformed polymer/surfactant complexes. A separate chapter is devoted to this important field with an emphasis on ''polysoaps''. Several more references to unionized, hydrophobically modified water-soluble polymers (''associative thickeners'') can be found in Chapter 10 on Applications.

Another recent development has been the detailed study, and the modeling, of the phase behavior of certain nonionic polymers, viz., those possessing cloud points, in combination with selected surfactants. Other developments include the determination, with an analysis, of detailed phase diagrams of selected polyion/surfactant counterion pairs and the demonstration of unusual rheological behavior of many of these systems. In this second category of complex, viz., that comprising a polyion and an oppositely charged surfactant, the notion of binding of the surfactant ion, reinforced by hydrophobic association, has persisted. Chapter 4, Part 2 presents an overview of this field, which is augmented in Chapter 5 by a description of new developments.

Finally, mention is made again of the new subject category which represents the final chapter of the book. This recognizes the growing practical and commercial importance of polymer/surfactant complexes under the title ''Applications''. It should be pointed out that this chapter does not include the category of proteins: in their case the utility and applications of their interactions with surfactants are so closely linked to the physical chemical aspects that they are woven directly into the text of Chapters 7 and 8.

A concluding word about the structure of the book, and the choices which seemed available, is offered in summary. One approach was, while maintaining the old format, to update and considerably expand the original review. The second was to augment it with new chapters involving new contributors. We, the authors, chose the second course. In this way the utility of the earlier review, with its extensive coverage of the early work, methods and principles, is retained while new perspectives, a broader foundation, and a comprehensive treatment of newer developments in this field have been made possible. We believe the overall compilation offered now represents a rather complete and up-to-date coverage of the field. One consequence, or ''penalty'', of the above approach is that some overlap between the chapters became inevitable. If this is considered a drawback we apologize. For our part we view this feature to be something of a positive attribute which can add to the utility, convenience, and readability of the work as a whole.

On a personal note I wish to express my appreciation to Union Carbide Corporation and Amerchol Corporation for sustained support of my research work, over many years, on the subjects dealt with in this volume.

REFERENCES

1. Lipo-Proteins, *Discuss. Faraday Soc.* No. 6, 1949. See introductory article by D. G. Dervichian.
2. **Goddard, E. D.,** Polymer-surfactant interaction. Part I. Uncharged water-soluble polymers and charged surfactants, *Colloids Surf.*, 19, 255, 1986.
3. **Goddard, E. D.,** Polymer-surfactant interaction. Part II. Polymer and surfactant of opposite charge, *Colloids Surf.*, 19, 301, 1986.

Chapter 2

SURFACTANT SOLUTIONS: ADSORPTION AND AGGREGATION PROPERTIES

K. P. Ananthapadmanabhan

TABLE OF CONTENTS

0-8493-6784-0/93/$0.00 + $.50

I. INTRODUCTION

Surfactants are characterized by the presence of two moieties in the same molecule, one polar and the other nonpolar. The polar group may carry a positive or negative charge, giving rise to cationic or anionic surfactants, respectively, or may contain ethylene oxide chains or sugar or saccharide-type groups, as in the case of nonionic surfactants. The nonpolar part of the molecule is generally a hydrocarbon chain, but may contain aromatic groups. In addition, surfactants with fluorocarbon, polypropylene oxide, or silicone(polydimethylsiloxane) hydrophobic groups are becoming increasingly popular.

The existence of groups with opposing characteristics is responsible for all the special properties of surfactants. The behavior of surfactants in aqueous solution is determined by their tendency to seclude their hydrophobic part from solution and expose their hydrophilic part towards the solution. This dual tendency is responsible for adsorption of surfactants at interfaces and for the formation of such aggregates as micelles.

A. SCOPE

The purpose of this chapter is to provide a brief review of the solution and interfacial properties of surfactants for readers interested in polymer- and protein-surfactant interactions. Since these interactions involve surfactant monomers and/or micelles, emphasis has been placed on the properties of monomers and micelles. Also, since some of these interactions are similar to surfactant adsorption at the solid-liquid interface, the latter subject is discussed here briefly. More emphasis is given to ionic surfactants, since nonionics do not interact strongly with proteins nor with ionic and nonionic polymers. By the same token, the class of surfactants referred to as "zwitterionics" will not be dealt with here. In the initial sections some general information on different classes of surfactants, including reference to relevant literature, is provided.

B. CLASSES OF SURFACTANTS

The structural formulas of various anionic, cationic, and nonionic surfactants are given in Table 1. For detailed reviews on surfactants, see References 1 to 8. Extensive listings of commercial surfactants arranged by their trade names and by chemical classifications can be found in annual editions of McCutcheon's publications[9] on detergents. Compilations of recent research papers presented at the biannual international symposia on Surfactants in Solution and dealing with such subjects as adsorption, micellization, solubilization, microemulsions, and so on, can be found in References 10 to 14.

The length of the hydrocarbon chain in most surfactants is in the range of C_8 to C_{18}. Surfactants with chain length less than eight are too soluble and above eighteen are of limited solubility.

1. Anionic Surfactants

Sulfates, sulfonates, carboxylates, and phosphates represent the most common anionic functionalities for this class of surfactant. The dissociation of the last two groups of molecules is dependent upon the pH. For example, the predominant species in carboxylate systems at pH values below pH 5 (pK = 4.95) is the undissociated acid molecule and above pH 10 is the fully ionized monomer. The unionized fatty acid molecules of long chain surfactants are relatively insoluble in water. Several studies have indicated that in the intermediate pH region where both the acid and the ionized molecules exist, formation of acid-soap complexes may occur.[15-17] These complexes have significantly higher surface activity than the components which make up the complex and this is manifested in their adsorption behavior under intermediate pH conditions.

In contrast to carboxylates, alkyl sulfates and sulfonates are salts of strong acids and therefore can be expected to remain fully ionized in the entire pH range. However, alkyl sulfates undergo a different type of slow hydrolysis in aqueous solutions at pH values below 3 to form an alcohol and bisulfite.[18]

$$ROSO_3Na + H_2O = ROH + NaHSO_3 \qquad (1)$$

TABLE 1
Structural Formulas of Various Classes of Surfactants

Type	Structure[a]	Ionization
Fatty acid soap	$R{-}C(=O){-}O^{-}\ldots Na^{+}$	Weak
Alkyl phosphate	$R{-}O{-}P(=O)(O^{-}\ldots Na^{+}){-}O^{-}\ldots Na^{+}$	Weak
Alkyl sulfate	$R{-}O{-}S(=O)_2{-}O^{-}\ldots Na^{+}$	Strong
Alkyl sulfonate	$R{-}S(=O)_2{-}O^{-}\ldots Na^{+}$	Strong
Primary amine salt	$R{-}N^{+}H_2{-}H\ldots Cl^{-}$	Weak
Secondary amine salt	$R{-}N^{+}(R')(H){-}H\ldots Cl^{-}$	Weak
Tertiary amine salt	$R{-}N^{+}(R')(R'){-}H\ldots Cl$	Weak

TABLE 1 (continued)
Structural Formulas of Various Classes of Surfactants

Type	Structure[a]	Ionization
Quaternary amine salt	$R-N^{+}(R')_3 \ldots Cl$ (R′ above and below N)	Strong
n-Alkyl xanthates	$R''-O-C(=S)-S^{-}Na^{+}$	Strong
n-Alkyl ethoxylated sulfate	$R-(O-CH_2-CH_2)_x-O-SO_2-O^{-}\ldots Na^{+}$	Strong
Na-dialkyl sulfosuccinate	$Na^{+} \ldots SO_3^{-}-CH(COO-R)-CH_2-COO-R$	
n-Alkyl dimethyl betaine	$R-N^{+}(CH_3)_2-CH_2-COO^{-}$	Zwitterionic
n-Alkyl phenyl polyoxyethylene ether	$R-C_6H_4O(CH_2CH_2O)_xH$	Nonionic
n-Alkyl polyoxyethylene ether	$R-(OCH_2CH_2)_xOH$	Nonionic

[a] R, Alkyl chain with eight or more $-CH_2-$ groups; R′, a short alkyl chain, usually a methyl group; R″, normally ethyl or amyl group.

These neutral alcohol molecules are significantly more surface active than the sulfates and are responsible for the minimum in surface tension vs. concentration plots of alkyl sulfates.[18] This aspect is examined in a later section. Alkyl sulfonates do not suffer from this instability.

In addition to the above, alkyl phosphates, phosphonates, alkyl ethoxylated sulfates, and dialkyl esters of sulfosuccinates are used widely as anionic surfactants. Another class of anionic surfactant, used mainly in mineral flotation, is referred to as thiosurfactants. Alkyl mono- and dithiocarbonates (xanthates), and mono- and dialkyl dithiophosphates belong to this category. A detailed discussion on these reagents can be found in Reference 19.

TABLE 2
HLB Ranges and Applications

Range	Application
3–6	Water in oil emulsifier
7–9	Wetting agent
8–15	Oil in water emulsifier
13–15	Detergent
15–18	Solubilizer

2. Cationic Surfactants

Alkyl amines with chain length C_8 to C_{18} are the most important surfactants in this category. Amines, i.e., alkyl derivatives of ammonia, can be obtained in primary, secondary, tertiary, or quaternary forms. While quaternary amines are fully charged over the entire pH range, all others undergo pH-dependent hydrolysis. The pK value of amines is in the range of 8 to 10. The undissociated molecules tend to exhibit low solubility. The ethoxylated variations of nonquaternary amines, however, exhibit excellent solubility under all pH conditions, but carry a full positive charge only in the acidic pH range.

As for carboxylates, it has been suggested that amines form half-neutralization (amine-aminium) complexes in the pH region where both uncharged and charged forms coexist in the system.[20-23]

Cationic surfactants with pyridinium and piperidinium groups are also of importance, especially in the pharmaceutical area.[1]

3. Nonionic Surfactants

The two main advantages of nonionic surfactants are their stability over the entire pH range and their compatibility with other surfactants. The amphiphilic nature of these surfactants is expressed in terms of an empirical HLB (hydrophilic-lipophilic balance) scale devised by Griffin.[24] This scale has been found to be extremely useful for selecting surfactants for emulsification purposes. In fact, HLB can be used to predict the potential use of a surfactant for a particular application. A broad classification of surfactants in terms of their HLB vs. properties is given in Table 2. Note that ionic surfactants also can be assigned an HLB value. Water-soluble surfactants have an HLB value higher than 13 and those with poor or no dispersibility in water have HLB values less than 6. A detailed discussion on nonionic surfactants and HLB can be found in Reference 4.

II. PROPERTIES OF SURFACTANT SOLUTIONS: ADSORPTION

As mentioned earlier, the properties of surfactants in solution are governed by their tendency to minimize the contact of their hydrophobic groups with

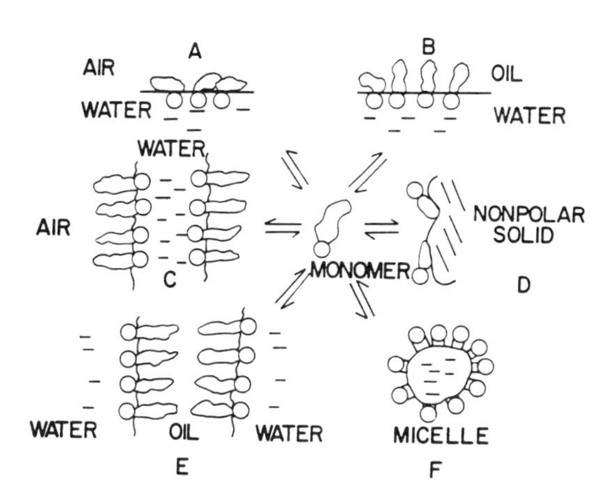

FIGURE 1. Schematic representation of surface activity and micelle formation. (a) air-water interface; (b) oil-water interface; (c) soap films; (d) adsorption onto nonpolar solids; (e) bilayers; (f) micelles. (From Mittal, K. L., Ed., *Micellization, Solubilization and Microemulsions,* Vol. 1, Plenum Press, New York, 1977, 12. With permission.)

water. This is accomplished by adsorbing at interfaces and association in solution. A schematic depiction of events[25,26] that occur because of the amphiphilic nature of surfactants is shown in Figure 1. Some of these properties, especially those related to adsorption at interfaces and micellization, will be examined in subsequent sections.

A. SURFACE ACTIVITY AT THE AIR-LIQUID INTERFACE

Surfactant molecules adsorb at the interface with their hydrophobic groups away from water and their hydrophilic groups in solution. A consequence of this is that some of the water molecules at the interface will be replaced by hydrocarbon or other nonpolar groups. Since the interaction force between water molecules and nonpolar groups is less than that between water molecules, adsorption of surfactants at the interface results in a reduction in the surface tension of the solution.

Surface tension vs. log concentration plots for sodium dodecylsulfate (SDS), dodecyl amine hydrochloride, sodium laurate, and myristyl trimethylammonium chloride are shown in Figure 2. Such plots, which are typical of surfactant solutions, exhibit a significant decrease with concentration initially, followed by a sharp break above which the surface tension remains essentially constant. The break is due to the formation of surfactant clusters referred to as micelles and the breakpoint is called the CMC (critical micelle concentration). Above this concentration, almost all of the added surfactant molecules are consumed in micelle formation and the monomer concentration does not increase appreciably. Since only the surfactant monomers adsorb at the interface, the surface tension remains essentially constant above the CMC. Thus, the surface tension can be directly related to the activity of monomers in the solution.

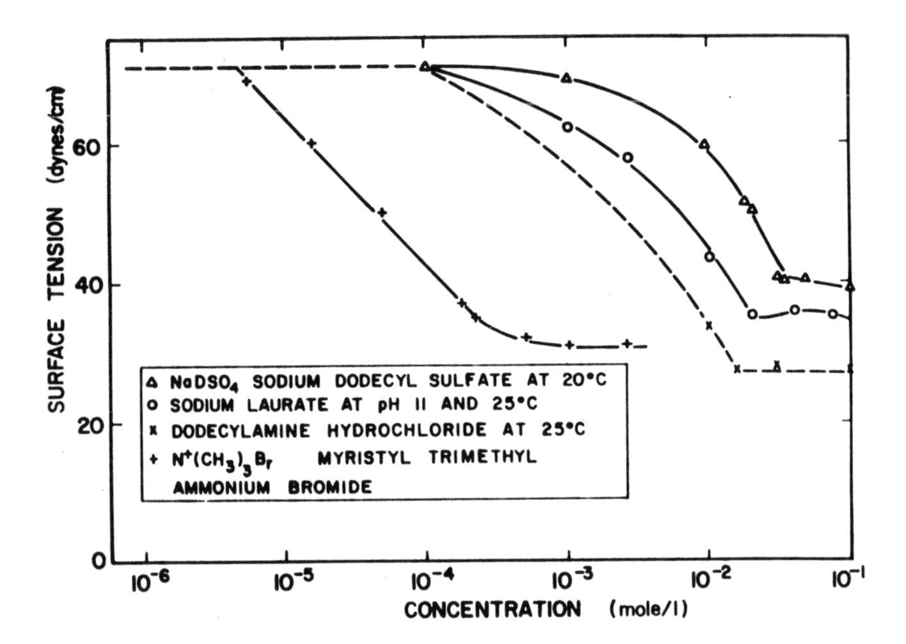

FIGURE 2. Surface tension vs. log concentration plots for sodium dodecylsulfate, sodium laurate, dodecylamine hydrochloride and myristyl trimethylammonium chloride solutions. (From Leja, J., *Flotation,* Plenum Press, New York, 1982, 270. With permission.)

1. The Gibbs Equation

This equation expresses the thermodynamic basis for the adsorption at interfaces and it provides a quantitative relationship between the activity of the molecules at the surface or interface and those in bulk solution. In its most general form the Gibbs equation for a system at constant temperature can be expressed as:[27,28]

$$d\gamma = -\sum \Gamma_i \, d\mu_1 \tag{2}$$

where $d\gamma$ is the change in surface tension of the solvent, Γ_i is the surface excess of species "i", i.e., it is the excess per unit volume present in the surface region over that in the same volume in the bulk solution, and $d\mu_i$ is the change in chemical potential of species "i". The surface excess per unit area is often referred to as Gibbs adsorption or simply as adsorption density. Note that according to this definition, the surface excess can be either positive, or negative when the solute for some reason is excluded or repelled from the interfacial region. Also, any solute which has a positive adsorption at the interface will decrease the surface tension. Similarly, a component which has a negative surface excess can actually lead to an increase in surface tension of the solvent. In general, inorganic electrolytes such as NaCl, which are known to have a negative surface excess, lead to a small increase in the surface tension of water, while surfactants which exhibit significant adsorption at the liquid-air interface decrease the surface tension substantially.

For solutions consisting of a single solute, Equation 2 can be rewritten as $d\gamma = -RT(\Gamma_0 \, d\ln a_0 + \Gamma_1 \, d\ln a_1)$ where subscripts 0 and 1 refer to the solvent and the solute, respectively. Accepting the general convention of fixing the Gibbs dividing surface such that $\Gamma_0 = 0$,[27,28] we may then write:

$$d\gamma = -RT \, \Gamma_1 \, d\ln c_1 \quad (3)$$

This equation applies to dilute solutions of nondissociating (e.g., nonionic) surfactants for which the activity coefficient can be taken to be unity.

For an ionic surfactant A^+B^- which dissociates completely into positive and negative ions,

$$d\gamma = -RT \, (\Gamma_{A+} \, d\ln a_{A+} + \Gamma_{B-} \, d\ln a_{B-}) \quad (4)$$

To maintain electrical neutrality at the interface, $\Gamma_{A+} = \Gamma_{B-} = \Gamma_1$, and we may set $a_{A+} = a_{B-} = a_1$, the mean activity. Hence for dilute solutions, neglecting activity coefficients, we have

$$d\gamma = -2RT \, \Gamma_1 \, d\ln c_1 \quad (5)$$

Note that compared to Equation 3, Equation 5 has a multiplying factor 2. In the presence of a constant amount of excess electrolyte with a common ion, the change in the activity of the counterion in Equation 5 will be negligible and, therefore, the corresponding equation will be:

$$d\gamma = -RT \, \Gamma_1 \, d\ln c_1 \quad (6)$$

The form of this equation for ionic surfactants in excess electrolyte solutions is the same as that for nonionic surfactant solutions, viz., Equation 3.

The surface excess, Γ, can be considered to be the same as the surface concentration and this information is readily obtained from the plot of γ vs. log c. Furthermore, the value of Γ under conditions of surface saturation, Γ_m, can be used to calculate the ultimate packing area of the surfactant molecule at the interface. For a typical surfactant such as SDS, the area per molecule in the absence of any added electrolyte is around 50 Å^2. An extensive tabulation of Γ_m, and its inverse, i.e., area per molecule, for a wide variety of surfactants can be found in Reference 2.

Since the changes in surface tension can be related to the activity of the monomeric surfactant species in solution, interactions of surfactant with other species/substrates which result in changes in monomer activity can be followed. The use of surface tension to monitor interactions between surfactants and polymers/proteins is described in some of the following chapters.

2. Surface Tension Minimum and the Gibbs Equation

Surface tension vs. concentration curves of a typically used anionic surfactant, SDS, often exhibit a minimum in the region of the CMC.[18,19] According to the Gibbs equation (see Equation 5), in the region of positive slope of the surface tension curve, i.e., at concentrations above the minimum, a negative surface excess is implied; i.e., no solute is present at the surface. If true, the surface tension of the solution should be the same as that of water instead of being the low value observed in the region of the minimum. It is now well known that the minimum in surface tension is due to the presence of a surface active impurity; in the case of SDS it is the lauryl alcohol. In many cases, the impurity results from hydrolysis. The explanation for the minimum is as follows. At concentrations below the CMC, the long chain alcohol adsorbs at the interface along with SDS molecules and decreases the surface tension to values lower than that of SDS alone. At concentrations above the CMC, the alcohol molecules are progressively solubilized in the micelles and therefore the surface tension increases to that of SDS alone. Thus, a minimum in surface tension vs. concentration plot of a surfactant can be taken as indication of the impure nature of the surfactant. Note, however, that the absence of a minimum does not always signify purity of the surfactant for the simple reason that addition of salt often eliminates the minimum by increasing the relative surface activity of the surfactant salt.[30]

3. Effect of Variables

a. Chain Length Effect and Traube's Rule

Solutions of surfactants belonging to a homologous series exhibit certain regularities in their surface tension behavior. Traube[28] observed that increasing the chain length of a surfactant by one $-CH_2-$ group resulted in a threefold decrease in the concentration of the surfactant required to obtain a certain surface tension reduction. The work "W" required to transfer a mole of solute from the bulk to the interface is given by:

$$W = RT \ln(c_s/c_b) \tag{7}$$

where c_s and c_b refer to concentration at the surface and in the bulk solution. Assuming that the surface concentration for a given surface tension reduction is the same, one can express the difference in work for surfactants of chain lengths n and n − 1 by:

$$W_n - W_{n-1} = RT \ln 3 = 640 \text{ cal/mol} \tag{8}$$

The value, 640 cal/mol, accordingly, represents the work required to bring one $-CH_2-$ group from bulk solution to the surface.

b. Effect of Ionic Strength

Increasing the salt concentration increases the surface tension reduction of ionic surfactant solutions. Typical results for the dependence of surface

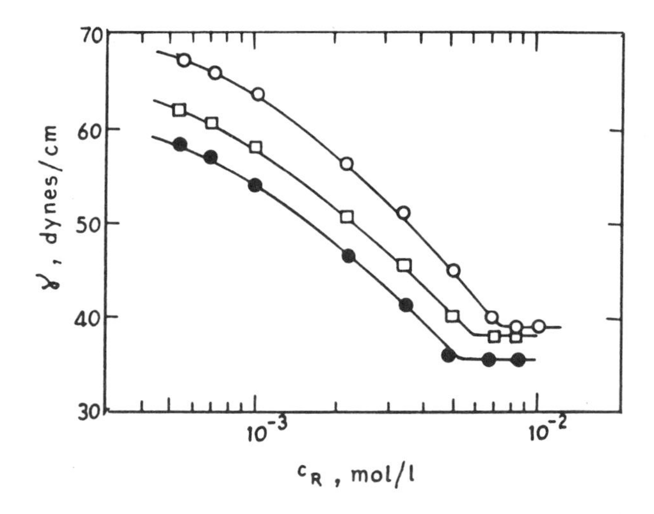

FIGURE 3. Dependence of surface tension of sodium dodecylsulfate on NaCl concentration at 25°C. ○, 1×10^{-3} *M* NaCl; □, 5×10^{-3} *M* NaCl; ●, 10×10^{-3} *M* NaCl. (From Chattoraj, D. K. and Birdi, K. S., *Adsorption and the Gibbs Surface Excess,* Plenum Press, New York, 1984, 118. With permission.)

tension of SDS on NaCl concentration are given in Figure 3.[31] The effect of electrolyte is essentially due to an increase in the activity of the surfactant salt. Also, the reduced electrical repulsion in the adsorbed monolayer will mean better packing of molecules at the interface.

c. *Effect of Temperature*

Generally, for conventional ionic surfactants, temperature does not have a major effect on adsorption at the liquid-air interface. Nonionic surfactants with oxyethylene groups, however, show an increase in adsorption with increase in temperature and this is because of the increase in their surface activity resulting from the progressive dehydration of the ethylene oxide groups.[1,4]

d. *Effect of pH*

The surface tension of solutions of hydrolyzable surfactants such as fatty acids and alkyl amines exhibits a marked pH dependence. These surfactants show maximum surface activity under pH conditions where the ionized and unionized species coexist. This has been shown to be due to the formation of ionomolecular complexes, such as acid-soaps, which are more surface active than either of the parent molecules.[15-17,20-23]

B. ADSORPTION OF SURFACTANTS AT THE SOLID-LIQUID INTERFACE

Adsorption of surfactants at the solid-liquid interface is governed not only by the solution properties of the surfactant, but also by the properties of the solid-liquid interface and interactions among the various dissolved species. The orientation of the surfactant at the solid-liquid interface is determined by the hydrophobic/hydrophilic character of the solid. On hydrophobic surfaces, surfactants adsorb with their hydrophobic tails towards the surface, making the surface hydrophilic in nature. On a hydrophilic surface, depending upon the surface chemical and electrokinetic properties, (i.e., charge characteristics) of the solid, surfactants will usually adsorb with their hydrophilic groups towards the surface. In some cases a double layer of adsorbed surfactant may form.

A number of reviews[32-37] on surfactant adsorption at the solid-liquid interface have appeared recently and therefore the subject will be reviewed only briefly here.

The free energy of adsorption of surfactants at the solid-liquid interface can be considered to be the sum of a number of contributing factors, such as hydrogen bonding, electrostatic interactions, hydrophobic interactions, and such specific interactions as covalent bonding.

$$\Delta G_{ads} = \Delta G_{elec} + \Delta G_H + \Delta G_{hydrophobic} + \Delta G_{specific} + \ldots \qquad (9)$$

Note that adsorption can occur even if some of the factors oppose the adsorption, as long as the net free energy change involved in adsorption is negative.

1. Electrostatic Forces

Almost all solids when immersed or suspended in a liquid assume an electrical charge and therefore electrostatic forces come into play in determining the adsorption properties of surfactants at the solid-liquid interface. Furthermore, the surface charge and the electrokinetic properties of the solid in an aqueous medium are markedly dependent upon such variables as pH, ionic strength, nature and concentration of specifically adsorbing species, etc.[32] The charge generation itself can be due to a number of interactions such as hydroxylation of the surface, as in the case of oxides, e.g., alumina; preferential dissolution of lattice species, as in the case of Agl; ion exchange-type mechanisms, as in the case of clay minerals; and dissolution/readsorption of lattice and hydrolyzed lattice species in sparingly soluble minerals such as calcite and apatite.[38,39] Most solids exhibit a positive charge at low pH values and a negative charge at high pH values and, therefore, carry a net zero charge at an intermediate pH, referred to as the point of zero charge (PZC). The PZC of a solid is a characteristic property which can indicate whether or not electrical forces will favor adsorption in a given system and vice versa. Obviously, electrical forces will help adsorption when the substrate and the surfactants are oppositely charged. PZC values for a number of typical solids can be found in Reference 40.

Analogous to solids and their PZC, proteins in solution and amphoteric polymers have an IEP (isoelectric point) which corresponds to the pH of zero

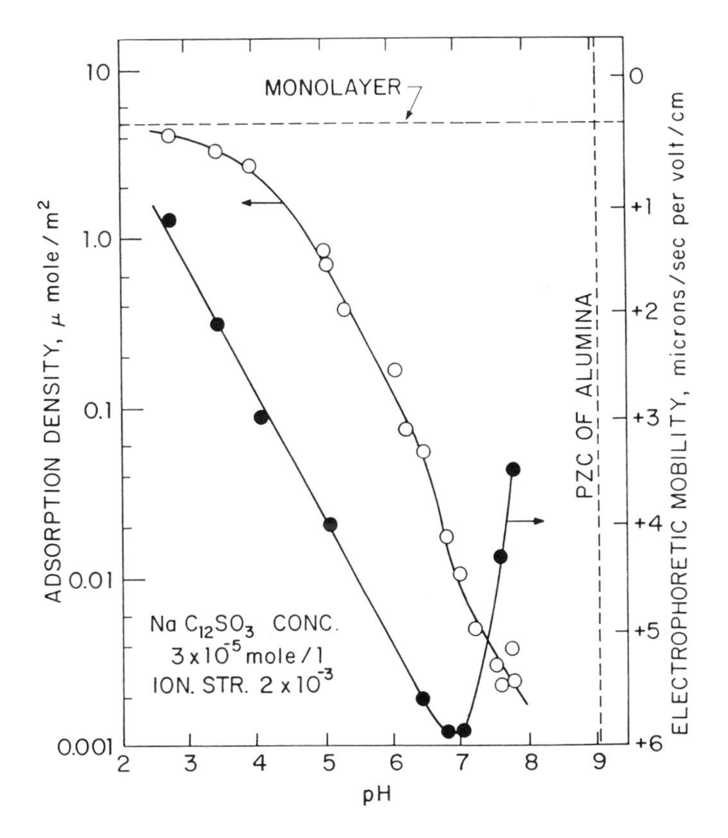

FIGURE 4. Adsorption of sodium dodecylsulfonate on alumina as a function of pH. (From Somasundaran, P. and Fuerstenau, D. W., *J. Phys. Chem.*, 70, 90, 1966. With permission.)

net charge of the molecule. At a given pH, whether or not electrostatic forces favor or oppose adsorption/binding of surfactants to proteins/amphoteric polymers is obviously determined by the relative value of this pH and the IEP.

On a typical hydrophilic oxide mineral such alumina, which has a PZC in the range of pH 7 to 9, anionic surfactants such as dodecylsulfate or dodecylsulfonate adsorb only below the PZC[41] (see Figure 4). Cationic surfactants, on the other hand, adsorb only at pH values above the PZC of the solid. Thus, electrostatic forces are primarily responsible for adsorption in these cases. At a given pH, once the adsorption density attains a certain critical value, hydrophobic chain-chain interaction among adsorbed surfactant molecules begins and this manifests itself in a sharp increase in the slope of the adsorption isotherm.[41] This aspect is examined below.

2. Hydrophobic Interactions

Typical adsorption isotherms for surfactants such as dodecyl sulfate on solids such as alumina exhibit four distinct regions[41,42] (see Figure 5). In region 1 adsorption occurs by a simple ion-exchange-type mechanism resulting

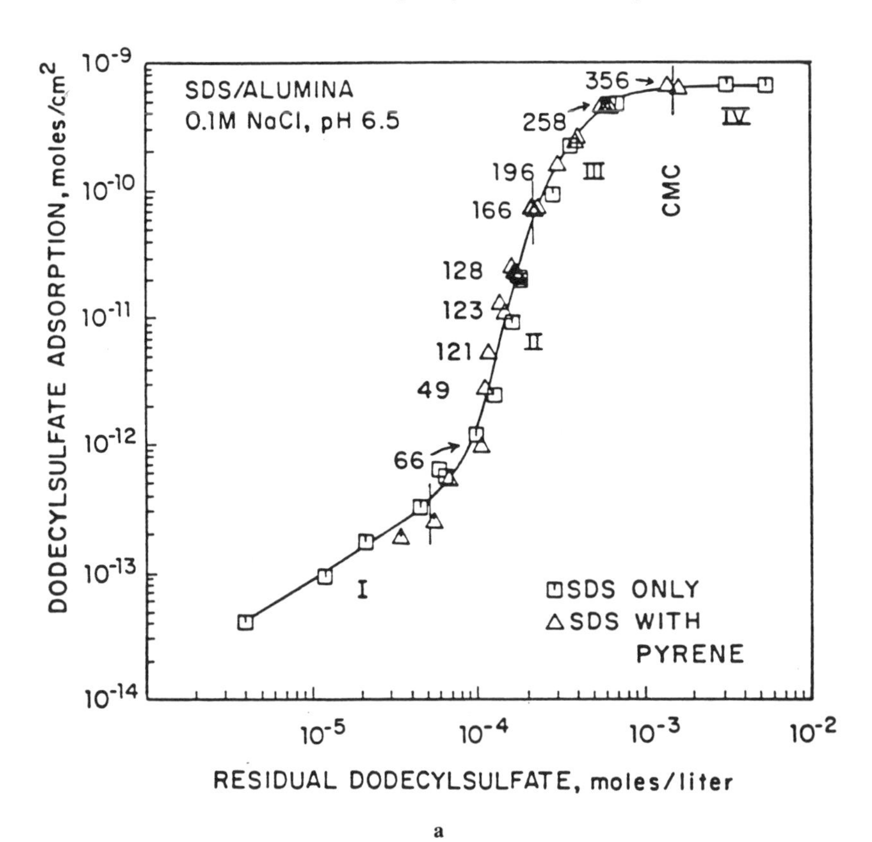

a

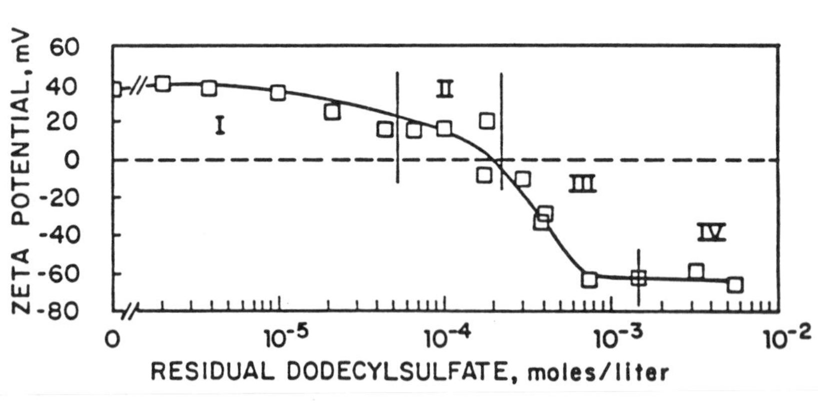

b

FIGURE 5. Adsorption of sodium dodecylsulfate on alumina at pH 6.5 at 0.1 *M* NaCl. Average surfactant aggregation number at each adsorption point is shown along the isotherm. (b) Zeta potential of alumina as a function of equilibrium concentration of SDS (designation of regions based on isotherm shape). (From Chandar, P. et al., *J. Colloid Interface Sci.*, 117(1), 31, 1987. With permission.)

from electrostatic interactions. In region 2, a marked increase in slope of the isotherm is observed. This often occurs at surfactant concentrations as low as two orders of magnitude below the CMC and has been attributed to the formation of micelle-like two-dimensional aggregates, so-called hemimicelles, at the solid-liquid interface.[41] In a recent study, Chandar et al.[42] have estimated the size of these aggregates at the solid-liquid interface using fluorescence spectroscopy (see Figure 5). The onset of region 3 corresponds to conditions of complete charge neutralization at the surface. Adsorption under these conditions occurs mainly by hydrophobic interactions and against electrostatic forces and the slope in this region is lower than that in region 2. It is speculated that surfactant molecules in this region adsorb with their hydrophilic groups towards the bulk solution and, indeed, the contact angle and hydrophobicity of particles exhibit a maximum at concentrations corresponding to the onset of region 3.[41,42] In region 4, the isotherm attains a plateau and this can be due to either surface saturation or formation of micelles in bulk solution. In the SDS-alumina system the plateau is evidently due to the formation of micelles.

In addition to the above-mentioned systems where primary electrostatic forces in combination with chain-chain interactions among adsorbed molecules result in enhanced adsorption, there are systems in which hydrophobic interactions between the substrate and the surfactant are the primary factor for surfactant adsorption. Typical hydrophobic surfaces, such as coal or carbon, are examples of such systems.

3. Other Interactions

Other interactions which can contribute to adsorption include hydrogen bonding, covalent bonding, and solvation/desolvation of adsorbate/adsorbent, and so on. The readers are referred to References 32 to 37 for additional information.

As is the case at the liquid-air interface, surfactant micelles themselves do not adsorb at the solid-liquid interface. They can, however, be considered a reservoir of monomers as they can break and reform to maintain the monomer-micelle equilibrium.

4. Effect of Variables

An increase in the chain length in a homologous series of surfactants increases the plateau adsorption and the plateau itself sets in at a lower equilibrium concentration.

In a homologous series of surfactants, the free energy change involved in the transfer of a $-CH_2-$ group from bulk aqueous solution into a hemimicelle has been estimated to be about 1.1 kT where k is the Boltzmann constant and T is the absolute temperature.[41] The latter value translates to about 650 cal/mol per $-CH_2-$ group, which is about the same as the value mentioned earlier for the transfer of a methylene group from aqueous solution to the air-water interface.

Adsorption of surfactants onto a solid-liquid interface can be altered by adding inorganic electrolytes to the system. So-called "indifferent electrolytes" which do not adsorb specifically at the solid-liquid interface and do not change the isoelectric point or PZC, reduce the surfactant adsorption in the electrostatic region and may increase the concentration at which hemimicelles would begin to form. This effect is due to the competition between the electrolyte and the surfactant for adsorption in the electrical double layer. Alternatively, it can be viewed as a nonspecific screening of the electrical charges on the solid, i.e., a reduction of its effective surface potential. Specifically adsorbing multivalent ions with the same sign of charge as the surfactant can reduce the adsorption of the surfactant. Oppositely charged multivalent ions, on the other hand, can increase surfactant adsorption.

Yet another variable which can have a significant effect on adsorption of surfactants at the solid-liquid interface is pH. The effect of pH, as discussed earlier, can be due to changes in the charge characteristics of the solid as well as the ionization characteristics of the surfactant. In general, adsorption of ionics is higher in the pH region where they are charged oppositely to the substrate. This can be seen clearly in the adsorption behavior of dodecylsulfonate on alumina (see Figure 6).[43]

III. MICELLE FORMATION IN AQUEOUS SOLUTIONS

As mentioned earlier, surfactant molecules in aqueous solutions at concentrations above their CMC aggregate to form fairly uniform sized structures referred to as micelles. The onset of micellization can be determined from a surface tension vs. log concentration plot which shows a sharp break at the CMC followed by an almost constant surface tension with increase in surfactant concentration. In fact, a number of other physical properties of surfactants, when plotted as a function of concentration, show a similar break and any of these can, in principle, be used to determine the CMC (see Figure 7). These include conductivity, viscosity, solubilization of an oil-soluble dye, and light scattering. Note that some of these are sensitive to monomer concentration and others to micellar concentration. A detailed discussion of various techniques and their reliability, together with possible variations in the measured CMC from technique to technique, can be found in Reference 44. CMC values for some of the common surfactants are given in Table 3.

All the physical properties mentioned above show a marked change over a narrow concentration region and for practical purposes a single concentration (CMC) can be defined at and above which aggregation occurs in solution. This is normally determined by an extrapolation procedure. The changes in solution behavior, generally, do not indicate the presence of any significant amounts of such premicellar aggregates as trimers and tetramers and above the CMC micelles have a definite size. Any experimentally observed deviations in solution behavior from that associated with monomers in the prem-

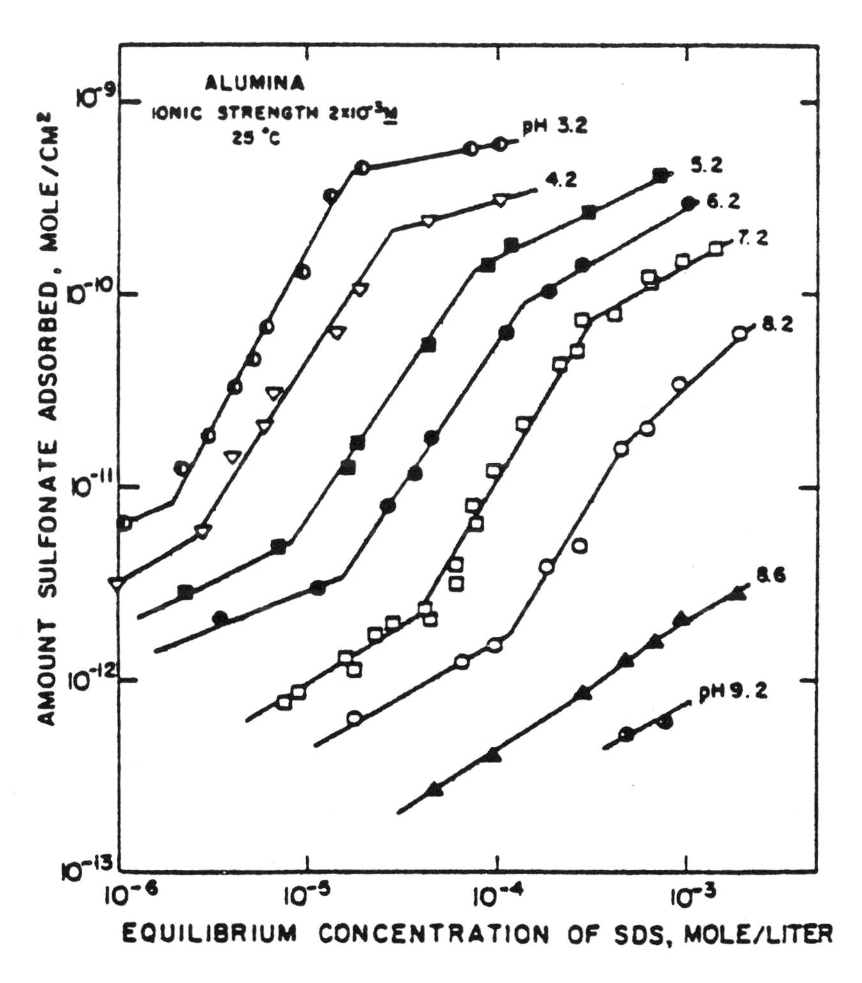

FIGURE 6. Adsorption isotherms of sodium dodecylsulfonate on alumina at 2×10^{-3} *M* NaCl at several pH values. (From Fuerstenau, D. W. and Wakamatsu, T., *Faraday Disc. Chem. Soc.*, 59, 157, 1975. With permission.)

icellar region have been attributed to the formation of dimers.[15,45] The topic of premicellar aggregation has been and still remains a controversial one, even today.[15,45-57] For all practical purposes, however, it can be considered that surfactants with a C_{12} hydrocarbon chain remain predominantly in the monomeric form up to the CMC and form micelles beyond it.

The driving force for micelle formation, as mentioned earlier, is the tendency of surfactant chains to minimize their water contact. Bringing several ionic groups together to form a micelle, however, involves overcoming the electrostatic repulsion in the head group region. Even in the case of nonionic surfactants, bringing hydrated hydrophilic groups together also involves repulsive interactions. On the basis of the ''iceberg model'' of water molecules surrounding a hydrocarbon chain,[48-50] micellization would involve release of

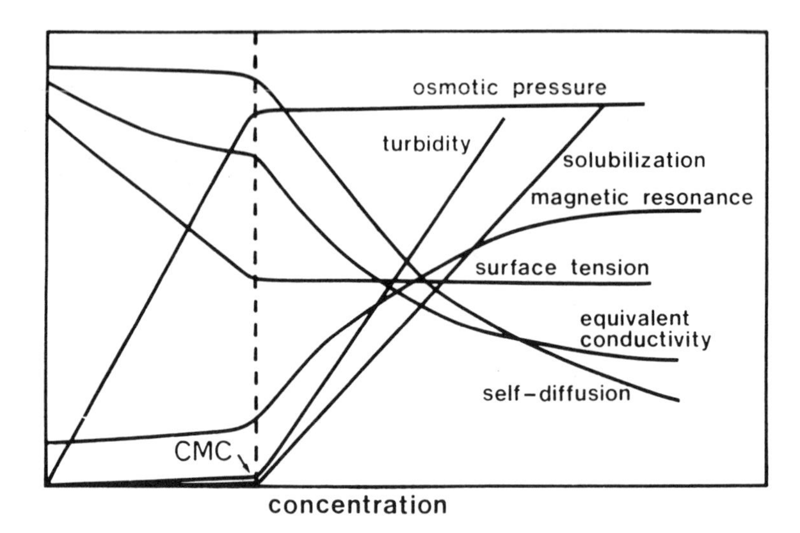

FIGURE 7. Changes in various properties of surfactant solutions around the CMC. (From Lindman, B., *Surfactants,* Tadros, Th. F., Ed., Academic Press, New York, 1984, 89. With permission.)

TABLE 3
CMC and Aggregation Number of Some of the Common Surfactants

Surfactant	CMC, mol/l T = 25°C	Aggregation number
Na octylsulfate	1.3×10^{-1}	27
Na decylsulfate	3.3×10^{-2}	41
Na dodecylsulfate	8.3×10^{-3}	64
Na tetradecylsulfate	2.1×10^{-3}	80
Dodecylammonium Cl	1.5×10^{-2}	
Decyltrimethylammonium Br	6.5×10^{-2}	39[a]
Dodecyltrimethylammonium Br	1.6×10^{-2}	55[a]
Dodecyltrimethylammonium Cl	2.0×10^{-2}	—
Hexadecyltrimethylammonium Br	9.3×10^{-4}	89[a]
Na octanoate	4×10^{-1}	—
Na decanoate	1.1×10^{-1}	—
Na dodecanoate	2.8×10^{-2}	—
$C_{12}H_{25}(OC_2H_4)_6OH$	8.7×10^{-5} [a]	400
$C_{12}H_{25}(OC_2H_4)_{12}OH$	14×10^{-5}	81

[a] Values at 20°C.

Data from References 2 and 44.

such structured water molecules, which should favor micellization. Thus, micellization essentially involves a delicate balance of several forces and interactions some of which favor and others of which oppose micelle formation. For a given surfactant, these tendencies can be expected to be dependent on its structure, chain length, charge characteristics, nature of the counterion, and so on. For example, nonionic surfactants which do not have to overcome electrical repulsion in the surface region would be expected to exhibit significantly lower CMC values than the corresponding chain length ionic surfactants and this is now well established.

The size, shape, and structure of micelles play an important role in governing their solution properties. These aspects have been investigated using a number of techniques such as fluorescence spectroscopy, light/laser/neutron/X-ray scattering, ESR, NMR and so on. An excellent review of the use of these techniques to study surfactant systems can be found in Reference 6. In the sections that follow, studies of micellar properties as investigated by various techniques is reviewed.

A. STRUCTURE OF MICELLES

Micelles in aqueous solution have essentially a "hydrocarbon-like" interior and hydrophilic groups on the outside. Well-studied micelles like those of SDS have a near-spherical geometry over a wide concentration range above the CMC. In most cases, there is no major change in shape until the surfactant approaches the solubility limit, where a liquid crystalline phase normally separates out. In certain cases, however, formation of larger micelles with increase in concentration above the CMC has been reported.

In the case of ionic surfactants, some of the counterions are bound strongly to the so-called "Stern layer" (see Figure 8) of the charged surface.[51] About 70% of the total micellar charge is neutralized by the ions in the Stern layer and the rest by the counterions in the Gouy-Chapman electrical double layer. The region within the micelle, but very close to the polar head, is often referred to as the palisade layer.

These general concepts of micellar shape and size have been around since the time of Hartley.[52] A number of recent studies have further probed into the finer details of size and shape of micelles in aqueous solutions. Some of these are examined in the following sections.

1. Nature of Micellar Core

Reference to the micellar interior as "hydrocarbon-like" stems from the observation that water-insoluble dyes and other oily materials can be effectively solubilized by micelles. Such materials often exhibit almost zero solubility in solutions of submicellar concentration, but show a marked increase in solubility with surfactant concentration, starting at the CMC. In fact, solubilization of certain standard dyes such as Orange OT is often used to determine the CMC of surfactants.[44,53] A number of recent spectroscopic studies indicate that the micellar interior is not totally devoid of water as

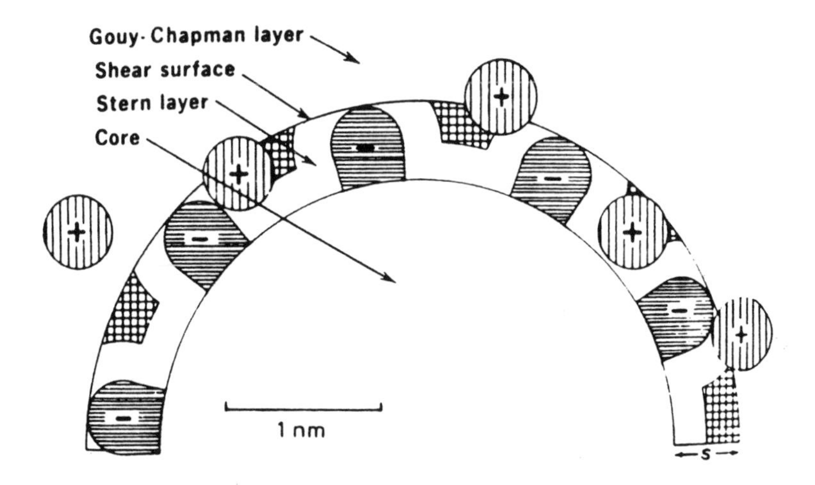

FIGURE 8. Partial cross-section of sodium dodecylsulfate micelle. Cross-hatched area in Stern layer is available to sodium ions. (From Stigter, D., *J. Colloid Interface Sci.*, 23, 379, 1967. With permission.)

implied by the description of "hydrocarbon-like" interior, in the sense that there may be some penetration of water molecules into the region of the first few carbon atoms, i.e., those closest to the head group.

An extension of the dye solubilization test to determine the CMC is the use of fluorescence probes to study micelles.[54-58] In a typical fluorescence test, for example, a relatively water-insoluble (solubility $<10^{-6}$ mol/l) probe, such as pyrene, is used to assess micellar properties. Pyrene exhibits a polarity-dependent characteristic spectrum with I_1/I_3 (the ratio of the intensity of the first peak at 373 nm to that of the third peak at 384 nm) as the indicator of polarity. The values of I_1/I_3 in water, alcohol, and hydrocarbon oil are 1.7, 1.0, and 0.65, respectively.

In a typical surfactant solution below its CMC, the I_1/I_3 value of dissolved pyrene is the same as that in water. Above the CMC, however, the probe partitions preferentially into the micelle, thereby providing a value of I_1/I_3 corresponding to that of the region where it is solubilized. In a typical SDS micelle, the above ratio has a value of 1.1 which is similar to its value in alcohol. Thus, in a plot of I_1/I_3 vs. surfactant concentration, a sharp drop in I_1/I_3 occurs at the CMC. Pyrene can, therefore, be used to determine the CMC of surfactants.[54] Another probe, pyrene carboxaldehyde, exhibits a solvent polarity-dependent emission maximum and this also can be used to determine the CMC of surfactants (see Figure 9A).[55] In such studies, it is important to ensure that the probe itself does not change the CMC (see Figure 9A) of the system and this aspect has been discussed in detail in Reference 55. In the latter study, the surface tension of surfactant solutions was measured in the absence and presence of pyrene and pyrene carboxaldehyde. The results obtained showed that the probes at levels normally used for CMC determination

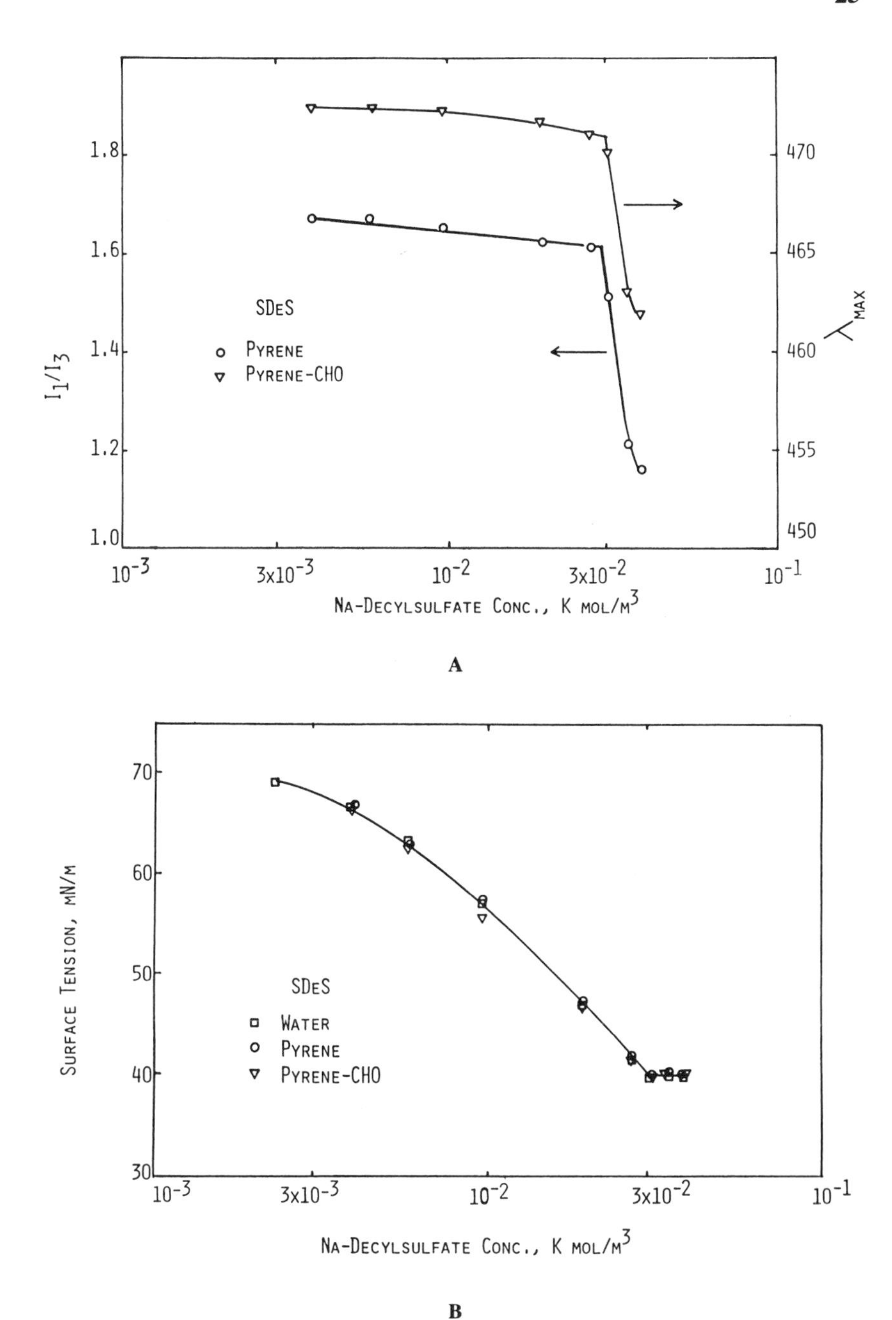

FIGURE 9. (A) Changes in the fluorescence characteristics of pyrene (I_1/I_3) and pyrene 3-carboxaldehyde (peak maximum) as a function of concentration of decylsulfate. (B) Surface tension of sodium decylsulfate in the absence and presence of pyrene and pyrene carboxaldehyde as a function of surfactant concentration. (From Ananthapadmanabhan, K. P. et al., *Langmuir,* 1(13), 352, 1985. With permission.)

did not change the surface tension behavior of the surfactant solution or the surface tension-derived CMC (see Figure 9B). This behavior is to be contrasted with the substantial effect found in the surface tension of SDS solutions upon incorporating the well-known micelle-indicator dye, pinacyanol chloride.[59] Interestingly, for dodecyltrimethyl ammonium bromide, pyrene 3-carboxaldehyde was a better probe than pyrene to reproduce the surface tension-derived CMC.[55]

As indicated above, in addition to CMC values, fluorescence probes can provide information on microstructural aspects of these aggregates. Pyrene, for example, indicates an alcohol-type environment in the region where it is solubilized. It is speculated that pyrene resides near/within the palisade layer of micelle — hence the medium value of the polarity it registers.

Probes such as pyrene have a sufficiently long lifetime in their excited state to interact with ground-state pyrene molecules to form excimers (dimers). This process is diffusion controlled and therefore the extent of formation of the excimer is inversely proportional to the viscosity of the medium. Thus, the ratio of the excimer to monomer peak (I_e/I_m) is inversely proportional to the viscosity.[57] This process, however, requires the presence of more than one pyrene per micelle and therefore the concentration of the probe has to be relatively high. A method to circumvent this problem is to use molecules which can form excimers intramolecularly. This was accomplished by Turro et al.,[58] by using dinaphthylpropane (DNP) as the probe. They reported a viscosity of 39 cP for hexadecyltrimethyl ammonium bromide micelles. Using a similar probe, viz. dipyrenal propane, Zachariasse[60] obtained a value of 19 cP for SDS micelles. In another study using a smaller probe, diphenyl propane, a much lower value (4 cP) was obtained for SDS micelles. These variations clearly indicate the difficulties involved in the quantitative assessment of the nature of the micellar core, but the estimates do show its fluid nature.

Fluorescence depolarization is another technique often used to study micelles.[61] By this approach, microviscosity of micellar core has been estimated to be in the range of 15 to 35 cP, which is in general agreement with the excimer measurements and confirms the fluid-like nature of the core.

A full description of the fluorescence technique as applied to polymer-surfactant interaction is presented in Chapter 9.

Another technique to study the micellar core involves the measurement of electron spin resonance of radicals such as nitroxides incorporated in the micelles.[62,63] The hyperfine splitting of the resonance spectrum is indicative of the polarity of the probe environment. The nitroxide radical can, for example, be attached to different positions on an alkyl chain to locate the position of that part of the chain in a micelle-like structure. Thus, stearic acid with a nitroxide group attached to carbons close to the carboxylic acid, in the middle of the chain and at the end of the chain can be used to obtain information near the head group, in the middle of the structure, as well as at the end of the chain. Again, the influence of the radical itself on the location of the probe should be considered when interpreting the observed results. Chandar

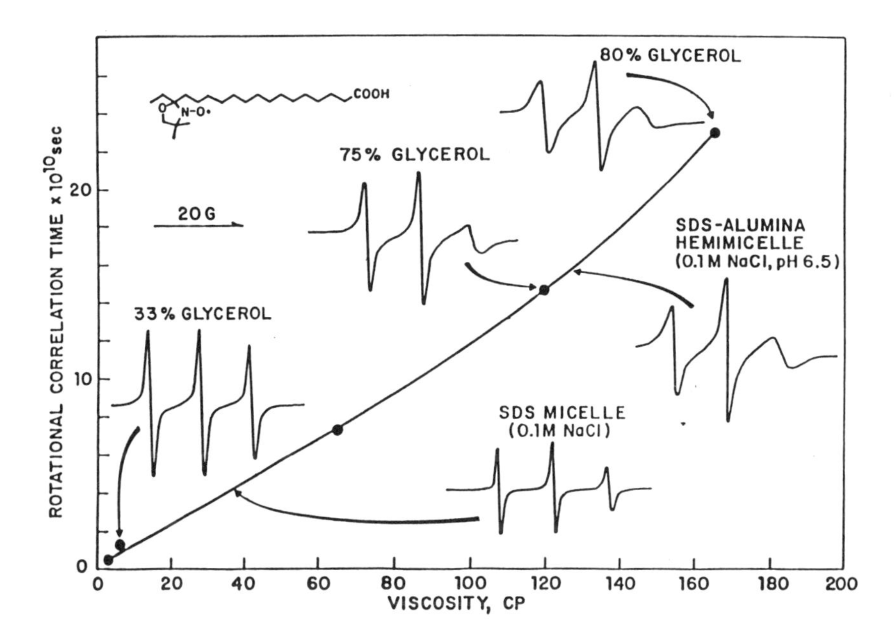

FIGURE 10. Comparison of ESR spectra of 16-doxylstearic acid in hemimicelles, micelles, and ethanol-glycerol mixtures and corresponding rotational correlation times and the viscosity. (From Chandar, P. et al., *J. Phys. Chem.*, 91, 150, 1987. With permission.)

et al.[64] have used this technique to show that the structure of SDS aggregates adsorbed onto an alumina surface is significantly more rigid and ordered than that of bulk solution micelles. Typical spectra of doxyl stearic acid in SDS micelles and hemimicelles, and the dependence of rotational correlation time with viscosity, are illustrated in Figure 10.

As mentioned earlier, one problem with the use of added probes to obtain information on microstructures is the possible perturbation of the structure because of the probe itself. One method which avoids such massive probes, yet provides molecular-level information, is NMR. The sensitivity of this technique, however, is relatively low. Use of deuterated or other labeled atoms (for example carbon, fluorine, oxygen, or nitrogen) may improve the sensitivity significantly.

A ^{13}C NMR study[65] of the spin lattice relaxation times of *n*-octyl trimethyl ammonium bromide micelles indicated a monotonic increase in methylene relaxation time from 0.9 s at the head group to 2.9 s at the tail. Thus, segmental motion appears to increase with the depth of penetration into the micelle. This is indicative of a liquid hydrocarbon-like nature of the micellar core and a fairly rigid structure near the head group.

The extent of water penetration into the micelle has been studied by several investigators using NMR.[66-68] Proton NMR[67] and spin relaxation studies[68] did not indicate significant water penetration into polyethylene oxide-based nonionic surfactant micelles. In contrast to this, solvent-dependent ^{13}C chemical shifts of carbonyl groups, either introduced as probes into the micelle as in the case of octanal into hexadecyltrimethyl ammonium bromide micelles or incorporated into the surfactant structure as in 8-ketodecyltrimethyl ammonium bromide, indicated water penetration up to as much as 7 carbon atoms from the head group.[69] However, this has been justifiably criticized because of lack of information on the exact location of the carbonyl groups within the micelle structure.[70] NMR studies of $CF_3(CH_2)_n$ surfactants also suggested significant water penetration into the micellar core.[71] Here again, the incompatibility of the fluorocarbon moiety with the hydrocarbon chains makes it difficult to clearly pinpoint the location of the CF_3 group.[72]

NMR has been used by Lindman and co-workers[73] to determine the hydration of micelles. Self-diffusion measurements of water using ^1H or ^2H, quadrupole relaxation using ^{17}O, and quadrupole splitting using ^2H are the three approaches employed by the above group to obtain information on micellar hydration. The self-diffusion data for ionic and nonionic surfactants were analyzed using a two-state model with free water molecules, and water molecules diffusing with micelles, and the results yielded hydration numbers in the range of 10 to 15 water molecules per surfactant molecule diffusing with the micelle.[74,75] This accounts essentially for the head group and counterion hydration, suggesting that significant hydration of alkyl chains does not occur in the micelle. Thus, water penetration into the micellar core, except for the first carbon atom, was ruled out by these authors. ^{17}O relaxation studies, which, according to these authors, eliminate some of the problems associated with the complex relaxation mechanism of water protons and the contributions from the exchange of the macromolecular ^1H and ^2H nuclei, also indicated exposure to water of fewer than two methylene groups of the hydrocarbon chain of the surfactant in micelles.

Note that some of the earlier micellar hydration studies,[74,76,77] using such techniques as viscosity and diffusion, also yielded hydration numbers in the range of 5 to 10 water molecules per surfactant in a micelle. These results, as well as the recent NMR results discussed above, appear to support the "hydrocarbon-like" nature of the micellar interior with no significant water penetration into the core.

2. The Micellar Surface and Counterion Binding

The ionic or ethylene oxide-type head groups of a surfactant are exposed to the aqueous phase in a micellar solution. In the case of ionic surfactants, a certain fraction of counterions is bound strongly to the micellar surface and in the process they stabilize the micellar structure. The remainder of the charges are balanced by the ion cloud in the Gouy layer. Thus, ionic micelles have a definite charge which can be measured using electrophoretic mobility. Typical zeta potential values for dodecylsulfate and dodecylammonium bromide micelles measured by tracer electrophoresis[78] are given in Figure 11.

Probe techniques have also been used to determine the electrical potential of micelles. It should be recognized that for an anionic surfactant solution,

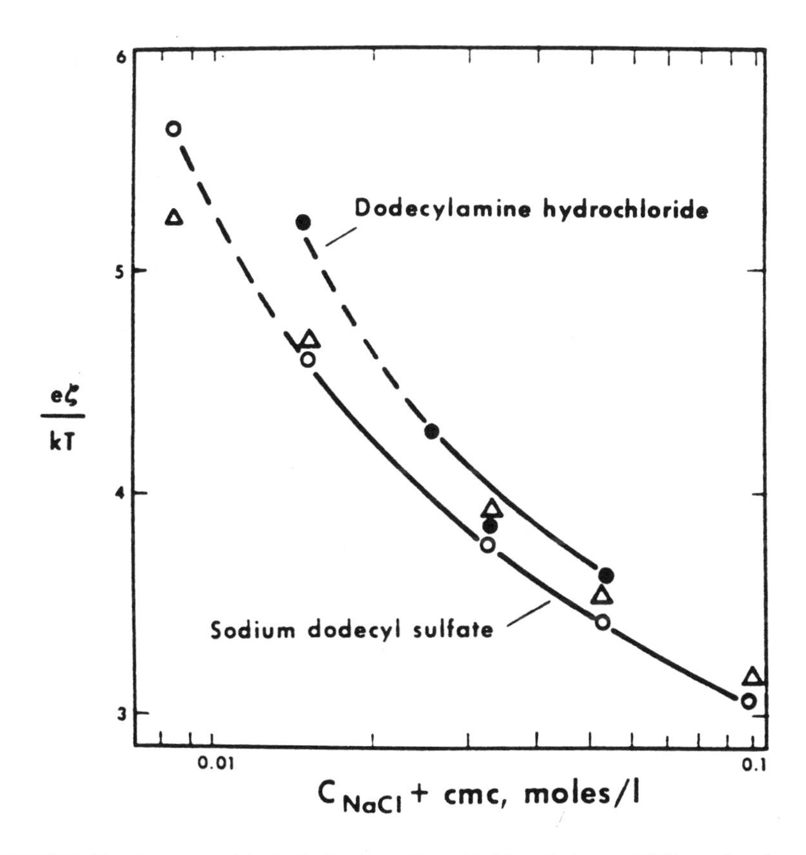

FIGURE 11. Zeta potential of micelles in sodium chloride solutions at 25°C as a function of the total counter ion concentration at the CMC. The dimensionless parameter, eζ/kT, is equal to unity when ζ = 25.7 mV at 25°C. (From Stigter, D., in *Physical Chemistry: Enriching Topics from Colloid and Surface Science,* van Olphen, H. and Mysels, K. J., Eds., Theorex, La Jolla, CA, 1975, 301. With permission.)

hydrogen ions, as well as the counterions of the surfactant, will be concentrated in the region of the micelle (or an adsorbed layer of surfactant such as the one at the liquid-air interface) and this will cause a lowering of pH in these regions.[79] Attempts have been made to determine the surface pH using probes and relate them to the surface potential. Mukherjee and Banerjee,[80] for example, using bromophenol blue and bromocresol green as indicators, determined the apparent pK of indicators when solubilized in ionic micelles and attributed the shift in the pK to the surface pH. Assuming a Boltzmann distribution of hydrogen ions at the surface, they estimated the potential at the surface. In such studies, as has been pointed out by Fernandez and Fromherz,[81] the shift in pK due to electrical effects should be separated from those due to dielectric changes near the micelle surface. Using this approach and estimating the dielectric effects from nonionic surfactant studies, the latter

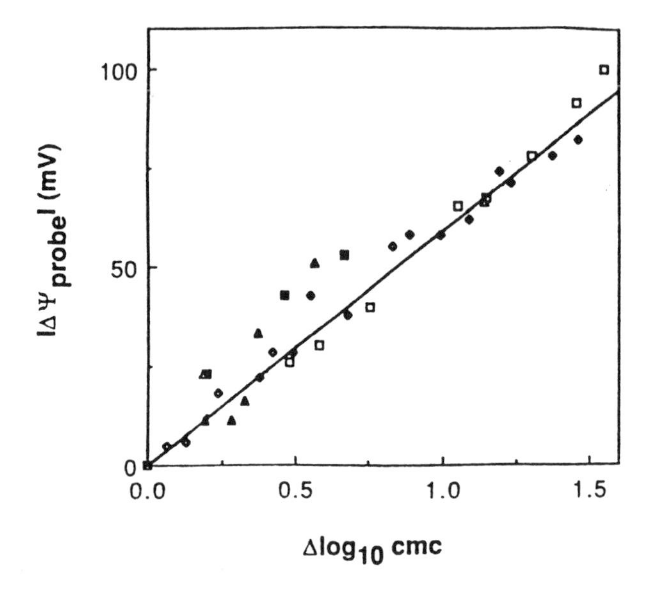

FIGURE 12. $|\Delta\Psi_{probe}|$ as a function of Δlog CMC for micellar DTAB/NaBr (□); DTAC/NaCl (◆); SDS/NaCl (◇); SDS/NaCl (Δ); SDS/$NaClO_4$ (■); and SDeS/$NaClO_4$ (▲). The line represents a gradient of 59 mV/unit change in Δlog CMC. (From Healy, T. W. et al., *Langmuir,* 6, 506, 1990. With permission.)

authors estimated a surface potential of -134 mV for SDS and $+148$ mV for hexadecyltrimethyl ammonium bromide micelles.

In a recent study, Healy et al.[82] have measured the surface potential of selected anionic and cationic surfactants in different electrolyte solutions using acid-base indicator probes (see Figure 12). The systems studied include dodecyltrimethyl ammonium bromide-NaCl/NaBr, sodium dodecylsulfate-NaCl/$NaClO_4$, and sodium decylsulfate-$NaClO_4$. Their results show that, for all of the systems studied, the potential at the micelle surface changes by 59 mV for a tenfold change in CMC, i.e., for a decade change in monomer concentration. This indicates that micelles behave like a bulk phase, or a membrane electrode, and that the surfactant monomer acts like a potential-determining ion for the surface. This interesting correlation allows one to make estimates of the surface potential of micelles in various salt solutions, provided the potential in water and the variation of CMC with the salt concentration are known.

The dielectric constant at the micelle surface can be expected to lie between that of water (78) and of hydrocarbons (2). Fluorescence measurements by Kalyanasundaram and Thomas[83] using pyrene carboxyaldehyde as the probe indicate values in the range of 15 to 50 for several ionic and nonionic surfactants.

Values of surface potential and of zeta potential are important in interactions of micelles with other species and substrates in the system. The extent of counterion binding to micelles, which determines the micellar potential, is an important and experimentally measurable parameter. A number of techniques, such as conductivity,[84] ion selective electrode measurements,[85] light scattering,[86] osmometry,[84] and NMR,[87] have been employed to determine it. These studies indicate that, for most cases, about 50 to 80% of the charges are neutralized by the counterions in the Stern layer and the remaining counterions are distributed in the Gouy-Chapman electrical double layer.

Note that the estimates of counterion binding can be expected to depend upon the technique used for measurements.[88] Surface potential measurements using probes represent values very close to the head group region, possibly at the Stern layer, as argued by Healy et al.[82] Values derived from zeta potential and conductivity measurements should be higher since they will include Stern layer counterions and also those up to the shear plane. Ion-selective electrode values on the other hand, will represent counterions in the Stern layer, shear plane, as well as some in the Gouy layer. Thus, the extent of binding as measured by different techniques will indicate the order to be: ion selective electrode measurements > zeta potential/conductivity > pK indicator/fluorescence/UV probes.

Counterion binding tends to show a minor, but measurable, decrease with increase in the concentration of the surfactant above the CMC. Interestingly, for alkyl trimethyl ammonium bromides (see Figure 13), the binding was found to increase, as indicated by a decrease in the degree of micelle ionization, α, with increase in chain length of the surfactant.[89] This has been attributed to increased hydrophobic interaction with increase in chain length and the consequently closer packing of the head groups. Increase in size of counterions reduced their binding in the case of dodecanoate micelles.[90] Lindman's results showed that binding of Na was more than that of Li for dodecylsulfate and this is in line with the larger hydrated radius of Li. Mukherjee et al.[91] demonstrated an increase in the CMC of various alkali dodecylsulfates with an increase in the size of hydrated alkali ion and found a greater degree of dissociation for Li than for Na and these are in harmony with the effects of various alkali metal salts on the CMC reported earlier by Goddard et al.[92]

Several investigators have reported that the degree of counterion binding to a micelle surface shows only a minor dependence on concentration of an added electrolyte.[51,88,93,94] This has been interpreted by Lindman et al. as suggesting that a counterion condensation type phenomenon described by Manning[95] for polyelectrolytes in solution occurs in micellar solution. According to this theory, developed for linear polyelectrolytes, counterion binding occurs to an extent that reduces the charge density to a certain critical value which is independent of temperature and added electrolyte. One of the limitations of this model is that it treats the charges as line charges and does not take into account their finite size. Gunnarsson et al.[96] and Jonsson and Wennerstrom[97] have recently developed a model for four different geometries taking into account the finite size of the charges. The counterions, however, are assumed to be point charges in this treatment. The model does predict

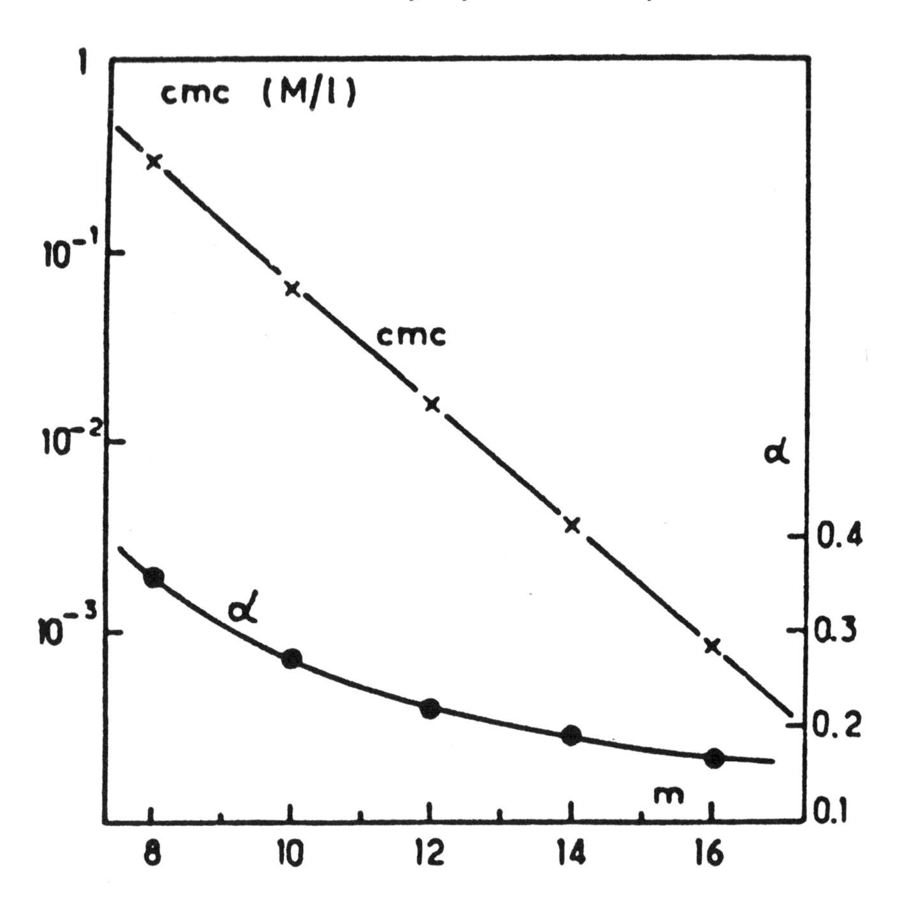

FIGURE 13. Variation in α (degree of micelle dissociation = 1 − fraction of counterions bound) and CMC for alkyltrimethyl ammonium bromides at 25°C. (From Zana, R. et al., *J. Colloid Interface Sci.*, 80, 208, 1981. With permission.)

the condensation type of behavior for micellar systems with no significant dependence of counterion binding with increase in electrolyte concentration.

Ion specificity effects among monovalent counterions are expected to be small when compared to analogous effects between ions having different levels of charge. For example, divalent counterions will generally displace monovalent ions from the surface of a charged micelle.

A factor with marked influence on the counterion binding is the size of the polar head of the surfactant. An increase in the size of the polar head reduces the counterion binding significantly.[89] For a series of anionic surfactants, Gustavsson and Lindman[98] report the shielding of Na^+ at the micellar surface to follow the order $-OSO_3^- >$ aryl-$SO_3^- >$ alkyl-$SO_3^- > -COO^-$ and this has been interpreted in terms of hydration sphere overlap and head group-induced polarization of the sodium ion hydration sphere. Since complete inversions of counterion sequences can be observed as regards their effect on surfactant properties such as CMC, Krafft point, surface tension lowering, etc., as the surfactant ion type is changed (e.g., $-SO_4^-$, $-PO_4^{2-}$, $-CO_2^-$) caution must be exercised in attempting to develop facile explanations of these counterion sequence effects. A summary of earlier work on this subject is provided in Reference 99.

B. MICELLAR SHAPE AND AGGREGATION NUMBER

McBain,[100] who coined the term "micelles" to describe the association structures in surfactant solutions, first proposed the shape of ionic micelles to be spherical and of "neutral" micelles to be lamellar. Hartley,[52] in his classical monograph, advanced the view that spherical-shaped micelles with aggregation numbers in the range of 50 to 100 adequately explain the properties of surfactant solutions. Recent studies and models[101] do indicate that the shape of micelles is not exactly spherical, but is fairly close to it in most cases. Evidence exists also for the formation of cylindrical and other shapes in certain systems.[102-105]

Note that micelles are essentially dynamic entities and, therefore, cannot be considered to have a rigid structure. It is, nevertheless, meaningful to discuss their average shape and size.

On the assumption that the core of micelles is hydrocarbon liquid-like without any "holes", theoretical models have been developed to predict their possible shapes. This restriction clearly puts a limitation on one of the micellar dimensions to be less than twice the length of the surfactant molecule. Schott,[106] for example, has shown that the range of theoretical aggregation numbers for spherical micelles of different chain length is restricted and that most of the experimentally determined values are above these calculated numbers, suggesting that micelles are not exactly spherical. To overcome this type of difficulty, other near-spherical shapes, such as globular or ellipsoidal, have been considered. Tanford[101] demonstrated that small changes in the ratio of minor to major axis (a_0/b_0) for an oblate or prolate ellipsoid can accommodate a significant increase in aggregation numbers over those estimated for spherical geometry. Tanford has argued in favor of an oblate ellipsoid over prolate ellipsoid for ionic surfactants.

Israelichvili et al.[107] and Mitchell and Ninham[108] developed a detailed theory of aggregation in surfactant solutions taking into account the shape and size of surfactants. They define a critical linear dimension l_c which is less than R, the length of hydrocarbon chain or the radius of a spherical micelle with no hole in the core. Their analysis leads to the conclusion that l_c is related to the shape, volume (v), and the surface area (a) and these relations can be expressed as:

Value of $v/(a\ l_c)$	Structure of micelle
$\leq 1/3$	Spherical in aqueous media
1/3–1 1/2	Cylindrical in aqueous media
1/2–1	Bilayers/vesicles in aqueous media
≥ 1	Reverse micelles in nonpolar media

For a linear chain, Tanford,[109] using the X-ray data of Reiss-Husson and Luzzati,[103] has obtained the following relationships for the hydrocarbon chain volume, v, and the critical length, l_c:

$$v = 27.4 + 26.9 \text{ nA}^3 \quad (12)$$

$$l_c \leq 1.5 + 1.265 \text{ nA} \quad (13)$$

where n is close to, but smaller than, the number of carbon atoms in the chain (usually taken as the number of carbon atoms in the chain minus one). For a fully extended saturated chain, l_c may be 80% of the chain length. The total volume of the aggregate (V) and surface area (A) are related by: $V/v = A/a = N$, the aggregation number of the micelle.

In short, surfactants with bulky hydrophilic groups and long, thin hydrophobic groups tend to form spherical/globular micelles in aqueous solutions. In contrast to this, surfactants with bulky hydrophobic groups and small hydrophilic groups tend to form lamellar or cylindrical micelles. While there is general agreement with regard to the general shapes and molecular dimensions of micelles, the exact shape continues to be an area of controversy and is providing challenges for continued research.

The shape of a micelle, as mentioned earlier, is related to its size. The size is usually expressed in terms of an aggregation number which can be determined in a number of ways. Some of the classical methods include light scattering,[6,110] and diffusion[111] and sedimentation rates in ultracentrifugation.[112,113] Other scattering techniques[6,114-117] such as X-ray scattering, quasi-elastic light scattering, and neutron scattering have also been used. NMR has been used successfully by Lindman[118,119] to determine aggregation numbers of ionic and nonionic surfactants. A relatively convenient technique which has become popular during the last 10 years is the use of fluorescence probes.[6,89,120-122] Both static and dynamic fluorescence techniques can be used to obtain aggregation numbers and the latter is considered to be superior to the former.[6] Equations allowing such calculations are described in Chapter 9. Zana[6] has recently reviewed the merits and demerits of various fluorescence techniques.

Note that fluorescence and NMR measurements involve determining the concentration or the number of micelles. This, along with the information on the total amount of surfactant in the micellar form, is used to estimate the aggregation number.[123] In contrast to this, scattering techniques provide information on the hydrophobic-hydrophilic interface and, therefore, an average volume occupied by the hydrocarbon core and a "dry radius". Quasi-elastic scattering techniques, on the other hand, yield a "wet radius". For most conventional surfactants with a single alkyl chain, the radius deduced from such measurements corresponds closely to the fully extended length of the surfactant chains.[123] The difference between the dry and the wet radius is rather small, indicating the absence of any long-range perturbing effect of micelles on the solvent.

Typical aggregation numbers obtained for commonly used surfactants are given in Table 3. Size and aggregation number of surfactants change markedly with such variables as ionic strength and temperature. Typical variation in N for sodium dodecylsulfate, and dodecylammonium chloride (DDACl) with salt concentration is given in Figure 14A. It is interesting that DDACl exhibits a very much larger increase in size with salt concentration than SDS with an identical number of $-CH_2-$ groups. The reasons for this are not clear at present, but it does demonstrate the important role of the polar group in the aggregation process. Variations in the aggregation number of several surfactants are given in Figure 14B.[124] Evidently, for ionic surfactants the aggregation number decreases and for nonionic surfactants it increases with increase in temperature. The reasons for this opposite behavior arise from entropic considerations and will be discussed in a later section.

C. FACTORS AFFECTING THE CMC AND AGGREGATION NUMBER

Micelle formation is accompanied by marked changes in the solution properties of surfactants and, in consequence, a great deal of work has been done to determine the CMC of surfactants[44] under a variety of conditions to establish the factors that influence the CMC and the micellization process. Chief among the factors affecting the CMC are the nature of the hydrophobic and hydrophilic groups, the electrolyte concentration, and temperature.

1. Hydrophobic Group

An increase in the length of the hydrocarbon chain in the unbranched portion of the chain decreases the CMC. For a homologous series, the dependence of CMC on hydrocarbon chain length, n, can be described by an equation of the type:[125]

$$\log(\mathrm{CMC}) = A - nB \tag{14}$$

where A and B are constants dependent on the polar group and the hydrocarbon chain, respectively. Note that this form of equation can be justified readily from the thermodynamic considerations of micelle formation and this aspect is discussed in a later section. B has a value approximately equal to 0.3 for ionic surfactants at 35°C.[2]

Nonionics and zwitterionics have a B value of about 0.5. Typical values of A and B for some of the commonly used surfactants are given in Table 4. These correlations can be approximated to make the following general conclusions regarding the variation of CMC with chain length.[2] An increase in chain length by two methylene groups results approximately in a decrease in CMC of ionic surfactants by a factor of 4 and of nonionic surfactants by 10. This relation holds up to a chain length of about 16 carbons. When the

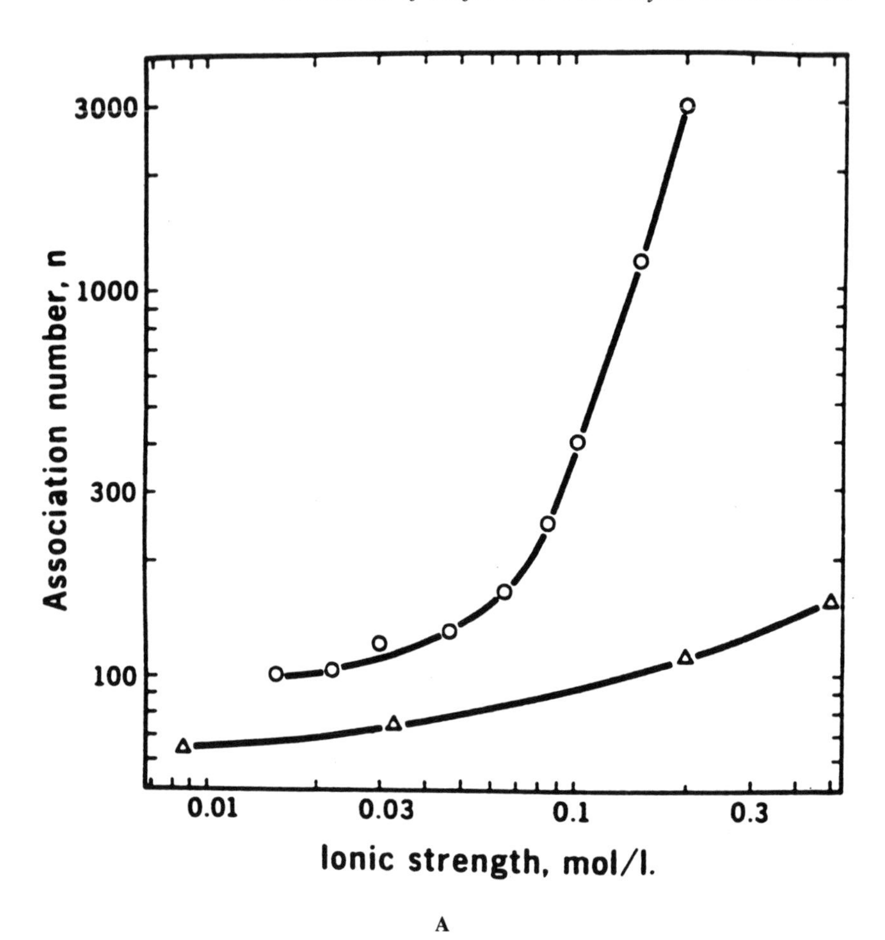

A

FIGURE 14. (A) Aggregation number of SDS and dodecylammonium chloride micelles in NaCl solutions as a function of ionic strength. Δ, SDS; ○, dodecylammonium chloride. (From Stigter, D., in *Physical Chemistry: Enriching Topics from Colloid and Surface Science,* van Olphen, H. and Mysels, K. J., Eds., Theorex, La Jolla, CA, 1975, 192. With permission.) (B) Temperature dependence of the micellar aggregation number, N: (A) effect of ionic head; (B) effect of surfactant chain length; (C) effect of counterion; (D) effect of nature of surfactant. Symbols: (●) Dodecylammonium chloride; (×) dodecylmethylammonium chloride; (○) dodecyldimethylammonium chloride; (+) dodecyltrimethylammonium chloride; (▲) tetradecyltrimethylammonium chloride; (□) cetyltrimethylammonium chloride; (△) tetradecyltrimethylammonium bromide; (▽) dodecyltrimethylammonium bromide; (◑) Triton X-100; (■) SDS; (◒) dodecyltrimethylammonium propane sulfonate. (From Malliaris, J. et al., *J. Phys. Chem.*, 89, 2709, 1985. With permission.)

hydrocarbon chain length exceeds 16 methylene groups, the CMC reduction is less rapid and this is speculated to be due to coiling of the chains and the consequent lower tendency for micellization. Formation of premicellar aggregates such as dimers also will be expected to increase the CMC of surfactants.

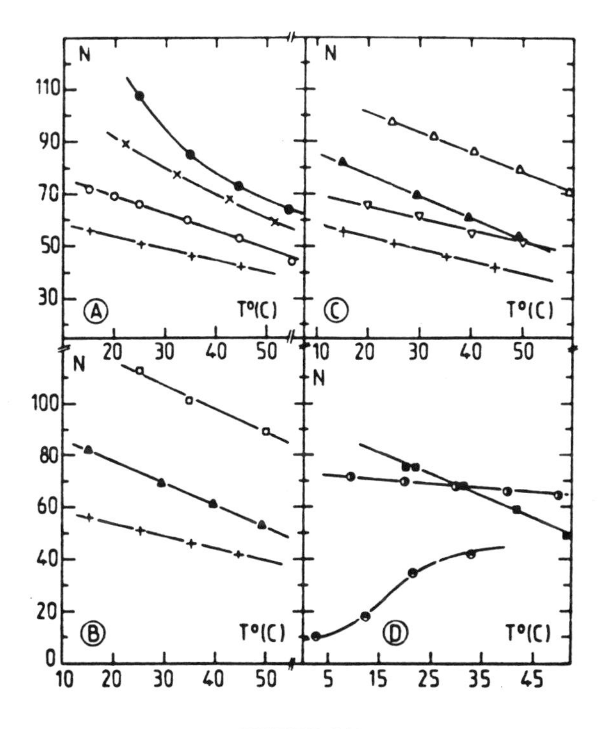

FIGURE 14B

TABLE 4
Values of A and B in Equation for the Dependence of CMC on Chain Length

Surfactant series	Temperature, °C	A	B
Na carboxylates	20	1.8	0.3
K carboxylates	25	1.9	0.29
Na(K) *n*-alkyl sulfates or sulfonates	25	1.5	0.3
n-Alkyltrimethyl ammonium bromides	25	2.0	0.32
n-Alkyltrimethyl ammonium chlorides (in 0.1 *M* NaCl)	25	1.2	0.33
n-Alkyl pyridinium bromides	30	1.7	0.31

Note: log CMC = A − nB.

From Rosen, M. J., *Surfactants and Interfacial Phenomena,* 2nd ed., John Wiley & Sons, New York, 1989. With permission.

Linear hydrocarbon surfactants tend to have a much lower CMC than their branched-chain counterparts. A methyl group in a branched hydrocarbon chain has about one half the effect of a methyl group in a straight chain in lowering the CMC.[2] The presence of a double bond in the chain increases the CMC. Similarly, introduction of polar groups such as –O– or –OH– into the chain normally increases the CMC.

The presence of a benzene ring in the hydrophobic group lowers the CMC and its effect is equivalent to that of about 3.5 methylene groups. For the same chain length, fluorocarbon chains exhibit a lower CMC than the corresponding hydrocarbon surfactant. For partially fluorinated chains the effect may be different. For example, substitution of the terminal $-CH_3$ group by a $-CF_3$ group was found to increase the CMC.[2] This has been attributed to incompatibility of fluorocarbons with hydrocarbons.

2. Hydrophilic Group

In aqueous media, ionic surfactants have a much higher CMC than nonionic surfactants with a corresponding chain length. This result is expected, since the aggregating nonionic molecules do not have to overcome the adverse effect of electrical repulsion during micelle formation. Correspondingly, as the number of ionizing groups increases in a surfactant molecule the CMC increases.

As the position of the ionic group changes from the terminal position to more central positions, the CMC increases. For a given chain length, the CMC increases as the ionic charge on the head group moves closer to the α carbon of the alkyl chain. This has been attributed by Stigter[126] to an increase in the self-potential of the surfactant ion and the consequent effects on the energetics of micellization. Geometrical effects in the packing of the chains are clearly involved. For a given chain length, variation of the actual ionic headgroup present in the surfactant can markedly affect the CMC. The CMC in alkyl ionic surfactants follows the order: aminium salts > carboxylates > sulfonates > sulfates.[126]

In general, the introduction of EO groups in a molecule makes it more hydrophilic and, in a homologous series of nonionic surfactants, the CMC will increase with the number of EO groups. However, in the case of ionic surfactants, the presence of the first few EO groups between the ionic group and the alkyl chain, as in ethoxy sulfate surfactants, actually reduces the CMC.[127] This may result from a combination of more facile packing of the aggregated molecules and more effective self-screening of the charged groups in the palisade layer. The latter effect possibly arises from the strong hydration of EO groups.

3. Effect of Counterions

As mentioned earlier, counterions adsorb at the ionic micelle surface and stabilize it and the extent of adsorption is dependent on the micelle charge density. The higher the adsorption of counterions, the lower the CMC. The extent of adsorption of counterions depends also on the properties of the counterions. For example, an increase in the valency, as well as in the polarizability of counterions, decreases the CMC. An increase in hydrated radius, on the other hand, increases the CMC. For dodecylsulfates, for example, the CMC decreases in the order:[91,92] $Li^+ > Na^+ > K^+ > Cs^+ > Mg^{2+} > Ca^{2+}$. For cationic dodecyltrimethyl ammonium and pyridinium salts the order of decrease in CMC[44] is $F^- > Cl^- > Br^- > I^-$.

4. Effect of Additives

As indicated previously, electrolytes have a significant effect on the CMC of ionic surfactants. An increase in the concentration of electrolytes progressively contracts the electrical double layer around the micelle and in the process stabilizes the micelle. This is manifested as a reduction in CMC with electrolyte concentration. Note, however, that an increase in the electrolyte level may not increase the amount of counterions bound in the Stern layer of the micelle.[88,123] The dependence of CMC on electrolyte level can be described using the following equation:[128]

$$\log \mathrm{CMC} = -A \log C_e + B \tag{15}$$

where A and B are constants and C_e is the total concentration of the counterions. This equation also can be derived from thermodynamic considerations, as shown in a later section. Zwitterionics and nonionics exhibit less dependence on electrolyte levels than ionic surfactants and the corresponding relationship is given by:[123]

$$\log \mathrm{CMC} = -K\, C_s + \text{constant} \tag{16}$$

where K is a constant for a particular surfactant-electrolyte system at a given temperature and C_s is the salt concentration. The effect on nonionic surfactants is essentially due to the salting out or salting in of the hydrophobic or hydrophilic groups by the electrolytes. These considerations stem from the observations that electrolytes lower the solubility of hydrocarbons[48] in water and may salt out or salt in the EO groups. Note that electrolytes have a much more significant effect on the cloud point than on the CMC of nonionic surfactants. In general, salts which lower the cloud point can be expected to lower the CMC, and those which raise the cloud point, to raise the CMC.

Organic additives with alkyl chains can also have a significant impact on the CMC of ionic and nonionic surfactants.[121,130] Short-chain alcohols (ethyl, propyl, butyl, etc.) tend to adsorb near the head group region, with the alcohol groups in between the ionic groups, and in the process reduce the electrical repulsion and lower the CMC. High concentrations of short-chain alcohols, on the other hand, influence the solvent characteristics of water and increase the surfactant solubility and therefore increase the CMC. Longer-chain alcohols can comicellize with ionic surfactants, causing a lowering of the CMC.

Addition of an oppositely charged surfactant in small amounts to a given ionic surfactant can also result in a significant decrease in the CMC.[2] Note, however, that addition of equimolar amounts of oppositely charged surfactants can result in precipitation and the region of homogeneous solution is restricted to very low total concentrations. Short-chain alcohols, such as methanol and ethanol, increase the CMC of nonionic surfactants.[131] This is attributed to their effect on the solvent properties of water and the weakening in the hydrophobic interactions involved. Long-chain alcohols, on the other hand, form mixed micelles and in the process lower the CMC, as is observed for ionic surfactants.

Other types of organic additives which affect micellization because of their influence on solvent properties include urea, formamide, ethylene glycols, and other polyhydric alcohol such as glycerol. These additives affect the CMC at relatively higher concentrations compared to the earlier discussed additives with alkyl chains. Water structure breakers such as urea increase the CMC[132] and structure makers such as fructose lower it. These effects can be rationalized as follows. It is known that surfactant dissolution in the premicellar region actually is accompanied by a decrease in entropy because of the "hydrophobic hydration" around the alkyl chains. Upon micellization the ordered water molecules are released and entropy is regained.[48-50] Therefore, one can expect water structure breakers to increase the CMC and correspondingly structure makers to lower it.

5. Effect of Temperature

Ionic surfactants typically exhibit a minimum in CMC with increase in temperature.[133] The initial lowering is attributed to a reduction in the hydration of surfactant monomers and the increase to a disruption in the structure of water around the hydrocarbon chains. In contrast to ionic surfactants, nonionic surfactants with EO-based hydrophilic groups exhibit a monotonic lowering in CMC with increase in temperature. In fact, this phenomenon at higher temperatures manifests as a phase separation, referred to as clouding, which is discussed below.

The process of micellization itself is dependent on temperature in a complex way. For example, cooling of a micellar solution of SDS below 12°C results in the precipitation of the surfactant. Here, the concentration of the surfactant in solution is equal to the CMC. This temperature, referred to as the Krafft temperature or Krafft, point is the temperature at which the solubility of the surfactant equals the CMC and above this temperature the total solubility of the surfactant increases markedly because of the formation of micelles, but below it only surfactant monomers exist and therefore, the total solubility is drastically limited. A typical phase diagram for an anionic surfactant is given in Figure 15.

The Krafft point of surfactants is dependent upon the chain length of the hydrophobe as well as the type, valency, and concentration of counterions. The Krafft point, in general, increases with alkyl chain length. Similarly, the

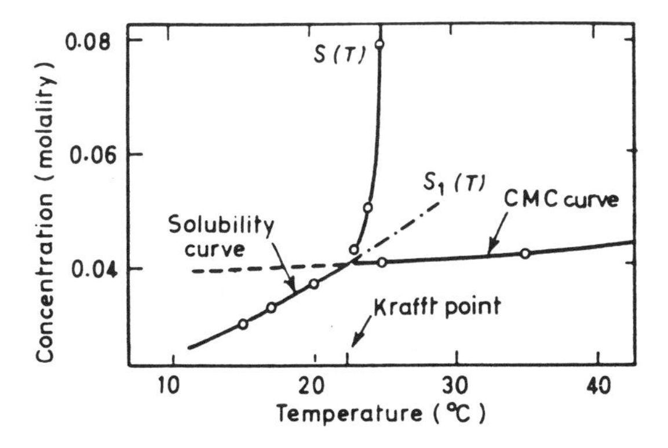

FIGURE 15. Solubility vs. temperature for sodium decylsulfonate showing the Krafft temperature. (From Atwood, D. and Florence, A. T., *Surfactant Systems, Their Chemistry, Pharmacy and Biology,* Chapman and Hall, New York, 1983, 7. With permission.)

Krafft point increases with the valency of the counterion. For example, calcium dodecylsulfate has a Krafft point around 50°C. The Krafft points of surfactants in mixed-counterion systems is dependent upon the relative concentrations of the counterions and the surfactant.

a. Clouding of Nonionic Surfactants

As mentioned earlier, the CMC of an EO-based nonionic surfactant in aqueous solution decreases with increase in temperature and this phenomenon at higher temperatures manifests as a phase separation referred to as clouding.[134,135] In fact, at temperatures just below the cloud point, nonionic surfactant micelles exhibit a marked increase in their size. In a sense, the clouded out phase can be considered as "giant micelles" which become visible, even to the naked eye.

The hydration of EO chains is responsible for the solubility of EO-based nonionic surfactants. An increase in temperature results in progressive dehydration of the EO groups, leading to a reduction in the CMC, increase in the micelle size, and eventually, clouding. The clouded out phase also contains substantial levels of water, indicating that complete dehydration of EO chains is not necessary for phase separation. In general, the amount of water associated with the clouded out phase decreases with increase in temperature beyond the cloud point. Note that ethylene oxide polymers such as polyethylene glycols or polyethylene oxides also exhibit clouding phenomenon at elevated temperatures[136] and therefore, formation of micelles is not a prerequisite for clouding. In fact, several other nonionic polymers which dissolve in water because of hydrogen bonding-type interactions exhibit clouding at elevated temperatures. This entropy-driven phenomenon is fairly general for polymers in aqueous and nonaqueous solutions and is characterized by their lower critical solution temperature (LCST).[137,138] (Note, in this regard, that the phase separation occurring in some polymer solutions upon lowering the temperature, characterized by an upper solution critical temperature [UCST], is enthalpic in origin.) The presence of an alkyl chain along with the EO group essentially lowers the cloud point of nonionic surfactants significantly and possibly renders definite structures to the clouded out phase.

For a given nonionic surfactant solution clouding occurs at a particular temperature and the latter is relatively, but not completely, insensitive to the surfactant concentration. An increase in the alkyl chain length lowers the cloud point. As can be expected, increase in the EO chain length increases the cloud point.

The cloud point of a surfactant can be altered significantly by such additives as electrolytes and surfactants.[4,134,135,139,140] The effectiveness of salts to lower the cloud point and salt out the surfactant follows the so-called lyotropic series or the Hofmeister series.[141] Thus, among Na salts, the tendency to depress the cloud point of a nonionic surfactant would follow the order:[134,136] phosphate > citrate > sulfate > chloride. Evidently, the higher the valency of the anion, the higher the tendency to depress the cloud point. The tendency of various metallic sulfates to depress the cloud point follows the order: $Al^{3+} > Na^{+} > Mg^{2+} > Zn^{2+} > Li^{+}$. Evidently among cations the depression effect does not follow the valency order. This has been attributed to the interactions of certain cations with the EO groups of the nonionic surfactant leading to a "salting in" rather than "salting out" of the surfactant.

The amount of water associated with the clouded out phase is dependent on the temperature at which separation occurs. For systems in which the cloud point is depressed markedly by addition of inorganic electrolytes, the separated phase may contain very high levels of water such that they can be regarded as two aqueous phases in equilibrium.[142,143] Note in this regard that the so-called aqueous two-phase systems[144,145] formed with polyethylene glycols and inorganic salts such as sodium sulfate or phosphate are essentially clouded out phases in which the cloud point is lowered to room temperature by addition of excess salt.[142,143]

The molecular mechanisms involved in the depression of cloud point of EO-based surfactants and polymers by inorganic electrolytes are beyond the scope of this review and can be found in References 4, 141 to 143, 146 and 147.

In general, water structure breakers increase the cloud point of nonionic surfactants and polymers.[148,149] Some of the additives which increase the cloud point include tetra-alkylammonium halides, where the alkyl group is methyl, ethyl, or propyl; urea; and such anions as SCN^-, I^-, and ClO_4^-.[148]

Ionic surfactants increase the cloud point of nonionic surfactants[139,140] and this has been shown to be due to the formation of mixed micelles. It is reasonable that the charged mixed micelles would not grow as rapidly as the nonionic micelles because of electrostatic repulsion and, consequently, the temperature range of stability of the micellar phase can be expected to expand with the addition of ionic surfactants.

D. THERMODYNAMICS OF MICELLIZATION

The two classical approaches which have been used to analyze the process of micellization are the phase separation approach and the mass action approach. In the phase separation approach, micelles are considered to form a separate phase at the CMC. In the mass action approach the micelles are treated as solution species. In both cases, they are in reversible equilibrium with the surfactant monomers. Micellization has been treated also using statistical thermodynamic techniques,[150,151] but these will not be considered here.

The phase separation model provides a simple treatment of the process and leads to almost identical mathematical expressions for the free energy micellization as those given by the mass action approach. It is, however, important to recognize that micelles do not strictly constitute a thermodynamic phase. An important drawback of the phase separation model is its prediction of constant activity of the monomer above the CMC which has been shown to be incorrect by surface tension[29] and potentiometric measurements.[47] It is, however, instructive to examine the utility and limitations of both models.

1. The Phase Separation Model

According to this model, the chemical potential of the monomer, μ_s, and of micelles, μ_m, should be the same at equilibrium. The standard state for the former is the solvated monomer at unit mole fraction referred to infinitely dilute solution, and for the micelle, the micelle itself is considered to be the standard state.

a. Nonionic Surfactants

The standard free energy of formation, ΔG^o_m, of nonionic micelles can be shown to be[1]

$$\Delta G^o_m = RT \ln x_{cmc} \tag{17}$$

where R is the universal gas constant, T the absolute temperature, and x_{cmc} the mole fraction of monomers at the CMC; x_{cmc} is essentially the CMC expressed as a mole fraction, and is defined as

$$X_{cmc} = \frac{n_s}{n_s + n_{H_2O}} \tag{18}$$

where n_s and n_{H_2O} represent the moles of surfactant and water in the system. Since n_s is small relative to n_{H_2O}, $x_{cmc} = n_s/n_{H_2O}$ and therefore,

$$\Delta G^o_m = 2.303 \ RT \ (\log CMC - \log w) \tag{19}$$

where w = moles per liter of water (55.4 at 20°C).

The standard enthalpy of micellization derived using the Gibbs-Helmoltz equation is given by:

$$\Delta H^{o}_{m} = R\left(\frac{\delta \ln x_{cmc}}{\delta(1/T)}\right) \tag{20}$$

The standard entropy of micellization can be obtained using the expression:

$$\Delta S^{o}_{m} = (\Delta H^{o}_{m} - \Delta G^{o}_{m})/T \tag{21}$$

b. Ionic Surfactants

In this case, the free energy calculations include the contributions from the counterions as well as the surfactant molecules. The expression for the free energy is given by:[1]

$$\Delta G^{o}_{m} = (2 - \alpha)\, RT \ln x_{cmc} \tag{22}$$

where $(1 - \alpha)$ is the fraction of charges on the micelles neutralized by the counterions. It is often assumed that the micellar phase consists of a charged phase with an equivalent number of counterions, in which case the above expression can be rewritten as:

$$\Delta G^{o}_{m} = 4.606\, RT\, (\log cmc - \log w) \tag{23}$$

2. The Mass Action Model

a. Nonionic Surfactants

Micellization in a nonionic system can be represented as:

$$nR \rightleftharpoons M \tag{24}$$

The equilibrium constant K_m under ideal conditions, where activity = mole fraction, is given by:

$$K_m = x_m/(x_R)^n \tag{25}$$

where x_m and x_R are the mole fractions of micelles and the free monomer. The standard free energy of micellization is given by:

$$\Delta G^{o}_{m} = -RT \ln K_m \tag{26}$$

Using this approach, Corkill et al.[152] have derived the following expression for the free energy of micellization:

$$\Delta G^{o}_{m} = RT[(1 - 1/n)\ln x_{cmc} + f(n)] \tag{27}$$

where

$$f(n) = \frac{1}{n}\left[\ln n^2 \left(\frac{2n - 1}{n - 2}\right) + (n - 1) \ln \frac{n(2n - 1)}{2(n^2 - 1)}\right] \tag{28}$$

If n is very large, the equation reduces to the form:

$$\Delta G^o_m = RT \ln x_{cmc} \tag{29}$$

Note that this expression is identical to Equation 17 derived using the phase separation action approach.

b. Ionic Surfactants

Consider an anionic micelle M with a charge of $-(n-p)$ where p is the number of counterions, A^+, associated with the micelle formed by n surfactant molecules. Therefore, the fraction of charges not neutralized by the counterions, α, is essentially equal to $(n-p)/n$.

$$nR^- + p\,A^+ \leftrightarrow M^{-(n-p)} \tag{30}$$

The equilibrium constant K_m is given by:

$$K_m = \frac{x_M}{(x_R)^n\,(x_A)^p} \tag{31}$$

The standard free energy of micellization per mole of monomeric surfactant is given by:

$$\Delta G^o_m = -(RT/n) \ln K_m \tag{32}$$

For large values of n and when conditions are near the CMC, the above expression can be reduced to the form:[1]

$$= (2 - \alpha)\,RT \ln x_{cmc} \tag{33}$$

This equation is the same as the expression (Equation 22) derived using the phase separation model.

It is clear that the phase separation model and the mass action model lead to identical expressions for the free energy under limiting conditions. The experimentally determined values of free energy of micellization, enthalpy, and entropy of micellization of some of the common surfactants are given in Table 5. It is clear that the negative values of ΔG^o_m results from the large values of ΔS^o_m. Clearly, as discussed earlier, micellization is an entropy-driven phenomenon.

TABLE 5
Thermodynamic Parameters of Micellization

Surfactant	Temp. °C	ΔG^o kJ/mol	ΔH^o kJ/mol	$T\Delta S^o$ kJ/mol
Na decylsulfate	25	−34.9	8	42.9
Na dodecylsulfate	25	−39.7	5	44.7
Dodecylpyridinium bromide	25	−38.2	−14	24.2
Dodecylpyridinium chloride	25	−37	—	—

The above equations derived using thermodynamic models can be used to show the theoretical basis for the empirical equations for the dependence of the CMC on surfactant chain length and electrolyte concentration.[1,53] Substituting for K_m from Equation 31, Equation 32 can be expressed as:

$$\Delta G_m^o = -(RT/n)\,[\ln x_m - n \ln x_R - p \ln A_x] \quad (34)$$

or

$$\log x_R = (-p/n) \log A_x + \Delta G_m^o/(2.303\ RT) + (1/n) \log x_M \quad (35)$$

If one makes the assumption that in micellar solutions, the monomeric surfactant concentration, x_R, is equal to the CMC, the above equation has the same form as Equation 15. Accordingly, the slope "A" of a log CMC vs. log electrolyte concentration plot is equal to $(-\ p/n)$ and the intercept B is equal to $\Delta G_m^o/(2.303\ RT) + (1/n) \log x_M$. For convenience of calculations, it is often assumed that, at the CMC, 2% of the surfactant is in the form of micelles. Thus, the theoretical basis of the constants A and B in Equation 15 is evident.

The free energy of formation of micelles, ΔG_m^o, can be considered to be a sum of the hydrophobic contribution, ΔG_h^o, and an electrostatic contribution, ΔG_{elec}^o. For a homologous series of surfactants:

$$\Delta G_m^o = n \cdot \Delta G^o_{-CH2-} + \Delta G^o_{-CH3} + \Delta G^o_{elec} \quad (36)$$

where ΔG^o_{-CH2-} and ΔG^o_{-CH3-} represent free energy contributions from the $-CH_2-$ groups and the terminal $-CH_3$ group, and n is the number of $-CH_2-$ groups in the alkyl chain. For a homologous series, the last two terms on the left-hand side can be taken as constants. Also using Equation 19, one can write:

$$2.303RT \log CMC = n \cdot \Delta G^o_{-CH2-} + \Delta G^o_{-CH3} + \Delta G^o_{elec} + 2.303\ RT\log w \quad (37)$$

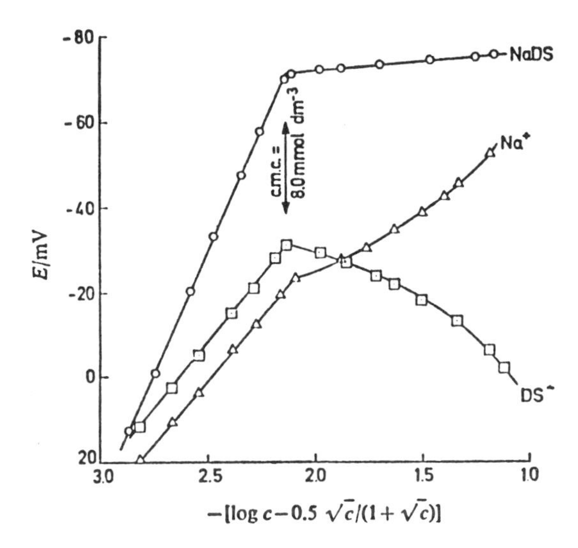

FIGURE 16. EMF vs. activity [= log c − $0.5\sqrt{c}/(1 + \sqrt{c})$] for SDS solutions. Cell I: electrode specific to DS^-/test solution/electrode specific to Na^+ and cell II: electrode specific to DS^- or Na^+/test solution/reference electrode. Thus, cell I gives mean activity of SDS ($a_\pm$) whereas cell II gives that of DS^- or Na^+. (From Cutler, S. G. et al., *J. Chem. Soc. Faraday Trans. 1,* 74, 1758, 1978. With permission.)

The above equation can be rearranged to have the same form as Equation 14 and it provides the theoretical basis for the empirical equation relating the dependence of CMC on surfactant chain length.

One difference between the phase separation model and the mass action model is that only the latter predicts a change in the monomer activity above CMC. Mysels[153] and Sexsmith and White[154] treated the micellization process using the mass action approach extensively and showed that the concentration of the monomer would actually decrease with surfactant concentration above the CMC.

3. Monomer Activity Above the CMC

Some of the experimental evidence which shows a reduction in the monomer activity above the CMC includes ion selective electrode and NMR measurements. Results of ion selective electrode measurements[47] for the SDS system are shown in Figure 16. It is clear that the activity of the counterions measured by this method increases and that of the monomers decreases with increase in surfactant concentration above the CMC. Similarly, Lindman[123] has shown using self-diffusion measurements of surfactant ions, counterions, water, and coions that the concentration of surfactant monomers decreases above the CMC (see Figure 17).

Note that the presence of a maximum in the surfactant monomer activity in itself is not an argument against the phase separation model, since the thermodynamic requirement for phase separation is only that the mean activity

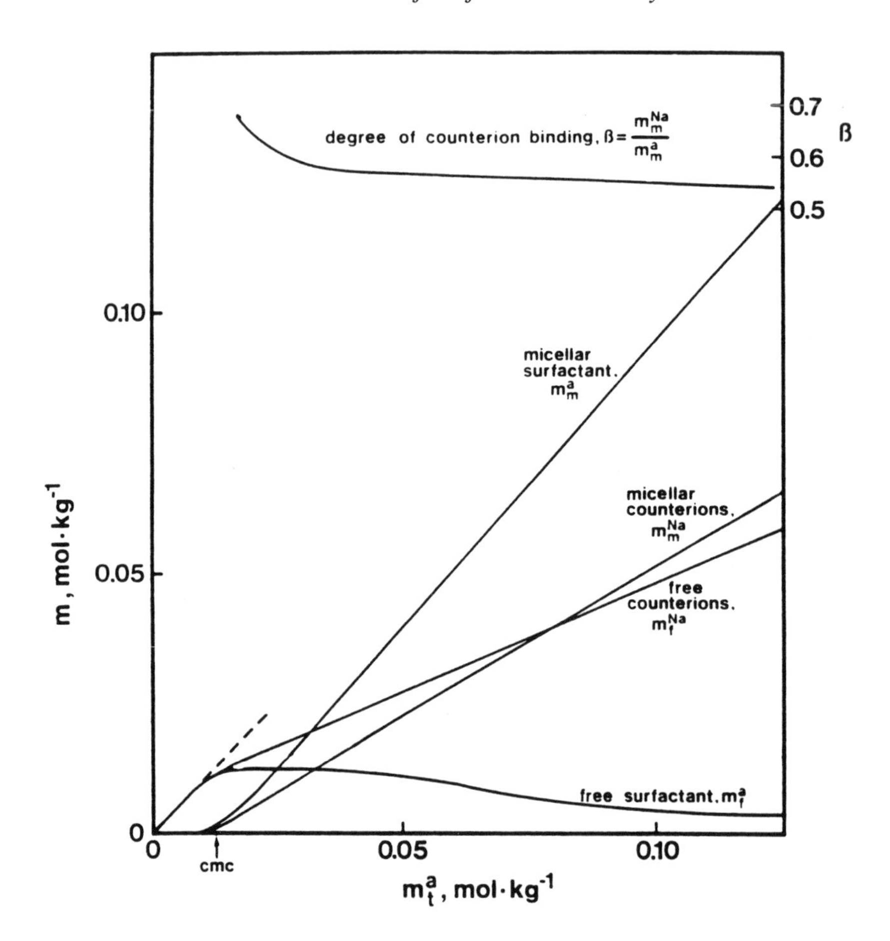

FIGURE 17. Concentration of free and micellized surfactant ions and counterions for solutions of sodium p-octylbenzenesulfonate. Additionally, the degree of counterion binding, β, is also given. (From Lindman, B., in *Surfactants,* Tadros, Th. F., Ed., Academic Press, New York, 1984, 83. With permission.)

of the surfactant (as opposed to that of the monomer or the counterion) should remain constant above the CMC. It is evident from Figure 16 that the mean activity of the surfactant in fact increases with increase in the concentration above CMC and this cannot be accounted for by the phase separation model. An increase in the mean activity of the surfactant is also reflected in a continued reduction in the surface tension of, for example, SDS above CMC.[29] These results are consistent with the consideration that the micelle is a solution species. Also, strictly speaking, micelles are not a phase since they cannot be separated from the monomers under thermodynamic conditions. For these reasons, the mass action approach is the more realistic one of the two considered.

A simple picture of what happens above the CMC is as follows. Since the micelles are only partially neutralized by the counterions, the activity of counterions continues to increase with increase in total surfactant concentration above the CMC. This preferential increase in counterion concentration can be viewed as leading to a reduction in the CMC of the system and this is reflected as an apparent reduction in the monomer activity.

4. Other Models

In addition to the two general representations of the micellization process, other models based on electrostatic[96,155] and geometric considerations[107-109] have been developed. These models in essence allow for taking into account the various energetic contributions from molecular interactions and in the process provide a clearer picture of the micellization process at the molecular level. These models are beyond the scope of this review and the reader is referred to References 94, 96, 107 to 109, 126, and 155 for further information. Some of the conclusions from the geometrical model were presented in an earlier part of the chapter in terms of the possible geometries of micelles as they depend on the size of the surfactant monomer.

5. Kinetics of Micellization

Micelles are flickering clusters in the sense that they continually form and break up into monomers. Recent advances have made it possible to follow such processes and obtain reliable kinetic information on them. An excellent review of methods to study micellar relaxation can be found in Reference 6. Some of the techniques which have been used include ultrasonic absorption,[156] temperature jump,[157] pressure jump,[158] and stopped flow measurements.[157]

There appears to be general agreement among researchers that there exist at least two relaxation processes in micellar systems. The fast process occurring at a time scale of 10 μsec or less has been attributed to monomers entering and leaving a micelle. The slow process, which has a relaxation time in the millisecond range, has been studied using the temperature jump, pressure jump, and stopped flow techniques.[6] The slow process has been attributed to the break up of a complete micelle into monomers. These two reactions can be represented by the following equations:

$$nR \rightleftarrows R_n \qquad \text{(slow)} \tag{38}$$

$$R_n \rightleftarrows R_{n-1} + R \text{ (fast)} \tag{39}$$

Theories of micellar kinetics developed by Aniansson and Wall[159] and subsequent extensions introduced by Kahlweit and Teubner[160] and Hall[161] account for most of the experimental observations. For details the reader is referred to References 6 and 159 to 161.

Some recent experimental values of the time constant for the slow process, τ_2, as a function of surfactant concentration for SDS solutions, are given in

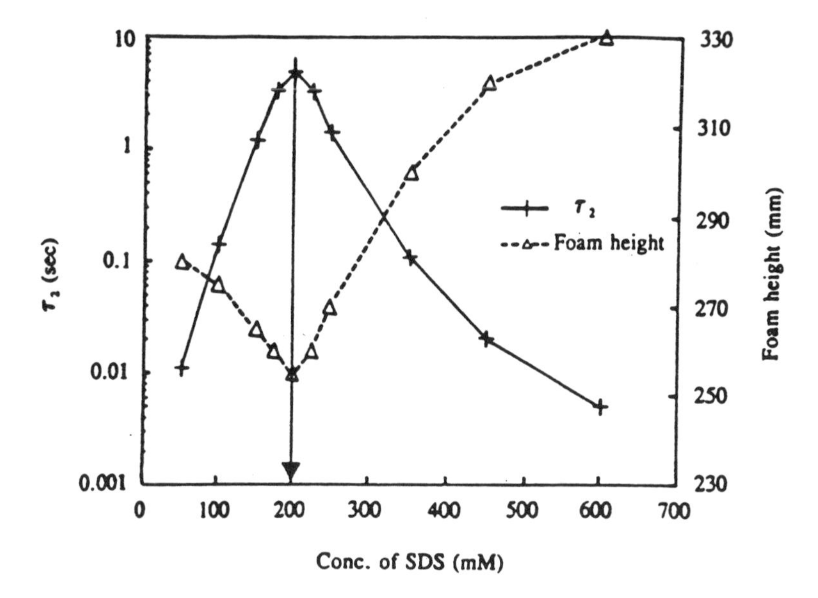

FIGURE 18. Effect of concentration of SDS on the micellar relaxation time and foaming ability. (From Oh, S. G. and Shah, D. O., *Langmuir*, 7, 1316, 1991. With permission.)

Figure 18.[162] It is clear that $1/\tau_2$ goes through a minimum at about 2.0×10^{-1} mol/l. Similar results have been reported for other anionic and cationic surfactants.[163] Theoretical models based on the assumption that the growth of the micelle occurs by addition of single monomers only ($nR \rightleftharpoons R_n$) could account for the reduction in $1/\tau_2$ with increase in concentration. At high counterion concentrations, however, growth can occur by reversible coagulation of submicellar aggregates:[163] $R_k + R_l \rightarrow R_n$ where $k + l = n$. This additional pathway could explain the increase in $1/\tau_2$ beyond a certain concentration.

Several practical applications of surfactants, e.g., foaming, wetting, detergency, and emulsification, depend on the dynamics of supply of surfactant monomer to the relevant interface(s). It has now been demonstrated that the kinetics of micellar breakdown can play a vital role in all these processes.[162,164,165] For example, results given in Figure 18 show that the minimum in foaming and the minimum in $1/\tau_2$ occur at the same surfactant concentration.

IV. CONCLUDING REMARKS

Significant advances have been made in our understanding of the molecular and microstructural aspects of surfactant aggregation in solution and at interfaces. This has been made possible because of major developments in a wide variety of methods, such as spectroscopy, microscopy, potentiometry, molecular modeling, and various scattering and kinetic techniques, along with the progress in our theoretical understanding of the subject. Important structure-performance relationships have been established in several cases which can be used to manipulate the geometry and design of molecules to tailor-make surfactants for particular applications. Kinetic aspects of monomer-micelle equilibrium and their implications as regards commercial applications continue to be pursued.

Most of the research activity is still restricted to the behavior of conventional surfactants. Research needs exist in developing a better understanding of unconventional surfactants, such as those based on silicones, fluorocarbons, saccharide and polymeric groups. One can also confidently predict that "bio" aspects, per se, including studies of bioderived surfactants and the biodegradability of surfactants in general, will be receiving ever-increasing emphasis.

REFERENCES

1. **Atwood, D. and Florence, A. T.,** *Surfactant Systems, Their Chemistry, Pharmacy and Biology,* Chapman and Hall, New York, 1983.
2. **Rosen, M. J.,** *Surfactants and Interfacial Phenomena,* 2nd ed., John Wiley & Sons, New York, 1989.
3. **Lucassen-Reyenders, E. H., Ed.,** *Anionic Surfactants, Physical Chemistry of Surfactant Action,* Surfactant Science Ser., Vol. 11, Marcel Dekker, New York, 1981.
4. **Schick, M. J., Ed.,** *Nonionic Surfactants: Physical Chemistry,* Surfactant Science Ser., Vol. 23, Marcel Dekker, New York, 1987.
5. **Richmond, J. M., Ed.,** *Cationic Surfactants,* Surfactant Science Ser., Vol. 34, Marcel Dekker, New York, 1990.
6. **Zana, R., Ed.,** *Surfactant Solutions,* Surfactant Science Ser., Vol. 22, Marcel Dekker, New York, 1987.
7. **Rieger, M. M., Ed.,** *Surfactants in Cosmetics,* Surfactant Science Ser., Vol. 16, Marcel Dekker, New York, 19.
8. **Tadros, Th. F., Ed.,** *Surfactants,* Academic Press, New York, 1984.
9. **McCutcheon, J. W.,** *McCutcheon's Emulsifiers and Detergents,* Vol. 1, M. C. Publishing, New Jersey, 1991.
10. **Mittal, K. L., Ed.,** *Micellization, Solubilization and Microemulsions,* Vol. 1 and 2, Plenum Press, New York, 1977.
11. **Mittal, K. L., Ed.,** *Solution Chemistry of Surfactants,* Vol. 1 and 2, Plenum Press, New York, 1979.
12. **Mittal, K. L. and Fendler, E. J., Eds.,** *Solution Behavior of Surfactants,* Vol. 1 and 2, Plenum Press, New York, 1982.
13. **Mittal, K. L. and Lindman, B. J., Eds.,** *Surfactants in Solution,* Vol. 1 to 3, Plenum Press, New York, 1984.
14. **Mittal, K. L. and Bothorel, K., Eds.,** *Surfactants in Solution,* Vol. 4 to 6, Plenum Press, New York, 1986; **Mittal, K. L., Ed.,** *Surfactants in Solution,* Vol. 7 to 10, Plenum Press, New York, 1989.
15. **Mukerjee, P.,** Dimerization of anions of long chain fatty acids in aqueous solutions and the hydrophobic properties of the acids, *J. Phys. Chem.,* 69, 2821, 1965.
16. **Lucassen, J.,** Hydrolysis and precipitation in carboxylate soap solutions, *J. Phys. Chem.,* 70(6), 1824, 1966.

17. **Kulkarni, R. D. and Somasundaran, P.,** Kinetics of oleate adsorption at the liquid-air interface and its role in flotation, in *Advances in Interfacial Phenomena on Particulate/Solution/Gas Systems, Application to Flotation Research,* Somasundaran, P. and Grieves, R. B., Eds., AIChE Symp. Ser. No. 150, 1975, 124.
18. **Miles, G. D. and Shedlovsky, L.,** Minima in surface tension of solutions of Na alcohol sulfates, *J. Phys. Chem.,* 48, 60, 1944.
19. **Leja, J.,** *Flotation,* Plenum Press, New York, 1982.
20. **Kung, H. C. and Goddard, E. D.,** Interactions of amines and amide hydrochlorides,
21. **Somasundaran, P.,** The role of ionomolecular complexes in Flotation, *Int. J. Miner. Process.,* 3, 35, 1976.
22. **Finch, J. A. and Smith, G. W.,** Dynamic surface tension of alkaline dodecylamine solutions, *J. Colloid Interface Sci.,* 45(1), 85, 1973.
23. **Ananthapadmanabhan, K. P., Somasundaran, P., and Healy, T. W.,** The chemistry of oleate and amine solutions in relation to flotation, *Trans. AIME,* 266, 2003, 1980.
24. **Griffin, W. C.,** Classification of surface active agents by ''HLB'', *J. Soc. Cosmet. Chem.,* 1, 311, 1949.
25. **Mukerjee, P. and Mittal, K. L.,** The wide world of micelles, in *Micellization, Solubilization and Microemulsions,* Mittal, K. L., Ed., Plenum Press, New York, 1977, 1.
26. **Shah, D. O. and Bansal,** Micellar solutions for improved oil recovery, in *Micellization, Solubilization and Microemulsions,* Mittal, K. L., Ed., Plenum Press, New York, 1977, 87.
27. **Adam, N. K.,** *The Physics and Chemistry of Surfaces,* 2nd ed., Dover, New York, 1968.
28. **Adamson, A. W.,** *Physical Chemistry of Surfaces,* 3rd ed., Interscience Publishers, New York, 1984.
29. **Elworthy, P. H. and Mysels, K. J.,** The surface tension of sodium dodecylsulfate solutions and the phase separation model of micelle formation, *J. Colloid Interface Sci.,* 21, 331, 1966.
30. **Mysels, K. J. and Florence, A. T.,** Techniques and criteria in the purification of aqueous surfaces, in *Clean Surfaces: Their Characterization for Interfacial Studies,* Marcel Dekker, New York, 1970, 227.
31. **Chattoraj, D. K. and Birdi, K. S.,** *Adsorption and the Gibbs Surface Excess,* Plenum Press, New York, 1984.
32. **Somasundaran, P. and Hanna, H. S.,** Physico-chemical aspects of adsorption at solid-liquid interfaces, Part I, Basic principles, in *Improved Oil Recovery by Surfactant and Polymer Flooding,* Shah, D. O. and Schechter, R. S., Eds., Academic Press, New York, 1977, 205.
33. **Moudgil, B. M., Somasundaran, P., and Soto, H.,** Adsorption of surfactants on minerals, in *Reagents in Mineral Technology,* Surfactant Science Ser., Vol. 27, Somasundaran, P. and Moudgil, B. M., Eds., Marcel Dekker, New York, 1988, 79.
34. **Harwell, J. H., Hoskins, J. C., and Schechter, R. S.,** Pseudophase separation model for surfactant adsorption: isomerically pure surfactants, *Langmuir,* 1, 251, 1985.
35. **Somasundaran, P. and Kunjappu, J. T.,** In-situ investigation of adsorbed surfactants and polymers on solids in solution, *Colloids Surfaces,* 37, 245, 1989.
36. **Koopal, L. K. and Keltjens,** Adsorption of ionic surfactants on charged solids, *Colloids Surfaces,* 17, 371, 1986.
37. **Giles, G. H.,** Surfactant adsorption at solid/liquid interfaces, in *Anionic Surfactants, Physical Chemistry of Action,* Surfactant Science Ser., Vol. 11, Lucassen-Reyenders, E. H., Ed., Marcel Dekker, New York, 1981, 143.
38. **Somasundaran, P. and Agar, G. E.,** The zero point of charge of calcite, *J. Colloid Interface Sci.,* 24, 433, 1967.
39. **Somasundaran, P.,** Zeta potential of apatite in aqueous solutions and its changes during equilibration, *J. Colloid Interface Sci.,* 27, 659, 1968.

40. **Somasundaran, P. and Ananthapadmanabhan, K. P.,** Bubble and foam separations — ore flotation, in *Handbook of Separation Process Technology,* Rousseau, R., Ed., Wiley-Interscience, New York, 1987, 775.
41. **Somasundaran, P. and Fuersteanau, D. W.,** Mechanisms of alkyl sulfonate adsorption at the alumina-water interface, *J. Phys. Chem.,* 70, 90, 1966.
42. **Chandar, P., Somasundaran, P., and Turro, N. J.,** Fluorescence probe studies on the structure of the adsorbed layer of dodecyl sulfate at the alumina-water interface, *J. Colloid Interface Sci.,* 117(1), 31, 1987.
43. **Fuerstenau, D. W. and Wakamatsu, T.,** Effect of pH on the adsorption of sodium dodecyl sulfonate at the alumina-water interface, *Faraday Discuss. Chem. Soc.,* 59, 157, 1975.
44. **Mukerjee, P. and Mysels, K. J.,** Critical Micelle Concentration of Aqueous Surfactant Systems, NSRDS-NBS 36, National Bureau of Standards, Washington, D.C., 1971.
45. **Somasundaran, P., Ananthapadmanabhan, K. P., and Ivanov, I. B.,** Dimerization of oleate in aqueous solutions, *J. Colloid Interface Sci.,* 99, 128, 1984.
46. **Rijnbout, J. B.,** Adsorption and dimerization of hexadecyl trimethylammonium bromide from surface tension measurements, *J. Colloid Interface Sci.,* 62, 81, 1976.
47. **Cutler, S. G., Meares, P. and Hall, D. J.,** Ion activities of sodium dodecyl sulfate solutions from electromotive force measurements, *J. Chem. Soc. Faraday Trans. 1,* 74, 1758, 1978.
48. **Goddard, E. D., Hoeve, C. A. J., and Benson, G. C.,** Heats of micelle formation of paraffin chain salts in water, *J. Phys. Chem.,* 61, 593, 1957.
49. **Nemethy, G. and Scheraga, H. A.,** Structure of water and hydrophobic bonding in proteins. II. Model for the thermodynamic properties of aqueous solutions of hydrocarbons, *J. Chem. Phys.,* 36, 3401, 1962.
50. **Nemethy, G. and Scheraga, H. A.,** Structure of water and hydrophobic bonding in proteins. III. The thermodynamic properties of hydrophobic bonds in proteins, *J. Phys. Chem.,* 66, 1773, 1962.
51. **Stigter, D.,** On density, shape and charge of micelles of sodium dodecyl sulfate and dodecyl ammonium bromide, *J. Colloid Interface Sci.,* 23, 379, 1967.
52. **Hartley, G. S.,** *Aqueous Solutions of Paraffin Chain Salts,* Hermann, Paris, 1936.
53. **Shinoda, K.,** The formation of micelles, in *Colloidal Surfactants,* Shinoda, K., Nakagawa, T., Tamamushi, B., and Isemura, T., Eds., Academic Press, New York, 1963, chap. 1.
54. **Kalyanasundaram, K. and Thomas, J. K.,** Environmental effects of vibronic band intensities in pyrene monomer fluorescence and their application in studies of micellar systems, *J. Am. Chem. Soc.,* 99, 2039, 1977.
55. **Ananthapadmanabhan, K. P., Goddard, E. D., Turro, N. J., and Kuo, P. L.,** Fluorescence probes for critical micelle concentration, *Langmuir,* 1(13), 352, 1985.
56. **Turro, N. J. and Kuo, P. L.,** Pyrene excimer formation in micelles of nonionic detergents and of water soluble polymers, *Langmuir,* 2, 438, 1986.
57. **Pownall, H. J. and Smith, L. C.,** Viscosity of hydrocarbon region of micelles measured by excimer fluorescence, *J. Am. Chem. Soc.,* 95, 3136, 1973.
58. **Turro, N. J., Aikawa, M., and Yekta, A.,** A comparison of intermolecular and intramolecular excimer formation in detergent solutions, *J. Am. Chem. Soc.,* 101, 772, 1979.
59. **Goddard, E. D. and Jones, T. G.,** Surface tension measurements and spectral dye method of determining critical micelle concentrations, *Res. Corresp.,* 8, A1, 1955.
60. **Zachariasse, K.,** Intramolecular excimer formation with diarylalkanes as a microviscosity fluorescence probe for SDS micelles, *Chem. Phys. Lett.,* 57, 429, 1978.
61. **Shiniyzky, M., Dianoux, A. C., Gitler, C., and Weber, G.,** Microviscosity and order in hydrocarbon region of micelles and membranes determined by fluorescence probes. I. *Biochemistry,* 10, 2106, 1971.

62. **Baglioni, R., Ottaviani, M. F., Martini, G., and Ferrani, E.,** ESR studies of spin labelled micelles, in *Surfactants in Solution,* Mittal, K. L. and Lindman, B., Eds., Plenum Press, New York, 1984.
63. **Povich, M. J., Mann, J. A., and Kawamoto, A.,** ESR spin label study of HDTAB micelles, *J. Colloid Interface Sci.,* 41, 145, 1972.
64. **Chandar, P., Somasundaran, P., Waterman, K. C., and Turro, N. J.,** Variation in nitroxide chain flexibility within sodium dodecyl sulfate hemimicelles, *J. Phys. Chem.,* 91, 150, 1987.
65. **Williams, E., Sears, B., Allerhand, A., and Cordes, E. H.,** Segmental motion of amphipathic molecules in aqueous solutions and micelles. Application of natural abundance carbon-13 partially relaxed FTNMR spectroscopy, *J. Am. Chem. Soc.,* 95, 4871, 1973.
66. **Clifford, J. and Pethica, B. A.,** Properties of micellar solutions. II. NMR chemical shifts of water protons in solutions of sodium alkyl sulfates, *Trans. Faraday Soc.,* 60, 1483, 1964.
67. **Podo, F., Ray, A., and Nemethy, G.,** Structure and hydration of nonionic micelles, a high resolution NMR study, *J. Am. Chem. Soc.,* 95, 6164, 1973.
68. **Clemett, C. J.,** Proton spin lattice relaxation times in some nonionic micelles, *J. Chem. Soc. A,* 2251, 1970.
69. **Menger, F. M., Jerkunica, J. M., and Johnston, J. C.,** The water content of a micelle interior. The Fjord vs. Reef Models, *J. Am. Chem. Soc.,* 100, 4676, 1978.
70. **Wennerstrom, H. and Lindman, B.,** Water penetration into surfactant micelles, *J. Phys. Chem.,* 83, 2931, 1979.
71. **Muller, N. and Simsohn, H.,** Investigation of micellar structure by fluorine magnetic resonance, Na perfluoroctanoate, *J. Phys. Chem.,* 75, 942, 1971.
72. **Mukerjee, P. and Mysels, K. J.,** in *Colloidal Dispersions and Micellar Behavior,* Mittal, K. L., Ed., ACS Symp. Ser. No. 9, American Chemical Society, Washington, D.C., 1975, 239.
73. **Lindman, B., Soderman, and Wennerstrom,** NMR Studies of surfactant systems, in *Surfactant Solutions,* Surfactant Science Ser., Vol. 22, Zana, R., Ed., Marcel Dekker, New York, 1987.
74. **Lindman, B., Puyal, M. C., Kamenka, N., Rymden, R., and Stilbs, P.,** Micelle formation of anionic and cationic surfactants from Fourier transform hydrogen-1 and lithium-7 NMR and tracer diffusion, *J. Phys. Chem.,* 88, 5048, 1984.
75. **Lindman, B., Wennerstrom, H., Gustavsson, H., Kamenka, N., and Brun, B.,** Some aspects of hydration of surfactant micelles, *Pure Appl. Chem.,* 52, 1307, 1980.
76. **Mukerjee, P.,** The hydration of micelles of association colloids, *J. Colloid Sci.,* 19, 722, 1964.
77. **Lindman, B. and Brun, B.,** Translational motion in aqueous Na n-octanoate solutions, *J. Colloid Interface Sci.,* 42, 388, 1973.
78. **Stigter, D.,** Electrostatic interactions in aqueous environment, in *Physical Chemistry: Enriching Topics from Colloid and Surface Science,* van Olphen, H. and Mysels, K. J., Eds., Theorex, La Jolla, CA, 1975, 181; **Stigter, D.,** Electrophoresis, in *Physical Chemistry: Enriching Topics from Colloid and Surface Science,* van Olphen, H. and Mysels, K. J., Eds., Theorex, La Jolla, CA, 1975, 293.
79. **Bujake, J. E. and Goddard, E. D.,** Surface composition of sodium lauryl sulphonate and sulphate solutions by foaming and surface tension, *Trans. Faraday Soc.,* 61, 190, 1965.
80. **Mukerjee, P. and Banerjee, K.,** A study of surface pH of micelles using solubilized indicator dye, *J. Phys. Chem.,* 68, 3567, 1964.
81. **Fernandez, M. S. and Fromherz, P.,** Lipid pH indicators as probes of electrical potential and polarity of micelles, *J. Phys. Chem.,* 81, 1755, 1977.

82. **Healy, T. W., Drummond, C. J., Grieser, F., and Murray, B. S.,** Electrostatic surface potential and critical micelle concentration relationship for ionic micelles, *Langmuir,* 6, 506, 1990.
83. **Kalyanasundaram, K. and Thomas, J. K.,** Solvent dependent fluorescence of pyrene 3-carboxyaldehyde and its applications in estimating the polarity of micelle solution interface, *J. Phys. Chem.,* 81, 2176, 1977.
84. **Cushman, A., Brady, A. P., and McBain, J. W.,** The osmotic activity and conductivity of aqueous solutions of some typical colloidal electrolytes, *J. Colloid Sci.,* 3, 425, 1948.
85. **Palepu, R., Hall, D. J., and Wyn-Jones, E.,** The use of ion-selective electrodes to determine the effective degree of micelle dissociation in tetradecylpyridinium bromide solutions, *J. Chem. Soc. Faraday Trans.,* 86(9), 1535, 1990.
86. **Mysels, K. J.,** Charge effects in the light scattering by association colloids, *Colloid Sci.,* 10, 507, 1955.
87. **Gust, H. and Lindman, B.,** NMR studies of the interaction between alkali ions and micellar aggregates, *J. Am. Chem. Soc.,* 97(4), 3923, 1975.
88. **Hartland, G. V., Grieser, F., and White, L. R.,** Surface potential measurements in pentanol sodium dodecyl sulfate micelles, *J. Chem. Soc. Faraday Trans. 1, 83, 591,*
89. **Zana, R.,** Ionization of cationic micelles: effect of the detergent structure, *J. Colloid Interface Sci.,* 78(2), 330, 1980.
90. **Brun, T. S., Hoiand, H., and Vikingstad, E.,** The fraction of counterions and singly dispersed amphiphiles in micellar systems from ion exchange membrane electrode measurements, *J. Colloid Interface Sci.,* 63, 590, 1978.
91. **Mukerjee, P., Mysels, K. J., and Kapauan, P.,** Counterion specificity in the formation of ionic micelles — size, hydration, and hydrophobic bonding effects, *J. Phys. Chem.,* 71, 4166, 1967.
92. **Goddard, E. D., Harva, O., and Jones, T. G.,** The effect of univalent cations on the critical micelle concentration of sodium dodecyl sulfate, *Trans. Faraday Soc.,* 49, 980, 1953.
93. **Wennerstrom, H., Lindman, B., Lindblom, G., and Tiddy, G. J.,** Ion condensation and NMR studies of counterion binding in lyotropic liquid crystals, *J. Chem. Soc. Faraday Trans. 1,* 75, 663, 1979.
94. **Stigter, D.,** Invariance of the charge of electrical double layers under dilution of the equilibrium electrolyte solution, *Prog. Colloid Polym. Sci.,* 65, 45, 1978.
95. **Manning, G. S.,** The molecular theory of polyelectrolyte solutions with application to properties of polynucleotides, *Q. Rev. Biophys.,* 11, 179, 1978.
96. **Gunnarsson, G., Jonsson, B., and Wennerstrom, H.,** Surfactant association into micelles. An electrostatic approach, *J. Am. Chem. Soc.,* 84, 3114, 1980.
97. **Jonsson, B. and Wennerstrom, H.,** Thermodynamics of ionic amphiphilic-water systems, *J. Colloid Interface Sci.,* 80, 482, 1981.
98. **Gustavsson, H. and Lindman, B.,** Alkali ion binding to aggregates of amphiphilic compounds studied by nuclear magnetic resonance chemical shifts, *Am. Chem. Soc.,* 100, 4647, 1978.
99. **Goddard, E. D., Kao, O., and Kung, H. C.,** Counterion effects in charged monolayers, *J. Colloid Interface Sci.,* 27, 616, 1968.
100. **McBain, J. W.,** Discussions on colloids and their viscosity, *Trans. Faraday Soc.,* 9, 99, 1913.
101. **Tanford, C.,** Micelle shape and size, *J. Phys. Chem.,* 76, 3020, 1972.
102. **Henriksson, U., Odberg, L., Erikson, J. C., and Westman, L.,** N^{14} relaxation in aqueous micellar solutions of n-hexadecyl trimethylammonium bromide and chloride, *J. Phys. Chem.,* 81, 76, 1977.
103. **Reiss-Husson, F. and Luzzati, V.,** The structure of micellar solutions of some amphipathic compounds in pure water as determined by absolute small angle X-ray scattering technique, *J. Phys. Chem.,* 68, 3504, 1964.

104. **Mazer, N. A., Benedek, G. B., and Carey, M. C.,** An investigation of micellar phase of SDS in aqueous NaCl solutions using quasielastic light scattering spectroscopy, *J. Phys. Chem.*, 80, 1075, 1976.
105. **Hayashi, S. and Ikeda, S.,** Micellar size and shape of SDS in NaCl solutions, *J. Phys. Chem.*, 84, 744, 1980.
106. **Schott, H.,** Are there spherical micelles?, *J. Pharm. Sci.*, 60(10), 1594, 1971.
107. **Israelachili, J. N., Mitchell, D. J., and Ninham, B. W.,** Theory of self-assembly of hydrocarbon amphiphiles into micelles and bilayers, *J. Chem. Soc. Faraday Trans. 2*, 72, 1525, 1976.
108. **Mitchell, D. J. and Ninham, B. W.,** Micelles, vesicles, and microemulsions, *J. Chem. Soc. Faraday Trans. 2*, 77, 601, 1981.
109. **Tanford, C.,** *The Hydrophobic Effect,* 2nd ed., John Wiley & Sons, New York, 1980.
110. **Debye, P. and Anacker, E. W.,** Micellar shape from light scattering measurements, *J. Phys. Chem.*, 55, 644, 1951.
111. **Vetter, R. J.,** Micelle structure in aqueous solutions of colloidal electrolytes, *J. Phys. Chem.*, 51, 262, 1947.
112. **Peri, J. B.,** The state of dispersion of detergent additives in lubricant oil and other hydrocarbons, *J. Am. Oil Chem. Soc.*, 35, 110, 1958.
113. **Dwiggins, C. W., Bolen, F. J., and Dunning, H. N.,** Ultra centrifugation determination of micellar characteristics of nonionic detergent solutions, *J. Phys. Chem.*, 64, 1175, 1960.
114. **Cabane, B., Duplessix, R., and Zemb, T.,** An introduction to neutron scattering of surfactant micelles in water, in *Surfactants in Solution,* Lindman, B. and Mittal, K. L., Eds., Plenum Press, New York, 1984.
115. **Bendedouch, D., Chen, S. H., and Koehler, W. C.,** Structure of ionic micellar solutions from SANS, *J. Phys. Chem.*, 87, 153, 1983.
116. **Hayter, J. B. and Penfold, J.,** Self consistent studies of structure and dynamics of concentrated micellar solutions, *J. Chem. Soc. Faraday Trans. 1*, 77, 1851, 1983.
117. **Corti, M. and Degiorgio, V.,** Investigation of aggregation phenomena in aqueous SDS solutions at high NaCl concentrations by quasielastic light scattering, in *Solution Chemistry of Surfactants,* Mittal, K. L., Ed., New York, 1979, 377.
118. **Lindman, B.,** Structure of micellar solutions of nonionic surfactants, NMR self diffusion and proton relaxation studies of PEO alkyl ethers, *J. Phys. Chem.*, 87, 1377, 1983.
119. **Lindman, B.,** Water self diffusion in nonionic surfactant solutions. Hydration and obstruction effects, *J. Phys. Chem.*, 87, 4756, 1983.
120. **Turro, N. J. and Yekta, A.,** Luminescent probes for detergent solutions. A simple procedure for determination of the mean aggregation number of micelles, *Am. Chem. Soc.*, 100(18), 1978.
121. **Zana, R., Yiu, S., Stsrazielle, C., and Lianos, P.,** Effect of alcohols on the properties of micellar systems, *J. Colloid Interface Sci.*, 80, 208, 1981.
122. **Atik, S. S., Nam, M., and Singer, C. A.,** Transient studies on intramicellar excimer formation. Useful probe of the host micelle, *Chem. Phys. Lett.*, 67, 75, 1979.
123. **Lindman, B.,** Structural aspects of surfactant micellar systems, in *Surfactants,* Tadros, Th. F., Academic Press, New York, 1984, 83.
124. **Malliaris, J., Moigne, J. L., Sturm, J., and Zana, R.,** Temperature dependence of the micelle aggregation number and rate of intramicellar excimer formation in aqueous surfactant solutions, *J. Phys. Chem.*, 89, 2709, 1985.
125. **Klevens, H. B.,** Structure and aggregation in dilute solutions of surface active agents, *J. Am. Oil Chem. Soc.*, 30, 74, 1953.
126. **Stigter, D.,** Micelle formation in ionic surfactants. II. Specificity of head groups, micelle structure, *J. Phys. Chem.*, 78, 24, 1974.
127. **Shinoda, K. and Hiral, T.,** Ionic surfactants applicable in the presence of multivalent cations. Physicochemical properties, *J. Phys. Chem.*, 81(19), 1842, 1977.

128. **Corrin, M. L. and Harkins, W. D.,** The effect of salts on critical concentration for the formation of micelles in colloidal electrolytes, *J. Am. Chem. Soc.,* 69, 684, 1947.
129. **Ray, A. and Nemethy, G.,** Effect of ionic protein denaturants on micellar properties of nonionic detergents, *J. Am. Chem. Soc.,* 93, 6787, 1971.
130. **Backlund, S., Rundt, K., Birdi, K. S., and Dalagaer, S.,** Aggregation behavior of ionic micelles in aqueous alcoholic solutions at different temperatures, *J. Colloid Interface Sci.,* 79, 578, 1981.
131. **Green, F. A.,** Interactions of a nonionic detergent. III. Further observations on hydrophobic interactions, *J. Colloid Interface Sci.,* 41, 124, 1972.
132. **Schick, M. J. and Gilbert, A. H.,** Effect of urea, guanidinium chloride and dioxane on the CMC of branched chain nonionic detergents, *J. Colloid Interface Sci.,* 20, 464, 1965.
133. **Goddard, E. D. and Benson, G. C.,** Conductivity of aqueous solutions of some paraffin chain salts, *Can. J. Chem.,* 35, 986, 1957.
134. **Durham, K.,** Properties of detergent solutions, amphipathy and adsorption, in *Surface Activity and Detergency,* Durham, K., Ed., MacMillan, London, 1961, 1.
135. **Nakagawa, T. and Shinoda, K.,** Physicochemical studies in aqueous nonionic surface active agents, in *Colloidal Surfactants,* Shinoda, K., Nakagawa, T., Tamamushi, B., and Isemura, T., Eds., Academic Press, New York, 1963.
136. **Bailey, F. E., Jr. and Callard, R. W.,** Some properties of polyethylene oxide in aqueous solutions, *J. Appl. Polym. Sci.,* 1, 56, 1959.
137. **Flory, P. J.,** *Principles of Polymer Chemistry,* Cornell Press, Ithaca, 1953.
138. **Napper, D. H.,** *Polymeric Stabilizers of Colloidal Dispersants,* Academic Press, London, 1983.
139. **Valaulikar, B. S. and Manohar, C.,** The mechanism of clouding in Triton X-100: the effect of additives, *J. Colloid Interface Sci.,* 108(2), 403, 1985.
140. **Maclay, W. N.,** Factors affecting the solubility of non-ionic emulsifiers, *J. Colloid Sci.,* 11, 272, 1956.
141. **Schott, H.,** Lyotropic numbers of anions from cloud point changes of nonionic surfactants, *Colloids Surfaces,* 11, 1984.
142. **Ananthapadmanabhan, K. P. and Goddard, E. D.,** Aqueous biphase formation in polyethylene oxide-inorganic salt systems, *Langmuir,* 3, 25, 1987.
143. **Ananthapadmanabhan, K. P. and Goddard, E. D.,** The relationship between clouding and aqueous biphase formation in polymer solutions, *Colloids Surfaces,* 25, 393, 1987.
144. **Albertsson, P.,** *Partition of Cell Particles and Macromolecules,* Wiley-Interscience, New York, 1971.
145. **Walter, H., Brooks, D. E., and Fisher, D.,** *Partitioning in Aqueous Two Phase Systems,* Academic Press, New York, 1985.
146. **Kjellander, R. and Florin, E.,** Salt effects on the cloud point of polyethylene oxide-water systems, *J. Chem. Soc. Faraday Trans. 1,* 77, 2053, 1981.
147. **Garvey, M. J. and Robb, I. D.,** Effect of electrolytes on solution behavior of water soluble macromolecules, *J. Chem. Soc. Faraday Trans. 1,* 75, 993, 1979.
148. **Saito, S.,** Salt effect on polymer solutions, *J. Polym. Sci. A,* 1, 7, 1789, 1969.
149. **Shinoda, K. and Takeda, H.,** The effect of added salts in water on the hydrophile-lipophile balance of nonionic surfactants: the effect of added salts on the phase inversion temperature of emulsions, *J. Colloid Interface Sci.,* 32, 642, 1970.
150. **Hoeve, C. A. J. and Benson, G. C.,** On the statistical mechanical theory of micelle formation in detergent solutions, *J. Phys. Chem.,* 61, 1149, 1957.
151. **Aranow, R. H.,** The statistical mechanics of micelles, *J. Phys. Chem.,* 67, 556, 1963.
152. **Corkill, J. M., Goodman, J. F., and Harrold, S. P.,** Theory of micellization of nonionic detergents, *Trans. Faraday Soc.,* 60, 202, 1964.
153. **Mysels, K. J.,** Charge effects in light scattering of association colloidal electrolytes, *J. Colloid Sci.,* 10, 507, 1955.

154. **Sexsmith, F. H. and White, H. J., Jr.,** The absorption of cationic surfactants by cellulosic materials. III. A theoretical model for the absorption process and a discussion of maxima in absorption isotherms for surfactants, *J. Colloid Sci.,* 14, 630, 1959.
155. **Ruckenstein, E. and Nagarajan, R.,** in *Micellization, Solubilization and Microemulsions,* Vol. 1, Mittal, K. L., Ed., Plenum Press, New York, 1977, 133.
156. **Rassing, J., Sams, P. J., and Wyn-Jones, E.,** Temperature dependence of the rate of micellization determined from ultrasonic relaxation data, *J. Chem. Soc.,* 69, 180, 1973.
157. **Tondre, T. and Zana, R.,** On the kinetics of micelle dissolution-formation equilibria in solutions of cationic detergents, *J. Colloid Interface Sci.,* 66, 544, 1978.
158. **Inoue, T., Tashino, R., Shibuya, Y., and Shinozawa, R.,** Kinetics of micelle formation of tetradecyl pyridinium salts, *J. Colloid Interface Sci.,* 73, 105, 1980.
159. **Aniansson, E. A. G. and Wall, S. N.,** On the kinetics of step-wise micellar association, *J. Phys. Chem.,* 78, 1024, 1974.
160. **Kahlweit, M. and Teubner, M.,** On the kinetics of micellization, *Adv. Colloid Interface Sci.,* 13, 1, 1980.
161. **Hall, D. G.,** Micellar kinetics of ionic surfactants, *J. Chem. Soc. Faraday Trans. 2,* 77, 1973, 1981.
162. **Oh, S. G. and Shah, D. O.,** Relationship between micellar life-time and formability of SDS and SDS/1-hexanol mixtures, *Langmuir,* 7, 1316, 1991.
163. **Kahlweit, M.,** Kinetics of formation of association colloids, *J. Colloid Interface Sci.,* 90, 92, 1982.
164. **Oh, S. G. and Shah, D. O.,** Effect of micellar life-time on the wetting time of cotton in sodium dodecylsulfate solutions, *Langmuir,* in press.
165. **Oh, S. G., Kelin, S. P., and Shah, D. O.,** Effect of micellar life-times on the bubble dynamics in SDS solutions, A. I. C. E. Jul., 38(4), 149, 1992.

Chapter 3

FUNDAMENTALS OF POLYMER SOLUTIONS

Matthew Tirrell

TABLE OF CONTENTS

0-8493-6784-0/93/$0.00 + $.50

I. OBJECTIVES IN THE STUDY OF POLYMER SOLUTIONS

A. UNIQUE FEATURES OF POLYMERS

Polymers mixed with solvents, including water, are the key elements of many systems occurring in nature and technology. Coatings, cosmetics, food products, lubricants, and biological systems are but a few examples where the key properties derive, in part, directly from the unique behavior of polymer solutions. The term polymer and macromolecule will be used here essentially synonymously to refer to molecules of which the mass exceeds a few thousand daltons. (Stricter definitions can be constructed, but they have little physical content.) This size places polymers over the range of size scales from atomic/

molecular to colloidal, and as such, they frequently function to manipulate colloidal characteristics based on the nature of the macromolecular structure.

Molecules of this size begin to exhibit interesting new properties owing to their *large spatial extent.* They can interact with other elements of the solution over length scales of tens, hundreds, or even thousands of angstroms, orders of magnitude larger than small molecules. This affects, *inter alia,* their thermodynamic phase behavior, the rheological characteristics they impart to the solution, and their ability to associate with other molecules in solution. A corollary effect of the large spatial extent of polymers is their *uncrossability* and propensity to *entanglement* as their molecular weight and concentration in solution get higher. This results in important features in the dynamic properties (diffusion, rheology, viscoelasticity) of sufficiently dense fluids containing long molecules.[1]

Also related to the large spatial extent of polymer molecules are phenomena deriving from *connectivity.* Macromolecules are covalently bonded strings of atoms; as such, they physically connect and interact with spatially separated regions in a solution, and can propagate effects (for example, stress and structure) from one region to another. While the emphasis in this chapter is on *solutions* of macromolecules, it is essential to realize that it is quite easy to create a different kind of structure, a polymer gel, from a polymer solution by increasing the degree of connectivity.[2] This can be done in several ways. Chemicals crosslinks can be made between different chain backbones, either by nonlinear polymerization or by chemical reactions between preformed polymers. In many of the applications of interest to this book (see Chapter 10 for some examples), macromolecules can associate with one another physically (as opposed to forming covalent bonds), owing to attractive interactions between certain regions on different macromolecules, producing many of the effects of network formation, sometimes known as ''physical gelation''. The degree of this association in a polymer solution can often be affected by the addition of a surfactant. A gel is different from a solution in that the three-dimensional connectivity confers on a gel the properties of a solid (albeit in many instances a soft, pliant one), whereas a true solution is a liquid. Solids often have yield stresses, deforming perfectly elastically at small strains, and in general, gels exhibit more elasticity than polymer solutions, even if the polymer solids content of the gel is as low as that of a dilute solution.

Macromolecules also can have special properties due to the *multiplicity of interactions* that can be built into the same molecule. Multiplicity of interactions in polymers has two aspects. One is what might be called *the interaction multiplier effect;* that is, for a macromolecule, a small interaction between the individual segments and other molecules can be multiplied into a large intermolecular interaction by virtue of the number of times the interaction is duplicated within the same macromolecule. Manifestations of this phenomenon, important in practice, include the fact that (1) the styrene and cyclohexane are miscible in all proportions at most temperatures of interest, while high-molecular-weight polystyrene and cyclohexane have a critical temperature for phase separation in the neighborhood of 35°C;[3] (2) polyethylene oxide (and other polymers) will adsorb effectively irreversibly from good solvents such as water, on surfaces where the binding energy per segment is small;[4] (3) two different polymers generally exhibit limited mutual miscibility.[5]

The second significant aspect of the multiplicity of interactions in which polymers can engage arises from the synthesis of *copolymers*. More than one kind of monomer can be incorporated into the same macromolecule, leading to a polymer that exhibits more than one kind of affinity or interaction. This characteristic can be manipulated to great advantage to optimize the properties of polymers, particularly through the ability to adjust thermodynamic or associative interactions, including adsorption. For example, subunits can be built into macromolecules that are attractive to some other constituents of the solution while being repulsive (nonadsorptive) to solid surfaces, or vice versa. Additional structural variations, beyond copolymerization, can be built into macromolecules through the degree and nature of branching. Nonlinearity in the backbone structure of a polymer affects its solution properties in several ways, owing essentially to the fact that branching makes polymers denser and more compact than their equal molecular weight linear counterparts.

B. MOLECULAR WEIGHT AND ITS DISTRIBUTION

Discussion of the synthesis of polymers in any depth is beyond the scope of this chapter. There are excellent alternative sources for that information.[6-8] However, certain generic features of polymer synthesis affect most polymer solution properties. All polymerization reactions involve elements of randomness in the rates of initiation, propagation, transfer, and termination reactions. This, in turn, leads to a distribution of molecular weights in all synthetic (and some biological) polymers. This distribution can be fairly narrow, as in the case of polymers synthesized by anionic polymerization, where the mole fraction of polymers of a particular chain length n, P_n, is given by a Poisson distribution:

$$P_n = x^n e^{-x}/(n - 1)! \tag{1}$$

where x is the average chain length. Chains experience few termination reactions in pure anionic polymerization conditions. More common is the geometric distribution of chain lengths,

$$P_n = \alpha^{n-1}(1 - \alpha) \tag{2}$$

resulting from a random distribution of lifetimes of growing chains during the polymerization, as is free-radical or condensation polymerization, where α is the probability that the chain continues to propagate as opposed to terminating at a particular chain length. This probability is controlled by several factors, such as stoichiometry and the reaction kinetics, depending on the type of polymerization. Polymerization reactions under direct control of the genetic code produce virtually perfectly uniform polymers with respect to molecular weight;[9] polymers produced by heterogeneous catalysis typically produce very broad distributions of size.[10] Exceptions to the molecular-weight uniformity of biological polymers are found among the polysaccharides.[11]

Since accurate experimental determination of the full distribution of molecular weights, as typified in Equations 1 and 2, is difficult, and since many properties are directly related to some average of these distributions, characterization of a polymer sample is done, first and foremost, by average chain lengths or molecular weights. These averages are defined in standard ways from the moments of the distributions, where m_k is the *k*th moment of the distribution:

$$m_k = \sum n^k P_n \tag{3}$$

The sum is taken over all possible values of the chain length. These moments are then used to calculate the average chain lengths or molecular weights:

$$\begin{aligned} \langle DP \rangle_n &= m_1 \text{ (number average chain length or degree of polymerization)} \\ \langle M \rangle_n &= M_o \langle DP \rangle_n \text{ (number average molecular weight)} \end{aligned} \tag{4}$$

$$\begin{aligned} \langle DP \rangle_w &= m_2/m_1 \text{ (weight average chain length or degree of polymerization)} \\ \langle M \rangle_w &= M_o \langle DP \rangle_w \text{ (weight average molecular weight)} \end{aligned} \tag{5}$$

where M_0 is the monomer molecular weight. (These equations can be extended in obvious ways for copolymers.) The standard measure of the breadth of a distribution of molecular weights is the polydispersity, Q, defined as the ratio of these two averages:

$$Q \equiv \langle DP \rangle_w / \langle DP \rangle_n \tag{6}$$

which becomes unity for a uniform, or monodisperse, polymer. For the common geometric distribution (Equation 2), Q is readily shown by the formulas above to be:

$$Q_{geometric} = (1 + \alpha)/(1 - \alpha) \tag{7}$$

converging to a value of 2 as the propagation probability becomes high. Polydispersity for the narrower anionic polymerization distribution (Equation 1) converges to 1 at high molecular weight.

Some specific classes of structures of common polymers are illustrated in Table 1. Synthetic polymers are most often made by addition polymerization of the carbon-carbon double bond, as in the cases of all the major commodity

TABLE 1
Structures of Some Common Polymers

Structure	Generic name	Examples	Common synthetic route
$-(-CH_2-\underset{X}{\underset{\vert}{CH}}-)-$	Vinyl polymer	X = H, polyethylene X = phenyl, polystyrene X = COOH, polyacrylic acid X = OH, polyvinylalcohol	Free radical, ionic, or catalyzed poly-merization
$-(-CH_2-CH-\underset{X}{\underset{\vert}{C}}-CH_2-)-$	Diene polymer	X = H, polybutadiene = CH_3, polyisoprene	Free radical, ionic, or catalyzed poly-merization
$-(-R_1-\overset{O}{\overset{\Vert}{C}}-\overset{H}{\overset{\vert}{N}}-R_2-)-$	Polyamide (Nylon)	$R_1 = C_2H_4$ $R_2 = C_3H_6$; polycaprolactam (Nylon 6)	Condensation poly-merization
$-(-R_1-\overset{O}{\overset{\Vert}{C}}-O-R_2-)-$	Polyester	$R_1 = -O-\overset{O}{\overset{\Vert}{C}}-C_6H_4-$ $R_2 = C_2H_4$ Polyethyleneterephthalate (Dacron, Mylar)	Condensation poly-merization
$-(-R_1-\overset{H}{\overset{\vert}{N}}-\overset{O}{\overset{\Vert}{C}}-O-R_2-)-$	Polyurethane		Step growth poly-merization

$-(-[CH_2-]_nO-)-$	Polyether	n = 2, polyethyleneoxide	Step growth polymerization
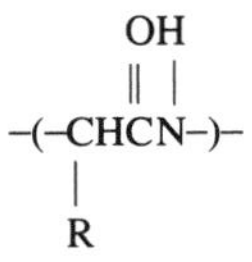 $-(-CHCN-)-$ with $=O$ and $-H$ on the C and N, and R hanging from the first C	Polypeptide	R = H, polyglycine R = $-(-CH_2-)_2\ N^+H_3$, polylysine R = CH_2-COO^-, polyaspartic acid (latter two are examples of polyelectrolytes)	Solid phase peptide synthesis, genetic engineering
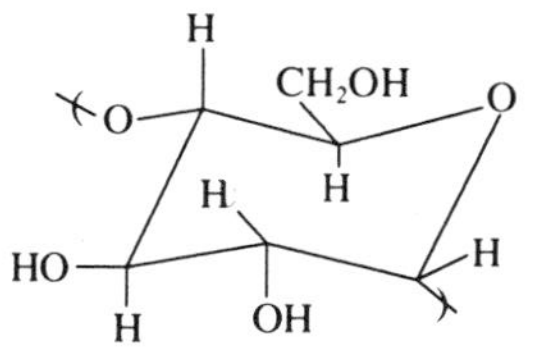 [structure of an amylose repeat unit: pyranose ring with O, CH_2OH, OH, HO and H groups]	Polysaccharide	Amylose (as illustrated)	Enzymatic synthesis

polymers (polyethylene, polypropylene, polystyrene, and polyvinylchloride), or by step growth condensation polymerization, as in the cases of the synthetic fiber-forming polymers (polyamides and polyesters). Polypeptides and polysaccharides are technologically useful polymers that occur in nature, or can be made synthetically, sometimes by genetic engineering techniques which rely on genetically modified organisms to synthesize the molecules.

C. AIMS

The aims of the study of the properties of polymer solutions are twofold. One aspect is the characterization of the polymer itself. Through understanding of the relationship of molecular structure to measurable solution properties, one can aspire to determine molecular-level characteristics such as molecular weight (and its distribution), chain configuration, the amount of charge on an ionizable polymer, or the degree of association in a polymer capable of that sort of interaction. The roles of these characteristics in determining the properties will change as the solution concentration changes.

The second general aim in the study of polymer solution properties cuts in the opposite direction. How can the properties of polymer solutions be manipulated by the changes that one can make in the characteristics mentioned above? As discussed in the opening paragraphs, the overriding key feature of polymer solutions is the large spatial extent of the macromolecules. We therefore begin this overview of the fundamental properties of polymer solutions with a discussion of the factors affecting the configurations of macromolecules in solution. This discussion will be directed almost exclusively toward solution of synthetic polymers, in particular, as protein solutions will be described in a later chapter.

II. FACTORS AFFECTING POLYMER CONFIGURATION

A. RANDOM WALK MODEL

1. Average Dimensions

The primary model of the liquid-state configuration of a polymer molecule is that of the random walk. We will consider only homopolymers here, containing just one sort of repeat unit; this repeat unit of length ℓ is the basic element of the model random walk. For vinyl polymers or polyolefins, this is the length of the -(–C–C–)- repeat unit, of order 2.5 Å, as illustrated in Table 1. Suitable definitions of the repeat units for more complex polymers such as polyamides or polyurethanes can be made.[12] The end-to-end vector, **r**, of the entire macromolecule is one convenient means to measure the characteristic size of the polymer configuration, that is, by measuring how much space it spans. This vector can be expressed as the sum of the individual repeat unit or segment vectors:

$$\mathbf{r} = \sum \ell_i \tag{8}$$

where the sum is taken from the first segment vector to the last ($i = 1$ to n). Since, in the fluid state, the macromolecule is in a constant state of thermal agitation, this quantity is not a useful one to deal with directly. This is due to the fact that this quantity is fluctuating and the means to measure instantaneous values of **r** are not readily available. Under these circumstances, we direct our attention to average values, and in systems undergoing Brownian motion, **r** will be continually reoriented on timescales of microseconds giving rise to an average value of zero. We therefore deal with the magnitude of **r**, or its square:

$$r^2 = \mathbf{r} \cdot \mathbf{r} = \sum\sum \ell_i \cdot \ell_j \tag{9}$$

which, when we average over the configurational possibilities, is related directly to the average projections of the individual segment vectors on one another:

$$\langle r^2 \rangle = \sum\sum \langle \ell_i \cdot \ell_j \rangle \tag{10}$$

For a completely random walk, there would be no correlation among the directions of the individual segment vectors so that the average projection of any bond vector on any other is zero ($\langle \ell_i \cdot \ell_j \rangle = 0$ for $i \neq j$) giving rise to the classical random walk result, known as the freely jointed chain[13] in polymer literature:

$$\langle r^2 \rangle_{\text{freely jointed}} = \sum \langle \ell_i \cdot \ell_i \rangle = n\ell^2 \tag{11}$$

Among the important features of this result is the fact that the characteristic size scale of a polymer in the fluid state, under conditions where the random walk model is appropriate, is anticipated to vary like the square root of molecular weight.

Real polymer chains are not random walks at the segment level since fixed bond angles and steric hindrances to free rotation about the backbone introduce correlations between the orientations of certain segments. Nonetheless, the random walk model need not be abandoned at this point. If the interactions producing the correlations among the segments occur between pairs of segments that are situated close to one another along the backbone (local interactions), then in many cases the random walk model retains utility. For example, if one introduces just fixed bond angles (θ) into the backbone of a formerly freely jointed chain, one finds that the average dimensions increase:[14]

$$\langle r^2 \rangle_{\text{freely rotating}} = n\ell^2(1 - \cos\theta)/(1 + \cos\theta) \tag{12}$$

by a factor of about two for the tetrahedral angle ($\theta = 109.5$). *However,* the important scaling characteristic of the random walk, $\langle r^2 \rangle^{1/2} \equiv r \sim n^{1/2}$, is retained.

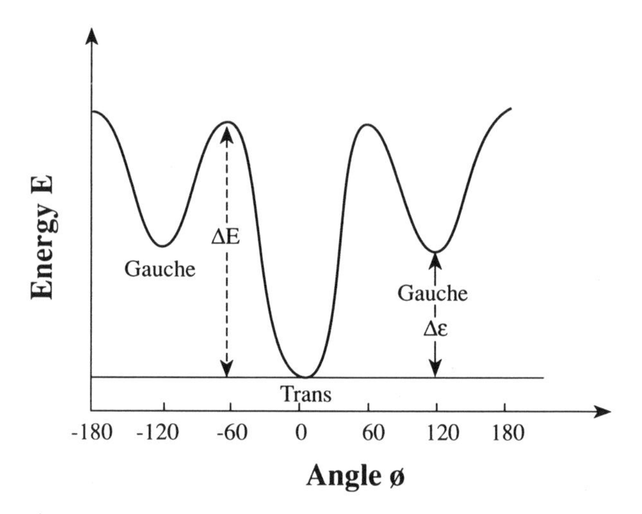

FIGURE 1. Potential energy diagram for the rotation about backbone bonds in a hypothetical, symmetrical polymer. Φ is the rotational angle, ΔE is the trans-gauche barrier, and $\Delta\epsilon$ is the difference in energy between trans and gauche states. (Adapted from de Gennes, P.-G., *Scaling Concepts in Polymer Physics,* Cornell University Press, Ithaca, 1979, chap. 8.)

This is also true when other local interactions along the chain are considered. For example, a typical, flexible, synthetic polymer will exhibit a potential energy surface with respect to rotations around the backbone segments such as that illustrated schematically in Figure 1. Benoit[15] and others[12] have shown that if one averages the rotational angles with respect to a Boltzmann weighting with a rotational potential energy as illustrated in Figure 1, an excellent approximation is

$$\langle r^2\rangle_{\text{hindered rotation}} = n\ell^2[(1 - \cos\theta)/(1 + \cos\theta)] \times [(1 + \langle\cos\Phi\rangle)/(1 - \langle\cos\Phi\rangle)] \quad (13)$$

adding an additional prefactor greater than unity to the basic random walk result, but not changing its scaling form. More advanced and accurate, though conceptually similar, calculations of the effects of molecular structure on $\langle r^2\rangle$ (and directly related properties such as the dipole moment, birefringence, and dichroism), are now being made on the basis of the rotational isomeric state model.[16-18] Commercial software is also available for this purpose.[19]

2. Effects of Chain Stiffness

When only local correlations such as those arising from bond angles and steric hindrance are considered, the final result for the dimensions of the molecule can always be expressed as:

$$\langle r^2\rangle_0 = Cn\ell^2 \quad (14)$$

defining a quantity known as the characteristic ratio, C, which physically is the ratio of the actual dimensions of a chain in the fluid state to what they would have been if the chain of n segments of length ℓ had performed a truly random walk. The subscript zero denotes that this formula is limited to local interactions, producing what is known in the polymer literature as the "unperturbed" dimensions of the macromolecules.

The characteristic ratio is a configurational property of polymers where the information on local stiffness, interactions, and ability to rearrange configurationally resides. C is higher for stiffer polymers and lower for more flexible ones. A typical range for values of C is from about 4 to 10, with chains comprising heterolinkages, such as the flexible ether bonds in the backbone of polyethylene oxide, occupying the low end of this range, and chains with bulky side groups (polystyrene) or rigid backbone links (polyamides or polypeptides) at the high end. Values of the characteristic ratio have been tabulated for many polymers.[12]

Another parameter which is useful in gauging the stiffness of a macromolecule, particularly charged macromolecules, is the persistence length, which, qualitatively, is a measure of the probability, determined by the Boltzmann weight of the energetic penalty, to bend a chain. For neutral polymers, the energy penalty to bend is essentially the energy difference between trans and gauche states, $\Delta\epsilon$ in Figure 1. The persistence length, p, then is roughly: $p \approx \ell \exp(\Delta\epsilon/kT)$.[20] This parameter varies from very high ($\geq n\,\ell$) when the penalty to bend is high, down to a few angstroms ($\approx\ell$) when the penalty is low. When p is a significant fraction of the total contour length of the entire chain ($n\ell$, which is a practical upper bound on p), the random walk model developed here is no longer valid; the chain is more rod-like in this regime. A useful model for the intermediate stiffness regime between flexible coil and rod is the Porod-Kratky wormlike chain,[21] for which the mean square dimensions are given by:

$$<r^2>_0/n\ell^2 = p/\ell - (p^2/n\ell^2)[1 - \exp(-n\ell/p)] \tag{15}$$

As the contour length becomes large relative to the persistence length (that is, as $n\ell/p \rightarrow \infty$), p is approximately ℓ, and $<r^2>_0$ is given by the random walk result (Equation 11). In the opposite limit p is approximately $n\ell$, and the chain is fully extended: $<r^2>_0 = n^2\ell^2$. Equation 15 interpolates between the two limits. This kind of formula is particularly useful in situations where a macromolecule experiences interactions which cause it to stiffen, such as by the increase of charge density along the chain, or by the binding onto the chain of other species in the solution (e.g., surfactants). Attention can be focused on the persistence length as the relevant parameter to characterize the configurational change.

3. Configurational Distribution

Returning to the random walk model, since all of the effects of the local interactions that we have discussed so far can be lumped into the parameter

C, it is possible to recover the simple random walk model in its entirety by redefining the segments of the random walk, that is, by defining an equivalent freely jointed chain with N segments of length a:

$$<r^2>_0 = Na^2 \tag{16}$$

where N and a refer to the number and size, respectively, of what have come to be called Kuhn statistical segments.[22] Obviously, by comparing Equations 14 and 16, and noting that $n\ell = Na$, we see that $N = n/C$ and $a = C\ell$. Equation 16 means that, unless we are interested in a local scale along the chain (less than about C monomers), then it is reasonable to treat the chain as a truly random walk. The chain is stiff over local scales but globally executes a random walk. There are two caveats to this: (1) that the molecular weight be high enough that it does not approach the total chain contour length (for typical values of C this is not an issue except for very short chains); (2) that only the local interactions along the chain that we have discussed so far come into play (longer-ranged interactions will be taken up shortly).

Thus far, we have dealt only with one measure of the average dimensions of the polymer chain, its mean square end-to-end distance. The establishment of the validity of the random walk model, however, opens up all of the power of the theory of random walks[13] to help us understand and model polymer configurations. Under these circumstances, the distribution of chain configurations is governed by a diffusion-type equation:[23-26]

$$\partial G(\mathbf{r},\mathbf{r}', N)/\partial N = (a^2/6)\, \nabla^2 G(\mathbf{r},\mathbf{r}', N) \tag{17}$$

where $G(\mathbf{r},\mathbf{r}', N)$ is the probability that a chain with one end at $\mathbf{r}'$ places its Nth segment at $\mathbf{r}$. For a single chain in dilute solution, it causes no loss of generality to place one end at $\mathbf{r}' = 0$, and to expect that, as $\mathbf{r}$ gets very large, G will go to zero, stipulations that lead to the solution of Equation 17 as a Gaussian distribution of end-to-end distances:

$$G(\mathbf{r},N) = (\beta/\pi^{1/2})^3 \exp(-\beta^2 r^2) \tag{18}$$

with $\beta = (3/2)^{1/2}/N^{1/2}a$. This is the origin of the common appellation ''Gaussian chains'' applied to random walk polymers.

Knowledge of this distribution function permits several additional insights. Computation of $<r^2>_0$ by multiplying Equation 18 by r^2 and integrating with respect to $\mathbf{dr}$ ($= 4\pi r^2 dr$) gives the expected result of Equation 16. With the distribution function in hand, we can also compute other configurational average properties. Foremost in importance among these is the radius of gyration, $<s^2>_0^{1/2}$, which is the root-mean-square distance of any segment (from now on, read ''segment'' as ''statistical segment'') from the center of mass of the chain. For a Gaussian chain, this is[27]

$$<s^2>_0 = Na^2/6 = <r^2>_0/6 \tag{19}$$

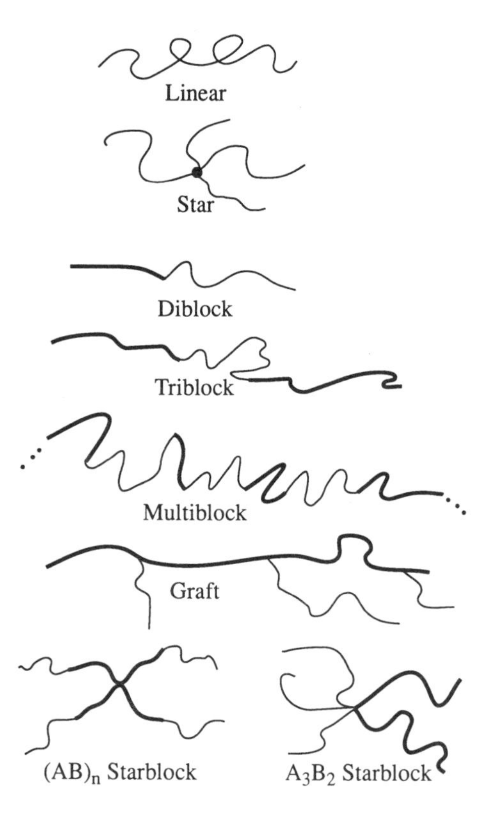

FIGURE 2. Schematic illustrations of various polymeric architectures.

Since there is this proportional relationship between mean radius of gyration and end-to-end distance for the Gaussian chain, the two quantities are used interchangeably to describe the sizes of linear Gaussian chains.

4. Branched Chains

For branched chains, the situation is different. End-to-end distance is no longer useful as a configurational measure. Many synthetic polymers, as well as polysaccharides, are used in branched form; some of the relevant architectures are illustrated in Figure 2. Yamakawa[18] and others[28,29] show how the radius of gyration can be calculated for a variety of nonlinear polymers. A common measure by which branching is characterized, both experimentally and theoretically, is by the ratio g, defined by:

$$g \equiv \langle s^2 \rangle_{0,b} / \langle s^2 \rangle_{0,l} \quad (20)$$

where the subscripts b and ℓ refer to branched and linear chains of the same total degree of polymerization, N. The branching ratio, g, is less than 1, in general; branched polymers are more compact objects than their linear counterparts of the same molecular weight. For example, for f-arm star macromolecules (Figure 2a) with arms of uniform molecular weight, $g = (3f - 2)/f^2$,[28] a reduction in radius of gyration of more than 20% for a three-arm star and almost 40% for a four-arm star. These effects are even larger when translated into the volumes of the molecules ($\sim \langle s^2\rangle_0^{3/2}$), which we will show subsequently is the principal determinant of the contribution that a dissolved macromolecule makes to the *viscosity* of a dilute polymer solution. Recent experimental work in the determination of g has been summarized and presented.[30,31]

5. Fluctuations of Chain Configurations

The distribution function can also be used to compute other moments and averages of the distribution. Higher-order moments (such as $\langle r^4\rangle_0 = (5/3)(Na^2)^2$ for the Gaussian chain) are useful in assessing the deviations of the population of configurations from the mean. The Gaussian distribution of Equation 18 is spherically symmetric, but it is important to realize that *the instantaneous configurations of any individual chain are not spherically symmetric*. In principle, this asymmetry could be determined by measuring some off-diagonal moments of the distribution of configurations;[32] in practice, this measurement is difficult, owing to the sensitivity and timescales of the configurational fluctuations involved.[33] Computer simulations are useful here and they show that the chain instantaneously is typically roughly ellipsoidal with a major axis aligned along the end-to-end vector.[34]

6. Effects of External Force on Chain Configuration

Force applied to a flexible object such as a polymer chain, such as occurs on stretching a polymer gel or shearing a polymer solution, will distort its configuration at the cost of a free energy penalty arising from the reduction in the number of configurations available to the N segments of the chain in the stretched state. If we take $G(\mathbf{r}, N)$ of Equation 18 as the partition function of the chain, then the Helmholtz free energy of the chain can be written:

$$A = A_o + kT \ln G(\mathbf{r},N) = K - kT\beta^2 r^2 = K - kT\beta^2(\mathbf{r}\cdot\mathbf{r}) \quad (22)$$

where A_0 and K are constants. An incremental change in $\mathbf{r}$ will produce an increment in the free energy of the chain given by:

$$dA = kT\, d[\ln G(\mathbf{r},N)] = 2kT\beta^2(\mathbf{r}\cdot d\mathbf{r}) \equiv \mathbf{F}\cdot d\mathbf{r} \quad (23)$$

Comparing the last two parts of this sequence of equations leads to:

$$\mathbf{F} = 2kT\beta^2\mathbf{r} \quad (24)$$

which has the form of a linear elastic spring, with effective spring constant $2kT\beta^2 = 3kT/Na^2$. Likewise, from Equation 22, we see that the energy of the chain due to stretching rises with the square of the extension:

$$A \approx 3kTr^2/2Na^2 \quad (25)$$

a result expected for linearly elastic springs. These are very useful formulas in constructing simple models of other important phenomena. Examples of where this kind of result is used in models of other effects include the bead-spring theories of polymer rheology,[35] swelling and stretching of polymeric gels and networks,[36] and the effects of excluded volume in swelling individual chains. This kind of simple model has several limitations which must at least be borne in mind in using these results. They are limited to relatively small deformations where the chain is not too strongly distorted from its Gaussian shape. Flory[37] and Bird et al.[35] discuss modifications to the force law more appropriate at higher extensions. No other possible contributions to the free energy of the chain (for example, segment stretching or bending, stretch-induced crystallization[37]) have been considered; these are seldom important in polymer solutions, except when the forces being applied to the chains are sufficiently high that chain breaking and degradation come into play as well.

B. LONG-RANGE INTERACTIONS WITHIN A SINGLE CHAIN

1. Excluded Volume

To this point, we have dealt only with effects on the global polymer configuration arising from short-range interactions along the chain. Solvation of polymers in good solvents introduces long-range interactions among the segments owing to the fact that segments distant from one another want to maximize their solvation and therefore avoid proximate locations in space. This self-avoidance characteristic, which is strong in good solvents, swells the chains from their Gaussian dimensions and destroys the validity of the random walk model. Of course, self-avoidance exists among the bare segments, irrespective of the solvent in which they are. However, this weak self-avoidance among the bare segments is rendered negligible in two situations where the ideal, Gaussian, random walk does match reality. The first is in a so-called *theta solvent,* which is a weak solvent in which the tendencies to swell due to self-avoidance are exactly compensated by a slight collapsing tendency induced in the polymer chain by the weak solvent. *Theta solvents are very important conceptually in polymer science and are useful for chain characterization, owing to this feature of adoption of ideal dimensions.* The second is the pure polymer fluid, the *polymer melt.* In this situation, the swelling tendency is negated by the intervention of segments coming from other polymers, screening the effects of self-avoidance. Chains in melts have ideal, Gaussian, global properties.

For chains in good solvents, self-avoidance must be recognized since it produces major swelling of the polymer chains. There are many sophisticated treatments of the phenomenon since the self-avoiding walk is a longstanding problem in the statistical physics of many-body interactions.[38,39] One theoretical approach is to modify Equation 17 by the introduction of a self-consistent field term which perturbs the random walk by making the chain pay a penalty in potential energy for proximal placement of segments. We will adopt a simple, nonrigorous approach, developed by Flory,[40] that is illuminating regarding the physics of the phenomenon. This approach regards the final swollen configuration to be dictated by a balance of two energies: F_{os} and F_{el}. The former is the contribution to the free energy of the isolated chain viewed as though it were an isolated cloud of unconnected polymer segments swarming in a certain volume, with a uniform concentration of segments within the swarm. At the level of binary interactions, which might be expected to be reasonable if the solvent were very good and the concentration of the swarm not too high, F_{os} (per chain) could be expressed as:

$$F_{os}/kT \approx vN(N/R^3) \tag{26}$$

where v is a dimensionless excluded volume parameter varying between the 1 for a very good (so-called, athermal) solvent and 0 for a theta solvent. R is the characteristic size of the coil (proportional to either the end-to-end distance or the radius of gyration, since we do not include the numerical constants necessary to make Equation 26 a strict equality). The term in parentheses is the average concentration of the swarm of segments. So, Equation 26 reads that the binary contribution to the osmotic free energy per chain is given by the strength of the binary interactions, times the number of segments per chain, times the average concentration within the domain of the coil.

If the chain swells under the action of this osmotic pressure of segments, then the chain will be stretched and this will induce an elastic energy in the chain, which we will take as being given by Equation 25:

$$F_{el}/kT \approx R^2/Na^2 \tag{27}$$

If one adds these two contributions to the free energy of the chain, then minimizes with respect to R to find the equilibrium swollen dimensions (taking the very good solvent limit of $v = 1$), the result is

$$R \approx aN^{3/5} \tag{28}$$

which, for usual chain lengths N, can be substantially greater than the random walk result of Equation 16, $\langle r^2 \rangle_0^{1/2} \equiv R_0 = aN^{1/2}$, that is, by a factor of $N^{1/10}$. In concrete terms, for $N \approx 1000$, this is an increase by a factor of two in the average linear dimensions of the chain over the random configuration, and an increase by a factor of eight in the volume pervaded by the coil. The swelling is sometimes expressed in terms of an expansion factor, $\alpha(N) \equiv R/$

R_0, or $R = \alpha(N)aN^{1/2}$, which can be calculated in terms of the strength of the excluded volume parameter.[18]

The validity of Equation 28 has been tested, both by computer simulation[41] and laboratory experiment, many times and is very reliable in its prediction of the 3/5 power dependence of the dimensions on chain length.[42] The problems with, and avenues to improve upon, the method of calculation have been discussed in great detail elsewhere.[38,39] For example, the procedure described is not self-consistent since the unswollen result (Equation 16) is used in the denominator of Equation 27, thus the stretching energy is overestimated. So, too, is the osmotic energy overestimated, as discussed subsequently in connection with Equation 48.[43] The advanced methods compute the individual contributions to the swelling more accurately and reliably, and are based on much more rigorous detailed physics; considerable new insight is gained, but they do not change the part of the results expressed in Equation 28 very much.

It is possible, through copolymerization or grafting, to create amphiphilic polymers, where the affinity of one part of the macromolecule for the solvent environment is strong while the other part of the polymer has strong antipathy for the solvent. Under these circumstances the solvent-loving part of the molecule will swell in a manner qualitatively similar to that described here. The part of the molecule that would prefer to avoid the solvent can either collapse into an isolated, minimum-surface-area globule, or, more likely at finite concentrations, associate with other such parts of other macromolecules, leading to micellization, or possibly to the formation of a three-dimensional association network, that is, a gel. This kind of association structure will affect the viscosity of a polymer fluid markedly. In practical applications, such as those discussed in Chapter 10, the degree of association, and consequently many of the physical properties of the fluid, can be manipulated over wide ranges in fluids such as these by addition of suitable surfactants which can disrupt the association structures.

2. Charged Polyelectrolytes

Repulsion between charged segments of a polymer can be of long range and can affect the configuration of a polymer even more strongly than excluded volume. Common polyelectrolytes are those based on sulfonate, phosphate, carboxylate, pyridinium, or charged peptide groups (for example, refer to polypeptide entry in Table 1) incorporated into the polymer. High charge density can effectively straighten the chain into a rod. The wormlike chain of Equation 15 is frequently a good model. The amount of dissolved salt in the solution (most frequently aqueous solution), which can screen the electrostatic repulsions, and the linear density of charged groups along the chain backbone are the principal determinants of the chain configuration.

The expansion of polyelectrolyte chains due to charge repulsion is most often described in terms of the persistence length, as defined in connection with Equation 15, here designated p_{el} to indicate its electrostatic character. The formula for the electrostatic persistence length:[44,45]

$$P_{el} = l_B/(4\kappa^2 b^2 \xi^2) \quad (29)$$

involves several factors. The quantity $l_B = e^2/\epsilon kT$ is the Bjerrum length at which two electronic charges of magnitude e, in a solvent of dielectric constant ϵ, interact with energy kT; κ^{-1} is the Debye screening length, given by $\kappa^2 = 8\pi l_B c_s$, with c_s the concentration of monovalent salt in solution (modifications to multivalent salt are straightforward[44]); b is the distance between charges along the chain (sometimes this is a statistical quantity); ξ is a number between 0 and 1 ($\equiv l_B/b$) that accounts for a phenomenon known as counterion condensation, discussed in the next paragraph. Equation 29 can be used in a wormlike chain model like Equation 15 to predict the configuration of the charged polymer. The papers of Odijk and co-workers[45,46] and of Fixman and Skolnick[47,48] provide more detailed accounts of the models and their nuances. We see from Equation 29 the main features embodied in this model, that the electrostatic persistence length decreases linearly with salt concentration and increases with the square of the charge density along the backbone.

Counterion condensation, advanced as a model and studied extensively by Manning[49-51] and others,[52-55] is a phenomenon that regulates the effective charge density on a polyelectrolyte chain. The maximum supportable charge density on the polyion is controlled by ξ; when the charge separations become less than l_B, counterions condense, or bind to the polyion, neutralizing some of the charge and reducing the effective charge density to the critical value of l_B. Counterion condensation has also been shown to play a role in regulating the charge on ionic micelles,[56-58] which might be thought of as spherical polyions.

The predicted variations in electrostatic persistence length and their effects on global configuration have been the subject of considerable experimental work.[44] Principal tools in these studies have been scattering of light and neutrons, as well as viscometric measurement. Some illustrative results will be given in subsequent sections which discuss these techniques. The model seems to be reasonably accurate in the limits of dilute polyelectrolyte and either extreme of salt concentration, high or low.[48] At high salt concentrations, the electrostatic interactions are screened and the chain regains the random (or self-avoiding) walk character discussed in connection with neutral chains, albeit with possibly a larger segment size owing to local electrostatic stiffening. At low salt concentrations, p_{el} increases rapidly with decreasing salt,[59] as predicted by Equation 29, stretching the polymer into a near-rodlike configuration. The situation at intermediate salt concentrations is presently somewhat unclear regarding the detailed correspondence between theory and experiment. The situation at higher polyelectrolyte concentration is even more complex and will be discussed further in the next section on effects of polymer concentration. The complexity arises because it is necessary, at higher polyion concentration, to account for interactions and screening among the polyions themselves and there is, as yet, no generally agreed upon way of doing this.[60-62]

C. FACTORS PARTICULAR TO WATER-SOLUBLE POLYMERS

Polymers exhibiting water solubility are not limited to polyelectrolytes but a diverse class of important structures, both ionic and nonionic, of both synthetic and biological origins.[63] In many cases, particularly in polymers of biological origin, the macromolecules are intricate, multicomponent copolymers. For proteins, there are 20 naturally occurring amino acid monomers to choose from; polysaccharides comprise more than 100 monomeric sugars and sugar derivatives.

Water plays a key role in determining the properties of aqueous solutions. Solvation of polymer chains may result from the interaction of ionic, polar, or hydrogen-bonded hydrophilic segments with water. Some water-soluble polymers contain monomers that have amphiphilic character themselves. In this sort of case, solvation of the polymer may involve the hydrophobic interaction, in which the local structure of water in the neighborhood of the hydrophobic portion of the segment is thought to play a role.[64] Some polymers, such as polyacrylic acid or polyacrylamide, precipitate from aqueous solutions when cooled, whereas others, such as polyethylene oxide, polypropylene oxide, or polymethacrylic acid, precipitate when heated (see Section III.B.1. for a more general discussion of polymer phase behavior). Addition of other solution components, such as salt, cosolvents, or surfactants, or changing the molecular weight of the polymer, can affect the solubility strongly.

Important nonionic, water-soluble polymers include polyethylene oxide, polyacrylamide, polyvinyl alcohol, and poly-*N*-vinylpyrrolidinone. In general, water solubility in nonionic polymers derives from strong polarity or hydrogen-bonding functionality in the monomer units, leading to favorable interactions with water molecules. Polyelectrolytes come in important anionic and cationic varieties. Associated with these polyions are counterions, which maintain electrical neutrality and whose characteristics of size and valency can affect the properties of the polyelectrolyte solution. Polyelectrolytes are frequently solvated in water in spite of, not because of, the intrinsic water solubility of the monomer segment. Rather, solvation takes place because of the ability of swelling in water to diminish the strong electrostatic repulsions among segments by separating them.[65] This has as a consequence that neutralization of the polyion, for example by changing pH, can lead to precipitation of the polymer. Anionic polyelectrolytes most often contain carboxylic acid or sulfonic acid functionality. Phosphoric acid polyelectrolytes are also possible. Whereas the pK_a of carboxylic acid polyions is in the neighborhood of 6, and therefore the degree of charge can be manipulated with pH, sulfonic acid groups are typically such strong acids that they are not neutralizable with increasing pH. Cationic polyelectrolytes come in three main varieties: ammonium, sulfonium, and phosphonium. The most important category of ammonium polymers includes amines and quaternary ammonium polymers, as well as polyimines and polyvinylpyridinium polymers. As these polymers frequently have organic groups in the ammonium moiety, these polymers can vary widely (and be manipulated to desired conditions) in their hydrophobicity. This can be used to control intermolecular associations.

Amphoteric polyelectrolytes contain both anionic and cationic charges. Such materials can be zwitterionic, with both positive and negative charges on the same pendant group, or ampholytic, where the groups are on the same backbone, but not the same segment. Zwitterionic amphoteric polymers are, per force, charge balanced; polyampholytes need not be. Amphoteric polymers exhibit a wide range of interesting configurational and association behavior, based on electrostatic attractions.[66] For example, if they embody the right balance of positive and negative charges, polyampholytes can exhibit behavior that is in its effects opposite from usual polyelectrolyte behavior. For example, polyampholytes can *expand on adding* salt and *contract on the removal* of salt, because electrostatic attractions contract the coils and are weakened (strengthened) on adding (removing) salt.

D. EFFECTS OF POLYMER CONCENTRATION ON CHAIN DIMENSIONS

As polymer concentration is increased in a solution of a polymer in a good solvent, intermolecular interactions become increasingly important. This causes the dimensions of individual polymer chains to shrink from their swollen, self-avoiding state toward the ideal, Gaussian state of the theta condition. The reason for this is understandable in terms of the scaling theory formulated by de Gennes.[43] In the dilute solution conditions we have discussed thus far, the characteristic length which has been the object of our concern is the average linear dimension of the polymer, R. As we put more macromolecules of size R into a solution, we come to a concentration where the polymer chains begin to overlap one another. This will happen at a concentration, c, of polymers in solution, where the solution concentration is approximately equal to the average concentration of segments within the domain of a single polymer coil ($\approx N/R^3$, see Equation 26). This defines the *overlap concentration, conventionally called c*,* expressed in terms of numbers of segments per unit volume, $\equiv N/R^3$, or the overlap volume fraction of polymer $\phi^* \equiv Na^3/R^3$. This concentration of segments is frequently not very high especially, as we shall see directly, for high molecular weight polymers. For this reason, this regime has come to be called the *semidilute regime*. Using Equation 28, we see that in good solvents,

$$\phi^* = N/(N^{3/5})^3, \text{ or } = N^{-4/5} \tag{30}$$

Clearly, for polymers of significant molecular weight, ϕ^* can be quite small.

Nothing dramatic happens in the properties of the solution at the overlap threshold, though some attempts have been made to measure it by fluorescence spectroscopy.[67] Viscosity measurements to be discussed later in this chapter provide a good operational, experimental definition. At polymer concentrations above ϕ^*, R gradually loses meaning as the important characteristic length scale governing the properties of polymer solutions. It is replaced by another, smaller length, ζ, which is the average distance between segments of different polymer chains. This length will decrease with increasing concentration from its value at the onset of overlap ($\zeta \approx R$) down to some very small value of the order of the segment size ($\zeta \approx a$) as the polymer solution becomes practically a pure polymer fluid. Its decrease with concentration in the regime just beyond overlap can be predicted by a scaling argument which begins with the idea that ζ should decrease as a power law with concentration above ϕ^*:

$$\zeta = R(\phi^*/\phi)^n \tag{31}$$

The exponent n is then deduced via the physically reasonable hypothesis that, above ϕ^*, ζ should be independent of N. In more plain words, once the macromolecules are overlapped, the characteristic length scale of the solution, the mean distance between monomers, is determined only by the concentration (or volume fraction) and not by the molecular weights of the polymers to which the monomers are attached. Since $R \sim N^{3/5}$ (in a good solvent) and $\phi^* \sim N^{-4/5}$, clearly $n = -3/4$ to make the molecular weight dependence of ζ nil, leading to:

$$\zeta/R \approx \phi^{-3/4} \tag{32}$$

Since this characteristic length scale can be picked up in scattering experiments, Equation 32 has been examined experimentally and found reasonably accurate.[68,69]

This result can be used to examine how chain dimensions change in good solvents as the concentration of polymer is increased. This shrinkage results from the screening of excluded volume interactions within a chain by the intervention of other chains mentioned in Section II.B.1. Above ϕ^*, excluded volume interactions are screened at length scales greater than ζ. A reasonably useful model for this can be developed by arguing that the smaller-scale elements of the chain remain swollen, but the overall dimensions of the chain can be thought of as a Gaussian chain of elements of size ζ (compare Equation 16):

$$R(\phi) = N_e(\phi)\ \zeta^2(\phi) \tag{33}$$

where $N_e(\phi)$ is the concentration-dependent number of these swollen elements along the chain. If these elements smaller than ζ remain swollen due to

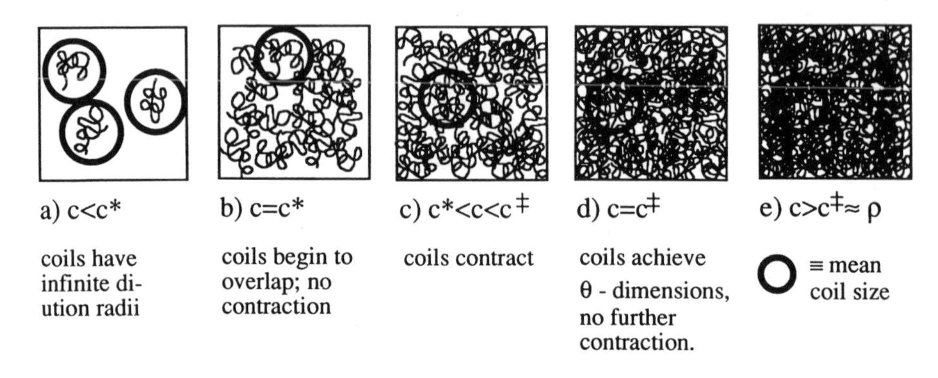
a) c<c*
coils have
infinite di-
ution radii
b) c=c*
coils begin to
overlap; no
contraction
c) c*<c<c‡
coils contract
d) c=c‡
coils achieve
θ - dimensions,
no further
contraction.
e) c>c‡≈ ρ
≡ mean
coil size

FIGURE 3. Schematic illustrations of polymer chains in solutions of increasing concentration.

excluded volume, their size ζ is related to the number of monomers in these elements, g, by Equation 28:

$$\zeta = ag^{3/5} \text{ or } g(\phi) = [\zeta(\phi)/a]^{5/3} \quad (34)$$

Since $N_e = N/g$, we can use Equations 32 to 34 to show that:

$$R(\phi)/R(\phi^*) = \phi^{-1/4} \quad (35)$$

This predicts that, in the semidilute regime, the size of the coil contracts from its excluded volume dimensions at ϕ^* down to its random walk dimensions at some higher concentration where all the excluded volume has been screened out. From there, the chain contracts no further as it has reached its unperturbed melt dimensions. This is illustrated schematically in Figure 3. Data testing Equation 35 are shown in Figure 4, where the radius of gyration of polystyrene is measured by light scattering in solutions of polymethylmethacrylate refractive index matched with ethyl benzoate.[70] Reasonably good agreement with Equation 35 is found here and by others.[68,71] Whereas the exponent of -0.25 is not always seen exactly, the numbers found experimentally are of that magnitude indicating, at least, the qualitative correctness and power of a very simplified theory. There are many attempts to improve on this calculation using more sophisticated methods.[72]

Concentrated solutions of polyelectrolytes pose more complex theoretical problems which are as yet incompletely understood.[73,74] One aspect of the complexity is the proper treatment of *electrostatic* screening among the polyions. The Debye length κ^{-1} is a function of polyelectrolyte concentration under these circumstances. Some aspects of these issues will be discussed in the subsequent section.

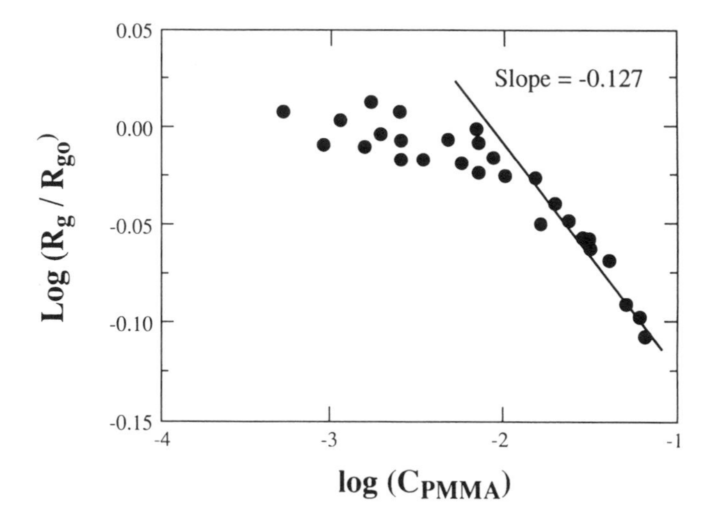

FIGURE 4. Radius of gyration of individual polymer coils as a function of concentration, measured by light scattering in isorefractive solutions. Data in plot are from Reference 70.

III. METHODS FOR DETERMINATION OF MOLECULAR WEIGHT, CONFIGURATION, SIZE, AND ASSOCIATION

A. GENERAL CONSIDERATIONS

The discussion to this point has been directed toward understanding the factors in control of the size and configuration of macromolecules in solution. These factors include chain length, stiffness, solvent quality, electrical charge, and polymer concentration. We focus in this section on methods for the measurement of some of these properties and their effects on polymer configurations. Configurations of polymers in solvents are determined by the balance of segment-solvent and segment-segment interactions. Description of these interactions is the aim of the study of the thermodynamics of polymer solutions, embodied in properties such as the osmotic pressure. Colligative properties, as osmotic pressure, are one of the most direct routes to the determination of the molecular weight of a polymer. In this section, we enumerate briefly the methods that can be used to determine the molecular weight, configuration, size, and degree of association of polymers in solution. This section is an introduction, not a comprehensive survey, for which the interested reader is referred elsewhere.[75,76] The Flory-Huggins theory to be discussed in the next section is the cornerstone of our understanding of the behavior of polymer solutions between the infinitely dilute solution in a good solvent and bulk polymer.

B. FLORY-HUGGINS THEORY, OSMOTIC PRESSURE, AND COLLIGATIVE PROPERTY MEASUREMENT

1. Phase Separation in Polymer-Solvent and Polymer-Polymer Mixtures

Macromolecules in a solvent form an asymmetric solution in the sense that the molecular volume of the solute is much greater than the molecular volume of the solvent.[77] This causes substantial deviations in the behavior of the solution from Raoult's law[78] characteristic of ideal, small molecule solutions. The Flory-Huggins theory, which, in its simplest form, considers the mixing of polymer segments and solvent molecules (of equal volume) on a lattice, has been very successful in rationalizing and predicting many features of polymer solutions. The Flory-Huggins theory considers all lattice sites to be occupied by one component or the other, in other words, incompressible solutions. The Sanchez-Lacombe theory[79] is the extension to lattices with empty sites, enabling the prediction of PVT and other behavior not comprised in the Flory-Huggins model.

The Flory-Huggins expression for the free energy of mixing (per lattice site) in a polymer-solvent mixture of polymer volume fraction ϕ is

$$(F_{mix}/kT)\big|_{per\ site} = \phi/N \ln\phi + (1 - \phi)\ln(1 - \phi) + \chi\phi(1 - \phi) \quad (36)$$

where χ is the interaction parameter measuring the relative energetics of contacts between polymer segments and solvent. The first two terms arise from the entropy of mixing. Notice that the first term, which is the polymer contribution to the entropy of mixing, is strongly reduced by the factor of 1/N; the connectivity of the polymer chain has reduced the number of ways of mixing the segments. This is one aspect of the reduced solubility of polymer relative to monomers mentioned in Section I.A. The other part of this effect comes from the energetics of contacts, or enthalpy of mixing, embodied in the third term. More specifically,

$$\chi = \chi_{ps} - (1/2)(\chi_{pp} + \chi_{ss}) \quad (37)$$

where the subscripts designate the interaction energy, in units of kT, for the indicated binary contact. In an "athermal" solution, $\chi = 0$, i.e., there is no penalty in replacing contacts between like species by heterocontacts. This limit is sometimes approached in polymer-polymer mixtures but never in real polymer-solvent mixtures.[80] Even a very good solvent for a polymer, as in the case of polystyrene in toluene, results in something like $\chi \approx 0.3$ to 0.4. Usually, χ is positive, owing to its typical origins in van der Waals attractions, where the individual χ_{ij}s in Equation 37 can be expressed as $\chi_{ij} = -k\alpha_i\alpha_j$, with the αs being the electronic polarizabilities (related to the refractive indices) of the two species; k is positive since the van der Waals forces are attractive. The net result is

$$\chi = (k/2)(\alpha_s - \alpha_p)^2 > 0 \quad (38)$$

If the polarizabilities are different, χ is positive. This simple picture is modified (χ can be negative for polymer-polymer interactions) if there are sufficiently strong specific interactions, such as hydrogen bonding, between solute and solvent.

The temperature dependence of χ is complex.[81] In the simplest case (van der Waals), the interactions are independent of temperature, so that χ (which is interaction energy/kT) varies like 1/T. This implies, *inter alia,* that there is phase separation of the homogeneous solution into dilute and concentrated polymer solution phases with decreasing temperature. The critical temperature, or more informatively, the critical value of χ, $\chi_c \approx 1/2 + 1/N^{1/2}$, and the critical composition occurs at a volume fraction, $\phi_c = 1/N^{1/2}$. Owing to the large size of polymers, the concentration at the critical point for phase separation is much lower than the critical *mole* fraction of 1/2 in ideal solutions. Monte Carlo simulations, conforming to the basic tenets of the lattice model, but avoiding the numerous assumptions leading to the derivation of Equation 36, have been used to determine the critical properties of the model. What is found, among other things, is that the critical temperature emerging from Equation 36 is too high, and the coexistence curve is considerably flatter in the region of the critical point than predicted by Equation 36.[82]

The same basic approach can be applied to modeling the thermodynamic interaction between different polymer species in a binary (or even multicomponent) mixture of polymers. The Flory-Huggins theory expression appropriate for the free energy of a mixture of two polymers (A and B) analogous to Equation 36 is

$$(F_{mix}/kT)|_{per\ site} = \phi_A/N_A \ln\phi_A + \phi_B/N_B \ln\phi_B + \chi\phi_A\phi_B \qquad (36')$$

Now the ability to mix is diminished even further by the reduced entropy of mixing terms of *both* the components. For this reason, it is very difficult to find miscible polymer pairs. The only real avenue toward enhancing miscibility is through the enthalpy of mixing. If χ is negative, then miscibility in all proportions between two polymers is possible. This can be synthesized into some polymers, especially by copolymerization, incorporating, onto different macromolecules, groups designed to interact attractively with one another. For a mixture of equal chain length polymers ($N_A = N_B = N$), the critical point for phase separation is at $\chi_c = 2/N$, and the critical composition occurs at a volume fraction, $\phi_c = 1/2$. The coexistence curve has the familiar parabolic shape varying like $\phi_A(1 - \phi_A)$. The molecular weight symmetry between the two large molecules produces a symmetric phase diagram.

The Flory-Huggins theory in this form predicts that polymers phase separate when the temperature is lowered, giving rise to so-called upper critical solution temperatures (UCST). In fact, for a variety of reasons, many polymer mixtures phase separate when the temperature is raised, as well, exhibiting lower critical solution temperatures (LCST). One common reason for this is differences in the thermal expansion coefficients of the two components. If the temperature is varied, the two components may be unable to find a mutually satisfactory density at which to form a homogeneous mixture. The Sanchez-Lacombe theory mentioned above is one means to adapt the Flory-Huggins approach to encompass these effects.[79] Some water-soluble polymers are known to manifest this behavior, as described in Section II.C.

The effect of mixing between different polymeric species persists even when they are both dissolved in a good solvent. Adding a solvent that is good for both the polymers diminishes any unfavorable interactions between different polymer species. It also contributes to the overall entropy of mixing in the solution. Therefore, immiscible polymers can frequently be dissolved at low concentration in a good mutual solvent, but phase separation will ensue on increasing polymer concentration.[83]

Micelles are also objects of reduced entropy of mixing, owing to the clustering of individual surfactant molecules in the aggregate. Chapter 5 discusses the mixing of polymers and micelles in solution from the point of view of Flory-Huggins theory.

2. Dilute Solutions and Molecular Weight Determination

While several important critiques can be leveled, and systematic improvements made in the model, no apology needs to be made for the Flory-Huggins theory, which has been the cornerstone of understanding polymer solution thermodynamics for the last 50 years. It serves as a very reasonable basis for our discussion of osmotic pressure and molecular weight measurement. In dilute solution, the last two terms of Equation 36 can be expanded in powers of ϕ to give:

$$(F_{mix}/kT)|_{per\ site} = \phi/N \ln\phi + (1/2)\phi^2(1 - 2\chi) + (1/6)\phi^3 + \ldots \quad (39)$$

If we rewrite this result on a per unit volume basis to escape the reference to an arbitrary lattice, taking the volume of a lattice site to be approximately that of a segment or solvent molecule, a^3, we obtain:

$$(F_{mix}/kT)|_{per\ volume}a^3 = \phi/N \ln\phi + (1/2)v\phi^2 + (1/6)\phi^3 + \ldots \quad (40)$$

where we have introduced the excluded volume parameter, v, of Equation 26, which here is seen to be essentially the (dimensionless) coefficient of the binary interaction contributions to the free energy. This second term of Equation 40 is the osmotic free energy (here on a per volume rather than per chain basis) used in Equation 26. Comparison between Equations 39 and 40 shows that $v = 1 - 2\chi$. For $\chi = 1/2$, or $v = 0$, the second term vanishes; this point of the elimination of the binary interaction term is the theta condition, as discussed in Section II.B.1. Referring back to the discussion under Equation 38, we see that the theta temperature is always greater than the critical temperature; the two converge at large N.

The osmotic pressure exerted by the macromolecules in solution may be deduced from standard thermodynamic operations as:

$$\Pi = -(\mu_1 - \mu_1^0)v_1 \tag{41}$$

In words, the osmotic pressure of the solution is equal to the difference in the chemical potential of the solvent between the solution and pure solvent states, divided by the molecular volume of the solvent. This is, in turn, easily shown to be[84,85]

$$\Pi = -(\partial F_{mix}/\partial V)_{T,P,n_2} \tag{42}$$

where V is the volume of the solution and n_2 refers to the number of polymer segments per unit volume of solution. This expression is the solution analog of the gas-phase relationship expressing the pressure as the derivative of the free energy with respect to volume for a fixed number of gas-phase species. In terms of the definitions of the variables we have been using to describe the solution, this expression is transformed into:

$$\Pi a^3 = -\phi^2[\partial(F_{mix}|_{per\ site}/\phi)\partial\phi]_{T,P,n_2} \tag{43}$$

Performing this operation on Equation 36 gives:

$$\Pi a^3/kT = \phi/N + \ln(1/1 - \phi) - \phi - \chi\phi^2 \tag{44}$$

For very dilute solutions, as ϕ goes to 0, we get:

$$\Pi a^3/kT = \phi/N \text{ or } \Pi/R_GT = c/M \text{ (as } \phi,c \rightarrow 0) \tag{45}$$

where R_G is the gas constant, M is the molecular weight of the polymer, and $c = \phi M/a^3N_{av}$ is the concentration of polymer in weight per unit volume (N_{av} is Avogadro's number).

Equation 45, which clearly has an ''ideal gas-like'' character, is the basis of molecular weight determination by osmotic pressure measurement. Extrapolation of Π/c to infinite dilution yields the molecular weight of the polymer; further information on some of the practical aspects of this determination will be given subsequently. Flory shows that measurements of osmotic pressure on a polydisperse mixture of molecular weights yield the number average molecular weight of Equation 4.[85]

3. Semidilute Solutions

At higher concentrations, $1/N \ll \phi \ll 1$, a series expansion comparable to Equation 39 can also be done, with the result:

$$\Pi a^3/kT = (1/2)(1 - 2\chi)\phi^2 + \ldots \tag{46}$$

since, as ϕ increases, the quadratic part dominates the ϕ/N term arising from the entropy of mixing. Here, we see clearly the interpretation of $v = 1 - 2\chi$ as the second virial coefficient. This is an expression that is anticipated to be valid in the semidilute regime, according to the Flory-Huggins, mean field picture.

A different conclusion about the concentration dependence of the osmotic pressure in the semidilute regime is reached by scaling arguments[86] which parallel those leading to Equation 32 for the concentration dependence of the molecular dimensions in Section II.D. In semidilute solutions beyond the macromolecular overlap threshold, we expect the osmotic pressure, just as the characteristic length ζ, to be independent of molecular weight and to increase from its dilute solution value at the overlap threshold as a universal function of the concentration of polymer segments alone. For example, we assume, following de Gennes,[86] that:

$$\Pi a^3/kT = (\phi/N)(\phi/\phi^*)^m \tag{47}$$

For Π to be independent of N, considering Equation 30, m must be equal to 5/4, or:

$$\Pi a^3/kT \sim \phi^{9/4} \tag{48}$$

in the semidilute regime.

This is different from the quadratic dependence predicted by Equation 46. Since $\phi < 1$, Equation 48 predicts that the osmotic pressure at some concentration in the semidilute regime is *smaller* by a factor of $\phi^{1/4}$ than the mean field prediction of Equation 46. The reason for this lower osmotic pressure is the correlation imposed on the placement of polymer segments by the long-range excluded volume effects discussed in Section II.B.1. Simply put, segments avoid one another and reduce the number of binary contacts, and therefore, the osmotic pressure. This sort of correlation is not accounted for in the Flory-Huggins theory. This deficiency becomes decreasingly important as polymer concentration increases and excluded volume becomes increasingly screened.

4. Experimental Observations

Data on osmotic pressure and its concentration dependence have been obtained traditionally either by direct measurement of the pressure resulting after equilibration of a polymer solution with pure solvent across a membrane impermeable to polymer and perfectly permeable to solvent, or, by an elastic light scattering measurement of the osmotic compressibility ($[\partial\Pi/\partial c]^{-1}$, to be discussed further in the next section). There are many technical points to consider in making an osmotic pressure measurement with a semipermeable membrane.[87] Adequate time must be allowed for equilibration, membranes must be leak-free, truly impermeable to polymer, and nonadsorptive, to enumerate a few. The method is therefore unsuitable for very low molecular weights, for which impermeability is hard to achieve, as it is for very high molecular weights, for which Equation 45 shows the osmotic pressure to be low, thus sensitivity is a problem.

Typical plots of Π/c vs. c will show curvature coming from the higher-order terms in concentration, affecting the accuracy of the extrapolation to 0 to get M. A useful practical approach to this situation is to extend the virial expansion of Equation 44 and write it in the following form:

$$\Pi/c = R_G T/M[1 + (1/2)(a^3 N_{av}/M)vc + (1/3)wc^2 + \ldots] \qquad (49)$$

The third virial coefficient can then be usefully approximated as proportional (with constant g)[85] to the square of the second, $w \approx g(a^3 N_{av}v/M)^2$, so that Equation 49 can be written as a perfect square:

$$\Pi/c \approx R_G T/M[1 + (1/2)(a^3 N_{av}/M)vc]^2 \qquad (50)$$

Thus, a plot of the square root of Π/c. vs. c is usually linear,[85] yielding molecular weight from the intercept and an estimate of the second virial coefficient, or excluded volume parameter, from the slope. (The statistical segment size, a, must be determined independently from measurement of the polymer configuration, in order to determine v.) The theta condition can be identified experimentally as the combination of solvent quality and temperature where the slope of such a plot is 0. Similar work has been done on polyelectrolyte solutions[88-90] where added salt concentration affects the interactions and is reflected in the second virial coefficient. Study of the osmotic pressure and its concentration dependence is therefore useful for the information it gives both on molecular weight and on interactions between macromolecules.

The concentration dependence of the osmotic pressure has also been studied in the semidilute solution regime.[91] Figure 5 shows data of Noda et al.[91] on toluene solutions of poly-α-methylstyrene. The data for different molecular weights conform quite well to the universal behavior predicted in Equation 48. Similar examinations have been made for semidilute solutions of polyelectrolytes, in the presence and absence of added salt. Figure 6 shows the data compiled by Wang and Bloomfield[92,93] on 16 different samples of sodium polystyrenesulfonate studied in different laboratories. With added salt, these data, too, conform to Equation 48, the scaling law developed for neutral polymers. This correspondence has been anticipated by Odijk and colleagues,[45,46,94] provided that the effects of charge could be taken into account via the salt-dependent, electrostatic persistence length of Equation 29.

Without added salt, the behavior is more complex. de Gennes et al.[95] have predicted that, above ϕ^*, the polyions screen one another electrostati-

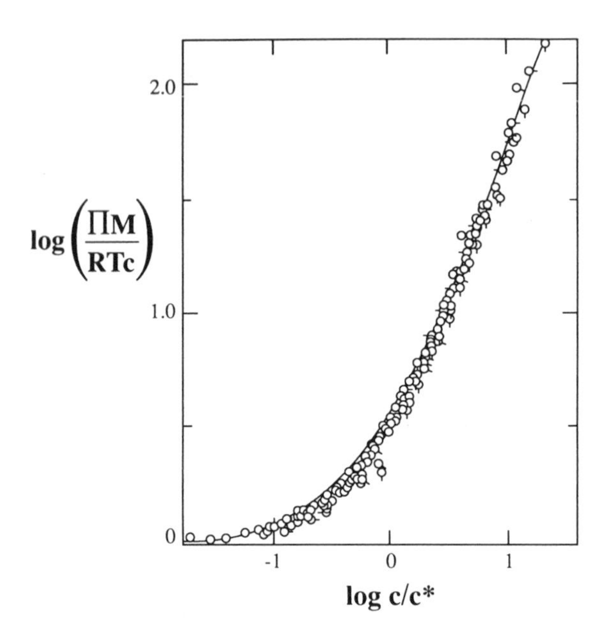

FIGURE 5. Experimental data of Noda et al.[91] on the osmotic pressure of toluene solutions of poly-α-methylstyrene, plotted according to the format suggested by Equations 47 and 48. Slope of plot at high concentration is close to 9/4.

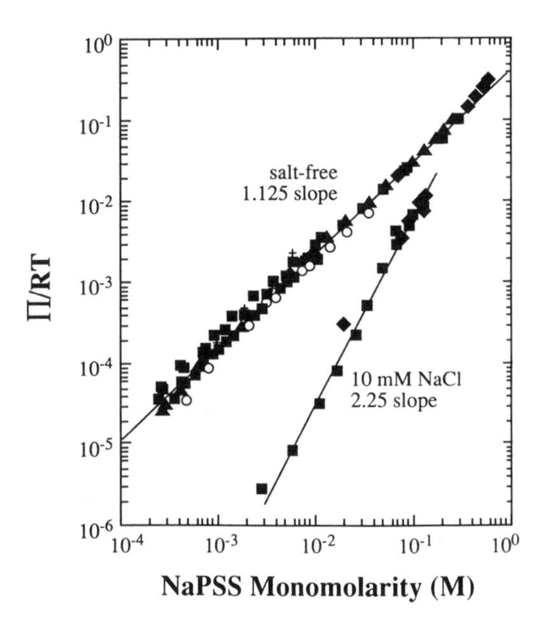

FIGURE 6. Data compiled by Wang and Bloomfield[93] on the osmotic pressure of sodium polystyrenesulfonate, as a function of polyelectrolyte concentration, in the absence and presence of salt, having slopes that agree quantitatively with expectations based on Equations 53 and 48, respectively.

cally, so that the Debye length is dependent on polyelectrolyte concentration under these conditions:

$$\kappa^2 = 8\pi l_B c \quad (51)$$

Odijk and co-workers have argued[45,46,94] that in the presence of electrostatic screening, the following relation should hold:

$$\Pi a^3/kT \sim (p_{el}/\kappa)^{3/4}(bc)^{9/4} \quad (52)$$

If the screening is coming from from other polyions, as in Equation 51, then κ depends on c. The electrostatic persistence length also depends on κ (Equation 29) and the net result for the dependence of osmotic pressure on polyelectrolyte concentration is

$$\Pi a^3/kT \sim c^{9/8} \quad (53)$$

which coincides well with the data of Figure 6 for salt-free conditions. There are indications[93] that at higher polyelectrolyte concentrations there is a crossover from the behavior of Equation 53 to that of Equation 48, presumably because strong mutual screening among the concentrated polyions, even with no added salt, causes them to behave in this respect as neutral polymers.

C. SCATTERING DETERMINATIONS OF MOLECULAR WEIGHT AND MOLECULAR DIMENSIONS

1. Elastic Scattering

a. Measurement of Molecular Weight and Osmotic Compressibility

The most useful tools in the study of radiation scattering by polymers in solution have been light and neutron scattering, owing to the ease with which contrast between solute and solvent can be developed. X-rays can be useful in certain solutions where the requisite electron density contrast is available and also in denser structures that may evolve in or from solution (such as gels); however, we deal only with light and neutrons here.

The experimental observable in an elastic scattering experiment, the total, time-averaged intensity of radiation scattered at a certain angle θ, $I(\theta)$, is usually represented in the following form known as the Rayleigh ratio (assuming unpolarized incident light), dividing out some of the quantities that depend only on the experimental set-up:

$$R_\theta \equiv r^2 I(\theta)/[VI^0(1 + \cos^2\theta)] \quad (54)$$

where r is the distance from the scatterer to the detector, V is the scattering volume and I^0 is the incident intensity. The Rayleigh ratio is closely connected, and directly proportional, to the turbidity of the system. In terms of the molecular characteristics of the solution, R_θ can be written, for light scattering, as:

$$R_\theta = KMc \tag{55}$$

where:

$$K = 8\pi^4 N_{av} \alpha^2 / \lambda^4 M^2 \tag{56}$$

and α is the excess polarizability of the solution over that of the solvent, proportional to the difference of solution and solvent dielectric constants, $\alpha = (\epsilon - \epsilon_0)/4\pi\rho$; ρ is the number density of the solute and λ is the wavelength of light. For experimental purposes, K is more conveniently rewritten in terms of the easily measurable refractive index increment at infinite dilution, $(\partial n/\partial c)$, which reflects the change in polarizability of the solution with increasing concentration:

$$K = (2\pi^2 n_0^2 / N_{av} \lambda^4)(\partial n/\partial c)^2 \tag{57}$$

where n_o is the refractive index of the solvent.

Equation 55 shows that, in principle, the molecular weight could be determined by light scattering from the infinite dilution value of Kc/R_θ. Flory[96] shows that the molecular weight determined by light scattering is the weight average molecular weight of Equation 5. While this is essentially the basis on which light scattering is used for molecular weight determination, there are some additional factors to consider. Principal among these arises from the angular dependence of the scattered radiation. While this slightly complicates molecular weight determination, as we shall see, it aids in molecular size determination. The Rayleigh ratio as defined above was considered to be independent of scattering angle, the factor of $(1 + \cos^2\theta)$ accounting only for the geometry of the scattering experiment. The scatterers were assumed until now to be points. For polymer molecules, the sizes of which approach $\lambda/10$ or even larger, there is additional angular dependence arising from mutual interference among the scattered waves. We postpone discussion of this angular dependence until after we describe further how molecular weight determination is done in the absence of this effect.

In order to obtain accurate molecular weights, the values of Kc/R_θ for Equation 55 must be obtained at infinite dilution. This necessitates extrapolation to zero concentration. Further information can be gained from this extrapolation by examining more deeply the origins of Equations 55 to 57. The basic physics embodied in these equations is that the scattering arises from the heterogeneity in excess polarizability introduced into the solution by the solvent. If this excess polarizability were perfectly homogeneous in its spatial distribution down to the smallest scales, there would be no scattering, since there would be no contrast among any of the elements of the scattering solution. Scattering arises, *inter alia,* from local fluctuations in concentration, which produce local variation in refractive index and polarizability.

The fluctuation theory of light scattering dating to Smoluchowski and Einstein shows how fluctuations in refractive index are related to the concentration dependence of the osmotic pressure, since it is against the osmotic pressure that the molecules must move to increase the concentration of the solution above the mean, locally.[96] These relationships are explained in most detail by Yamakawa.[97] The final result is

$$Kc/R_\theta|_{\theta=0} = (R_G T)^{-1}(\partial\Pi/\partial c) = [(1/M) + (a^3 N_{av}/M)vc + wc^2 + \ldots] \quad (58)$$

where the caveat $\theta = 0$ has been added in view of the questions of angular dependence discussed above. Thus, light scattering can be used to determine the molecular weight and the second virial coefficient, owing to the relationship between scattering and osmotic compressibility. In practice, for the purposes of achieving a linear extrapolation, a manipulation exactly analogous to Equation 50 is useful in the practical analysis of the concentration dependence of light-scattering data.

The discussion of light scattering to this point applies to dilute solutions from the point of view of molecular weight determination. However, Equation 58, relating scattering to osmotic compressibility, is valid and useful well into the semidilute regime. At higher concentrations, further complications may develop. There is evidence in many concentrated solutions for anomalously high scattering arising persumably from clusters, associations, or other long-lived heterogeneities formed in the solution.[98,99] In the absence of as-yet-ill-defined phenomena such as these, the scattered light intensity would be expected to increase with increasing concentration up to the point that the solution is approaching bulk polymer in its concentration, where the intensity would decline again due to the increasing homogeneity of the solution.

Neutron scattering has become an increasingly important tool in the study of polymer solutions, especially nondilute solutions, owing to the possibility to create contrast among different constituents of the solution by selective isotopic labeling, usually with deuterium. Essentially all the phemomenology of scattering introduced above for light scattering applies as well to neutron scattering,[100] except for the mechanism of generating scattering contrast. Whereas electromagnetic waves interact with the electrons in a scattering sample, neutrons are scattered by the atomic nuclei. The quantity that plays the role of $(\partial n/\partial c)$ in a neutron-scattering experiment is the neutron-scattering contrast length density, $(\mathbf{b}/\boldsymbol{\nu})_{neut}$, having the dimension of reciprocal length squared. The data on which these contrast lengths are based, the scattering length densities for individual nuclei, $(\mathbf{b}/\boldsymbol{\nu})_n$, are tabulated for every nucleus and can be calculated for any type of solvent or polymer segment. To make the analogy complete, the contrast length density for light scattering, $(\mathbf{b}/\boldsymbol{\nu})_{light}$,

TABLE 2
Comparison of Contrast Length Densities for Light and Neutron Scattering

System (polymer/solvent)	(b/ν_{light}) (cm^{-2})	$(b/\nu)_{neut}$ (cm^{-2})
Polystyrene(H)/cyclohexane(H)	2.94×10^8	1.74×10^{10}
Polystyrene(D)/cyclohexane(H)		7.12×10^{10}
Polystyrene(H)/cyclohexane(D)		-5.46×10^{10}
Polyisoprene(H)/toluene(H)	1.41×10^8	-0.69×10^{10}
Polyisoprene(D)/toluene(H)		4.36×10^{10}
Polyisoprene(H)/toluene(D)		-5.60×10^{10}
Polyisoprene/bromobenzene	-0.65×10^8	

Note: Values given are temperature dependent and are given here for 20°C. H and D refer to fully protonated or fully deuterated materials, respectively. Scattering intensity is determined by the magnitude of the absolute value of the contrast length density.

can be calculated exactly, using the Lorentz-Lorenz formula, in terms of the refractive index increment, to give:

$$(\mathbf{b}/\boldsymbol{\nu})_{light} = \{9n/[2\pi(n^2 + 2)^2]\}(k^2M_i/N_{av}V_i)(\partial n/\partial c) \qquad (59)$$

where k is the wave number of the radiation ($\sim 1/\lambda$), and M_i and V_i are the molecular weight and volume of species i, which could be either polymer segments or solvent.

The contrast length density is a good basis on which to compare the capabilities of neutron and light scattering. Des Cloiseaux and Jannink[101] have provided some of the relevant numbers, shown in Table 2.

The data of Table 2 show that the contrast possible in neutron scattering can be two orders of magnitude higher than for light scattering, but this is not the principal advantage of neutron scattering. The radiation flux available in neutron beams (of order 10^7 cm^{-2} s^{-1}) is considerably less than the photon flux from a laser (of order 10^{19} cm^{-2} s^{-1}), sometimes necessitating long data acquisition times. The contrast advantage only makes neutrons competitive with light in many applications, but, more importantly, facilitates experimental contrast variation in a way *impossible* for light. As illustrated for the polyisoprene/bromobenzene example of Table 2, with judicious choice of solvent (or by mixing solvents) it is possible to vary light-scattering contrast to negative values or zero. However, this sort of variation *changes the solution under study;* neutron scattering affords the opportunity to vary contrast without varying the polymer-solvent interactions extensively (though isotopic substitution can affect solvent quality, it is, in general, not nearly as large an effect as completely changing the chemical nature of the solvent, as must be done to change the light-scattering contrast).[102] The contrast variation possible with neutrons is particularly useful in concentrated or multicomponent solutions where single molecules or one component can selectively be made to have contrast. Sites for neutron scattering exist in Europe, Japan, and the U.S., either as continuous flux sources from nuclear fission reactors, or as pulsed sources, accelerators in which metals are bombarded with protons; fast neutrons are among the nuclear fragments. Typical wave numbers for neutrons of use in the study of polymer solutions are of order 10 nm^{-1}.

b. Measurement of Average Dimensions and Structure in Solution

Scattering from molecules that are not infinitesimal in size relative to the wavelength of the radiation being used depends not only on the molecular weight but also on the size and shape of the scattering molecule. While this can complicate molecular weight determination slightly, it is a very useful tool to study configurational aspects of polymers. The angular dependence for molecules of polymeric size comes from mutual interference by radiation scattered from different parts of the macromolecule. This is expressed in a quantity known as the intramolecular interference factor, which is proportional to the scattered intensity at the scattering vector $\mathbf{k} = \mathbf{k}_f - \mathbf{k}_i$ ($\mathbf{k}_f$ and $\mathbf{k}_i$ being the wave vectors of the scattered and incident beams):

$$G(\mathbf{k}) = \sum\sum \alpha_n \alpha_m \exp[i\mathbf{k} \cdot (\mathbf{R}_n - \mathbf{R}_m)] \tag{60}$$

where α_n and $\mathbf{R}_n$ are the polarizability and position of the *n*th scattering element or segment, respectively, and the sums are taken from n,m = 1 to N. This quantity replaces α^2 in Equation 55 and 56, which are appropriate to point scatterers. Equation 60 expresses the effects of intramolecular interference arising from the structure (spatial distribution of segments) in the polymer. Since a macromolecule is a fluctuating object, the time-averaged, total intensity of scattered light is proportional to the average of this quantity, which is usually expressed as the (chain length-independent) structure factor:

$$g(\mathbf{k}) \equiv (1/N) \sum\sum \langle \exp[i\mathbf{k} \cdot (\mathbf{R}_n - \mathbf{R}_m)] \rangle \tag{61}$$

which expresses all the information about the placement of the segments of the chain. The averaging indicated in Equation 61 is done with respect to the distribution function of chain configurations appropriate to the type of polymer in question. For flexible chains, the Gaussian distribution of Equation 18 may be appropriate if the excluded volume swelling is not too large; analogous calculations have been done for branched and ring-shaped flexible polymers, as well as rigid spheres and rods.[97] For Gaussian chains, we recognize that $(\mathbf{R}_n - \mathbf{R}_m)$ is the end-to-end distance for the stretch of chain between the *n*th and *m*th segment. Therefore, it is distributed as in Equation 18, which can be inserted into Equation 61, with $|n - m|$ playing the role of N:

$$\langle \exp[i\mathbf{k} \cdot (\mathbf{R}_n - \mathbf{R}_m)] \rangle = \exp[(-a^2\mathbf{k}^2/6)\,|n - m|] \tag{62}$$

The double summation of Equation 61 is worked out in detail in Yamakawa;[97] alternatively, for long enough chains, n and m can be considered continuous variables and Equation 18 can be evaluated as a double integral.[103] The result is known as the Debye function:

$$g(\mathbf{k}) = N(2/x^2)(e^{-x} - 1 + x) \tag{63}$$

where $x = \mathbf{k}^2\langle s^2\rangle_o$. This can be expanded in powers of x, for small x, with the result that:

$$g(\mathbf{k}) = N[1 - (1/3)\mathbf{k}^2\langle s^2\rangle_0 + \ldots] \tag{64}$$

In fact, Equation 64 is a general result, irrespective of the configurational distribution function of the polymer, and so can be used to determine the radius of gyration of a particle, polymer, aggregate, or micelles by extrapolation of the scattered intensity to zero angle. Since accurate extrapolation requires data at low values of x (the so-called Guinier regime), light-scattering apparatus optimized to measure low-angle scattering are particularly important for large macromolecules. Incorporation of this angular dependence changes Equation 58 to read:

$$Kc/R_\theta\big|_{c=0} = \{[N/Mg(\mathbf{k})] + \ldots\} \tag{65}$$

$$\approx [(1/M) + (16\pi^2/3\lambda^2 M)\langle s^2\rangle_o \sin^2(\theta/2) + \ldots] \tag{66}$$

where the $c = 0$ caveat indicates that this is only strictly true at infinite dilution.

Taken together, Equations 58 and 66 indicate that a double extrapolation to infinite dilution and to zero scattering angle is necessary to determine M and $\langle s^2\rangle_o$ accurately. The standard procedure for handling experimental data in this way is known as a Zimm plot.[104] Takahashi and co-workers[89] illustrate how this procedure can be applied to polyelectrolytes in measuring the radius of gyration of sodium polystyrenesulfonate as a function of added NaCl concentration. Daoud et al.[68] pioneered the measurement of dimensions of individual macromolecules in semidilute solutions by neutron scattering, using essentially the methods described in this section and the last one.

2. Quasi-Elastic or Dynamic Scattering

Dynamic light scattering has become an extraordinarily important tool in the study of polymer solutions over the last decade. The rationale for this is at least twofold. One reason is that it provides data on the molecular mobility and diffusion coefficients of macromolecules which are of increasing conceptual and practical importance. A second is that, under proper circumstances, these data can be interpreted directly in terms of the size and configuration of the polymer molecule, in a way that sometimes provides an easier route to this information than total intensity elastic scattering.

Dynamic scattering measures the dynamic structure factor (compare with Equation 61):

$$g(\mathbf{k},t) \equiv (1/N) \sum\sum <\exp[i\mathbf{k} \cdot [\mathbf{R}_n(t) - \mathbf{R}_m(0)]> \quad (67$$

In words, this function measures the time evolution of correlations in the positions of different segments of the polymer. For dilute solutions, the behavior of $g(\mathbf{k},t)$ has two limiting regimes.[103,105,106] The most common is $kR_g << 1$ ($R_g \equiv <s^2>_o^{1/2}$), meaning that one is exploring distances large compared to the macromolecule, where $g(\mathbf{k},t)$ can be directly rewritten as:

$$g(\mathbf{k},t) \equiv (1/N) \sum\sum <\exp[i\mathbf{k} \cdot [\mathbf{R}_g(t) - \mathbf{R}_g(0)] + \underline{i\mathbf{k}} \cdot \underline{[\mathbf{R}_n(t) - \mathbf{R}_g(t)]} - \underline{i\mathbf{k}} \cdot \underline{[\mathbf{R}_m(0) - \mathbf{R}_m(0)]}]> \quad (68)$$

where $\mathbf{R}_g$ denotes the position of the center of gravity of the chain. The underlined terms, each of which refer to the position of a segment relative to the center of gravity (at *equal* instants in time) are both of the order of kR_g, which is much less than one, so:

$$g(\mathbf{k},t) \approx (1/N) \sum\sum <\exp\{i\mathbf{k} \cdot [\mathbf{R}_g(t) - \mathbf{R}_g(0)]\}> \quad (69)$$

As t gets large, the distribution of $\mathbf{R}_g(t) - \mathbf{R}_g(0)$, which is the excursion of the center of mass of the chain from its starting point, becomes Gaussian with the classical diffusion result that the variance, $<[\mathbf{R}_g(t) - \mathbf{R}_g(0)]^2> = 2Dt$, where D is the diffusion coefficient of the center of mass, or more simply put, of the chain as a whole. If $[\mathbf{R}_g(g) - \mathbf{R}_g(0)]$ has a Gaussian distribution, the average given in Equation 69 can be readily evaluated analytically (see Doi and Edwards[103] for further details) to give:

$$g(\mathbf{k},t) \approx N\exp(-D\mathbf{k}^2 t) \quad (70)$$

This function is proportional to the time-decaying correlation in intensity of scattered light from one instant to the next. This can be readily measured with commercial correlators which sample the scattered light at controllably spaced periods in time to compute the time correlation function.[106] In a dilute solution of a single, well-dissolved macromolecular species (or micelle), measurement of the initial decay of the correlation function gives the diffusion coefficient directly:

$$\Gamma \equiv \{-d \ln [g(\mathbf{k},t)]|_{t=0}\}/dt = D\mathbf{k}^2 \quad (71)$$

In practice, there is not always but one mode of the decay of the correlation function,[98,99] and the modes of decay are not always proportional to $\mathbf{k}^2$, the signature of a diffusion process.[107] Multiple modes can sometimes be an indication of aggregation or clustering in solution[108] or some other form of interchain dynamic coupling. For example, in polyelectrolyte solutions there is also the dynamic coupling between the polyion and the cloud of counterions to consider.[44,105,109] Lack of proportionality to $\mathbf{k}^2$ could indicate a nondiffusive relaxation process such as convective motion or viscoelastic relaxation.[110,112] Micellization of amphiphilic polymers can be detected by dynamic light scattering,[112] through the observation of a slow diffusive mode of decay.

If the observed decay *is* $\mathbf{k}^2$ dependent, and therefore diffusive, Equation 71 provides the experimental route to the measurement of the diffusion coefficient (or diffusion coefficients, if there is more than diffusive mode). The translation of these data to inform on the size of the diffusing objects is made, in dilute solution, via the Stokes-Einstein equation:

$$D = kT/6\pi\eta_s R_H \tag{72}$$

where η_s is the viscosity of the solvent and R_H is the average hydrodynamic radius of the macromolecule in solution. This hydrodynamic radius is different from the radius of gyration because the hydrodynamic resistance felt by a macromolecule diffusing in solution averages over the positions of the various segments of the chain in a manner different from the averaging embodied in Equations 18 and 19. Speaking a bit roughly, R_g is determined from the second moment of the configurational distribution ($\langle r^2 \rangle^{1/2}$), whereas R_H is determined from the first reciprocal moment ($\langle r^{-1} \rangle^{-1}$). However, these two measures of size are typically quite comparable in magnitude for flexible coils, and have similar dependence on chain length. A reasonable hydrodynamic model of a macromolecule[113] leads to a calculated theoretical ratio of $R_g/R_H \approx 1.48$. Figure 7 shows recent data of Hodgson and Amis[114] on the molecular weight dependence of the infinite dilution diffusion coefficient of the polyelectrolyte, protonated poly-2-vinylpyridine, in aqueous solution, determined by dynamic light scattering. This can be taken to be the reciprocal of the molecular weight dependence of the hydrodynamic radii of the molecules.

In the limit of $kR_g \gg 1$, Γ is not expected to depend on $\mathbf{k}^2$ since one is looking at the internal modes of motion of the macromolecule. In this limit, it is anticipated that:

$$\Gamma = Dk^3 R_g \tag{73}$$

which has, indeed, been observed to within about 15 to 20% accuracy.[115] The k^3 dependence arises from effects of hydrodynamic interaction, the fact that the solvent flow around one segment is altered by the wakes arising from flow around neighboring segments. Equation 73 can be rationalized very easily

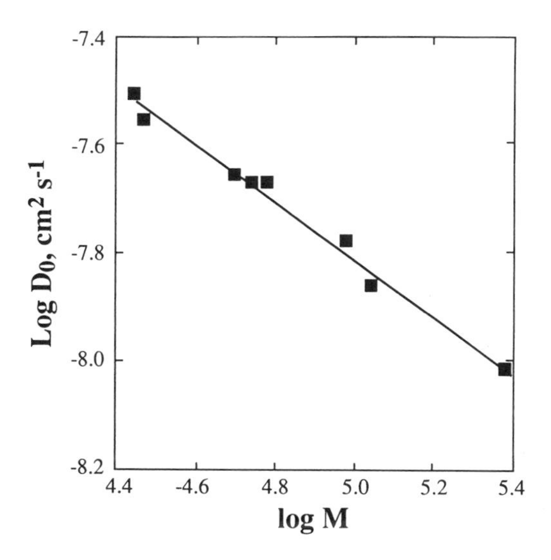

FIGURE 7. Diffusion coefficient of polyvinylpyridinium ion as a function of molecular weight. (Adapted from Hodgson, D. F. and Amis, E. J., *J. Chem. Phys.*, 95, 7653, 1991.)

by a scaling or dimensional analysis argument[116] which asserts that Γ must, in general, have the following dependence:

$$\Gamma = DR_g^{-2}(kR_g)^x \tag{74}$$

and, as we reach $kR_g \gg 1$, Γ, representing the time decay rate of fluctuations from internal modes of the macromolecule, will become independent of molecular weight, or of R_g; that is, dependent only on the local friction and on the amplitude of the local motion involved. Since from Equation 72, $D \sim R_g^{-1}$, x must be equal to 3 to make Γ independent of R_g, leading directly to Equation 73. For $kR_g \approx 1$, Equations 71 and 73 are identical, indicating that by varying **k**, one moves smoothly from probing the center of mass motion to probing the internal dynamics.

The decay modes observed in dynamic light scattering can only be attributed to diffusive motion of individual macromolecules in the limit of infinite dilution. Analogously, internal modes of individual macromolecules can be seen at finite polymer concentration only if $k\zeta \gg 1$. The concentration dependence of dynamic light scattering from polymer solutions is a subject which is too complex and is emerging (and therefore evolving) too rapidly to be adequately reviewed here. Some simple generalizations may, nonetheless, be useful in lieu of thorough review. In a concentrated solution, $\mathbf{k}^2$ modes, as suggested in Equation 72, can be seen but the D they are proportional to is the *mutual* (sometimes called *cooperative*) diffusion coefficient, which controls the rate of decay of concentration gradients in solution, not the molecular self-diffusion coefficient. These two are only equal in the limit of infinite dilution (in a binary mixture). At low but finite concentrations, this diffusion coefficient can usually be represented as a series expansion in concentration:

$$D = D_0[1 + kc + \ldots] \tag{75}$$

where the coefficient k depends, in general, on the friction in, or the viscosity of, the solution and on the osmotic compressibility (see Equation 58) which gauges the chemical potential driving forces involved in the creation and destruction of concentration gradients. Therefore, k is another measure of the quality of a solvent for a polymer and is generally positive in good solvents and 0 or negative in weaker solvents. Refractive index matching, discussed in connection with Equation 35 and Figure 4, has frequently been used as a method to observe single macromolecule dynamics in concentrated solution,[117] for example, the concentration dependence of the diffusion coefficient.[118] Inelastic neutron scattering has also been applied to study dynamic behavior in concentrated polymer solutions.[119]

D. INTRINSIC VISCOSITY AND CHROMATOGRAPHY MEASUREMENTS

Viscosity and chromatographic measurements of the size and configuration of macromolecules are, in some respects, similar to diffusion measurements by dynamic light scattering in that they are both sensitive mainly to the hydrodynamic volume pervaded by the macromolecule in solution. Neither provides an absolute measurement of the molecular weight of the polymer.

1. Intrinsic Viscosity

The analysis of intrinsic viscosity of dilute polymer solutions is built directly on the Einstein theory for the viscosity of dilute suspensions of hard spheres, the result of which can be written:

$$(\eta - \eta_s)\eta_s = 2.5\phi \tag{76}$$

where η is the viscosity of the suspension of spheres and ϕ is their volume fraction. This volume fraction is given by $(n/V)V_{sph} = (cN_{av}/M)V_{sph}$, where (n/V) and c are the number and weight concentrations of polymer, respectively, and V_{sph} is the effective volume of the coil in solution. One anticipates that V_{sph} would be proportional to R^3:

$$V_{sph} = k'R^3 \tag{77}$$

with R^3 given by an expression like that of Equation 16 or 28 with an exponent for the molecular weight dependence between 1/2 and 3/5, depending on the quality of the solvent.

The definition of the intrinsic viscosity of a polymer solution is

$$[\eta] \equiv \lim_{c \to 0} (\eta - \eta_s)/\eta_s c = 2.5k'N_{av}R^3/M \tag{78}$$

The constants $2.5k'N_{av}R^3$ are usually lumped into one constant called Φ to give what is known as the Flory-Fox equation:[120]

$$[\eta] = \Phi R^3/M \tag{79}$$

Φ is experimentally found to be nearly constant for many synthetic polymers[121] with a value between 2.0 and 2.6×10^{23} mol^{-1}. The best theory predicts Φ to be asymptotically constant at 2.25×10^{23} mol^{-1}.[121] Equation 78 shows that the intrinsic viscosity is a measure of the pervaded volume of the macromolecule per unit mass. It measures the spatial extent of the molecule and since, for linear coiling polymer R grows more rapidly than $R^{1/3}$, the intrinsic viscosity is generally an increasing function of molecular weight. (Exceptions to this might occur for very highly branched macromolecules.)

Notice that Equation 79 is, apart from the numerical constant, essentially the reciprocal of Equation 30 for the overlap concentration c^*. In fact, the reciprocal intrinsic viscosity is a good empirical estimate of the overlap concentration. The product $c[\eta]$ measures the degree of overlap among macromolecules in solution.

The experimental molecular weight dependence of a polymer sample is usually expressed in the empirical form known as the Mark-Houwink-Sakurada equation:

$$[\eta] = KM^a \tag{80}$$

Values of the empirical constants K and a are extensively tabulated.[42] Significantly, the exponent a is found to vary between 0.5 (in theta solvents) and 0.8 (in good solvents), exactly as anticipated based on the theoretical considerations of the effects of solvent quality on chain swelling embodied in Equations 16, 28, and 79.

In experimental practice, the intrinsic viscosity is determined from a measurement of the solution viscosity extrapolated to zero concentration (and zero shear rate), based on the following series expansion of the low concentration viscosity:

$$\eta = \eta_s(1 + [\eta]c + k_h[\eta]^2c^2 + \ldots) \tag{81}$$

where the coefficient k_h is known as the Huggins constant.[122] Normally, for neutral polymers, the extrapolation is smooth and monotonic; the Huggins constant is found (as with other coefficients giving low-order corrections to infinite dilution polymer solution properties) to depend systematically on solvent quality ($0.30 \sim 0.40$ in good solvents; $0.50 \sim 0.80$ in theta solvents).

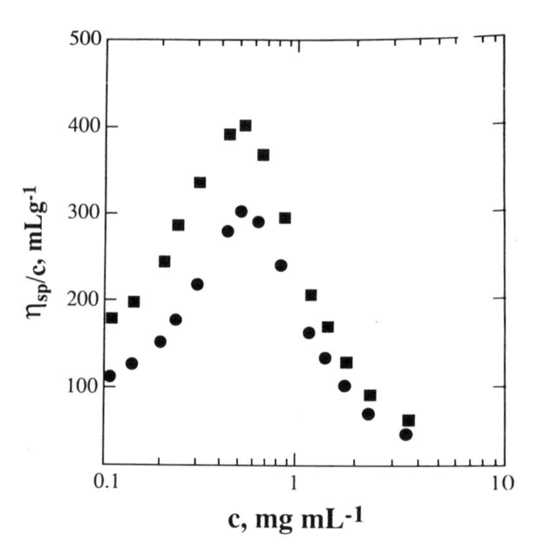

FIGURE 8. Intrinsic viscosities of two different molecular weights of quaternized polyvinyl-pyridine, in salt-free solutions, as functions of the polyion concentrations. (Data obtained from Reference 114.)

This is also true in solutions of polyelectrolytes containing added salt. In salt-free polyelectrolyte solutions, the extrapolation is more complex, owing to the fact that mutual screening of electrostatic interactions varies as the polyion concentration varies. This leads to behavior such as that illustrated in Figure 8.[114] In very dilute solutions, the viscosity of the solution rises with increasing polymer concentrations because of increasing interactions among the polyions, which are very extended, perhaps nearly rodlike, in these salt-free, dilute solutions. At higher concentration, electrostatic screening sets in among the polyions, reducing the Debye length and the viscosity.

2. Chromatography

Liquid-phase chromatography applied to polymers (frequently called gel permeation chromatography or GPC) operates, as does intrinsic viscosity, on a method that is sensitive to the hydrodynamic size of the macromolecule. Polymer dissolved in a very dilute solution phase flows through a column packed with a porous solid, sometimes literally a gel of crosslinked polymer, now more frequently a microporous silica (with a surface treatment to minimize polymer adsorption). The pore sizes in the column must be comparable to, but a bit larger than, the polymers in solution for the column to be effective for a particular sample.

The sample to be studied is injected into pure solvent flowing through the column in a narrow pulse near the entrance to the column. If a polymer having a very well-defined molecular weight (narrow molecular weight distribution) is injected. It will flow through the column, sampling that portion of the pore space which is physically accessible to it because of its size. Generally speaking, the solvent has access to all the pore space, so the polymer pulse elutes from the column at a residence time (or elution time) which is smaller than the mean residence time of the solvent. (The peak is also broadened by diffusion and possible effects of flow inhomogeneity in the column, effects for which correction must be made in accurate analysis of a GPC experiment.) Similarly, if a polydisperse sample of polymer is injected, the larger molecules elute first as they sample less of the pore space. Thus, after correction for instrument broadening, the shape of the elution pattern of a polydisperse sample for a GPC is a reflection of the molecular weight distribution of the polymer. GPC is the most common and convenient means to determine the polydispersity of a polymer sample. Thorough treatments of the practical aspects of GPC applied to polymers are available.[123]

The elution time (or volume of solvent flow before elution) for a particular polymer from a particular column is controlled by the partition coefficient, K, in the following way:

$$V_e = V_i + KV_p \tag{82}$$

where the subscripted Vs refer to the elution, interstitial, and pore volumes in the column, respectively. Theoretical calculations of the partition coefficient have been made by Cassasa, based on Equation 17, Solving Equation 17 subject to boundary conditions representing the confinement of macromolecules to pores of various geometries, Cassasa calculated the number of configurations available to macromolecules in pores of different sizes and shapes. This number, divided by the same quantity in an unrestricted, bulk solution, gives the partition coefficient. That the partition coefficient is controlled empirically by the hydrodynamic volume of the molecule is not only physically reasonable but has clear experimental support, as shown in Figure 9. The elution volumes of a large number of different macromolecules were measured on the same columns and in the same solvent by Grubisic et al.[124] One sees clearly in Figure 9 that there is an excellent correlation between elution position (or K) and hydrodynamic volume, which we demonstrated in Equation 79 is proportional to $[\eta]M$.

E. FLUORESCENCE METHODS APPLIED TO POLYMER SOLUTIONS

Fluorescence measurements have become in recent years an increasingly important source of information on the structure and dynamics of polymer solutions.[125] They are being used to determine two general kinds of features: (1) the proximity and association between different groups in a polymer solution (structure) and (2) the rate of spatial reorientation of certain groups (dynamics). While many aromatic groups occur in synthetic polymer structures, most investigative work employing fluorescence on polymers uses probes or labels that must be synthetically attached to the polymer or mixed

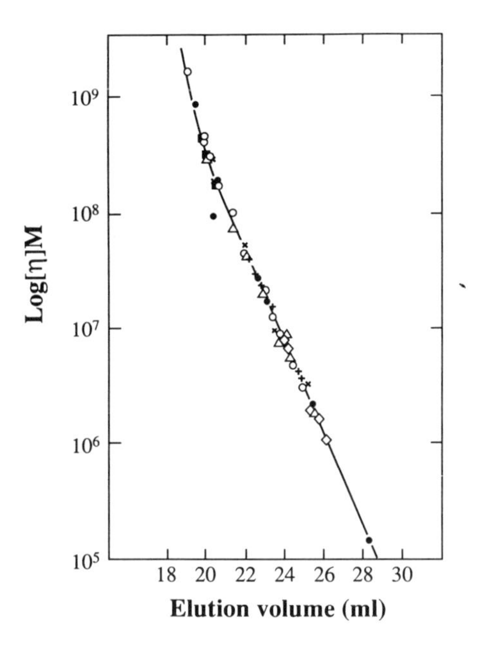

FIGURE 9. Illustration of universal calibration of gel permeation chromatography, as a plot of hydrodynamic volume as a function of elution volume. (After Grubisic et al., *J. Polym. Sci. Polym. Lett. Ed.*, B5, 753, 1967.)

into the polymer solution as an additional component. Pyrene is by far the most commonly used fluorescent probe. The use of fluorescence methods in the study of polymer-surfactant interactions is discussed in detail in Chapter 9.

Structural studies using fluorescence rely on observation of the change in the fluorescence emission of the reporting probe depending on its local environment. One useful and fluorescently active change which can be seen is excimer formation. Excimers are excited dimers of the fluorescent probe, which form after the individual probe species have absorbed the incident photons. The ability to form such a dimer in the excited state depends on the number of other probes in the vicinity of the excited molecule.

Another category of structural studies of polymer solutions using fluorescence relies on dissolution of free pyrene (or other suitable probes, such as dyes) in the solution. In aqueous solutions, pyrene exhibits a characteristic value of the ratio of fluorescence intensities in its first and third vibronic peaks when the pyrene finds itself in a polar, aqueous local environment. If there are less polar (for example, hydrophobic) domains in the solution (for example, regions of association between hydrophobic side chains, micelles) into which the pyrene can be locally solubilized, the fluorescence emission ratio of these two peaks will drop significantly.

The typical dynamic fluorescence experiment on polymer solutions measures the reorientation of an intrinsic or extrinsic (label) chromophore on the polymer. The high sensitivity of optical techniques allows measurements of chain motion, even with but one chromophore per chain of 1000 segments and at a chain concentration of 0.1%. The concept of this category of experiment is old and recent developments have been reviewed by Ediger.[126] We merely recapitulate the basic idea here and refer the interested reader elsewhere[126] for more details. When a chromophore with a strongly polarized electronic transition is excited by polarized light, those chromophores with transition dipoles in the direction of the excitation polarization are preferentially excited. If these chromophores emit before they significantly reorient, the fluorescence will also be preferentially polarized along this direction. The polarization of the fluorescence depends on the comparison of the time important timescales in the problem: the fluorescence lifetime τ_f and the reorientation time τ_r. If τ_r is much less than τ_f, the emission will be depolarized and this leads to the measurement and quantification of the reorientation time.

F. STUDIES OF THE INTERFACIAL CONFIGURATIONS OF POLYMERS ADSORBED FROM SOLUTION

Adsorption and interfacial properties of polymers are subjects worthy of books in themselves, and several review articles are available.[127-132] Here, we only point out some of the principal features of polymer adsorption from solution, without extensive background or review, making connection with our discussion of polymer configurations in solution and pointing out the experimental techniques used to study adsorbed layers when they are immersed in solvent.

The interaction multiplier effect mentioned in Section I.A plays a very important role in polymer adsorption. Polymer molecules tend to adsorb nearly irreversibly to solids even when the binding energy of individual segments is weak compared to kT. This can be true since, with many possible binding segments per molecule, the binding energy *per molecule* can be strong while the binding energy *per segment* is weak. This situation, which is the most commonly studied one in investigations of polymer adsorption, is frequently called (somewhat paradoxically) the ''weak'' adsorption limit. The reason for this is that a corollary of the weak binding per segment is that the polymer chain adsorbs in a rather loose configuration with many segments off the surface, extending into solution in ''loops'' and ''tails''. (''Trains'' has become the accepted terminology for the stretches of chain segments on the surface.) This situation results in a restricted equilibrium where the configurations of the adsorbed polymers can reach equilibrium, since segments can rearrange, but polymers desorb with difficulty, even under strong chemical potential driving forces to do so. On the other hand, exchange between molecules in adsorbed layers and those in solution has been documented.[133,134] Other dynamic aspects of adsorbed layers, such as kinetics of adsorption and exchange,[135-137] competitive adsorption,[138] and fluid mechanical effects[139-141] are subjects of active current investigation.

To date, the study of adsorbed layers has aimed mainly at understanding the configurations of adsorbed polymers in detail. Complete characterization of an adsorbed layer requires the determination of several quantities and the employment of several different experimental methods. Adsorbed amount, that is, the weight or number of adsorbed polymers per unit area, is the most basic information necessary. This sort of data is best obtained by a method which directly measures the adsorbed amount in some absolute terms. Radiolabeling,[142] direct weighing using an electrobalance, quartz crystal microbalance, or other mass sensitive methods,[143] precise measurement of solvent-free thickness,[138,140] and quantitative determination of a spectroscopic signal[144] are all methods that have been employed for quantitative, absolute determination of adsorbed amounts. Other methods may be calibrated precisely against these absolute methods, including: measurement of the attenuation of a spectroscopic signal from the substrate via x-ray Photoelectron Spectroscopy (XPS),[142] measurement of the refractive index of an adsorbed layer,[145,146] and secondary ion mass spectrometry.[147] Still other methods can also be used to infer the absorbed amount, without calibration against some other method (though they can sometimes be made more accurate by such cross-calibration). Ellipsometry[148] is in this category. Some of these methods are also suitable for measurement of adsorption at fluid-fluid interfaces, where the measurement of interfacial tension,[149] via spinning drop or Wilhelmy plate methods, has also been used to infer adsorbed amounts. The adsorbed amount is found experimentally to vary approximately as $M^{1/2}$ for a series of homologous, chemically identical homopolymers.[128] All other factors being equal, the adsorbed amount will increase when the quality of the solvent for the polymer diminishes.

Charge, both on the polymer and on the surface, is obviously a key determinant of polyelectrolyte adsorption. For surfaces and polymers bearing the same sign of charge, adsorption is inhibited and can be accomplished only if there is another driving force to adsorb (for example, hydrophobic portions of the polymer and surface which can associate). For opposite charges on the polymer and surface, in salt-free water, the adsorption is very strong and the polyelectrolyte would be expected to have an equilibrium configuration where there is an optimal match between polyion and surface charges. This would lead to a chain which lies essentially flat on the surface. The surface density of chains would be close to σ/Ne, where σ is the charge of the bare surface. The kinetics by which this flattening might occur could be very long since, in this case, the individual segment binding is strong; consequently, surface rearrangements could be slow.

Average layer thickness, L, is another basic quantity in the characterization of adsorbed polymer layers. Ellipsometry and surface force measurements have emerged as the principal means to these data, along with the measurement of hydrodynamic layer thicknesses, either by adsorption on the inside of pore walls through which solvent is caused to flow[140] or by the adsorption on the exterior of small particles and monitoring the change in hydrodynamic radius by dynamic light scattering.[150] Measured layer thicknesses in the weak adsorption regime are comparable to, but smaller than, the free solution dimensions of the polymer and are also seen to vary approximately as $M^{1/2}$ ($L \sim M^{0.4}$ seems more accurate and has some theoretical support for adsorbed layers in good solvents[151,152]). The contributions of loops and tails to this measured thickness are difficult to uncover experimentally, but it is generally believed on theoretical grounds that tails make a very substantial contribution to the overall layer thickness, especially to measurements of it by hydrodynamic means.[140,150] The theory of configurations of adsorbed polymers has mainly been constructed on the basis of the self-consistent mean field theory outlined in Section II.B.1. A lattice version of this model has been most thoroughly explored.[153,154]

The most detailed information one can obtain on an adsorbed layer is the complete profile of segments on, and extending away from, the surface. Neutron scattering has been used very effectively here,[155] though it is a delicate experiment since the scattering length density must be chosen to maximize the polymer-solvent contrast and eliminate the solution-solid contrast, conditions that are not always well achieved. There is no general agreement on whether the mean field theory accurately models the profiles or if correlations due to excluded volume, as discussed in Section II.B.1, are important.[156] What is clear is that, for uniform homopolymer adsorption, the density of segments drops off monotonically in a roughly exponential fashion (scaling theories predict a power law fall-off[152] away from the surface. Neutron reflectivity[157] has recently been brought to bear on this problem and is potentially very powerful. In this method, neutrons are reflected from the solid surface (eliminating the need to contrast match the solid) and the effect of the adsorbed layer on the reflectivity as a function of reflection angle can be interpreted in terms of the profile of segments.

Most of this discussion has focused on the adsorption of homopolymers, but it is important and interesting to realize that macromolecular architecture can play a significant role in all aspects of the structure of adsorbed layers. Some of these possibilities are sketched in Figure 10. Copolymers bring to macromolecules the possibility of amphiphilic or surfactant-type behavior akin to that found in small-molecule surfactants, but on length scales of macromolecular size. The opportunities to use these molecules to manipulate surface and solution properties are enormous. This is particularly true when the polymer solution is the continuous phase for a suspension of dispersed, solid particles. In this case, polymers, particularly surface-active, amphiphilic polymers, can be very useful in controlling the properties of the suspension.[130] Hydrophobically modified polymers, the structures discussed in Chapter 6, are often used in this manner in aqueous-based dispersions. This sort of hydrophobe-induced adsorption can be manipulated to advantage by the addition of surfactant, as described in Chapter 10 on applications.

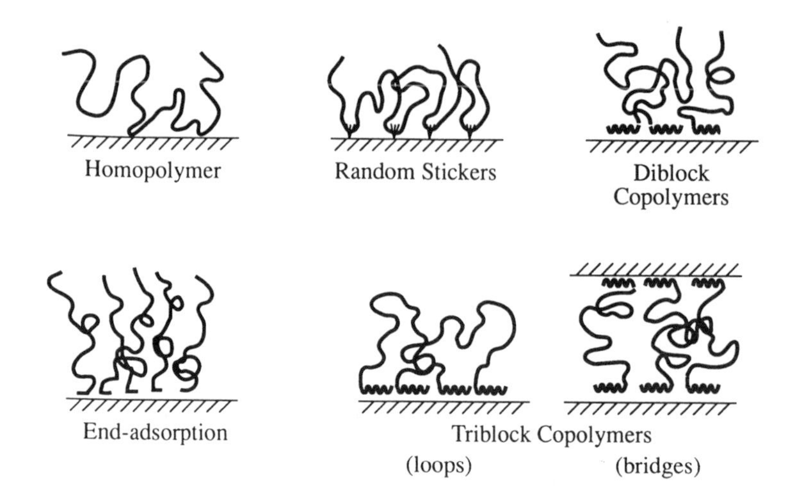

FIGURE 10. Schematic illustrations of various adsorbed configurations of macromolecules.

IV. RHEOLOGICAL PROPERTIES OF POLYMER SOLUTIONS

Rheology, the study of the flow and viscoelastic properties of polymers, provides very important information on the structure and behavior of solutions, beyond the molecular weight determination by intrinsic viscosity discussed in Section III.D.1. Rheological measurements are the source of information on how solutions respond to stresses imposed on them in various ways. Thorough treatment of the rheology of solutions is well beyond the scope of this chapter, but several excellent, recent sources are available.[35,158,159] We begin by discussing some of the rheological behavior exhibited by polymer solutions beyond the zero shear rate, infinite dilution, steady-shear limit appropriate for intrinsic viscosity measurement. We conclude with a very brief introduction to the theoretical principles useful in interpreting this behavior.

A. OBSERVATIONS OF RHEOLOGICAL BEHAVIOR IN POLYMER SOLUTIONS

1. Constant Shear Rate and Constant Stress Rheometry

Rheological measurements can be made in a multitude of ways. The reader is referred to other sources for detailed information on the execution of rheological measurements in the laboratory.[160,161] In one generic route, the polymer sample (solution for our purposes) is subjected to a predetermined, applied strain and the stress response is measured. This is the basic situation we will deal with in this section, but it is important to realize that considerable useful information can be obtained by performing the complementary experiment of applying a predetermined, constant stress and observing the strain or flow response. An example of the latter procedure is capillary viscometry,

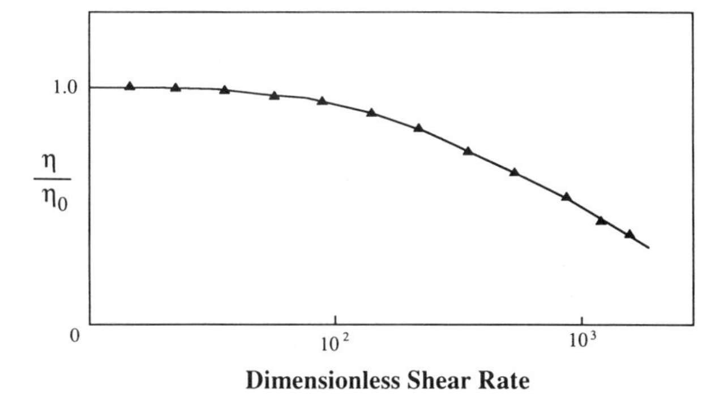

FIGURE 11. Data on viscosity as a function of shear rate of a 5% solution of polystyrene in Aroclor. (Adapted from Graessley, W. W., *Adv. Polym. Sci.*, 16, 1, 1974.)

where a constant pressure is applied and the flowrate measured. For simple fluids, equivalent information can be obtained by either route. However, for fluids containing interesting microstructures, such as many polymer solutions, one or the other route may be more convenient or informative. This is particularly true in solutions in which the macromolecules associate with one another to form aggregates or gel structures, the properties of which one wants to study (see Chapter 5 and 10 for examples of how gel structures can form in mixtures of interacting polymers and surfactants). Materials such as this often exhibit yield stresses, characteristic of solids, when subjected to low applied loads. Constant stress rheometry is very useful here in that it enables the yield stress to be approached from below, and therefore, precisely determined. Furthermore, the elastic properties of the microstructure can be studied at stresses low enough to avoid disrupting it.

Some of the principal rheological properties of interest in characterizing a polymer solution are viscosity, η, the ratio of shear stress to shear rate in a steady shear experiment; storage, or elastic, modulus, G', the ratio of the in-phase component of the stress response to the amplitude of a sinusoidal applied strain; and $\tan \delta \equiv G''/G'$, where G'' is the ratio of the out-of-phase component of the stress response to the amplitude of a sinusoidal applied strain. The first quantity measures the dissipation of energy due to friction in flow, the second gauges the elasticity of the material, and the third is the relative importance of these two. All of these properties may depend on shear rate or on the frequency of the imposed deformation.

2. Shear Rate Dependence of Viscosity

Polymer solutions generally exhibit a kind of shear rate dependence of the viscosity, known as shear-thinning, in which the viscosity decreases, from a low shear rate Newtonian plateau, with increasing shear rate. This is illustrated in Figure 11.[162] It is important to realize that, as in this case, shear-thinning can occur even in very dilute solutions; intermolecular interactions are not a prerequisite for shear-thinning. Though a theory which is completely satisfactory quantitatively is not available, a qualitative explanation of shear-thinning can be given. Viscosity measures the friction, or transfer of momentum, perpendicular to the shear planes. The apparent viscosity will be reduced in a polymer solution when the polymer contribution to friction or momentum transfer in this direction is reduced. This can happen in a polymer solution subjected to flow when the alignment and orientation of macromolecules induced by flow diminish the average spatial extent of the polymer in the direction normal to the shear planes. In other words, the average dimensions of a macromolecule in a flowing solution are no longer spherically symmetric, even on average. The shape of the macromolecule is anisotropic, with the principal axis of the radius of gyration (which is now properly considered to be a tensor quantity) oriented along the principal axis of the flow, with a corresponding contraction of the average dimensions in the perpendicular direction. This is, roughly speaking, what causes shear-thinning in dilute solutions. Considerable effort is being expended to achieve a more complete theory of this phenomenon.[22,35] Practically, this is very important in many uses of polymer solutions, where they are applied or must perform at high shear rate.

A different viscosity can be measured in a polymer solution subjected to a stretching, rather than a shearing, deformation. This so-called extensional viscosity, η_e, is the ratio of the stretching stress to the stretching rate. For simple liquids, this is directly proportional to the shear viscosity. For polymers, this quantity can be considerably greater than η since the elasticity and finite extensibility of macromolecules resists strong stretching. These are naturally important data to have concerning applications of polymer solutions that involve stretching.

3. Concentration Dependence of Viscosity

Increasing the concentration of a polymer solution increases its viscosity. For concentrations of polymer in the range between very dilute solutions and about 20%, an empirical equation, known as the Martin equation,[163] works well to correlate concentration dependence of viscosity:

$$(\eta - \eta_s)/c\eta_s = [\eta]\exp(k_m c[\eta]) \tag{83}$$

where k_m is an empirical constant to be determined by fitting the data but, were it equal to k_h, Equation 81 could be viewed as the low concentration series expansion of Equation 83. With the coil overlap parameter $c[\eta]$ being the appropriate dimensionless concentration, we see that, in this regime, up to the order of 20%, it is the degree of overlap among polymers that controls the concentration and molecular weight dependence of viscosity.

Solutions of higher concentration do not conform to the Martin equation in their concentration and molecular weight dependences. As concentration, molecular weight, and the degree of overlap among chains in solution increase, a different kind of dynamic intermolecular interaction sets in, known as

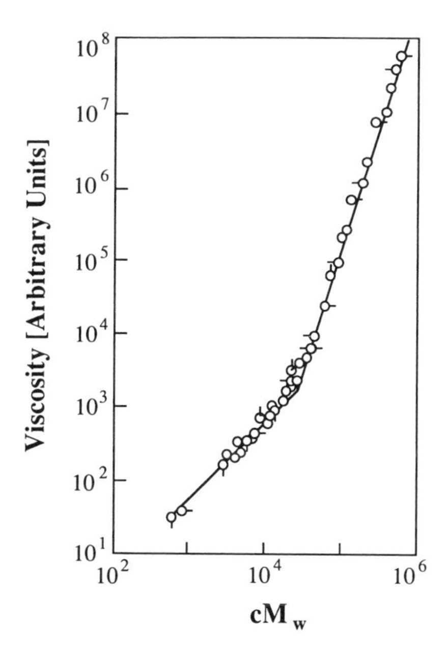

FIGURE 12. Viscosity data on a concentrated polystyrene solution illustrating good correlation with entanglement density. (Adapted from Graessley, W. W., *Adv. Polym. Sci.*, 16, 1, 1974.)

entanglement. Entanglement has several characteristic signatures in the effects that it has on the rheological properties of polymer fluids. One of these is shown in Figure 12. Above a certain threshold of concentration and molecular weight, the *viscosity becomes much more strongly dependent* on these quantities than it was at lower concentrations. Experimentally, the break in concentration and molecular weight dependence is so distinct that it can be used to define M_c empirically, the critical molecular weight for the onset of entanglement in the viscosity (of the undiluted polymer melt). The fact that the viscosity depends on the product $c\langle M\rangle_w$, rather than on $c[\eta]$ as it does at lower concentration, signals that the viscosity is now being controlled by the density of binary interactions between segments of macromolecules (which is proportional to $c\langle M\rangle$), rather than by coil overlap per se (which is proportional to $c[\eta]$).

While there is no theory capable of predicting the onset of entanglement, the fact that it happens at some high degree of overlap is reasonably well understood to result from the uncrossability of chains mentioned in the first section. At some concentration, the cooperativity required for these long, anisotropic objects to move around one another takes over and slows down the dynamic processes, leading to the enhanced concentration and molecular weight dependence. Entanglement also has striking effects on the dynamic mechanical response of polymer fluids to oscillatory deformation. G', which for dilute solutions is a monotonically increasing function of frequency of oscillation, exhibits a marked plateau (i.e., $\approx$ constant elastic modulus), above some threshold, cM_e, where M_e is the critical molecular weight for the appearance of entanglement in the elastic modulus, and is frequently empirically found to be $\approx (1/2)M_c$.[160] This appearance of a nearly constant modulus that does not relax over certain timescales is behavior like that of a solid or a permanently crosslinked rubber, leading to one common conception of ''entanglements'' as being temporary crosslinks. Entanglements enhance the elastic response of polymer fluids.

B. THEORETICAL INTERPRETATION OF RHEOLOGICAL BEHAVIOR IN POLYMER SOLUTIONS

The simplest useful, theoretical picture of the dynamics of a single polymer chain is given by the Rouse model.[164] It represents the chain as a linear series of N beads of diameter b (which should not be thought of as identical with the segments of the polymer defined by Equation 16) connected by springs, that is, as a coupled harmonic oscillator (anharmonic springs can also be introduced). The beads interact frictionally with the surrounding medium and are assigned a friction coefficient, f. The springs are usually taken to be linearly elastic with a force law given by Equation 24. For this reason, dynamic units of the Rouse model should be thought of as short stretches of the chain, larger than individual segments and large enough to exhibit the elasticity embodied in Equation 24.

To get dynamic information from the Rouse model, one writes down the differential equations of motion for each bead in terms of their positions in space, accounting for the frictional and spring forces acting on them. The solution of this set of equations yields a discrete set of exponential mode correlation functions and a corresponding set of mode relaxation times, τ_j, for the internal modes of motion of the polymer chain:

$$\tau_j = fN^2b^2/3\pi kTj^2; \; j \geq 1 \tag{84}$$

The $j = 0$ mode corresponds to the center of mass motion and gives the diffusion coefficient of the Rouse chain:

$$D_{Rouse} = kT/fN \tag{85}$$

This is not the same as Equation 72, since hydrodynamic interactions among the beads are left out here. Zimm showed how to modify the eigenvalue problem to include hydrodynamic interaction and give the proper molecular weight dependence ($D \sim R_H^{-1}$ as in Equation 72, not $D \sim N^{-1}$). The longest relaxation time ($j = 1$) of the Rouse model, with the Zimm modification for hydrodynamic interaction, is given by a modification of Equation 84:

$$\tau_{1,Zimm} = \eta_s(N^{1/2}b)^3/(3\pi)^{1/2}kT \tag{86}$$

This is the quantity that controls the rate of internal dynamics, relaxation, and viscoelasticity of polymers in unentangled solutions. Forcing the polymer molecules at rates faster than this fundamental relaxation time will provoke an elastic response, as well as nonlinear phenomena such as shear-thinning in the viscosity. Shear-thinning is to be expected at shear rates greater than $\tau_{1,Zimm}^{-1}$, since the macromolecule will be unable to relax sufficiently fast from the shear-induced orientation that produces shear-thinning.

The theoretical picture for entangled polymer solutions is much less clear. In entangled polymer *melts,* there is considerable experimental support for the reptation model,[165] which views the effects of mutual uncrossability among the macromolecules to cause the polymers to move like ''snakes'' moving mainly along their own backbone contour. The reptation model predicts that the viscosity *should* become much more strongly concentration and molecular weight dependent (for example, $\eta \sim M^3$ in an entangled solution, compared with the weaker dependence embodied in Equation 80). Experimentally, however, there is still considerable disagreement over the accuracy of the predictions of the reptation model for polymer solutions.[159] Support for the correctness of the reptation idea is seen in the comparison of the diffusion coefficients of linear and branched macromolecules of the same size in entangled solutions. The fact that branched chains diffuse much more slowly is a reasonable indication that linear polymers have a mode of diffusion available to them (reptation) which is unavailable to branched polymers.

V. POLYMER INTERACTIONS WITH SURFACTANTS

We finish this chapter with an introduction to polymer-surfactant interactions from a polymer perspective as a prelude to the detailed treatment in the remainder of the book and as a pointer to future possible research emphasizing the role of the polymer.

Polymers and surfactants are often used together in industrial formulations to take advantage of their characteristically different properties. When present together they can interact to provide beneficial properties or cause unwanted problems. Such interactions can occur in both aqueous and nonaqueous systems. Owing to their industrial importance, aqueous systems have been studied much more in detail than nonaqueous systems. As discussed in more detail later in the book, polymer-surfactant combinations in aqueous solutions which interact with each other can be broadly classified into three categories, namely uncharged polymer-ionic surfactant, oppositely charged polymer-surfactant and hydrophobe-modified polymer-surfactant systems. While the dominant forces responsible for interactions in the latter two systems can be readily identified, the reasons for the interaction between nonionic polymer and charged surfactant are less clear. In all cases, the interaction of polymers with surfactants, unlike that with other small molecules is made complex by the key role played by the self aggregation/micellization properties of the surfactant.

It is known, but still largely only in a qualitative sense, that there are several properties of polymers which influence their interaction with surfactants (see Chapters 4 and 5). In charged polymer-oppositely charged surfactant systems, in general, the interactions are significantly stronger than those in uncharged polymer-charged surfactant systems because of the dominant electrostatic forces. But even here it is quite apparent that it is not only the charge density of the polymer (polyelectrolyte) which is important, but also the disposition of the charges, e.g. on the backbone or on the side chains, as well as their specific nature. In this respect a list of the more commonly investigated polyelectrolytes is presented in Table 3, and it has been quite evident to investigators that major differences in their reactivity (towards oppositely) surfactants exist, as judged for example by the wide range of critical association concentrations which the latter exhibit (Chapter 4). This has also been evident to workers studying similar interactions with proteins, who find considerable variation in reactivity from protein to protein (Chapter 8).

Another molecular property of major importance, but of varying level from polymer to polymer, is their flexibility. Polymers are able to adopt many conformations to promote association structures in surfactant systems. This is an important factor which is involved in the interaction of both charged and uncharged polymers. Synthetic linear polymers tend, in general, to be more flexible than most of the biopolymers, and direct evidence has been obtained of appreciable differences in the rheology of polymer/surfactant systems depending on the backbone of the polyelectrolyte employed. (Chapter 4, Part II).

Turning to nonionized polymers, one can identify two important factors, in addition to the specific chemical structure of the polymer, relating to their reactivity towards ionic surfactants. The first is molecular weight. Although limited information exists in a general sense there are clear indications for the polyethylene oxide (PEO) family that a minimum molecular weight exists for interaction with sodium dodecylsulfate. It is of interest that this minimum molecular weight is reduced considerably if PEO is replaced by its close relative polypropylene oxide. In the latter case, the polymer has a decidedly more hydrophobic character, a factor pointed out many years ago by Breuer and Robb[166] to promote reactivity (Chapter 4). Goddard et al.[167,168] have suggested that a measure of this factor is the extent of lowering of the surface tension of water observed for solutions of the polymers. The effect is illustrated, qualitatively, by the data compiled in Table 4.

All the concepts concerning the reactivity of polymer surfactant mixtures which have been raised briefly above are elaborated upon in subsequent chapters.

TABLE 3-A
Typical Anionic Polymers Investigated for their Interactions with Oppositely Charged Surfactants

Anionic polymer	Monomer unit
Polyacrylic acid	$-[CH-CH_2-]$ with pendant $COOH$
Polymethacrylic acid	$-C-CH_2$ with pendant CH_3 and $COOH$
Polyvinyl-sulfonate	$-CH-CH_2-$ with pendant $SO_3^-...Na^+$
Polystyrene sulfonate	$-CH-CH_2-$ with pendant $C_6H_4SO_3^-...Na^+$
Dextran sulfate	
Carboxymethyl-cellulose	
Alginate	
Pectic acid	

ACKNOWLEDGMENTS

I would like to acknowledge, with appreciation, financial support during the course of this writing from the National Science Foundation (NSF/CTS-9107025, Interfacial Transport and Separations Program [CTS], and Polymers Program [DMR]), and also from the Center for Interfacial Engineering, an

TABLE 3-B
Typical Cationic Polymers Investigated for their Interactions with Oppositely Charged Surfactants[a]

Cationic polymer	Monomer unit	Commercial name
Polyquaternium-11 (copolymer of vinyl pyrrolidone and dimethyl aminoethyl-methacrylate modified with diethyl sulfate)	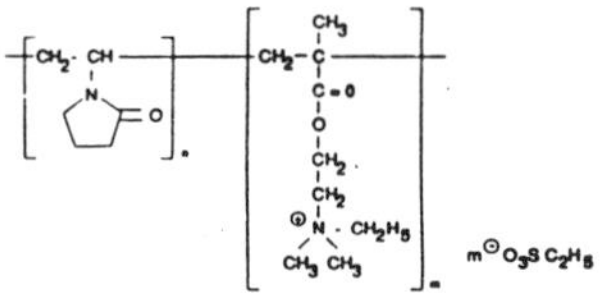	Gafquat
Polyquaternium-10 (hydroxyethyl cellulose reacted with a trimethyl ammonium substituted epoxide)	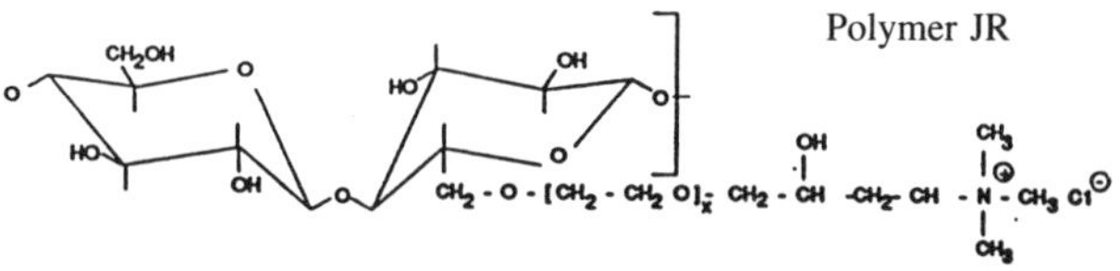	Polymer JR
Polyquaternium-24 (hydroxyethyl cellulose reacted with a lauryl dimethylammonium substituted epoxide)	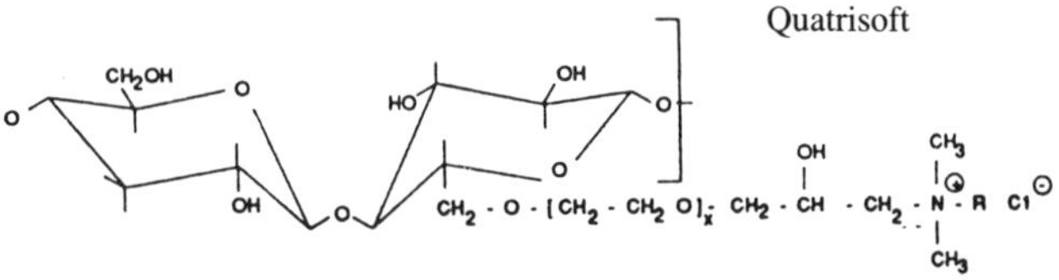	Quatrisoft
Polyquaternium-6 (polymer of dimethyl diallyl ammonium chloride)	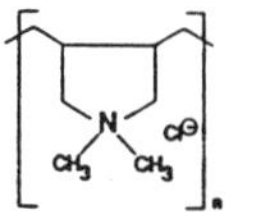	Nalquat6-20 Merquat-100
Polyquaternium-7 (copolymer of dimethyl diallyl ammonium chloride and acrylamide)	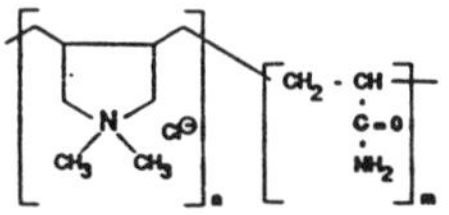	Merquat-550
Chitosan	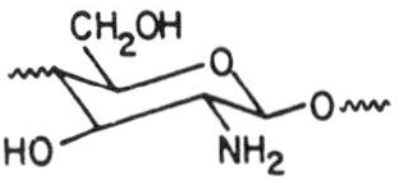	

[a] Structures and commercial names of various polyquaternium polymers can be found in CTFA (Cosmetic, Toiletry and Fragrance Association) International Cosmetic Ingredient Dictionary, 4th Ed., Pub. CTFA, Washington, D.C., 1991.

TABLE 4
Nonionic Polymers which Interact with SDS

Polymer	Monomer unit	Interaction with SDS	Effect on H_2O surface tension
Polyacrylamide	$-CH-CH_2-$ (with $CO-NH_2$ on CH)	No	Feeble/none
Poly-n-isopropyl acrylamide	$-CH-CH_2$ (with $CO-NHC_3H_7$ on CH)	Yes	Significant
Dextran	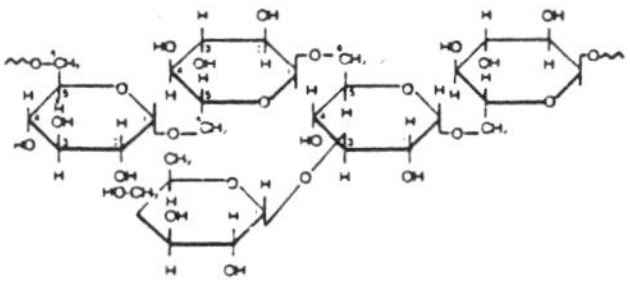	No	Feeble/none
Hydroxyethyl cellulose	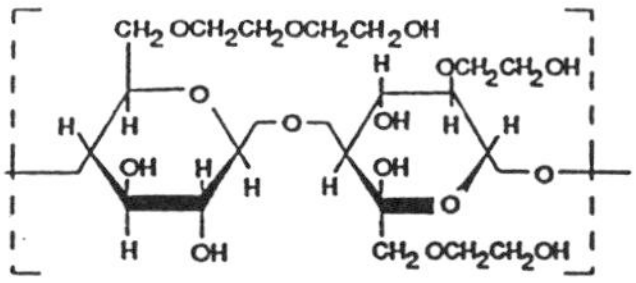	No	Feeble/none
Polyethylene glycol	$-[CH_2-CH_2O]$	Yes	Moderate
Polypropylene glycol	$-[O-CH-CH_2]$ (with CH_3 on CH)	Yes	Significant
Polyvinyl alcohol	$-[CH-CH_2]$ (with OH on CH)	Yes	Moderate
Polyvinyl pyrrolidone	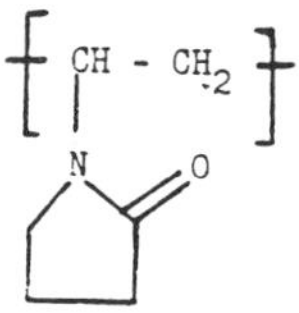 $[CH-CH_2]$ (with pyrrolidone ring N on CH)	Yes	Moderate

NSF-supported Engineering Research Center at the University of Minnesota. The writing was completed while I was the holder of a Sherman Fairchild Scholar Fellowship at the California Institute of Technology. For that, and the stimulating hospitality in the Division of Chemistry and Chemical Engineering at Caltech, I am very grateful. Thanks are also due to Julie Kornfield for access to her excellent personal library, and Enrico Westenberg for vital help with references.

REFERENCES

1. **de Gennes, P.-G.,** *Scaling Concepts in Polymer Physics,* Cornell University Press, Ithaca, 1979, chap. 8.
2. **Clark, A. H. and Ross-Murphy, S. B.,** Structural and mechanical properties of biopolymer gels, *Adv. Polym. Sci.,* 83, 57, 1987.
3. **Schultz, A. R. and Flory, P. J.,** Phase equilibria in polymer-solvent systems, *J. Am. Chem. Soc.,* 74, 4760, 1952.
4. **Klein, J. and Luckham, P. F.,** Forces between two adsorbed poly(ethylene oxide) layers immersed in a goal aqueous solvent, *Nature,* 300, 429, 1984.
5. **de Gennes, P.-G.,** *Scaling Concepts in Polymer Physics,* Cornell University Press, Ithaca, 1979. chap. 4.
6. **Lenz, R. W.,** *The Organic Chemistry of Synthetic High Polymers,* Wiley-Interscience, New York, 1967.
7. **Odian, G.,** *Principles of Polymerization,* Wiley-Interscience, New York, 1981.
8. **Rempp, P. and Merrill, E. W.,** *Polymer Synthesis,* Huthig and Wepf, Basal, 1986.
9. **Alberts, B., Bray, D., Lewis, J., Raff, M., Roberts, K., and Watson, J. D.,** *Molecular Biology of the Cell,* Garland Publishing, New York, 1989.
10. **Galván, R. and Tirrell, M.,** Molecular weight distribution predictions for heterogeneous Ziegler-Natta polyerization using a two-site model, *Chem. Eng. Sci.,* 41, 2385, 1986.
11. **Roehrig, K. L.,** *Carbohydrate Biochemistry and Metabolism,* AUI Publishing, Westport, CT, 1984.
12. **Flory, P. J.,** *Statistical Mechanics of Chain Molecules,* John Wiley & Sons, New York, 1969.
13. **Chandrasekar, S.,** Stochastic problems in physics and astronomy, *Rev. Mod. Phys.,* 15, 3, 1943.
14. **Wall, F. T.,** Statistical lengths of rubberlike hydrocarbon molecules, *J. Chem. Phys.,* 11, 67, 1943.
15. **Benoit, M. H.,** Sur la statistique des chaines avec interactions et empêchements stériques, *J. Chim. Phys.,* 44, 18, 1947.
16. **Volkenshtein, M.,** *Configurational Statistics of Polymeric Chains,* Interscience, New York, 1963.
17. **Birshtein, T. M. and Ptitsyn, O. B.,** *Conformations of Macromolecules,* Interscience, New York, 1966.
18. **Yamakawa, H.,** *Modern Theory of Polymer Solutions,* Harper & Row, New York, 1971, chap. 2.
19. **BIOSYM Technologies, Inc.,** *Polymer Modules,* San Diego, 1992.
20. **de Gennes, P.-G.,** *Scaling Concepts in Polymer Physics,* Cornell University Press, Ithaca, 1979, chap. 2.
21. **Kratky, O, and Porod, G.,** X-Ray investigation of dissolved chain molecules, *Recl. Trav. Chim. Pays. Bas.,* 68, 1106, 1949.
22. **Doi, M. and Edwards, S. F.,** *The Theory of Polymer Dynamics,* Clarendon Press, Oxford, U.K., 1986, 11.
23. **Edwards, S. F.,** The statistical mechanics of polymers with excluded volume, *Proc. Phys. Soc. London,* 85, 613, 1965.
24. **Casassa, E. F., and Tagami, Y.,** An equilibrium theory for exclusion chromatography of branched and linear polymer chains, *Macromolecules,* 2, 14, 1969.
25. **Freed, K. F.,** Self-consistent field theories of the polymer excluded volume problem. I. Edwards' functional integral approach, *J. Chem. Phys.,* 55, 3910, 1971.
26. **Helfand, E. and Tagami, Y.,** Theory of the interface between immiscible polymers. II, *J. Chem. Phys.,* 56, 3592, 1972.

27. **Flory, P. J.,** *Principles of Polymer Chemistry,* Cornell University Press, Ithaca, 1953, 430.
28. **Zimm, B. H. and Stockmayer, W. H.,** The dimensions of chain molecules containing branches and rings, *J. Chem. Phys.,* 17, 1301, 1949.
29. **Orofino, T. A.,** Branched polymers, II. Dimensions in non-interacting media, *Polymer,* 2, 305, 1961.
30. **Burchard, W.,** Static and dynamic light scattering from branched polymers and biopolymers, *Adv. Polym. Sci.,* 48, 1, 1983.
31. **Tsukahara, Y., Mizuno, K., Segawa, A., and Yamashita, Y.,** Study on the radical polymerization behavior of macromonomers, *Macromolecules,* 22, 1546, 1989.
32. **Gobush, W., Šolc, K., and Stockmayer, W. H.,** Statistical mechanics of random flight chains. V. Excluded volume expansion and second virial coefficient for linear chains of varying shape, *J. Chem. Phys.,* 60, 12, 1974.
33. **Pusey, P. N.,** Calculation of the fourth-order correlation function of a polymer coil (as measured by cross-correlation light scattering), *Macromolecules,* 18, 1950, 1985.
34. **Šolc, K. and Stockmayer, W. H.,** Shape of a random-flight chain, *J. Chem. Phys.,* 54, 2756, 1971.
35. **Bird, R. B., Curtiss, C. F., Armstrong, R. C., and Hassanger, O.,** *Dynamics of Polymeric Liquids,* Vol. 2, *Kinetic Theory,* 2nd ed., Wiley-Interscience, New York, 1987.
36. **Mark, J. E. and Erman, B.,** *Elastomeric Polymer Networks,* Prentice-Hall, Englewood Cliffs, N.J. 1992.
37. **Flory, P. J.,** *Principles of Polymer Chemistry,* Cornell University Press, Ithaca, 1953, 427.
38. **Freed, K. F.,** *Renormalization Group Theory of Macromolecules,* John Wiley & Sons, New York, 1987.
39. **des Cloiseaux, J. and Jannink, G.,** *Polymers in Solution: Their Modelling and Structure,* Oxford Science Publishers, Oxford, U.K., 1990, chap. 11.
40. **Flory, P. J.,** *Principles of Polymer Chemistry,* Cornell University Press, Ithaca, 1953, 520.
41. **Kremer, K., Baumgärtner, A., and Binder, K.,** Collapse transition and crossover scaling for self avoiding walks on the diamond lattic, *J. Phys. A: Math. Nucl. Gen.,* 15, 2879, 1982.
42. **Brandrup, J. and Immergut, E. H.,** *Polymer Handbook,* John Wiley & Sons, New York, 1975, chap. 4.
43. **de Gennes, P.-G.,** *Scaling Concepts in Polymer Physics,* Cornell University Press, Ithaca, 1979, chap. 3.
44. **Schurr, J. M. and Schmitz, K. S.,** Dynamic light scattering studies of biopolymers: effects of charge, shape and flexibility, *Annu. Rev. Phys. Chem.,* 37, 271, 1986.
45. **Odijk, T. and Houwaart, A. C.,** On the theory of the excluded-volume effect of a polyelectrolyte in a 1-1 electrolyte solution, *J. Polym. Sci. Polym. Phys. Ed.,* 16, 627, 1978.
46. **Odijk, T.,** Polyelectrolytes near the rod limit, *J. Polym. Sci. Polym. Phys. Ed.,* 15, 477, 1977.
47. **Fixman, M. and Skolnick, J.,** Polyelectrolyte excluded volume paradox, *Macromolecules,* 11, 863, 1978.
48. **Fixman, M.,** The flexibility of polyelectrolyte molecules, *J. Chem. Phys.,* 76, 6346, 1982.
49. **Manning, G. S.,** The molecular theory of polyelectrolyte solutions with applications to the electrostatic properties of polynuycleotides, *Q. Rev. Biophys.,* 11, 179, 1978.
50. **Manning, G. S.,** Limiting laws and counterion condensation in polyelectrolyte solutions. IV. The approach to the limit and the extraordinary stability of the charge fraction, *Biophys. Chem.,* 7,95, 1977.

51. **Manning, G. S.,** Limiting laws and counterion condensation in polyelectrolyte solutions. VII. Electrophoretic mobility and conductance, *J. Phys. Chem.,* 85, 1506, 1981.
52. **Fixman, M.,** The Poisson-Boltzmann equation and its application to polyelectrolytes, *J. Chem. Phys.,* 70, 4995, 1979.
53. **Anderson, C. F. and Record, M. T.,** Polyelectrolyte theories and their applications to DNA, *Annu. Rev. Phys. Chem.,* 33, 191, 1982.
54. **Wilson, R. W., Rau, D. C., and Bloomfield, V. A.,** Comparison of polyelectrolyte theories of the binding of cations to DNA, *Biophys, J.,* 30, 317, 1980.
55. **Klein, B. K., Anderson, C. F., and Record, M. T.,** Comparison of Poisson-Boltzmann and condensation model expressions for the colligative properties of cylindrical polyions, *Biopolyers,* 20, 2263, 1981.
56. **Chao, Y. S., Sheu, E. Y., and Chen, S. H.,** Experimental test to a theory of dressed micelles: the case of the monovalent counterion, *J. Phys. Chem.,* 89, 4862, 1985.
57. **Sheu, E. Y., Wu, C. F., Chen, S. H., and Blum, L.,** Applications of rescaled mean spherical approximation to strongly interacting ionic micellar solutions, *Phys. Rev. A.,* 32, 3807, 1985.
58. **Zimm, B. H. and LeBret, M.,** Counter-ion condensation and system dimenionality, *J. Biomol. Struct. Dyn.,* 1, 461, 1983.
59. **Maret, G. and Weill, G.,** Magnetic birefringence study of the electrostatic and intrinsic persistence length of DNA, *Biopolymers,* 22, 2727, 1983.
60. **Witten, T. A. and Pincus, P.,** Structure and viscosity of interpenetrating polyelectrolyte chains, *Europhys. Lett.,* 3, 315, 1987.
61. **Khoklov, A. R.,** On the collapse of weakly charged polyelectrolytes, *J. Phys. A.,* 13, 979, 1980.
62. **Joanny, J. F. and Leibler, L.,** Weakly charged polyelectrolytes in a poor solvent, *J. Phys. Paris,* 51, 545, 1990.
63. **McCormick, C. L., Bock, J., and Schultz, D. N.,** Water-soluble polymers, in *Concise Encyclopedia of Polymer Science and Technology,* Kroschwitz, J. T., Ed., Wiley-Interscience, New York, 1990.
64. **Cramer, C. and Truhlar, D. G.,** An SCF solvation model for the hydrophobic effect and absolute free energies of aqueous solvation, *Science,* 256, 213, 1992.
65. **Ross, R. S. and Pincus, P.,** The polyelectrolyte brush: poor solvent, *Macromolecules,* 25, 2177, 1992.
66. **McCormick, C. L. and Johnson, C. B.,** Water-soluble polymers. 28. Ampholytic copolymers of sodium-2-acrylamida-2-methylpropyl)dimethylammonium chloride: synthesis and characterization, *Macromolecules,* 21, 686, 1988.
67. **Torkelson, J. M., Lipsky, S., Tirrell, D. A., and Tirrell, M.,** Fluorescence and absorbance of polystyrene in dilute and semidilute solutions, *Macromolecules,* 16, 326, 1983.
68. **Daoud, M., Cotton, J. P., Farnoux, B., Jannink, G., Sarma, G., Benoit, H., Duplessix, R., Picost, C., and de Gennes, P. G.,** Solutions of flexible polymers. Neutron experiments and interpretation, *Macromolecules,* 8, 804, 1975.
69. **Wiltzius, P., Haller, H. R., Cannell, D. S., and Schaefer, D. W.,** Universality for static properties of polystyrenes in good and marginal solventsx, *Phys. Rev. Lett.,* 51, 1183, 1983.
70. **Kent, M. S., Tirrell, M., and Lodge, T. P.,** Measurement of coil contraction by total intensity light scattering from isorefractive, ternary solutions, *Polymer,* 43, 314, 1991.
71. **King, J. S., Boyer, W., Wignall, G. D., and Ullman, R.,** Radii of gyration and screening lengths of polystyrene in toluene as a function of concentration, *Macromolecules,* 18, 709, 1985.
72. **des Cloiseaux, J. and Jannink, G.,** *Polymers in Solution: Their Modelling and Structure,* Oxford Science Publishers, Oxford, U. K., 1990, chap. 13.

73. **Higgs, P. G. and Raphael, E.,** Conformation changes of a polyelectrolyte chain in a poor solvent, *J. Phys. I,* 1, 1, 1991.
74. **Raphael, E. and Joanny, J. F.,** Annealed and quenched polyelectrolytes, *Europhys. Lett.,* 13, 623, 1990.
75. **Fujita, H.,** *Polymer Solutions,* Elsevier, New York, 1990.
76. **Yamakawa, H.,** *Modern Theory of Polymer Solutions,* Harper & Row, New York, 1971, chap. 4.
77. **des Cloiseaux, J. and Jannink, G.,** *Polymers in Solution: Their Modelling and Structure,* Oxford Science Publishers, Oxford, U. K., 1990, 149.
78. **Hildebrandt, J. H. and Scott, R. L.,** *Solubility of Nonelectrolytes,* 3rd ed., Reinhold, New York, 1950.
79. **Sanchez, I. C.,** Bulk and interface thermodynamics of polymer alloys, *Annu. Rev. Mater, Sci.,* 13, 387, 1983.
80. **Brandrup, J. and Immergut, E. H.,** *Polymer Handbook,* John Wiley & Sons, New York, 1975, chap. 3.
81. **de Gennes, G.-G.,** *Scaling Concepts in Polymer Physics,* Cornell University Press, Ithaca, 1979, chap. 4.
82. **Sariban, A. and Binder, K.,** Critical properties of the Flory-Huggins lattice model of polymer mixtures, *J. Chem. Phys.,* 86, 5859, 1987.
83. **Anathapadmanabhan, K. P. and Goddard, E. D.,** Aqueous biphase formation in polyethylene oxide-inorganic salt systems, *Langmuir,* 3, 25, 1987.
84. **de Gennes, P.-G.,** *Scaling Concepts in Polymer Physics,* Cornell University Press, Ithaca, 1979, 74.
85. **Flory, P. J.,** *Principles of Polymer Chemistry,* Cornell University Press, Ithaca, 1953, 531.
86. **de Gennes, P.-G.,** *Scaling Concepts in Polymer Physics,* Cornell University Press, Ithaca, 1979, chap. 3.
87. **Bonnar, R. U., Dimbat, M., and Stross, F. H.,** *Number Average Molecular Weights,* Interscience, New York, 1958.
88. **Takahashi, A., Kato, N., and Nagasawa, M.,** The osmotic pressure of polyelectrolyte in neutral salt solutions, *J. Phys. Chem.,* 74, 944, 1970.
89. **Takahashi, A., Kato, N., and Nagasawa, M.,** The second virial coefficient of polyelectrolytes, *J. Phys. Chem.,* 71, 2001, 1967.
90. **Mattoussi, H., O'Donohue, S., and Karasaz, F.,** Polyion conformation and second virial coefficient dependences on the ionic strength for flexible polyelectrolyte solutions, *Macromolecules,* 25, 743, 1992.
91. **Noda, I., Kato, N., Kitano, T., and Nagasawa, M.,** Thermodynamic properties of moderately concentrated solutions of linear polymers, *Macromolecules,* 14, 668, 1981.
92. **Wang, L. and Bloomfield, V. A.,** Osmotic pressure of semidilute solutions of flexible, globular and stiff-chain polyelectrolytes with added salt, *Macromolecules,* 23, 194, 1990.
93. **Wang, L. and Bloomfield, V. A.,** Osmotic pressure of polyelectrolytes without added salt, *Macromolecules,* 23, 804, 1990.
94. **Odijk, T.,** Possible scaling relations for semidilute polyelectrolyte solutions, *Macromolecules,* 12, 688, 1979.
95. **de Gennes, P. G., Pincus, P., Velasco, R. M., and Brochard, F.,** Remarks on polyelectrolyte conformation, *J. Phys. Paris,* 37, 1461, 1976.
96. **Flory, P. J.,** *Principles of Polymer Chemistry,* Cornell University Press, Ithaca, 1953, 291.
97. **Yamakawa, H.,** *Modern Theory of Polymer Solutions,* Harper & Row, New York, 1971, chap. 5.
98. **Amis, E. J. and Han, C. C.,** Cooperative and self-diffusion of polymers in semidilute solutions by dynamic light scattering, *Polymer,* 23, 1403, 1982.

99. **Balloge, S. and Tirrell, M.,** The QELS "slow-mode" is a sample-dependent phenomenon in poly(methyl methacrylate) solutions, *Macromolecules,* 18, 817, 1985.
100. **Marshall, W. and Lovesey, S. W.,** *Theory of Thermal Neutron Scattering,* Oxford University Press, Oxford, 1971.
101. **des Cloiseaux, J. and Jannink, G.,** *Polymers in Solution: Their Modelling and Structure,* Oxford Science Publishers, Oxford, U.K., 1990, chap. 7.
102. **Bates, F. S., Muthukumar, M., Wagnall, G. D., and Fetters, L. J.,** Thermodynamics of isotropic polymer mixtures: significance of local structural symmetry, *J. Chem. Phys.,* 89, 535, 1988.
103. **Doi, M. and Edwards, S. F.,** *The Theory of Polymer Dynamics,* Clarendon Press, Oxford, U.K., 1986, 23.
104. **Zimm, B. H.,** The scattering of light and the radial distribution function of high polymer solutions, *J. Chem. Phys.,* 16, 1093, 1948.
105. **Berne, B. J. and Pecora, R.,** *Dynamic Light Scattering with Applications to Chemistry, Biology and Physics,* Wiley-Interscience, New York, 1976.
106. **Chu, B.,** *Laser Light Scattering,* Academic Press, New York, 1974.
107. **Eisele, M. and Burchard, W.,** Slow-mode diffusion of poly(vinylpyrolidone) in the simidilute regime, *Macromolecules,* 17, 1636, 1984.
108. **Drifford, M. and Dalbiez, J. P.,** Effect of salt on sodium polystyrene sulfonate measured by light scattering, *Biopolymers,* 24, 1501, 1985.
109. **Sedlák, M. and Amis, E. J.,** Concentration and molecular weight-regime diagram of salt-free polyelectrolyte solutions as studied by light scattering, *J. Chem. Phys.,* 96, 826, 1992.
110. **Sedlák, M. and Amis, E. J.,** Dynamics of moderately concentrated, salt-free, polyelectrolyte solutions: molecular weight dependence, *J. Chem. Phys.,* 96, 817, 1992.
111. **Du Bois-Violete, E. and de Gennes, P. G.,** Quasi-elastic scattering by dilute, ideal, polymer solutions. II. Effects of hydrodynamic interactions, *Physics, New York,* 3, 181, 1967.
112. **Xu, R., Winnik, M. A., Reiss, G., Chu, B., and Croucher, M. D.,** Micellization of PS-PEO block copolymers in water. V. Test of star and mean field models, *Macromolecules,* 25, 644, 1992.
113. **Zimm, B. H.,** Dynamics of polymer molecules in dilute solution: viscoelasticity, flow birefringence and dielectric loss, *J. Chem. Phys.,* 24, 269, 1956.
114. **Hodgson, D. F. and Amis E. J.,** Dilution solution behaviour of cyclic and linear polyelectrolytes, *J. Chem. Phys.,* 95, 7653, 1991.
115. **Lodge, T. P., Han, C. C., and Akcasu, A. Z.,** Temperature dependence of dynamic light scattering in the intermediate momentum transfer region, *Macromolecules,* 16, 1180, 1983.
116. **Doi, M. and Edwards, S. F.,** *The Theory of Polymer Dynamics,* Claredon Press, Oxford, U.K., 1986, 107.
117. **Wheeler, L. M., Lodge, T. P., Hanley, B., and Tirrell, M.,** Translational diffusion of linear polystyrenes in dilute and semidilute solutions of poly(vinyl methylether), *Macromolecules,* 20, 1120, 1987.
118. **Tirrell, M.,** Polymer self-diffusion in entangled systems, *Rubber Chem. Technol.,* 57, 523, 1984.
119. **Richter, D., Baumgartner, A., Binder, K., Ewen, B., and Hayter, J. B.,** Dynamics of collective fluctuations and Brownian motion in polymer melts, *Phys. Rev. Lett.,* 47, 109, 1981.
120. **Flory, P. J. and Fox, T. G.,** Treatment of intrinsic viscosities, *J. Am. Chem. Soc.,* 73, 1904, 1951.
121. **Yamakawa, H.,** *Modern Theory of Polymer Solutions,* Harper & Row, New York, 1971, chap. 6.

122. **Huggins, M. L.,** The viscosity of dilute solutions of long chain molecules. IV. Dependence on concentration, *J. Am. Chem. Soc.*, 64, 2716, 1942.
123. **Janca, J., Ed.,** *Steric Exclusion Liquid Chromatography,* Marcel Dekker, New York, 1984.
124. **Grubisic, Z., Rempp, P., and Benoit, H.,** A universal calibration for gel permeation chromatography, *J. Polym. Sci. Polym. Lett. Ed.*, B5, 753, 1967.
125. **Winnik, F. M., Tamai, N., Yonezawa, J., Nishimuru, Y., and Yamuzuki, I.,** Temperature-induced transition of pyrene-labeled (hydrotypropyl) cellulose in water: picosecond fluorescence studies, *J. Phys. Chem.*, 96, 1967, 1992.
126. **Ediger, M. D.,** Time-resolved optical studies of local polymer dynamics, *Annu. Rev. Phys. Chem.*, 42, 225, 1991.
127. **Patel, S. S. and Tirrell, M.,** Measurement of forces between surfaces in polymer fluids, *Annu. Rev. Phys. Chem.*, 40, 597, 1989.
128. **Kawaguchi, M. and Takahashi, A.,** Polymers adsorption at solid-liquid interfaces, *Adv. Colloid Interface Sci.*, 37, 219, 1992.
129. **Stamm, M.,** Polymer interfaces on a molecular scale, *Adv. Polym. Sci.*, 100, 357, 1992.
130. **Halperin, A., Tirrell, M., and Lodge, T. P.,** Tethered chains in polymer microstructures, *Adv. Polym. Sci.*, 100, 31, 1992.
131. **Russell, T. P.,** The characterization of polymer interfaces, *Annu. Rev. Mater. Sci.*, 21, 249, 1991.
132. **Ploehn, H. J. and Russel, W. B.,** Intreractions between colloidal particles and soluble polymer, *Adv. Chem. Eng.*, 15, 137, 1990.
133. **Pefferkorn, E., Carroy, A., and Varoqui, R.,** Adsorption of polyacrylamide on solid surfaces. Kinetics of the establishment of adsorption equilibrium, *Macromolecules,* 19, 944, 1986.
134. **de Gennes, P. G.,** Polymers at an interface; a simplified view, *Adv. Colloid Interface Sci.*, 27, 189, 1987.
135. **Leermakers, F. A. M. and Gast, A. P.,** Block copolymer adsorption studied by dynamic scanning angle reflectometry, *Macromolecules,* 24, 718, 1991.
136. **Motschmann, H., Stamm, M., and Toprakcioglu, C.,** Adsorption kinetics of block copolymers from a good solvent: a two-stage process, *Macromolecules,* 24, 3681, 1991.
137. **Frantz, P. and Granick, S.,** Kinetics of polymer adsorption and desorption, *Phys. Rev. Lett.*, 66, 899, 1991.
138. **Watanabe, H., Patel, S. S., and Tirrell, M.,** Displacement of homopolymer from a surface by a block copolymer, in *New Trends in Physics and Physical Chemistry of Polymers,* Lee, L. H., Ed., Plenum Press, New York, 1989.
139. **Klein, J., Perahia, D., and Warburg, S.,** Forces between polymer-bearing surfaces undergoing shear, *Nature,* 352, 143, 1991.
140. **Webber, R. M., Anderson, J. L., and John, M. S.,** Hydrodynamic studies of adsorbed diblock copolymers in porous membranes, *Macromolecules,* 23, 1026, 1990.
141. **Frederickson, G. H. and Pincus, P. A.,** Drainage of compressed polymer layers, Dynamics of a 'squeezed sponge', *Langmuir,* 7, 786, 1991.
142. **Parsonage, E. E., Tirrell, M., Watanabe, H., and Nuzzo, R.,** Adsorption of poly(2-vinyl pyridine-polystyrene), block copolymers from toluene solutions, *Macromolecules,* 24, 1987, 1991.
143. **Terashima, H., Klein, J., and Luckham, P. F.,** The adsorption of polymers onto mica: direct measurements using microbalance and refractive index techniques, in *Adsorption from Solution,* Ottewill, R. H., Rochester, C. H., and Smith, A. L., Eds., Academic Press, New York, 1983.
144. **Guzonas, D. A., Boils, D., Tripp, C. P., and Hair, M. L.,** Role of block size asymmetry on the adsorbed amount of polystyrene-*b*-poly(ethylene oxide) on mica surfaces from toluene, *Macromolecules,* 25, 2434, 1992.

145. **Almog, Y. and Klein, J.**, Interactions between mica surfaces in a polystyrene-cyclopentane solution near the θ temperature, *J. Colloid Interface Sci.*, 106, 33, 1985.
146. **Azzam, R. M. A. and Bashara, N. M.**, *Ellipsometry and Polarized Light*, North-Holland, Amsterdam, 1977.
147. **Briggs, D., Rance, D. G., and Briscoe, B. J.**, Surface properties in *Comprehensive Polymer Science*, Allen, J. and Bevington, J. C., Eds., Pergamon Press, Oxford, 1989.
148. **Char, K., Gast, A. P., and Frank, C. W.**, Fluorescence studies of polymer adsorptioon. I. Rearrangement and displacement of pyrene-terminated poly(ethylene glycol) on collodial silica particles, *Langmuir*, 4, 989, 1988.
149. **Wu, S.**, *Polymer Interfaced and Adhesion*, Marcel Dekker, New York, 1982, chap. 3.
150. **Garvey, M. J., Tadros, T. F., and Vincent, B.**, A comparison of the adsorbed layer thickness obtained by several techniques of various molecular weight fractions of poly(vinylalcohol) on aqueous polystyrene latex particles, *J. Colloid Interface, Sci.*, 55, 440, 1976.
151. **Takahasi, A. and Kawaguchi, M.**, The structure of macromolecules adsorbed on interphases, *Adv. Polym. Sci.*, 46, 1, 1982.
152. **de Gennes, P. G.**, Polymer solutions near an interface, I. Adsorption and depletion layers, *Macromolecules*, 14, 1637, 1981.
153. **Schuetjens, J. M. H. M. and Fleer, G.**, Statistical theory of the adsorption of interacting chain molecules. II. Train, loop and tail size distribution, *J. Phys. Chem.*, 84, 178, 1980.
154. **Scheutjens, J. M. H. M. and Fleer, G.**, Interaction between two adsorbed polymer layers, *Macromolecules*, 18, 1882, 1985.
155. **Cosgrove, T., Health, T. G., Ryan, K., and van Lent, B.**, The conformation of adsorbed poly(ethylene oxide) at the polystyrene/water interphase, *Polym. Commun.*, 28, 64, 1987.
156. **Auvray, L. and Cotton, J. P.**, Self similar structure of an adsorbed polymer layer: comparison between theory and scattering experiments, *Macromolecules*, 20, 202, 1987.
157. **Satija, S. K., Majkrzak, C. F., Russell, T. P., Sinha, S. K., Sirota, E. B., and Hughes, V.**, Neutron reflectivity study of block copolymers adsorbed from solution, *Macromolecules*, 23, 3860, 1990.
158. **Pearson, D. S.**, Recent advances in the molecular aspects of polymer viscoelasticity, *Rubber Chem. Technol.*, 60, 439, 1987.
159. **Lodge, T. P., Rotstein, N. A., and Prager, S.**, Dynamics of entangled polymer liquids: do linear chains reptate?, *Adv. Chem. Phys.*, 79, 1, 1990.
160. **Ferry, J. D.**, *Viscoelastic Properties of Polymers*, 3rd ed., John Wiley & Sons, New York, 1980.
161. **Macosko, C. W.**, *Rheological Measurements*, VCH Verlagsgesellschaft, Weinheim, Germany, 1992.
162. **Graessley, W. W.**, The entanglement concept in polymer rheology, *Adv. Polym. Sci.*, 16, 1, 1974.
163. **Graessley, W. W.**, Entangled linear, branched and network polymer systems. Molecular theories, *Adv. Polym. Sci.*, 47, 67, 1982.
164. **Rouse, P. E.**, A theory of the linear viscoelastic properties of dilute solutions of cooling polymers, *J. Chem. Phys.*, 21, 1272, 1953.
165. **Kausch, H. H. and Tirrell, M.**, Polymer interdiffusion, *Annu. Rev. Mater. Sci.*, 19, 341, 1989.
166. **Breuer, M. M. and Robb, I. D.**, Interactions between macromolecules and detergents, *Chem. Ind.*, 530, 1972.
167. **Goddard, E. D. and Leung, P. S.**, Interaction of cationic surfactants with a hydrophobically modified cationic cellulose polymer, *Langmuir*, 8, 1499, 1992.
168. **Goddard, E. D.**, Polymer-surfactant interactions. Part I. Uncharged water soluble polymer and charged surfactants, *Colloids Surf.*, 19, 255, 1986.

Chapter 4

POLYMER-SURFACTANT INTERACTION

PART I.

UNCHARGED WATER-SOLUBLE POLYMERS AND CHARGED SURFACTANTS

E. D. Goddard

TABLE OF CONTENTS

I. BACKGROUND

The groundwork for today's research on polymer-surfactant systems was laid initially in the early part of this century in studies of proteins associated with natural lipids,[1] and later in the 1940s and 1950s, in studies of their association with synthetic surfactants. These systems involving proteins are not treated here, but will be in ensuing chapters. Also not treated here are the extensive studies[2] on the interactions which can occur between polyoxyethylated nonionic surfactants and polycarboxylic acids; these are akin to the well-known interpolymer association reactions which occur between polyethylene oxide and polycarboxylic acids and are therefore not a unique feature of surfactants.

In this chapter we will be concerned with association reactions in systems comprising an uncharged (synthetic) water-soluble polymer and a charged synthetic surfactant (usually anionic). While emphasis will be placed on more recent work, for historical perspective it has seemed appropriate to describe the techniques, results, and conclusions of earlier work as well. To this extent, some overlap with the earlier reviews of Breuer and Robb[3] and Robb[4] has been unavoidable.

As regards the subject matter of this chapter, it is appropriate first to mention the name of S. Saito, who can properly be termed the father of this field of research. His extensive early work has already been summarized in the article of Breuer and Robb[3] and this will not be repeated here. However, frequent references will be made to Saito's work in the body of this chapter, which, overall, will be divided into four sections, viz. investigative methods, factors influencing the association reaction, interaction models, and, finally, the raison d'être of the association complexes.

Our starting point is 1967 when a publication by Jones[5] appeared on the properties of mixed polyethylene oxide (PEO)/sodium dodecyl sulfate (SDS) systems which has had a major impact on this field. In this paper Jones formalized the concept, in a system of fixed polymer concentration and increasing amounts of surfactant, of two critical concentrations, viz. T_1 and T_2, of the surfactant. T_1 represents the concentration at which interaction between the surfactant and the polymer first occurs, and T_2 the concentration at which the polymer becomes saturated with surfactant. Although refinements concerning the interpretation of these critical concentrations have since been made, this concept has dominated thinking in this field ever since, and is well illustrated by the surface tension method employed by Jones.

II. INVESTIGATION METHODS

A. SURFACE TENSION

Jones illustrated the elegant simplicity of the surface tension method when applied to a mixture of a highly surface-active species, the surfactant, and a feebly surface-active species, the polymer. (Although a boundary tension, viz. oil/water interfacial tension, method had been employed previously by Cockbain[6] for a study of bovine serum albumin/SDS interaction, such measurements are not as facile as the surface tension method and, though qualitatively similar, lacked the resolution of Jones' surface tension measurements: T_1 was less well defined, the boundary tension differences with and without polymer were small, and the protein itself was quite surface active.) If one makes the reasonable assumption that the surface tension is a sensor of the free surfactant in solution, i.e., the surface activity of the surfactant molecule considerably exceeds that of the polymer/surfactant "complex," a method to monitor the concentration changes of uncomplexed surfactant in the mixed systems is thus provided. Since T_1 is, in general, less than the critical micelle concentration (CMC) of the surfactant, it can be deduced that the "adsorbed" or "aggregated" state of the SDS on/with the polymer represents a more favorable energy condition for the surfactant molecules than do regular micelles. Of course, when the sites in this state are saturated, it is expected that the monomer concentration will once more build to the point beyond which regular micelles form: this occurs at T_2. Several investigators, including Jones, have demonstrated that T_1 is only weakly dependent on the amount of polymer in solution, i.e., the concentration of surfactant for "aggregate" formation is mainly a function of surfactant concentration for a particular polymer. On the other hand, the values of T_2, which represent "saturation" of the polymer sites, are directly proportional to the concentration of polymer, as was found by Cockbain.[6]

Schwuger[7] investigated the influence of PEO molecular weight on the surface tension behavior of SDS solutions. For a mol wt of 600, association

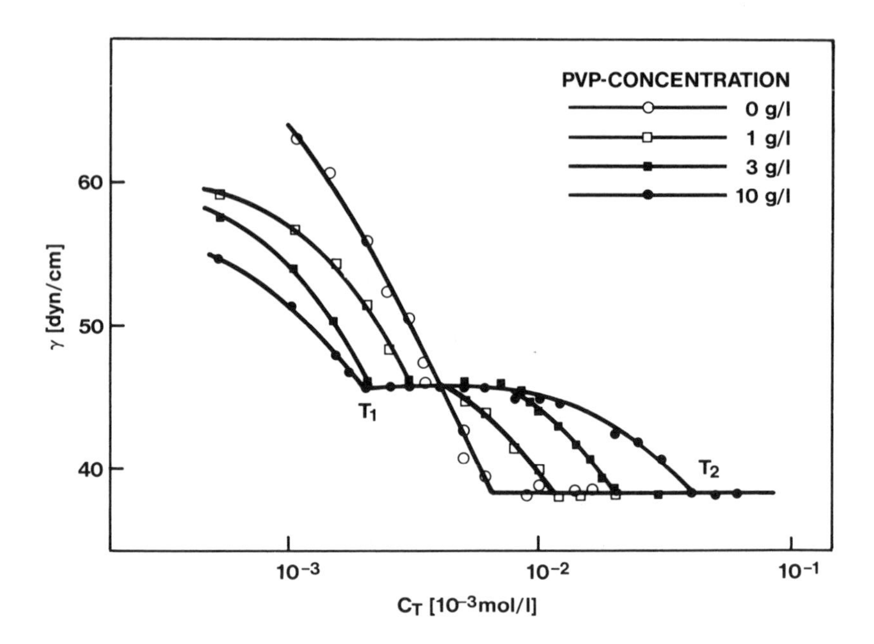

FIGURE 1. Surface tension (γ)/concentration plot of SDS in the presence of PVP at various concentrations; "*T*" assignments follow Jones.[3] (Reproduced from Breuer, M. M. and Robb, I. D., *Chem. Ind.*, 530, 1972. With permission.)

was relatively weak, but for a mol wt of 1550 it was pronounced, and beyond 4000, it was found to be strong and largely independent of molecular weight. For polypropylene oxide (PPO) where one is limited to molecular weights below about 1000 for water-solubility reasons, the interaction was pronounced for the member of mol wt 600, but the high surface activity of this polymer complicates interpretation of surface tension data.[8]

Another system which has been widely studied by the surface tension method is polyvinyl pyrrolidone (PVP)/SDS as illustrated by the work of Lange,[9] Schwuger and Lange,[10] and of Arai and co-workers.[11,12] The essential features of the surface tension plots were similar to those obtained with PEO and are summarized as follows.

1. T_2 increases directly as the amount of polymer increases while, except for low levels of polymer, T_1 varies very little.[9-11] These trends can be inferred from the data of Figure 1.
2. The presence of salt markedly decreases T_1, i.e., association occurs at a much lower concentration of surfactant.
3. In the presence of added salt, variation of T_1 with polymer concentration is smaller than in its absence.
4. Constancy of surface tension, implying constant SDS activity, in the initial part of the binding region, and indicative of high affinity or

"cluster"-type adsorption, is more evident at higher polymer concentrations.[9]

Note that although the surface tension method is readily applicable to a study of these mixed systems, the usual prerequisite of using high-purity surfactants and, in general, components free of surface-contaminating impurities, always applies. If the polymer itself is surface active, the results are more difficult to interpret as in the case of PPO/SDS, already discussed, and in the studies of partialy hydrolyzed polyvinyl acetate (PVAc)/dodecylbenzenesulfonate (DDBS) by Tadros,[13] although clear indications of interaction in such systems can be still be obtained.

B. ELECTRICAL CONDUCTIVITY

Jones[5] reported changes in the specific conductance/concentration plot of SDS occasioned by the presence of PEO which were consistent with the surface tension characteristics referred to above, viz., the appearance of a premicellar breakpoint where the plot departs from that of SDS alone, and a second, postmicellar breakpoint where the plot rejoins that of SDS. These corresponded to, and their values were in good agreement with, T_1 and T_2 from the surface tension method (see Figure 2). Thus, the first breakpoint was independent of, or only weakly dependent on, polymer concentration, while the second breakpoint varied linearly with it. These effects were confirmed by Moroi et al.,[14] who ascribe complex formation in the case of PEO/SDS directly to an interaction between the ionic headgroup and the EO chain. Qualitatively, the lower absolute conductivity, vis a vis the polymer-free system, observed in the T_1/T_2 concentration range of SDS, can be explained by a loss of free ions of the surfactant from solution, either by adsorption on, or cluster formation with, the macromolecule. Similar results were obtained for PEO/SDS by Schwuger.[7] Reference is also made to Gravsholt's work[15] on PEO/SDS and PVP/SDS. Tokiwa and Tsujii[16] demonstrated a minimum molecular weight requirement of the PEO to achieve the indicated type of behavior; it was observed for PEO 6000, but not for PEO 220, mixtures with sodium octylbenzenesulfonate.

Similar independence of behavior over a very wide range of molecular weight (6000 to 900,000) in the PEO/SDS system has recently been found by Francois et al.[17] These authors report, in addition, the existence of a transitional concentration T_1', intermediate between T_1 and T_2, which increases, but much more slowly than T_2, with polymer concentration. A similar intermediate transition in the specific conductivity/SDS concentration plot, showing the same weak dependence on polymer concentration, has been reported by Fadnavis and co-workers[18,19] for mixed SDS solutions with PVP and with two grades of polyvinyl alcohol (PVOH; 83% and 74% hydrolyzed PVAc). The demonstrated existence of this initial interaction zone (viz., T_1 to T_1') invites speculation as to molecular mechanisms involved (e.g., does it correspond to site binding of surfactant with limited aggregation at low coverage

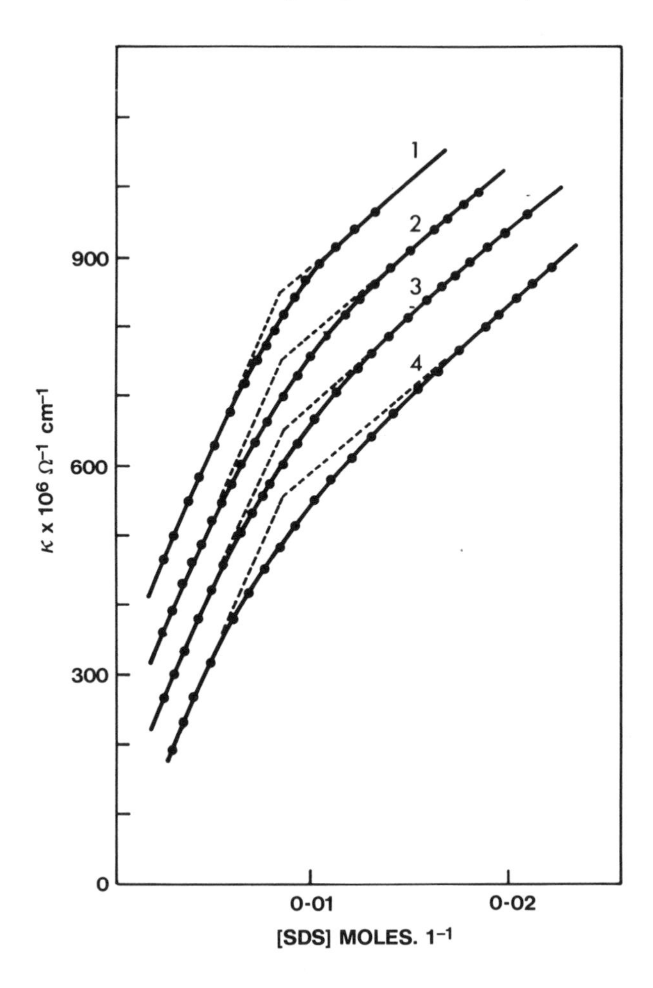

FIGURE 2. Specific conductance (κ)/concentration plot of SDS in the presence of PEO: (1) 0.025% PEO; (2) 0.05% PEO; (3) 0.065% PEO; (4) 0.090% PEO. Scale is for curve 4; curves 3, 2, and 1 raised by 100, 200, and 300 units, respectively. Dotted line, pure SDS solution. (Reproduced from Jones, M. N., *J. Colloid Interface Sci.*, 23, 36, 1967. With permission.)

of the polymer, prior to more substantial aggregation at higher coverage?) and will undoubtedly lead to more probing in this area.

There is evidence of a higher degree of ionic dissociation (α) of SDS in the "complex" than of regular surfactant micelles. Some support for this may be seen in the equivalent conductance curves that Muto et al.[20] obtained for mixtures of SDS and EO/PO copolymers. Also, Zana et al.[21] interpreted their conductance curves as indicating a pronounced increase in α for SDS "clusters" in the complexes formed with both PVP and PEO to values of 0.85 and 0.65, respectively, as compared with ~0.35 for SDS micelles.

Similar trends, but with much lower absolute values of α, viz., 0.2 and 0.3, respectively, for SDS micelles and "clusters" on PEO, were obtained by Francois et al.[17] by analysis of their specific conductance/SDS concentration data.

C. VISCOSITY, ELECTROPHORESIS, AND ULTRACENTRIFUGATION

In a 1957 paper, Saito[22] showed that considerable increases in the viscosity of methylcellulose (MeC) and PVP solutions can be brought about by the addition of SDS, presumably due to a combination of electrical charging and attendant conformational effects. At about the same time, Isemura and Imanishi[23] found that certain uncharged insoluble polymers, like PVAc, which can be solubilized in solutions of anionic surfactants, like SDS, acquire a negative charge under these conditions, as does the soluble PVOH in mixed solution with SDS. These charging effects were demonstrated by direct measurements of electrophoretic mobility. Related work may be found in papers by Saito[24] and Saito and Mizuta.[25] Nowadays, the interplay of polymer charging, related conformational changes, and viscosity increases is well recognized. In the present context, Jones[5] reported a steady increase in the relative viscosity of PEO 19,000 solutions on adding increasing amounts of SDS. A clear change in viscosity occurred at a concentration about T_2, but no slope change was present near T_1. In the same way, Lange[9] observed an increase in viscosity of PVP 40,000 on addition of SDS; in this case the effect was recorded as an increase in specific viscosity which occurred in the vicinity of T_1 for this system. These results clearly imply a change in polymer conformation, viz., an expansion of the polymer coils, on association with the charged surfactant, as reported also by Takagi et al.[26] and Nagarajan and Kalpakci[27] for PEO/SDS systems, and by Lewis and Robinson[28] and Tadros[13] for PVOH/SDS and PVAc/DDBS systems, respectively. In a very recent paper, Prud'homme and Uhl[29] infer from their rheological data that some coil collapse, and a viscosity decrease, occur on first addition of SDS to a PEO solution; coil expansion and higher viscosity follow at higher SDS concentrations. It should be pointed out that Bocquenet and Siffert[30] have also reported a decrease in viscosity — in this case of polyacrylamide (PAAm) in the presence of DDBS. A systematic investigation of the variation of the viscosity of PEO solutions, covering a range of molecular weight and concentration, as a function of added SDS concentration, has been described recently by Francois et al.[17] Their plots of reduced viscosity show a sudden increase at a concentration (T_1) of SDS, which was independent of polymer molecular weight and concentration (in the ranges studied), and a levelling off at a concentration (T_2) which increased with polymer concentration, see Figure 3. The polyelectrolyte nature of the complex was further confirmed in this work by the progressive reduction in reduced viscosity observed on adding salt. The (comparatively low) charge density of the "polyelectrolyte" complex is effectively reduced on adding salt,[17] but a complicating factor is that the binding ratio ($\sim$0.3 mol SDS per mol EO) increases with added salt. Nonetheless, a plot of reduced viscosity vs. the reciprocal of the ionic strength

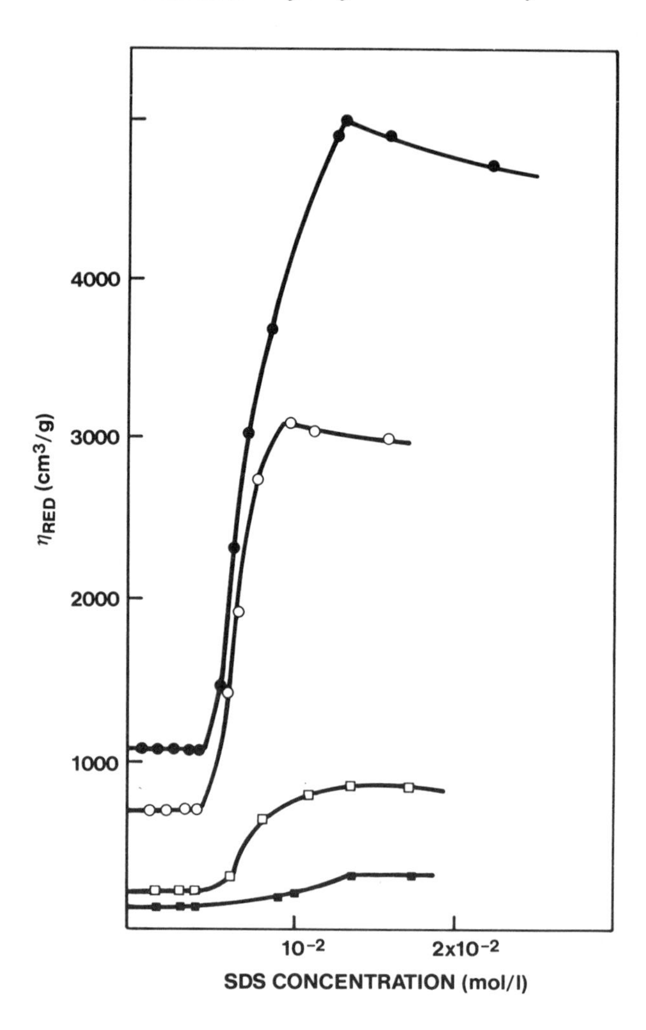

FIGURE 3. Reduced viscosity of PEO as a function of SDS concentration: (●) mol wt = 2 × 10^6, concentration = 6 × 10^{-1} g l^{-1}; (○) mol wt = 10^6, concentration = 6 × 10^{-2} g l^{-1}; (□) mol wt = 2 × 10^5, concentration = 5 × 10^{-1} g l^{-1}; (■) mol wt = 7 × 10^4, concentration = 5 × 10^{-1} g l^{-1}. (Reproduced from Francois, J. et al., *Eur. Polym. J.*, 21, 165, 1985. With permission.)

was linear (characteristic of polyelectrolytes) and led to an extrapolated value equal to that of (uncharged) PEO. An unexplained feature was the initial dip in reduced viscosity on addition of the first amounts of SDS if salt was present at sufficiently high concentration, viz. 0.1 to 0.2 *M* NaCl.

Work on the influence of surfactants on the solution viscosity of a high-molecular-weight PEO has also been reported by Kucher et al.[31] Increases in viscosity were noted both with SDS and fatty acid soaps.

Shirahama et al.[32] carried out electrophoretic experiments on the PVP/SDS system and found that the "complexes" have a substantial negative charge and a mobility value of the same order as that reported by Isemura and Imanishi[23] for PVAc/SDS. An interesting, additional finding was that the measured mobility was independent of the molecular weight of the PVP in the range 24,500 to 700,000. This result is in line with the behavior of polyelectrolytes. In this paper, Shirahama et al. made another important contribution: they invoked the "necklace" model of the complex, in which the "beads" are SDS clusters along the polymer "string" or chain. The conformation they envisage is statistically elongated to lessen the electrostatic free energy. According to these authors the individual clusters, i.e., the charge centers, control the mobility of the complex and they demonstrated that the measured value is comparable to that of free SDS micelles, actually about two thirds of the latter value. Finally, Shirahama et al. point out that the necklace model is fully consistent with the studies of binding of SDS to PVP of Arai et al.,[11] where cluster formation would account for the cooperative nature of the binding.

Ultracentrifugation experiments were carried out by Francois et al.[17] on a high-molecular-weight PEO specimen, in the presence and absence of SDS. Introduction of the latter, up to the saturation concentration (T_2), led to progressively faster sedimentation of the polymer, i.e., a higher molecular weight of the complex, only one peak being evident in the sedigram. Beyond T_2, a second peak corresponding to free SDS micelles was observed. Extrapolation of the sedimentation coefficients (reciprocal) to zero polymer concentration yielded values of 11 and 4 svedbergs for the saturated complex and the free PEO, respectively. From these data a ratio, viz., 3.46, of the molecular weights of the complex and PEO was calculated. This corresponds to a binding ratio of 0.38 mol SDS per mole EO, in rather good agreement with that obtained by Shirahama,[33] viz., 0.41, from binding data for the same concentration (0.1 *M*) of NaCl employed.

D. CLOUD POINT ELEVATION AND SOLUBILITY

A phenomenon, related to the "polyelectrolyte effect," is the enhancement of solubility in water of a polymer when it associates with an ionic surfactant. Most uncharged polymers owe their solubility to the presence of polar groups, such as ether, hydroxyl, carboxyl, amide, etc., which will hydrate in the presence of water. This hydration, of the ether group in particular, can diminish progressively with temperature and the critical balance governing solubility can be upset at a specific temperature ("cloud point") at which the polymer comes out of solution. If the polymer acquires charges, e.g., by ionization of acidic or basic groups or by adsorption of a charged species, such as a surfactant, enhanced solubility or elevation of the cloud point can be expected. Such effects can be seen in the data of Tadros[13] in which the cloud point of PVOH (88% hydrolyzed PVAc) was raised 20°C by a 0.01% concentration of DDBS. Similar elevation of cloud point is well known in the case of MeC. It is to be expected that for polymers with hydrophobic character, conferred through either a hydrophobic backbone or side-chain alkyl groups, the mechanism of cloud point elevation would very likely involve the adsorption of surfactant ions on these sites; see results of Saito et al.[34] for the PVAc (70% hydrolyzed)/SDS system. The work of Lewis and Robinson[28] has given direct evidence that the binding of SDS to two "hydrophobic" polymers, viz., MeC and 80% hydrolyzed PVAc, takes place in two steps, the first involving deaggregation of the polymer, i.e., the development of more hydrophilic entities in solution.

It is not readily possible to examine the influence of ionic surfactants upon the cloud point of PEO, since the clouding temperature exceeds 100°C. However, the phenomenon can clearly be seen by employing a related polyalkylene oxide, viz., PPO, which, by dint of hydrophobic methyl side groups, has much lower water solubility. PPO's of mol wt 1025 and 2000 have cloud points at about 40 and 20°C, respectively. Pletnev and Trapeznikov[8] have shown that these cloud points are raised to 90°C by added SDS in the range of 1 to 2×10^{-2} *M*. The anionic surfactant, NaDDBS, is even more active, a fact which these authors ascribe, at least in part, to enhanced interaction of the aromatic group of this surfactant with the alkylene oxide groups of the polymer. Of course, another way to depress the cloud point of a PEO chain compound is to introduce a single hydrophobic group, as in the alkyl group of conventional nonionic surfactants. The ability of ionic surfactants to increase the cloud point of such "block copolymers" is now a fact of common experience.

A point of interest is that the elevation of solubility can be mutual, i.e., the polymer can increase the solubility of the surfactant. Thus, Schwuger and Lange[10] report, for example, that the presence of PVP at the level of 0.6% in water can depress the Krafft point of sodium hexadecyl sulfate from 31 to 24°C.

E. GEL FILTRATION

Sasaki et al.[35] have pointed out the advantages of the gel filtration method to obtain information on the binding of surfactant to polymers. As compared, for example, to equilibrium dialysis, it is rapid and does not require the addition of electrolyte needed to suppress the Donnan effect in the dialysis method. Following earlier work[36] on the interaction between PVAc and SDS, these authors have recently reported on the PEO 6000/SDS system using a Sephadex G100 column and electrical conductance to obtain the elution curve. Elution volumes, corresponding to decreasing molecular volumes, increased in the order complex < micelles < single surfactant ions. Complex formation was found to occur above a concentration (T_1) of SDS of 4×10^{-3} *M* which was independent of PEO concentration (except at the lowest value studied, viz., 0.025% PEO, for which T_1 increased to 6×10^{-3} *M*). Illustrative elution plots are given in Figure 4 for SDS concentration (1) below T_1; (2) above T_1, but below T_2; (3) above T_2. It is of interest that for condition (2) only one elution peak was observed, suggesting that only one polymeric species is present (i.e., that the SDS is "shared" equally among the PEO

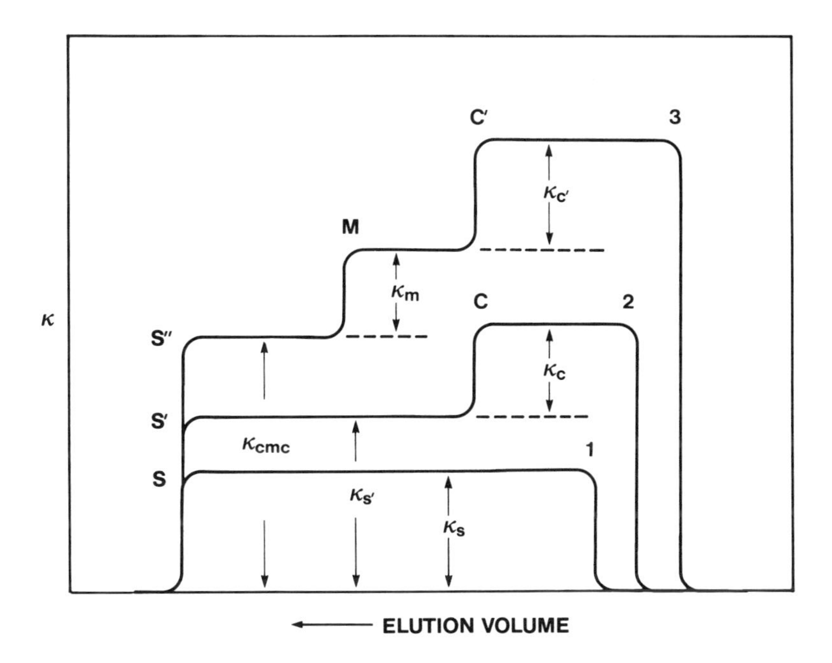

FIGURE 4. Elution volumes vs. specific conductance (κ) for PEO/SDS systems: (1) single ions (S) present only; (2) single ions and complex (C); (3) single ions, complex, and micelles (M). (Reproduced from Sasaki, T. et al., *Bull. Chem. Soc. Jpn.*, 53, 1864, 1980. With permission.)

molecules), thus anticipating results obtained by Cabane and Duplessix using small-angle neutron scattering (SANS);[72] see below. Another point of interest is that the molecular weight of the saturated complex decreased as the concentration of PEO increased, e.g., from 21,000 at 0.025% PEO to 15,000 at 0.2% PEO.

A somewhat different approach to the use of the gel permeation method by Szmerekova et al.[37] was to compare elution curves of PEO 1200 with water and 0.5% SDS as eluents. A lower elution volume of PEO, coupled with the existence of a negative vacant peak in the refractive index plot, was taken as indication of polymer/surfactant complex formation in the latter case. The intensity of the vacant peak corresponding to a "deficit" of SDS, allowed calculation of the binding ratio, viz., 0.30 mol SDS per mole EO, which is in good agreement with other reported values. A weaker, but finite, degree of binding to PEO was found for the nonionic surfactant, nonylphenol polyethylene glycol.

In view of the relative simplicity of the gel filtration method, it can confidently be expected that more use will be made of it to study polymer/surfactant systems. Reference to related use of the technique may be found in the above articles.

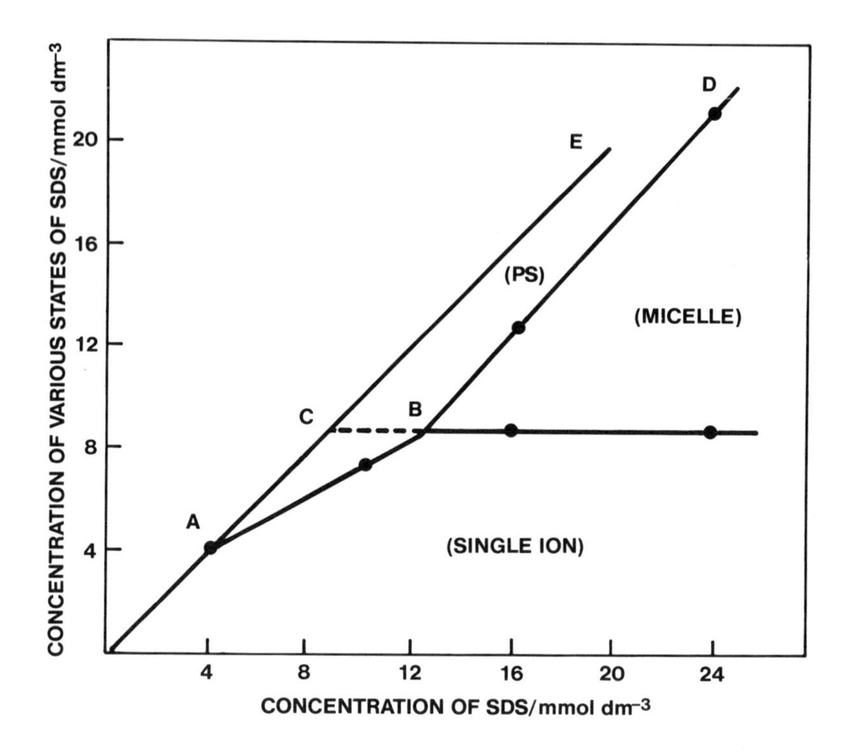

FIGURE 5. Phase diagram of PEO/SDS system (0.05 wt% PEO); PS, polymer/surfactant complex. (Reproduced from Sasaki, T. et al., *Bull. Chem. Soc. Jpn.*, 53, 1864, 1980. With permission.)

From the conductance vs. concentration relations for SDS, it is possible to construct a phase diagram for each PEO concentration condition. A typical diagram is shown in Figure 5.[35] The shape of the complex (ECABD) region suggests that the polymer only becomes saturated with SDS at the "B" (T_2) surfactant concentration condition, and this is in accord with the observation that the elution volume of the complex was found to decrease steadily as the SDS concentration was increased from A to B, and beyond that remained constant. The binding ratio determined from these studies was ~0.33 mol SDS per base mole PEO, corresponding to a molecular weight of the complex of ~20,000, which was found to decrease slightly with increased PEO concentration.

F. ADSORPTION ON SOLIDS

Since both polymer and surfactant can adsorb on solid surfaces, there has been interest in determining what effect each has on the adsorption properties of the other. Chibowski,[38] for example, demonstrated that when PAAm was preadsorbed on calcite, enhanced adsorption of SDS resulted at pH > 8. On the other hand, when present together in solution the polymer depressed the adsorption of surfactant. Chibowski and Szczypa[39] have recently extended their investigations to include sodium laurate (NaL) and dodecylammonium hydrochloride (DA). Some reduction in adsorption of PAAm on clay on addition of DDBS has been reported by Bocquenet and Siffert.[30] All these effects appear to involve polymer/surfactant interaction. Saunders[40] showed that the adsorption of MeC on polystyrene could be inhibited by added SDS. He ascribed the effect to preferential adsorption of SDS on the latex particles, but quite clearly the formation in solution of MeC/SDS-associated species has to be considered in the interpretation of these data. Tadros[13] studied adsorption effects on silica of the mixed pairs PVOH/DDBS and PVOH/cetyltrimethylammonium bromide (CTAB). Although, as would be expected, there are pronounced pH effects with this substrate, several conditions exist under which preadsorbed surfactant enhances the adsorption of the polymer and vice versa. Likewise, substantially increased adsorption of PVOH occurred at several pH values when DDBS was present concurrently with the polymer. Mutual enhancement of adsorption on calcite of sodium oleate/starch pairs was also found by Somasundaran.[41] Clear indications of synergism in adsorption on titania were obtained by Ma[42] for the strongly interacting pair, PVP/SDS, so long as the concentration of SDS was below $4 \times 10^{-4}\, M$, i.e., $\sim T_1$. To explain the results, this author invoked the formation of a type of surface complex between PVP and SDS which can occur on titania at a bulk concentration as low as $1 \times 10^{-3}\, M$ SDS. The formation in solution of higher SDS concentrations of "conventional" PVP/SDS complexes resulted in lowered adsorption of both compounds since the complexes are relatively surface inactive.

Gebhardt and Fuerstenau,[43] while confirming the virtual absence of interaction between PAAm and sodium dodecanesulfonate ($SDSO_3$) in solution, demonstrated that the presence of preadsorbed PAAm on hematite had no effect on the adsorption of $SDSO_3$ (except at high concentrations of the latter) nor on the electrophoretic mobility of hematite as a function of $SDSO_3$ adsorption density. On the other hand, Hollander[44] determined that each ingredient in a PAAm/DDBS system had a depressing effect on the adsorption of the other on kaolinite, and such an effect was found by Moudgil and Somasundaran[45] for adsorption of the $SDSO_3$/PAAm pair on hematite, but in this case the polymer had a small content of anionic groups. The aforementioned data again imply a possible role of polymer/surfactant complexes in the adsorption process, but further analysis of the individual systems is beyond the scope of this review.

G. SURFACTANT BINDING STUDIES

1. Dialysis Equilibrium

This has been a traditional method for the study of polymer/surfactant interactions, and was first used in the early work involving proteins. Fishman and Eirich[46] applied it to the PVP/SDS system. Their results support the pattern developed so far, namely, no identifiable interaction below a critical concentration (T_1) of surfactant, followed by a zone of interaction, and then a region corresponding to the formation of micelles. The values of T_1 and of the maximum binding ratio of SDS ($r = 0.3$ mol SDS per base mole PVP) of Fishman and Eirich are, according to their interpretation, in reasonable agreement with values derived from surface tension data.

A feature of such binding studies is that, within the interaction zone, a steep uptake of surfactant by the polymer (in the inner compartment of a dialysis cell) is generally observed. The steepness of the uptake has led most investigators to the conclusion that a high level of cooperativity is involved in the binding of the surfactant. These features are evident in studies of the above system which were extended by Arai et al.[11] to include three homologous sodium alkyl sulfates (C_{12}, C_{11}, C_{10}). The form of the binding curves was similar inasmuch as uptake occurred at a concentration close to that observed as "T_1" by surface tension measurements; it levelled off as the equilibrium concentration approached the CMC. This work was carried out in the presence of 0.1 *M* NaCl to minimize Donnan membrane effects. In consequence, both T_1 and the CMC were depressed but, interestingly, salt elevated the binding ratio which, in the case of the C_{12} homolog, was 0.9 mol SDS per base mole PVP. Not unexpectedly, both T_1 and T_2, for a given amount of polymer, decreased as the surfactant chain length increased, in much the same way as does the CMC in a homologous series of surfactants. Shirahama[33] carried out dialysis equilibrium measurements on PEO/SDS in the presence of 0.1 *M* NaCl and obtained results similar to the aforementioned, see Figure 6. Later, the work was extended by Shirahama and Ide[47] to include the C_{14} and C_{11} alkyl sulfates. The results were in accord with predictions based on chain length effects but, in the case of the C_{11} sulfate, binding was found to be limited by the onset of micellization. For the SDS system, the limiting binding ratio was 0.4 mol SDS per base mole PEO, a value somewhat higher than that observed in salt-free systems by the surface tension method, viz., ~0.2 to ~0.3:1.[5,7,48] Shirahama stressed the cooperative nature of the binding; his theoretical treatments to describe the binding of surfactant "clusters" will be referred to later.

In the studies by Smith and Muller[49] of the binding of sodium 12,12,12-trifluorododecyl sulfate (F_3SDS) by PEO of various molecular weights, it was shown that the interaction pattern was similar for PEO 20,000 and PEO 7000, but that little interaction occurred with PEO 1500. Binding started at a surfactant concentration of 10 m*M* (2/3 of the CMC) and the saturation value of binding corresponded to 0.3 mol F_3SDS per base mole PEO. Muller developed a simple model of cooperative binding of the surfactant by the polymer which will be presented later. Finally, we refer to work by Lewis and Robinson[28] on the binding of SDS by two relatively hydrophobic, surface-active polymers, viz., MeC and partially (80%) hydrolyzed PVAc. In these systems "two-step" binding isotherms were obtained, as referred to above. The extent of binding found was higher, reaching 1.5 mol SDS per base mole polymer, in the case of MeC. We mention here that for a 100% hydrolyzed PVAc system, i.e., PVOH/SDS, the binding was similar to that obtained for PEO/SDS, viz.,

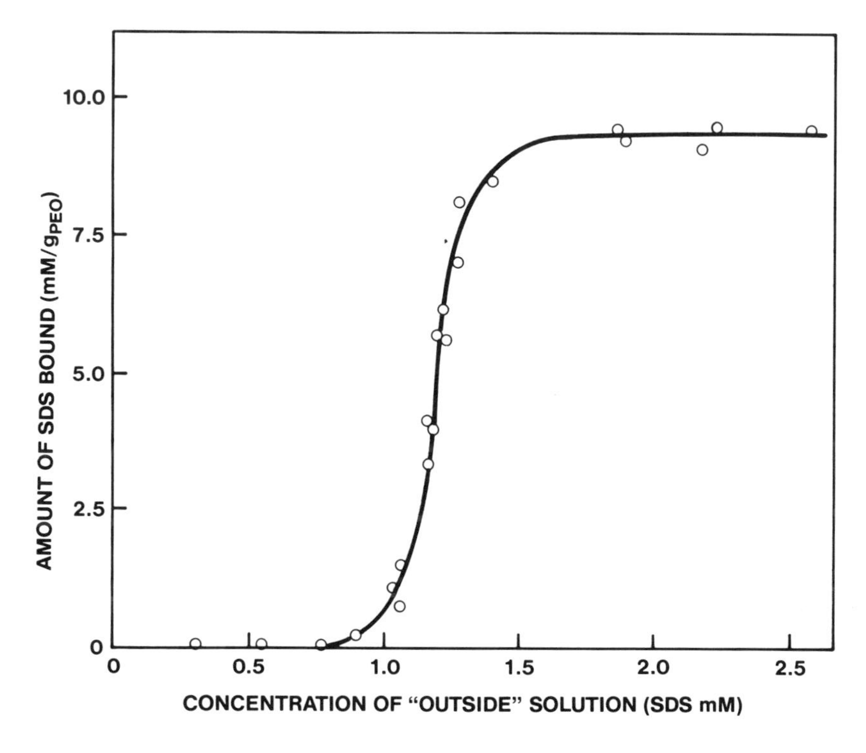

FIGURE 6. Binding isotherm of PEO/SDS system in 0.1 *M* NaCl: solid line from Hill equation.[99] (Reproduced from Shirahama, K., *Colloid Polym. Sci.*, 252, 978, 1974. With permission.)

a one-step process and a binding ratio corresponding to 0.3 mol SDS per base mole polymer.

2. Use of Ion-Specific Electrodes

Since the development of membrane electrodes specific for dodecyl sulfate[50] and other surfactant ions, their use as a convenient method to study binding of such ions has become increasingly popular (see also Part II of this review[108]). Kresheck and Constantinidis[51] applied the method to PVP/sodium alkyl sulfate systems, employing the decyl (SDecS) and octyl (SOS) homologs. For the electrode employed, a plot of e.m.f. against the logarithm of the surfactant concentration was linear over about 2 decades of concentration and showed a slope change at the CMC. In the presence of polymer two new slope change points straddling the CMC, and consistent with T_1 and T_2 transitions, were observed. These authors deduced from their data a saturation binding ratio of SDecS to PVP of 0.34 to 1, in good agreement with alkyl sulfate values previously referred to, and independent of the molecular weight of PVP in the range 10,000 to 340,000. It should be pointed out that for this deduction the T_2–T_1 value was not employed; rather a function T_2–CMC′ was used, where CMC′ is a hypothetical or effective CMC in the presence of polymer,

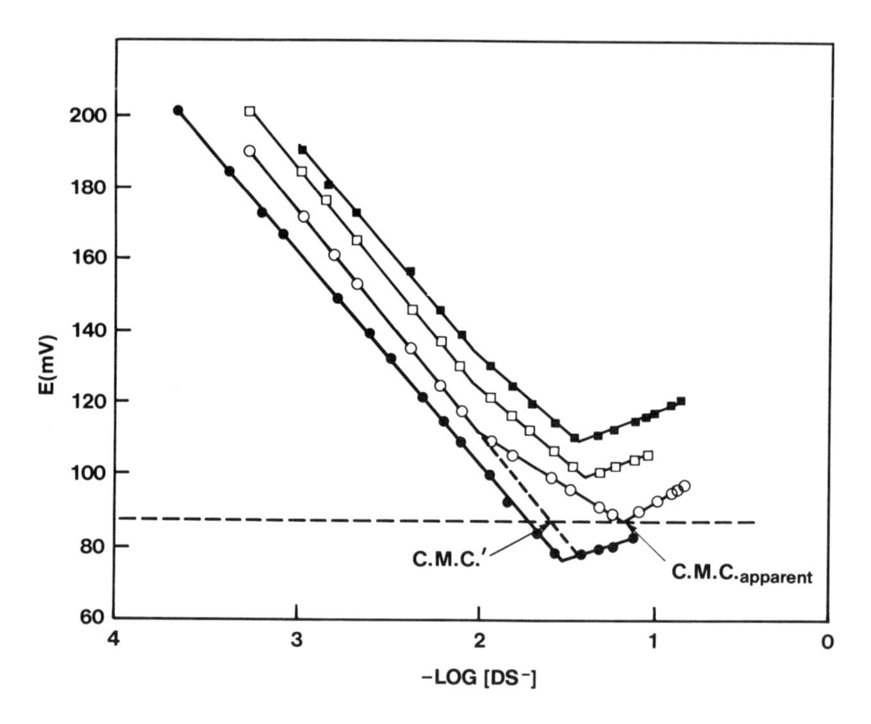

FIGURE 7. E.m.f./concentration plots of SDS in the presence of PVP: (○) 1.2% PVP; (□) 0.5% PVP; (■) 0.25% PVP; (●) 0% PVP. Each upper curve is raised progressively by 10 mV. For CMC designations see text. (Reproduced from Kresheck, G. C. and Constantinidis, I., *Anal. Chem.*, 56, 152, 1984. With permission.)

obtained by extrapolation of the low-concentration e.m.f. data to an e.m.f. value corresponding to that at the "CMC apparent," or T_2, value (see Figure 7). (Strictly speaking, such an approach should also be used in estimating binding ratios from surface tension data, as illustrated in Figure 8. The amount bound is evidently T_2'–T_1, or, what should be equivalent, T_2–CMC′ rather than T_2–T_1. Of course, in both cases the reliability of the extrapolation would have to be considered. Actually, evidence increasingly indicates that, except at high polymer levels, the extent of the true plateau condition T_2'–T_1 is rather restricted. Aside from the possibility that the complex itself is acquiring increasing surface activity as binding approaches saturation, this means that, unlike the conditions corresponding to micelle formation, the activity of DS monomers increases steadily as more of the complex forms. See, for example, the phase diagram shown in Figure 5. This means in turn that the free energy of formation of complex, expressed per mole of SDS, also decreases as the polymer becomes more saturated. Another way to view this is that the degree of cooperativity in complex formation is lower than in regular micelle formation, i.e., the cluster size of surfactant molecules is smaller and this aspect has been pointed out by Muller (see below) in developing his model of complex formation.) A feature of the e.m.f./concentration plots of Kresheck and Con-

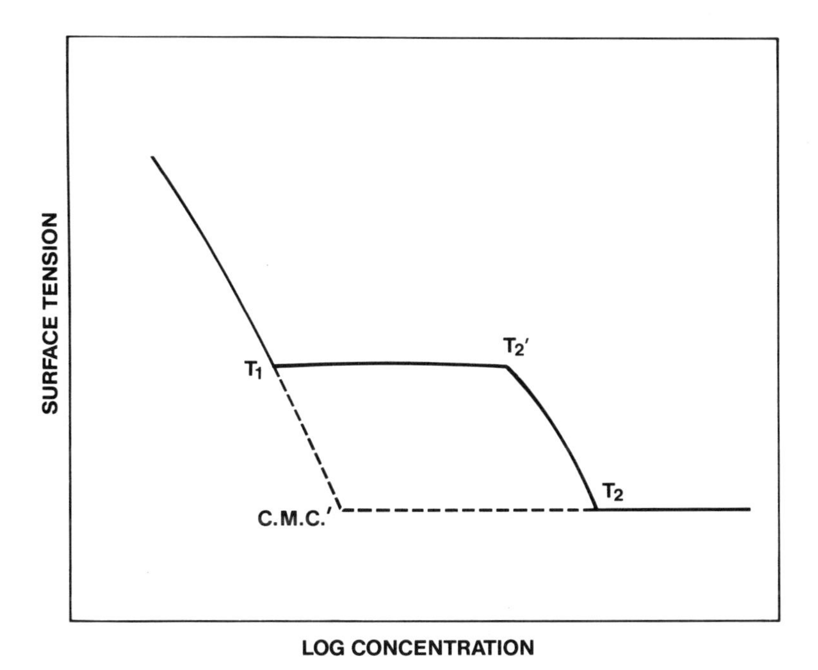

FIGURE 8. Diagrammatic surface tension/concentration plot of a surfactant in the presence of a complexing polymer. See text.

stantinidis,[51] indeed, is that there is a pronounced slope between T_1 and T_2, implying a steady increase in surfactant ion activity prior to micellization. This slope did, however, decrease as the polymer concentration, and hence T_2–T_1, increased. This effect is similar to that already mentioned in respect of a higher level of constancy of surface tension in the T_1 to T_2 concentration range as the concentration of PVP is increased.[9] It is of interest that earlier data of Birch et al.,[50] using a ''DS'' electrode, showed an e.m.f. plateau in this region with PVP but no such plateau region when the polymer was PVOH. This difference probably reflects stronger binding of SDS to PVP.

Another, somewhat less direct, method to obtain information about the binding of an anionic surfactant is to meter the change in counterion (e.g., Na^+) activity during binding, using an electrode specific to said counterion.[52,53] This method is especially useful if used in conjunction with a surfactant-ion electrode, as has been done by Birch et al.[50] and by Gilanyi and Wolfram.[54] The latter authors carried out a rather complete experimental and theoretical investigation of the binding of surfactant by polymers; the systems chosen were PVOH/SDS and PEO/SDS. The binding models developed, which are described in a later section, treat the binding as a cooperative process analogous to micellization and of similar free energy. A typical set of binding isotherms for the PEO/SDS system is shown in the section ''Interaction Models'' presented later. One interesting point is that Gilanyi and

Wolfram deduced that the number (n) of surfactant ions in a cluster is derivable from the dependence of T_1 on the polymer concentration. Values arrived at are $n = 12$ for PVOH in water, and $n = 40$ to 50 for PVOH and PEO in 0.1 M NaCl.

Recently, Francois et al.[17] reported the use of a sodium ion electrode to study the interaction of SDS and 0.2% PEO 20,000. Linear sections of a log–log plot of sodium ion activity vs. SDS concentration, intersecting at T_1 and T_2, were obtained, resembling the data of Gilanyi and Wolfram[54] for PVOH/SDS. Using the well-known expression derived by Botré et al.[52] for the observed activity of the counterion,

$$a_{Na} = \gamma[C_{DS^-} + \alpha(C - C_{DS^-})]$$

Francois et al. obtained a value for α, the degree of dissociation of the SDS clusters, of about 0.25. Here γ refers to the activity coefficient of the counterion.

H. SOLUBILIZATION OF OIL-SOLUBLE ADDITIVES (DYES)

The ability of surfactant micelles to solubilize various oil-soluble materials is well recognized. This property can also be found in surfactant assemblies in the adsorbed state, e.g., on solid surfaces.[55,56] In his 1957 paper, Saito[22] demonstrated that a number of water-soluble polymers can enhance the capacity of various surfactants to solubilize the oil-soluble dye Yellow OB. This enhancement was manifested as an initiation of solubilization activity before the formal CMC of the surfactant. In his 1958 paper, Saito[57] extended the studies to other dyes. This field has since been explored by a number of other workers. For example, Jones[58] investigated the solubilization capacity, towards the dye Orange OT, of PEO 5400/SDS mixtures. In this case the data showed excellent agreement between the point of initiation of solubilization and the T_1 value derived from conductivity and, likewise, between the concentration at the point of coincidence with the solubilization curve of the polymer-free, micellar system and the T_2 value from conductivity. Agreement between T_1 and T_2 values by the dye method and those obtained by other methods is also evident in the results of Arai et al.[11] for solubilization of Yellow OB by PVP/SDS mixtures. The solubilization efficiency of this polymer-surfactant combination is notable, and in this case the micellar solubilization plots with and without polymer did not come into coincidence, T_2 being assigned to the concentration point where the two plots became parallel. Data for solubilization of Orange OT by mixtures of PVP 40,000 and a series of alkyl sulfates (C_{10}–C_{16}) were presented by Lange.[9] In one diagram, Lange's data provide a useful illustration of two effects, which show that while the T_1 values are less than the corresponding CMCs they systematically follow the same dependence pattern that CMCs do on surfactant chain length (see Figure 9). These results provide a clear indication that the bound surfactant must exist in some kind of cluster and that there is a similarity between the

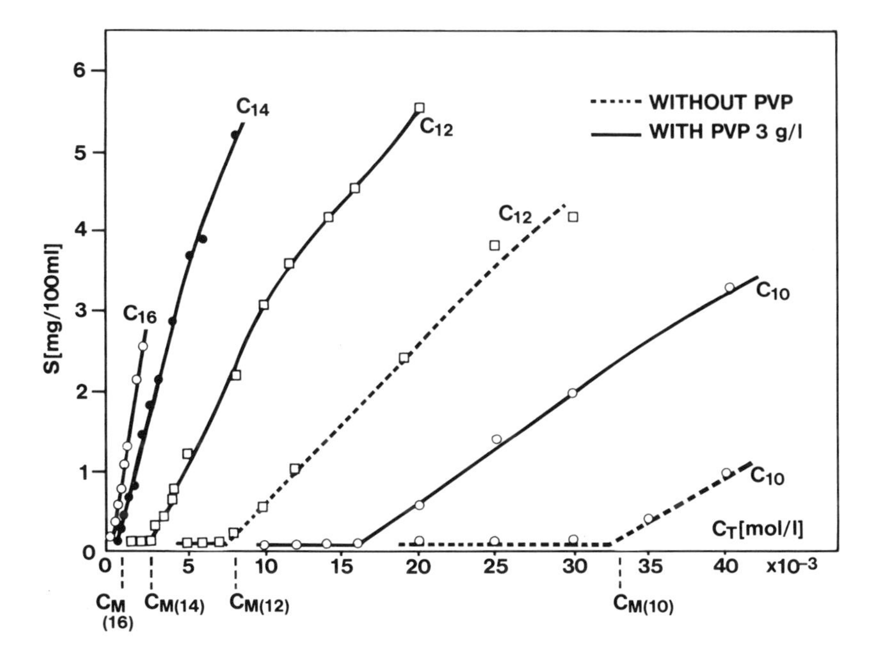

FIGURE 9. Effect of PVP on the solubilization/concentration plots of sodium alkyl sulfates of different chain lengths and indicated CMC (C_M). Dye is Orange OT. (Reproduced from Lange, H., *Kolloid Z. Z. Polym.*, 243, 101, 1971. With permission.)

process of "binding" and micellization. From studies of the solubilization of the Yellow OB by PEO/SDS solutions, Shirahama and Ide[47] adduced evidence from the red shift of the spectrum, as compared to that of simple micellar SDS solutions, that there must be contact of the solubilized dye with EO moieties in the "complex". Some support for this conclusion came from Tokiwa and Tsujii,[16] who reported a red shift for PEO 6000/octylbenzenesulfonate (but not seen with the PEO 220 counterpart). From a detailed study of the effect of molecular weight of PEO in these systems, these authors concluded that a transition in solubilization character occurs for the PEO member with a molecular weight of 600.

I. FLUORESCENT PROBES

The use of fluorescent probes to obtain information about micellar environments is comparatively new.[59] By this technique one can obtain a measure of the effective polarity of that portion of the micelle where the fluorescer is located. For pyrene, the most widely used probe for such studies, this measure is provided by the ratio of the 1st and 3rd fluorescent peaks. Turro and coworkers[60] investigated both PVP/SDS and PEO/SDS systems using pyrene. Their data for the former are illustrated in Figure 10. We comment first on the characteristics of SDS alone, for which there is a low concentration region ($<10^{-3}$ *M*) with I_I/I_{III} values of 1.9, corresponding to the polarity of water,

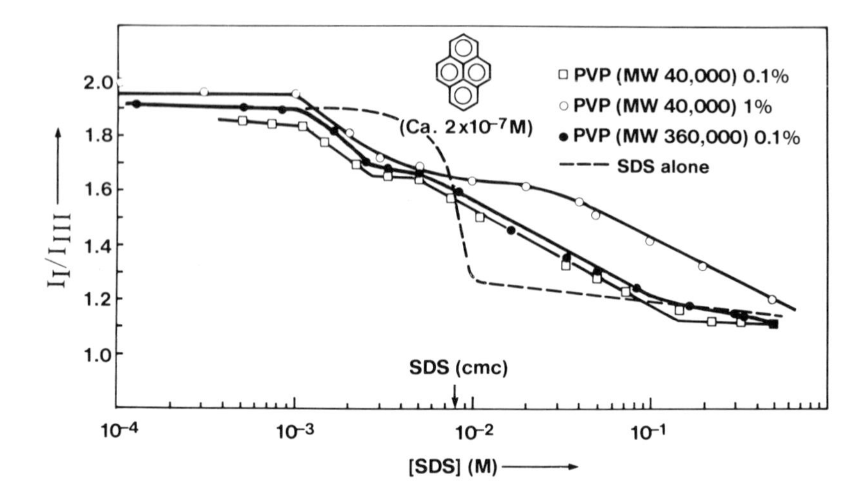

FIGURE 10. Plot of characteristic fluorescence parameter I_I/I_{III} of pyrene vs. SDS concentration in the presence of PVP. (Reproduced from Turro, N. J. et al., *Macromolecules*, 17, 1321, 1984. With permission.)

a high concentration region ($\geq 10^{-2}$ *M*) with I_I/I_{III} values of 1.3 to 1.1, signifying a polarity corresponding to that of ethanol, and a transition region in between. Introduction of PVP yielded most interesting changes in the fluorescence characteristics, namely the appearance of a plateau of I_I/I_{III} values in, roughly speaking, the intermediate or transition region. At first sight the results seem to offer another confirmatory illustration of the behavior of mixed polymer/surfactant systems which are in conformance with those obtained by the surface tension method, with the two transition concentrations in this case being identified by the two ends of the plateau as the beginning and end of the interaction zone. For example, the first one has about the same value as T_1 observed by other methods, and the length of the plateau increases with polymer concentration yielding a binding ratio of ~0.3 mol SDS per base mole PVP, consistent with that obtained by other methods. However, several additional comments are in order.

1. For the 0.1% PVP system, the second transition concentration is lower than the CMC of SDS. Hence, it is more appropriately designated as T_2' rather than T_2 (see previous discussion).
2. There is no explanation for the rather sharp reduction in the I_I/I_{III} value, starting at very low SDS concentration ($\sim 10^{-3}$ *M*), i.e., lower than T_1. Such a transition is not detected by other methods, perhaps because of lack of sensitivity (or insufficient probing of the low-concentration range), and there is a possibility that the presence of the dye itself may be involved in its existence.

3. The "with" and "without" polymer curves do not come into coincidence until very high SDS concentrations, which would yield unrealistically high values of T_2.

The results for the PEO/SDS system differed in an important respect in that, except possibly for the lowest polymer level (0.02%), there was no plateau: after T_1 the points were more or less colinear with the I_I/I_{III} vs. SDS concentration plot of micellar (i.e., polymer-free) SDS. In other words, the polarity of the "complex" in this case was about the same as that of SDS micelles. The plateau observed in 0.02% PEO/SDS solutions terminates at a concentration of SDS which far exceeds the CMC, implying an unrealistically high degree of binding of SDS. Therefore, it does not correspond to "T_2."

Complementary data to the above have recently been published by Zana et al.[21,61] on the PEO/SDS and PVP/SDS systems. Like Turro et al.,[60] these authors comment on the apparently higher polarity zone, vis a vis regular SDS micelles, encountered in the SDS clusters formed in the presence of these polymers, especially the PVP. They actually speculate that an interaction is involved between pyrene and the polymer at the micellar surface.

While certain refinements may be needed in the detailed interpretation of the above data (for example recognition that the probe is measuring a different property from that measured by the other techniques and/or that there is possibly perturbation of the system on addition of the probe even at the very low concentrations employed, viz 10^{-7} M), their value is obvious for providing information on the microenvironment of the surfactant clusters and on the dependence of the latter on the particular associating polymer employed. These data extend those of Bloor and Wyn-Jones,[62] who first pointed out the relatively polar environment of the PVP/SDS aggregates and the close similarity of PEO/SDS aggregates in this respect to regular SDS micelles.

Fluorescent decay curves, obtained under conditions where the probe to surfactant micelle, or probe to surfactant aggregate, mole ratio is at about unity, and where excimer formation of the fluorescer molecules occurs, can provide information concerning the size of the micelle or aggregate.[63] Using this approach, Zana et al.[21,61] have found the size of SDS clusters which form in the presence of PEO or PVP to be at about 20 to 50 monomers in the T_1–T_2 concentration region. This is to be compared with values in the range 60 to 80 monomers for regular SDS micelles by this technique. Their data clearly show that the cluster size increases in the T_1–T_2 concentration range and that the clusters which form first (i.e., at SDS concentrations near T_1) are actually quite small (~20 monomers). Beyond T_2, there is an increase in (average) size of SDS aggregates due to increasing dominance of regular SDS micelles. A point of considerable interest is that very good agreement with Zana's cluster size can be seen in the fluorescence decay results of Francois et al.[17] for comparable concentration values (wt/vol.): these authors employed a high-molecular weight (~10^6) PEO while the PEO specimen of Zana was of mol wt 20,000, leading them to conclude that molecular weight has little influence on the aggregation number of SDS clusters. Another noteworthy result is their demonstration of an increase of cluster size of the bound SDS aggregates on introducing salt (0.1 *M*) into the PEO/SDS system. The potential value and importance of this information are obvious, constituting one of the few possible approaches which allow direct estimates of the size of the surfactant clusters in the "complexes" to be obtained. Equally important is that these estimates are in line with the predictions of various models of binding that have been proposed.

Aggregation numbers of SDS clusters in PEO and PVP complexes have also been reported by Lissi and Abuin,[64] in this case using a fluorescence method based on the quenching of ruthenium tris(bipyridyol)chloride by 9-methylanthracene in mixed solution of the polymer and surfactant. The aggregates were estimated to comprise 35 ± 5 monomers in the presence of PEO, and 28 ± 6 in the presence of PVP. The aggregation numbers were unchanged for surfactant/polymer ratios in the range from about 40% of "coverage" of the polymer to saturation, and were independent of molecular weight, from 6×10^3 to 1×10^5 Da for PEO and 4×10^4 to 4×10^5 Da for PVP. In the presence of added salt (0.1 *M* NaCl) the size of the SDS clusters in mixed solution with PEO increased to 45 ± 5.

J. ELECTROOPTIC OR KERR EFFECT

The Kerr effect refers to the birefringence produced when an electric field, usually in the form of a pulse, is applied across an isotropic fluid. In employing this technique for a study of polymer/ionic surfactant systems, Rudd and Jennings[65] have pointed out various restrictions: (1) the high conductivity of the surfactant and of the polyelectrolyte-like complex when it forms with the polymer limits the concentration that can be employed; and (2) no adequate theory exists for the birefringence of a flexible polymer in solution.

Despite these restrictions and limitations as regards detailed interpretation, the measurements yielded extremely interesting results. Rudd and Jennings employed 2% PVP solutions containing SDS up to 2.5×10^{-2} *M*. The Kerr constant (B) for concentrations up to 3×10^{-3} *M* was extremely small (Region I). Between 3×10^{-3} *M* ("T_1") and 9×10^{-3} *M* there was a 50-fold, linear increase in B, clearly pointing to the creation of a new charged species in solution (Region II). What is not clear is why saturation was observed beyond this point (Region III): if the saturation point corresponds to "T_2" then it is obviously very considerably lower than T_2 values ($\gg 10^{-2}$ *M*) expected for this concentration of polymer by other methods, including the fluorescent probe. More work using this technique on polymer/surfactant systems would be very desirable.

K. FAST KINETICS MEASUREMENTS

The kinetics of micelle formation have been studied and discussed by a number of authors. Nowadays there is wide agreement that there are two characteristic relaxation times — a short one, τ_1, corresponding to the exchange of a surfactant monomer with a preformed aggregate (micelle) and a longer one, τ_2, corresponding to micelle formation/disintegration. Roughly speaking, τ_1 is measured in microseconds and τ_2 in milliseconds. Wyn-Jones and co-workers, in a series of papers,[62,66-68] have studied the kinetics of surfactant aggregation in the presence and absence of added water-soluble polymers. For the PVP/octylsulfate system ultrasonic measurements[66] revealed the existence of a premicellar zone in which relaxation rates ($1/\tau_1$) rose to a value 10-fold greater than that at the CMC of the surfactant solutions alone, for which no premicellar relaxations were observed. Beyond the CMC, the characteristics of the solutions with and without the polymer were much the same, each having a single relaxation time.

In general, relaxation rates corresponding to monomer-micelle exchange are lower as one increases the chain length in a homologous series of surfactants. Using a pressure jump method to measure $1/\tau_1$ for the relatively slow monomer/aggregate exchange process of sodium hexadecyl sulfate, Wyn-Jones[68] found that the presence of PVP resulted in an increase of relaxation rate of 2.5-fold near the CMC of this surfactant; the faster rate persisted up to a surfactant concentration roughly double the CMC. Pressure jump measurements were also employed for τ_2 measurements on solutions of the shorter chain length homolog SDS in the presence and absence of polymer. For both PVP/SDS and PEO/SDS, relaxation rates were substantially higher in the presence of polymer;[62,67] see Figure 11. By determining the temperature dependence of τ_2, activation energies (ΔE) could be obtained for the surfactant aggregation/deaggregation process. In the concentration region in which the aggregated surfactant was chiefly associated with polymer, i.e., there were no free micelles, ΔE was reduced by almost two thirds compared with the value in the polymer-free system. In all cases a single relaxation time was observed. These results clearly imply that the formation/disintegration process involving single surfactant ions and the adsorbed/aggregated surfactant on the polymer chain is faster and much more facile than between single ions and ordinary surfactant micelles. Similar results were observed for the PVP/SDS and PEO/SDS systems. However, certain differences in behavior of the two polymers as evidenced by different characteristics of $1/\tau_2$ and of ΔE, absolutely and in relation to the polymer saturation or T_2 line, indicate that further work using these techniques would be desirable.

Determinations of τ_2 have been made by Tondre[69] on mixtures of SDS with dioxane and PEO (two grades), utilizing a temperature jump technique and measuring the concomitant changes in absorption of acridine orange when the micellar equilibria are perturbed. Whereas addition of PEO 1000 or dioxane to a micellar solution of SDS led to continuous changes in relaxation rate ($1/\tau_2$) and signal amplitude, added PEO 10,000 beyond a certain concentration brought about a sharp change in the relaxation rate and absorption amplitude curves; see Figure 12. The value of these "transition concentrations" of polymer increased linearly with SDS concentration, and in fact Tondre points

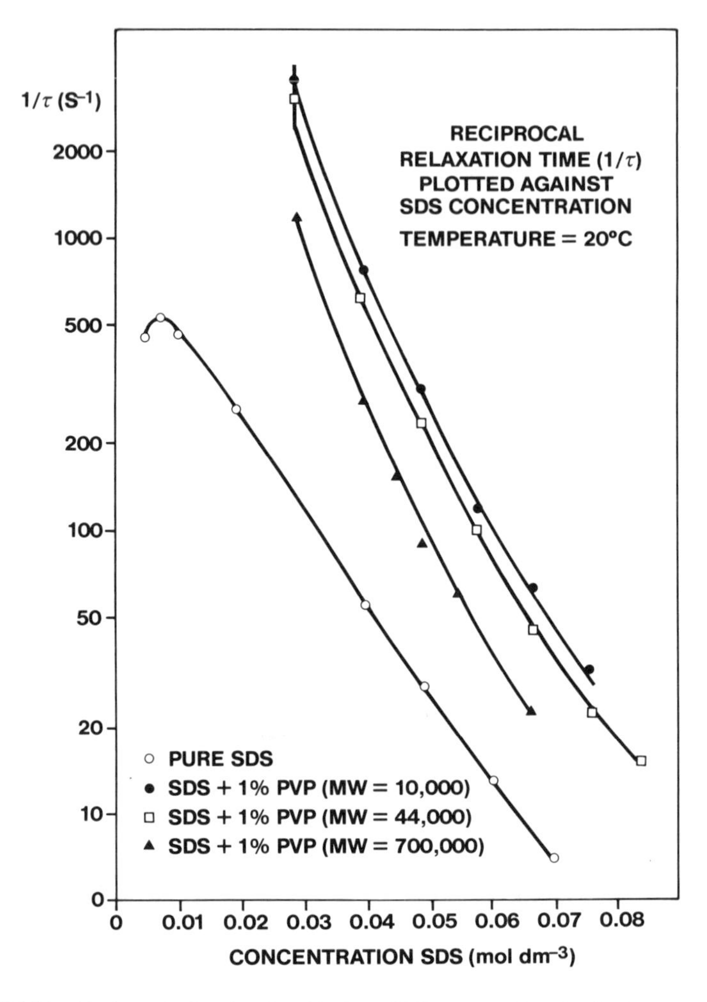

FIGURE 11. Reciprocal relaxation time ($1/\tau_2$) vs. SDS concentration plot, in the presence and absence of PVP. (Reproduced from Bloor, D. M. et al., in *Techniques and Applications of Fast Reactions in Solution,* NATO Adv. Study Inst. Ser., Ser. C [NASCDSO], VC50, D. Reidel, Dordrecht. With permission.)

out that the values lead to T_2 values of SDS at given concentrations of polymer which are in good agreement with those obtained by other methods. The implication, then, is that for solutions containing a mixture of regular micelles and saturated polymer/surfactant complexes, the relaxation rate remains relatively low and constant; once into the region of unsaturated complexes, i.e., to the right of the transition points of Figure 12, the relaxation rate increases rapidly as the degree of saturation of the polymer decreases. Although explanations of these observations are still tentative, it is obvious that the results are very significant. Another point of interest is that Tondre, like Wyn-Jones, deduced from temperature studies that the activation energy for the surfactant

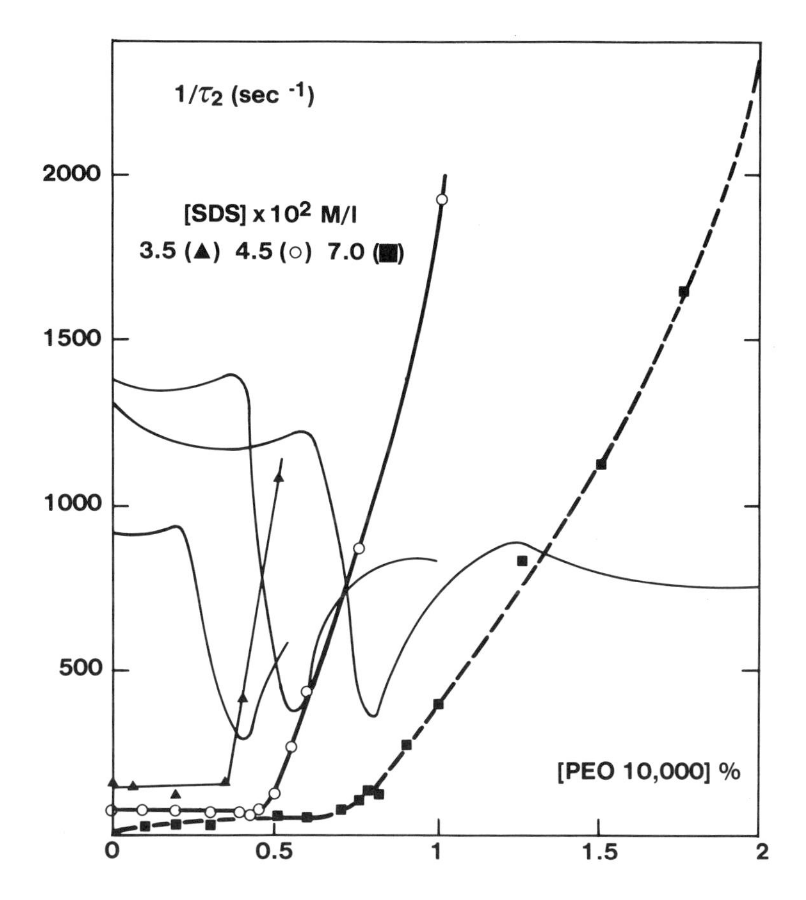

FIGURE 12. Influence of PEO on the reciprocal relaxation time ($1/\tau_2$) at different concentrations of SDS. The continuous lines refer to corresponding relaxation signals in arbitrary units. (Reproduced from Tondre, C., *J. Phys. Chem.*, 89, 5101, 1985. With permission.)

aggregation/deaggregation process is considerably reduced in the presence of the complexing polymer.

L. NUCLEAR MAGNETIC RESONANCE

The early work of Muller and Johnson[70] on F_3SDS revealed a characteristic NMR shift (δ) of the F atom which accompanied the micellization of this surfactant, as the terminal $-CF_3$ group experienced a change from an aqueous environment to that of the micelles. This work was extended to include mixed PEO/F_3SDS systems.[49] The results obtained are in good agreement with the interaction patterns revealed by other methods, e.g., surface tension, conductivity, and dialysis. Thus, at constant PEO level and low surfactant concentration, the chemical shift was the same as the value characteristic of submicellar F_3SDS solutions. At a certain concentration ("T_1") lower than the CMC of F_3SDS, a slope change in the plot of δ against the reciprocal of the surfactant concentration was observed. At a second concentration (''T_2'') above the CMC the slope again changed, becoming the same as or close to that of polymer-free F_3SDS micellar solutions. T_1 was, furthermore, found to be independent of polymer concentration while T_2 increased with it. This allowed calculation of the binding ratio which was ~0.25 mol F_3SDS per base mole PEO, in good agreement with the values these workers obtained by direct binding studies. Results were essentially the same for PEO 7000 and PEO 20,000 but interaction was feeble with PEO 1500. The similar values of δ for the polymer-bound surfactant and micellized surfactant confirmed again the concept of the surfactant molecules existing as aggregates in the former state. An important additional point is that each solution had a single fluorine NMR signal, indicating a rapid exchange of surfactant ions between all possible sites.

Cabane[48] carried out an extensive NMR study of the PEO 20,000/SDS system. He investigated the chemical shifts in the ^{13}C-NMR spectra of the various carbons of the dodecyl chain accompanying the addition of increasing amounts of PEO to micellar (2%) solutions of SDS. With the carbon closest to the sulfate headgroup of SDS termed ''C_1'' and the terminal carbon ''C_{12}'', virtually no shifts were observed for carbons C_4 through C_{12}. However, substantial shifts occurred for carbons C_1 through C_3, see Figure 13. Furthermore, the shift was pretty well linear up to a concentration of added PEO of about 1%. This represents the T_2 condition, i.e., where the polymer is just saturated with surfactant. Further addition of polymer, generating conditions where the polymer is not saturated, and presumably where available SDS is shared among the polymer molecules, resulted in little or no further ^{13}C line shifts. Nuclear relaxation rates for the surfactant counterions, viz., ^{23}Na, which are sensitive to their degree of binding to the sulfate groups, similarly underwent a pronounced and linear change up to the T_2 condition; thereafter, the change with PEO addition was much weaker. When the experiment was done in reverse, i.e., SDS was added in increasing amounts to a fixed-concentration (1%) solution of PEO, a linear shift in the ^{13}C line of PEO occurred up to a concentration close to the T_2 value for this system. Thereafter, little further shift was observed.

Cabane makes the following deductions from his results: first, the internal environment experienced by carbons C_4 through C_{12} of the hydrocarbon chains of the surfactant in the polymer/surfactant aggregates is indistinguishable from that in regular micelles, suggesting that the aggregates are themselves micelles. On the other hand, the shifts experienced by the C_1, C_2, and C_3 carbons in the polymer/aggregate state indicate that these carbon atoms encounter a different environment and Cabane suggests that EO groups replace water in the outer region of the ''micelles,'' i.e., the zone comprising the headgroup and first few carbons. If such a configuration is favorable, one has an explanation of why aggregation of the surfactant is more favorable in the presence of the polymer, i.e., why T_1 is less than the CMC. Cabane's picture of the ''complex'' or aggregate is that of a micellar system with the polymer

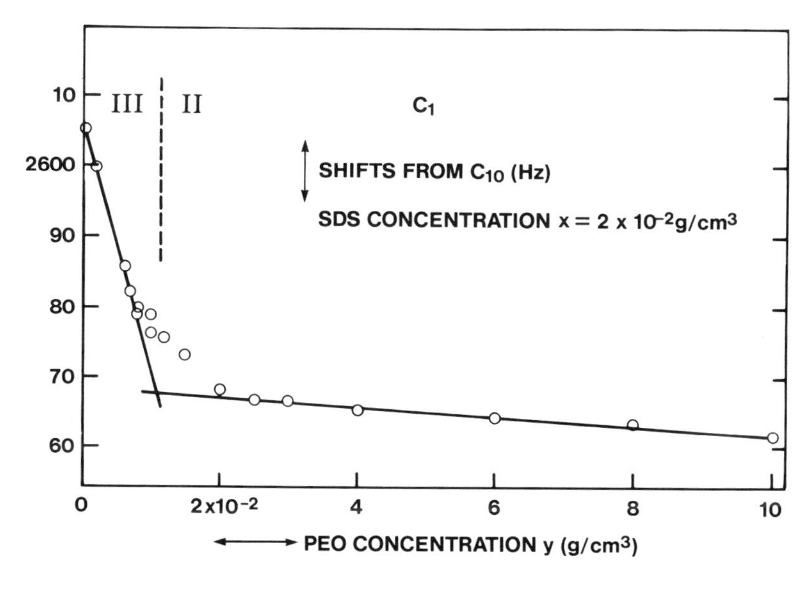

a

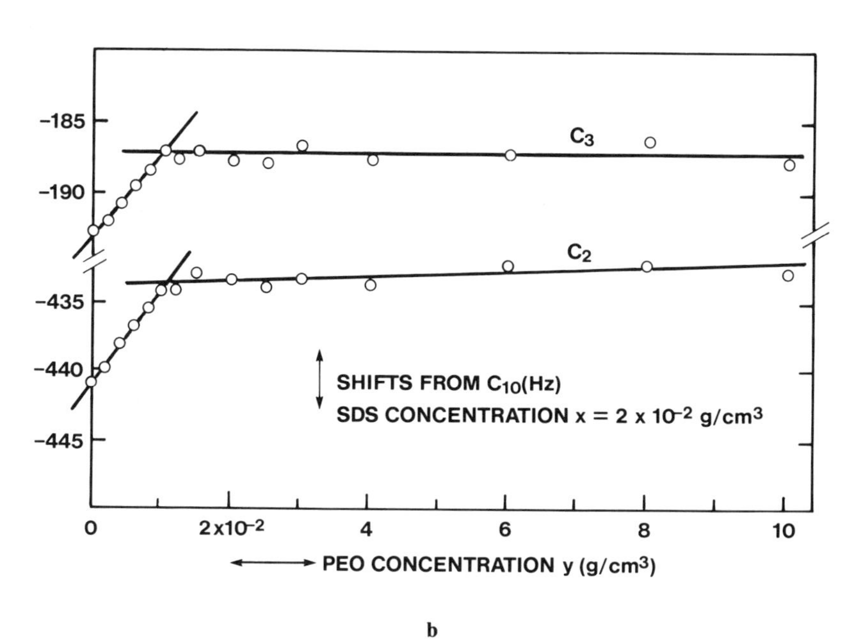

b

FIGURE 13. (a) Effect of PEO on the difference in chemical shift between carbons C_1 and C_{10} of the SDS chain. (b) Difference of chemical shifts between carbons C_2 and C_{10} of the SDS chain, and between C_3 and C_{10}. (Reproduced from Cabane, B., *J. Phys. Chem.*, 81, 1639, 1977. With permission.)

chains wrapped around the micelles, but with only a fraction of the polymer monomers actually in contact with the micellized surfactant and the rest in the form of loops.

M. SMALL-ANGLE NEUTRON SCATTERING (SANS)

The power of SANS in its own right to provide information and to augment that obtained from other scattering techniques is becoming increasingly recognized. In particular, it offers the feature, through contrast matching, of yielding information separately about two components, such as two solutes, dissolved in a liquid phase, like water. In such a case, one solute is obtained in hydrogenous form and the other in deuterated form, and the liquid is a mixture of H_2O and D_2O in such a ratio as to "blank out" one or other of the solutes. This approach makes use of the large difference in scattering length density of H and D atoms to neutrons; clearly, it seems well suited to the subject of interest, viz., polymer/surfactant mixed solutions. Cabane and Duplessix[71] have reported a detailed study of the system PEO/SDS in aqueous NaBr solutions of various ionic strengths employing a wide range of scattering angles and of solution composition and concentration. The molecular weight of the PEO was also varied. The PEO/SDS compositions were all chosen to be stoichiometric, representing polymer just saturated with SDS, i.e., along the T_2 composition line. Scattering measurements for such compositions were carried out in three scattering vector (Q) ranges. The details of the measurements, theory, and derivations used to interpret the results are outside the scope of this article and only the main conclusions are given here:

1. The Guinier range ($3 \times 10^{-3} < Q < 5 \times 10^{-2}$ Å^{-1}). Zimm plots were constructed from the data from which the dimensions, molecular weight, and composition of the "complexes" were deduced. Each macroaggregate, comprising a single polymer molecule and surfactant "micelles," was shown to have a radius of gyration, R_g, comparable to that of a coil of the free PEO parent. While at first sight this result is surprising, the data do refer to high ionic strength conditions where "polyelectrolyte" expansion effects due to surfactant binding will have been diminished.
2. The asymptotic range ($0.1 < Q < 1$ Å^{-1}). The SDS molecules of the macroaggregate are clustered in micelle-type units of 20 Å radius. These resemble regular micelles of surfactant in solutions of low ionic strength. Unlike the latter, these surfactant clusters do not grow as the ionic strength is increased.
3. The intermediate range ($9 \times 10^{-3} < Q < 0.1$ Å^{-1}). The SDS clusters adsorbed on the PEO strands resemble a string of beads: the distance between the beads changed from 90 to 60 Å as the salt concentration was increased from 0 to 0.8 M, but was independent of polymer molecular weight. So far as the PEO is concerned, the decay of the intensity $I(Q)$ followed a $Q^{-\alpha}$ law. The values of the exponent α show that EO

monomers of pure PEO repel each other (i.e., water is a good solvent) more so than when they are associated with SDS.

In another paper Cabane and Duplessix[72] considered the scattering behavior of a series of PEO/SDS solutions, not along the stoichiometric T_2 line, but along a line of constant content of PEO (mol wt 135,000) and varying SDS concentration. In these experiments the solvent, viz., 82% H_2O and 18% D_2O, was chosen to match the scattering length density of the macromolecule, so that the scattering was produced exclusively by the surfactant. In this case the intensity of scattering at low wave vector (Q) values is given by:

$$I_{Q\to 0} = (\rho_m - \rho_s)^2 \cdot \mu_m^2 \cdot n \cdot M^2$$

where n refers to the number of scattering particles of mass M, specific volume μ_m and average density of scattering length ρ_m; ρ_s refers to the latter parameter for the solvent. The experiment was designed to establish whether, as the concentration of SDS was increased from T_1, one type of polymer/surfactant macroaggregate formed, i.e., the SDS was shared among the macromolecules, or a series of saturated macroaggregates was progressively formed, leaving unassociated polymer molecules in diminishing number until final attainment of the T_2 condition when all the polymer molecules will have become saturated. In the former case, from the above equation, the mass M of surfactant per macroaggregate should be proportional to the surfactant concentration and hence the scattering should be proportional to the square of the surfactant concentration. In the latter case, an increase in SDS concentration should increase the number of saturated PEO/SDS macroaggregates at the expense of the free PEO molecules, while M would remain constant and hence the scattering intensity should be proportional to the surfactant concentration. The data showed conclusively that the former case is valid, i.e., surfactant is shared among all macromolecules in the T_1–T_2 range. From the calculated value of M, and assigning the clusters of adsorbed SDS a micellar molecular weight of 26,900, Cabane and Duplessix[72] concluded that a saturated macroaggregate formed from PEO 135,000 contains about 27 micelles, i.e., there is 1 micelle for every 114 monomer units of EO, constituting a PEO subsegment weight of ~5000. This is consistent with Schwuger's[7] finding that the aggregation characteristics of SDS with PEO become independent of molecular weight starting with PEO 4000 and the results indicate a molar ratio about 0.8 of SDS to EO in the complexes formed in the strong salt solutions employed.

N. CALORIMETRY, ENTHALPY TITRATION, ENERGETICS

Shirahama and Ide[47] estimated the enthalpy of transfer of SDS from the micellar form to PEO/SDS aggregates by direct calorimetry. The magnitude of ΔH was small, although these authors admit the value they obtained, viz., 0.6 kcal mol^{-1} SDS at 30°C, is subject to considerable error. Kresheck and

Hargraves[73] carried out enthalpy titrations of micellar SDS into solutions of PVP. They deduced that the binding of SDS clusters to the polymer is essentially athermal, but adduced evidence that the initial binding of the surfactant to the polymer, i.e., at low equilibrium concentrations, involved SDS monomers. Small endothermic effects were found for decyl and octyl sulfates: for these lower homologs the concentrations for binding were obviously much higher.

It is evident from frequent references in this chapter that several authors consider, with validity, the interaction between ionic surfactant and nonionic polymer to involve a surfactant aggregation process akin to micellization. It is well known that the heat of micellization of SDS at room temperature is small:[74] low values of its heat of aggregation in the presence of polymer may therefore be expected. Indirect estimates of the heat of aggregation, ΔH, of SDS in the presence of PVP, i.e., the heat of interaction of SDS and PVP, were made by Murata and Arai[12] through temperature dependence determinations of T_1 and application of the equation

$$\Delta H = -RT^2 (\partial \ln T_1/\partial T)_p$$

where R is the gas constant, T the absolute temperature, and P the pressure.

Over the temperature range of 18 to 30°C, the values obtained approximate zero, as is the case for the heat of micellization of SDS. In the same way as the CMC of this surfactant increases gradually as the temperature is raised above 25°C, so does T_1; in other words both aggregation processes become exothermic, and the ΔH values for both processes are comparable and small. (Following the analogy of micelle formation, the free energy [ΔG] value for the binding of surfactant should approximate that for micellization but be slightly larger, reflecting a slightly more favorable process.[12])

O. MISCELLANEOUS TECHNIQUES

1. Micellar Catalysis/Inhibition

It is well known that the presence of micelles can either catalyze or inhibit reactions in solution, especially hydrolysis. Fadnavis and Engberts[18] examined the influence of added PVP on solutions of SDS (whose micelles are known to inhibit the neutral hydrolysis of 1-benzoyl-1,2,4 triazole [B1,2,4T]) containing B1,2,4T. Addition of the polymer, which is itself without effect on the hydrolysis reaction, decreased the level of micellar inhibition. The effect is ascribed to the higher polarity of the binding sites for B1,2,4T on the PVP/SDS clusters than on regular SDS micelles. More work in the area of catalysis can be expected.

2. Electron Spin Resonance (ESR)

Shirahama et al.[75] have examined the ESR spectra of the hydrophobic spin probe 2,2,6,6-tetramethyl-4-tetradecanoyl-piperidine-*N*-oxide in simple micellar solutions of SDS and in solutions where SDS is present in complexes with PEO or PVP. The line width of a peak at high field was found to be considerably less broad in the presence of the polymers. Varying the molecular weight of PEO from 20,000 to 300,000 had no effect on the results. Lessening of the peak broadening, which may be identified with a greater level of separation of the spin probe moles, is taken by these authors as an indicator that the SDS aggregates in which the probes are solubilized are smaller in the presence of the polymers.

Recently, Witte and co-workers[76] have examined the properties of the "spin" probe di-terbutylnitroxide (DTBN) solubilized in regular SDS micelles and in the aggregates of SDS in complexes with PEO and PVP. The spectral characteristics suggest a more hydrophilic environment in the SDS aggregates than in the regular micelles. Detailed analysis of the characteristic three-line ESR spectrum allowed calculation of correlation times and these were plotted as a function of SDS concentration. The concentration profile is rather complex. Changes were detected at the lowest concentrations of SDS chosen ($\sim 2.5 \times 10^{-3}$ M, near the T_1 value). Between the CMC value of SDS and T_2, the inverse correlation times of the complex rose rapidly, leading to a sharp maximum at an SDS concentration $c \approx T_2$ for the PVP system, and a broader maximum at $c \approx 0.5T_2$ for the PEO system. Thereafter, i.e., at higher SDS concentrations, the inverse correlation times dropped fairly gradually, crossing and eventually falling below those of simple SDS micellar solutions. The faster tumbling rates in the region of the maxima are interpreted by the authors as indicating less compact, smaller aggregates of SDS in the complex as compared to regular SDS micelles.

3. X-Ray Diffraction

Shaginyan and co-workers[77] have obtained X-ray diffraction patterns of liquid crystals composed of sodium pentadecylsulfonate and PVP or PVOH. Their data indicate a mosaic structure containing aggregates of the surfactant. The domains were enlarged in the presence of PVOH, suggesting its localization in the lamellae; contractions found with PVP are interpreted as indicating its presence in the hydrophobic regions of the aggregates.

III. FACTORS INFLUENCING THE ASSOCIATION REACTION

At several points in this chapter, reference has been made to various factors which have a direct influence on polymer/surfactant association. In most cases it is clear that several factors influence the association in the same way as they do micellization of the surfactant, confirming the similarity of the two processes.

A. TEMPERATURE

This has been alluded to above: increases above room temperature generally make association less favorable, as evidenced by an increase in T_1. It is well known that the CMC of ionic surfactants changes in the same way.[74]

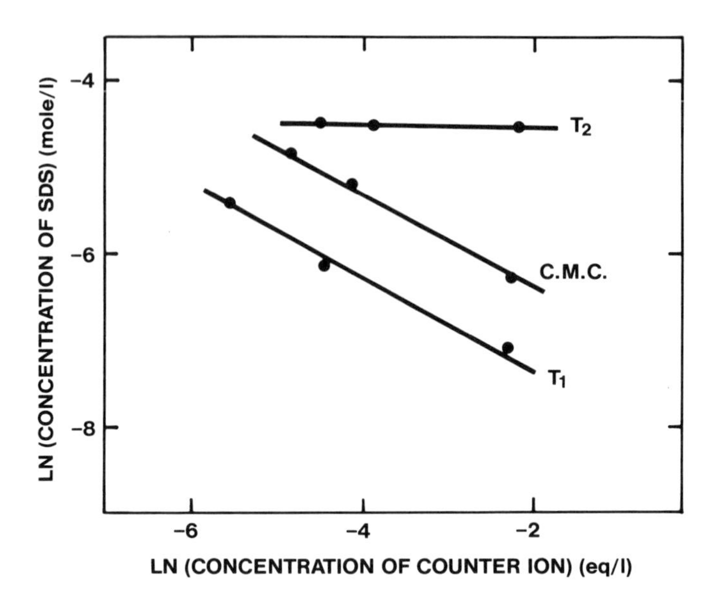

FIGURE 14. Effect of salt on the transition concentrations of 0.1% PVP/SDS systems; corresponding salt effect on CMC of SDS included. (Reproduced from Murata, M. and Arai, H., *J. Colloid Interface Sci.*, 44, 475, 1973. With permission.)

B. SALT

The addition of inorganic electrolyte generally depresses the T_1 values, i.e., promotes the formation of complexes. For example, Murata and Arai[12] found a linear relationship in a log–log plot of T_1 against sodium ion concentration for the PVP/SDS association, see Figure 14. Interestingly, the slope of this plot was exactly the same as that of the corresponding CMC/sodium ion concentration plot for this surfactant. Addition of salt also increases the binding ratio of surfactant to polymer, i.e., extends the T_1 to T_2 range. For example, in the case of PVP/SDS, addition of 0.1 *M* NaCl increased the ratio to 0.9 mol SDS per base mole PVP from the 0.3/1 ratio observed in water. A similar effect occurs with PEO/SDS, and this has been systematically studied by Cabane and Duplessix,[71] who report an increase in binding ratio of SDS to EO of 0.25 at zero salt to a limit of 0.85 at 0.4 ionic strength (0.4 *M* NaBr).

Singh et al.[78] studied the interaction of a series of alkaline earth (Mg, Ca, Sr) valerates with PVOH by the electrical conductance method. Breakpoints corresponding to T_1 and T_2 were identified. Although corresponding data for the alkali metal valerates were not obtained, it is safe to assume that, in view of the much lower surface activity of these surfactants, complex formation with PVOH would have been weak or absent. Saito and Kitamura[79] have, however, drawn attention to another property influencing the interaction, viz., the ability of the added counterion to affect the structure of water. Saito and co-workers have concluded that structure-breaking cations tend to increase the tendency of a surfactant anion to associate with a polymer, and structure-making cations do the reverse. Since some of the counterions studied in this work were themselves large organic ions, the possibility of their bonding directly to the polymer cannot be excluded.[3] The reader is referred to Saito's papers for detailed information.[34,80] His work on counterion effects with cationic surfactants is referred to below.

C. SURFACTANT CHAIN LENGTH

In a homologous series, the initial binding concentration, T_1, decreases with increasing chain length of the alkyl sulfate present with the binding polymer.[9,11,47] A linear relation between log T_1 and n, the number of alkyl chain carbons, exists, as has been found between log CMC and n. For results obtained with PVP/SDS mixtures in 0.1 M NaCl solution, Arai et al.,[11] using the relation $\ln T_1 = nw/kT + \text{constant}$, found a value for w of $-1.1\ kT$. This corresponds to the free energy change per CH_2 group on transferring the surfactant from the unassociated state in water to the complex, and is comparable to the value for the analogous transfer of the surfactant to a micelle.

D. SURFACTANT STRUCTURE

The surfactant used in most polymer/surfactant studies has been SDS. Outside of comparisons of its performance with members of its own homologous series, i.e., chain length variations, studies of variations in surfactant structure have been limited. However, the work of Tokiwa and Tsujii[16] clearly shows that sodium alkylbenzenesulfonates interact with PEO in much the same way as SDS, as does NaL with this polymer.[81] On the other hand, modification of the alkyl sulfate structure by insertion of ethylene oxide groups has been shown by Saito[82] to enfeeble the interaction with PVP considerably: no viscosity increase of PVP was noticed with $SD(EO)_4S$ and $SD(EO)_{10}S$, and only the former's solubilization capacity towards the dye Yellow OB was improved in the presence of PVP. It should be pointed out that the CMC of these dodecylethoxysulfates is considerably lower than that of the parent SDS. In other words, the free energy for conventional aggregate, i.e., micelle, formation is already more favorable in this case. On the necklace, or "polymer-wrapped-around-micelle" model, one could perhaps argue that the hydrophilic chain groups provided by a polymer like PEO are already "in place" in these surfactants, affecting both the degree of contact of the hydrocarbon micellar core with the external water phase and also the degree of repulsion of the sulfate headgroups. In the latter respect, Schwuger[83] has recently reported higher degrees of ionic dissociation of dodecylethoxysulfate micelles than of dodecyl sulfate micelles, in the same way as has been found for the latter micelles on addition of PEO by Zana et al.[21] (These points are elaborated upon in the conclusion of this review.)

E. CATIONIC VS. ANIONIC SURFACTANTS

Several authors allude to the fact that interaction between uncharged water-soluble polymers is much more facile with anionic surfactants than with cationics (Breuer and Robb,[3] Kresheck and Hargraves,[73] Gravsholt,[15] Schwuger,[7] Zana et al.,[21] Moroi et al.,[14] Saito and co-workers[57,84-86]). Recently, Nagarajan and Kalpakci[27] found only feeble indications of interaction of dodecylammonium chloride (DA^+Cl^-) and dodecylammonium bromide (DA^+Br^-) with PEO, as judged by viscosity criteria. Saito and Yukawa[85,86] have demonstrated that interaction between a polymer and a cationic surfactant can, however, be promoted when a strongly interacting counterion is present. Thus DA^+CNS^- interacts quite strongly with PVAc, as judged by reduced viscosity data, whereas the corresponding hydrochloride shows comparatively weak interaction. Parenthetically, the thiocyanate ion (CNS^-) also shows more marked interaction with cationic surfactant micelles. Saito and co-workers extended the studies to PVP, PEO, and PVOH,[79,80] and dye solubilization and cloud point data reinforced the conclusions reached from viscosity measurements. Their results also demonstrate that the "right" headgroup of the cationic surfactant is a requirement since interaction of dodecylpyridinium thiocyanate with PVOH was weak.[85] For a listing of the extensive work carried out by Saito and co-workers on polymer/cationic surfactant systems, the reader is referred to the article of Breuer and Robb.[3] Attention is also drawn to Tadros' work[13] involving CTAB and a relatively hydrophobic PVOH sample (88% hydrolyzed PVAc). He obtained clear evidence of interaction by surface tension, viscosity, conductance, and cloud point techniques. In particular, the marked increase in reduced viscosity of this polymer in the presence of CTAB at low polymer concentration represents an excellent demonstration of the "polyelectrolyte effect" as polymer binds surfactant.

F. NONIONIC SURFACTANTS

There is little indication of reactivity of polyoxyethylated nonionic surfactants with uncharged water-soluble polymers. One exception is the strong interaction between these surfactants and polyacrylic acid.[2,87,88] However, such interaction is well known between nonalkylated polyoxyethylenes and polyacrylic acid, i.e., surfactant properties are not a requirement of the former reactant for such activity. Recently reported data of Szmerekova et al.[37] on interaction activity between a nonylphenol polyethylene glycol and PEO may well be due to an affinity of the aromatic phenol moiety and PEG, as suggested previously by Pletnev and Trapeznikov.[8] We include in this category similar activity between this type of surfactant and hydroxyethylcellulose seen from the light scattering results of Boscher et al.[89] Nagarajan and Kalpakci,[27] however, found no evidence of any octylphenol ethoxylate and PEO interaction by viscometric methods.

G. MOLECULAR WEIGHT OF POLYMER

A minimum molecular weight of polymer is apparently required to ensure "complete" interaction with the surfactant. Several authors have placed this molecular weight for PEO and PVP to be ~4000. Below values of ~1500 the interaction tendencies with these polymers are restricted.

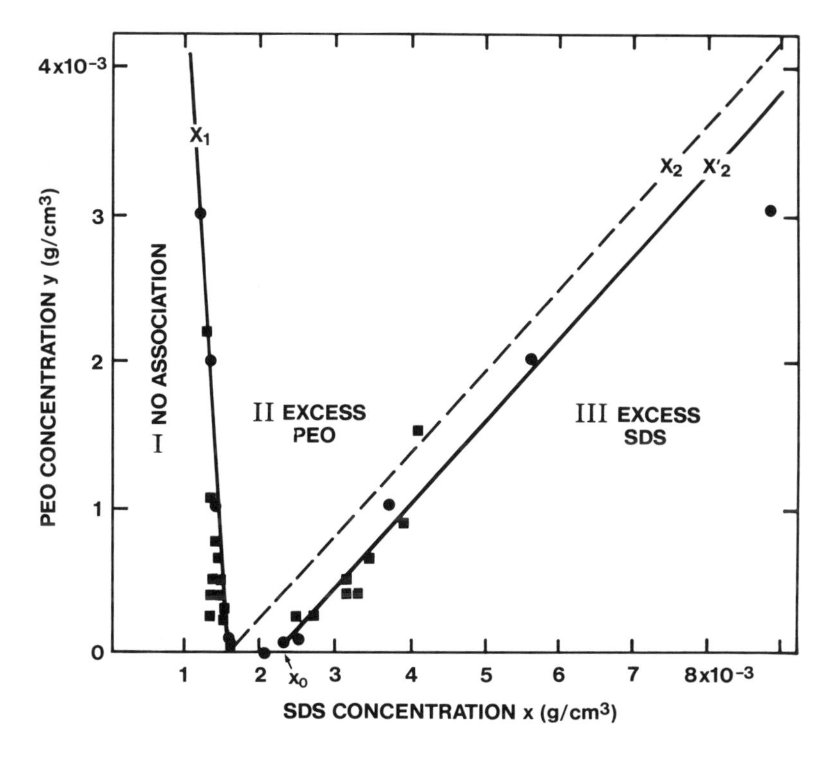

FIGURE 15. "Phase diagram" of PEO/SDS, according to Cabane and Duplessix.[71] The term x_0 refers to the CMC of SDS; see text for explanation of other terms. (Reproduced from Cabane, B. and Duplessix, R., *J. Phys. (Paris),* 43, 1529, 1982. With permission.)

H. AMOUNT OF POLYMER

Although there are some discrepancies in available data, the value of T_1 seems to be insensitive to, or to decrease slightly with, polymer concentration; T_2, on the other hand, increases linearly with it. These effects are best visualized in phase diagrams of the type developed by Jones,[5,58] refined by Cabane and Duplessix,[48,71,72] and lately used by Francois et al.[17] In the most recent diagram of Cabane and Duplessix (Figure 15) allowance is made for the area between saturation of the polymer (line x_2) and subsequent increase of surfactant monomer concentration to the value required to form micelles (line x_2'). In the usage adopted in this review, x_1 would correspond to T_1, x_2 to I_2', x_2' to T_2, and x_0 to the CMC. This diagram illustrates the equivalence of the two ways of estimating the amount of bound surfactant at a given concentration of polymer, namely $x_2' - x_0$ and $x_2 - x_1$. Discrepancies would occur at higher polymer levels because of the slight reduction of x_1 with polymer concentration, i.e., negative tilt of the x_1 line. Points of refinement to the phase diagram which can be expected in the future are inclusion of the T_1' intermediate concentration (seen in electrical conductivity), the "sub-T_1" concentration (see in fluorescence experiments), i.e., unless these turn out to be specific to these experimental methods, and the substantial variation in T_1 which occurs at very low polymer concentration.[10]

I. POLYMER STRUCTURE AND HYDROPHOBICITY

Several uncharged polymers now identified with the ability to form complexes with charged surfactants are PEO, PVP, and PVOH. The lack of activity of other polymers in this respect, such as hydroxyethylcellulose (HEC) may well be due to a lower level of macromolecular flexibility,[27] but even the more flexible polysaccharide dextran shows little tendency to interact with SDS[50] or DDBS.[27] The relative inertness of PAAm has not been satisfactorily explained.[43,90] "Reactivity" can sometimes be conferred upon a polymer structure by introducing hydrophobic sites in the macromolecule, as in MeC vs. HEC, in low-molecular-weight polyalkyleneoxides in which propylene oxide replaces ethyleneoxide, and in PVOH specimens prepared from, but still containing residual amounts of, PVAc.[91] In fact, the "reactivity" of polymers does seem to be correlatable with a kind of "HLB index." For anionic surfactants, Breuer and Robb[3] list polymers in the following order of increasing reactivity: PVOH $<$ PEO $<$ MeC $<$ PVAc $\leq$ PPO $\sim$ PVP; and for cationic surfactants: PVP $<$ PEO $<$ PVOH $<$ MeC $<$ PVAc $<$ PPO.

The position of PVP in the two series is noteworthy. This polymer is known to be weakly cationic; its slight residual positive charge promotes interaction with anionic surfactants and does the reverse with cationic surfactants. This explanation is not inconsistent with the data of Fadnavis and Engberts[18] concerning effects on the CMC of SDS of the added model "monomer" for PVP, viz., *N*-isopropylpyrrolidone. Slight charge effects[7] may also be operative with PEO in view of the reported cationic nature of its ether groups due to oxonium ion formation.[92]

Support for the postulated role of polymer hydrophobicity comes from a study[93] of the interaction of a series of nonionizable polypeptides (poly-DL-alanine and three derivatives of poly-L-glutamine) with SDS. Strongest interaction as judged, for example, by the lowest T_1 value (from dye solubilization and electrical conductivity measurements) and highest ionic dissociation value (α, from sodium ion activity measurements), were obtained with the most hydrophobic member, viz., poly-DL-alanine, and least interaction with the most hydrophilic member, viz., poly-*N*-(2 hydroxyethyl)-L-glutamine. Lastly, we refer again to the ability of surfactants such as SDS and NaL to solubilize certain polymers such as PVAc which are sufficiently hydrophobic as to be normally water insoluble.[23]

The influence of macromolecular flexibility previously referred to is illustrated by the relatively strong interaction of SDS with amylose (which can undergo a helix-coil transition) and the relatively weak interaction of SDS with amylopectin (which cannot).[3,94] An interesting feature of these polymers (both derived from starch) was reported by Fishman and Miller,[95] who found that reactivity towards a cationic surfactant could be induced in highly alkaline solution (pH $\geq$ 12). Development of precipitates in certain concentration ranges of the mixtures suggested the formation of polyelectrolyte/surfactant complexes (see Part II),[108] the polymer becoming "ionized" by "adsorption of hydroxyl ions". Previous workers,[96] who had observed the same phenomenon with insulin, dextran, and starch, ascribed the negative charge of the polymer at high pH to ionization of its hydroxyl groups (see also Reference 97).

IV. INTERACTION MODELS

Smith and Muller[49] developed a simple model to explain (1) their binding data for the F_3SDS/PEO system, which were not consistent with a Langmuirian process, and (2) their NMR data, which indicated that every "bound" surfactant molecule experienced the same environment. The latter finding suggested that surfactant molecules might be bound cooperatively in micelle-like clusters, but smaller in size than regular micelles since the observed binding isotherm was insufficiently steep to be consistent with a ~50-molecule "micelle". These authors made the assumption that each polymer macromolecule consists of a number of "effective segments" of mass M_s, and total concentration $[P]$, that act independently, an implication being that M_s represents a minimum molecular weight of the polymer for interaction. Each segment was seen as being able to bind a cluster of n surfactant anions, D^-, in a single step, the binding equilibrium being represented as

$$P + nD^- \rightleftarrows PD_n^{n-}$$

and the equilibrium constant being given by

$$K = [PD_n^{n-}]/[P][D^-]^n$$

K is obtained from the half saturation condition, viz.,

$$K = [D^-]_{1/2}^{-n}$$

Assuming trial values of n and using experimental values of other parameters, one has all the information to calculate a series of experimental isotherms. The best fit yields the parameters M_s, n, and K. The data indicated the cluster size, n, to be about 15, and M_s to be 1830, which explains the experimental finding of these authors that PEO 1500 is relatively ineffective for surfactant binding whereas higher-molecular-weight PEO members are effective. The free energy of binding is obtained from the expression

$$\Delta G^0 = -RT \ln K^{1/n}$$

and the value obtained, viz., -5.07 kcal mol^{-1}, is close to that of micelle formation for the surfactant employed (F_3SDS), again suggesting that binding and micellization are related processes.

This model has been further developed by Gilanyi and Wolfram[54] and by Nagarajan,[98] who include micelle formation as well as complex formation in their basic mass balance equation. The former authors treated the same case of an anionic surfactant binding to a neutral polymer but allowed, in addition, for (a different degree of) binding of the surfactant counterion, in an analogous way to treatments of micellization of such surfactants. This was done by considering the degree (α) of ionic dissociation of the bound-surfactant aggregate. Their expression for the free energy of transfer of surfactant (K^+D^-) from solution to the bound state, as complex, is

$$\Delta G^0 = RT \ln [D^-] + RT (1 - \alpha)\ln[K^+] - \frac{RT \text{ [complex]}}{[\text{P}_0] - [\text{complex}]}$$

where $[P_0]$ is the initial total concentration of "active" polymer sites. Gilanyi and Wolfram also presented mass balance and free energy equations for a treatment of the aggregation process similar to that used in the phase separation treatment of micellization. In addition, they discuss the applicability of a Langmuir adsorption of surfactant clusters on the "adsorption sites" of the polymer. The analogy of the process of transfer of surfactant molecules to polymer-bound clusters with that to surfactant micelles is stressed by these authors, who state that this process provides the essential driving force for formation of the complex.

Shirahama,[33] in his earlier paper, had pointed out that his data for the binding of SDS to PEO fitted a Langmuir adsorption equation if provision were made that the binding was accompanied by cluster formation of the surfactant molecules ($n \approx 20$ monomers). The equation then took the form

$$\Theta = KC^n/(1 + C^n)$$

where Θ is the degree of binding, n is an empirical exponent, C is the equzzilibrium concentration of bound species, and K is a constant, and had actually been proposed by Hill[99] in 1910. Shirahama was also able to account for the binding results using a statistical mechanical approach. Reference will be made again to this in Part II of this chapter.[108]

Nagarajan[98] has recently presented a theoretical description of the aggregation of surfactant in the presence of water-soluble polymer in a treatment which is essentially similar to that of Gilanyi and Wolfram. We summarize below the main elements of Nagarajan's model, its predictions, and a comparison with experimental data of these authors.

The aqueous solution of surfactant and polymer is assumed to contain both free micelles and "micelles" bound to the polymer molecule. The total surfactant concentration X_t is thus partitioned into singly dispersed surfactant,

X_1, surfactant in free micelles, X_f, and surfactant bound as aggregates, X_b, in the mass balance equation

$$X_t = X_1 + g_f(K_f X_1)^{g_f} + g_b\, n\, X_p \left[\frac{(K_b X)_1{}^{g_b}}{1 + (K_b X_1)^{g_b}}\right] \quad (1)$$

In Equation 1, the second and third terms represent X_f and X_b, respectively; g_f is the average aggregation number of the free micelles; and K_f is the intrinsic equilibrium constant for their formation. Furthermore, each polymer molecule is assumed to have n binding sites for surfactant aggregates of average size g_b. K_b is the intrinsic equilibrium constant for the binding of the surfactant on the polymer. It can also be visualized as the intrinsic equilibrium constant for the formation of polymer-bound micelles. X_p is the total concentration of polymer molecules in solution. The polymer influences Equation 1 through the term nX_p, its effective mass concentration, which is independent of polymer molecular weight.

Conformational changes attending polymer-micelle complexation are assumed not to affect K_b and g_b, nor n. The relative magnitudes of K_b, K_f, g_b, and g_f determine whether or not complexation with the polymer occurs. If $K_f > K_b$, and $g_b \approx g_f$, the formation of free micelles occurs in preference to complexation. If $K_f < K_b$, and $g_b \approx g_f$, complexation/aggregation on the polymer takes place first and upon saturation of the polymer, free micelles form. If $K_f < K_b$, but g_b is much smaller than g_f, then formation of free micelles can occur even prior to saturation of the polymer. A first critical surfactant concentration will be observed close to $X_1 = K_b^{-1}$ and a second critical concentration will occur near $X_1 = K_f^{-1}$.

The calculated relations[98] between X_1 and X_t for different polymer concentrations are shown in Figure 16. The values chosen for the parameters K_b, K_f, g_b, and g_f are those for the SDS/PEO system. Note that in this system, $K_b > K_f$ and g_b is slightly smaller than g_f.

In the region from O to A, the surfactant molecules remain singly dispersed. In the region from A to B, the formation of polymer-bound micelles occurs. Note the monomeric surfactant concentration X_1 increases very little in this region. This is a consequence of the relatively large size, g_b, of the polymer-bound micelle and reflects the cooperative nature of formation of bound micelles. In contrast, if g_b is small (say 10), then X_1 would increase more significantly in the region AB. If the mass concentration of the polymer is very small (nX_p is small), then the region AB may escape detection: if nX_p is very large, practically speaking the saturation point B may not be reached. Expressing this result differently, the total surfactant concentration X_t at C (i.e., T_2) depends directly upon the mass concentration (nX_p) of the polymer. In contrast, the first critical concentration A (i.e., T_1) is independent of the concentration of polymer in solution.

The features calculated above for the PEO/SDS system were verified qualitatively by the specific ion activity results obtained by Gilanyi and

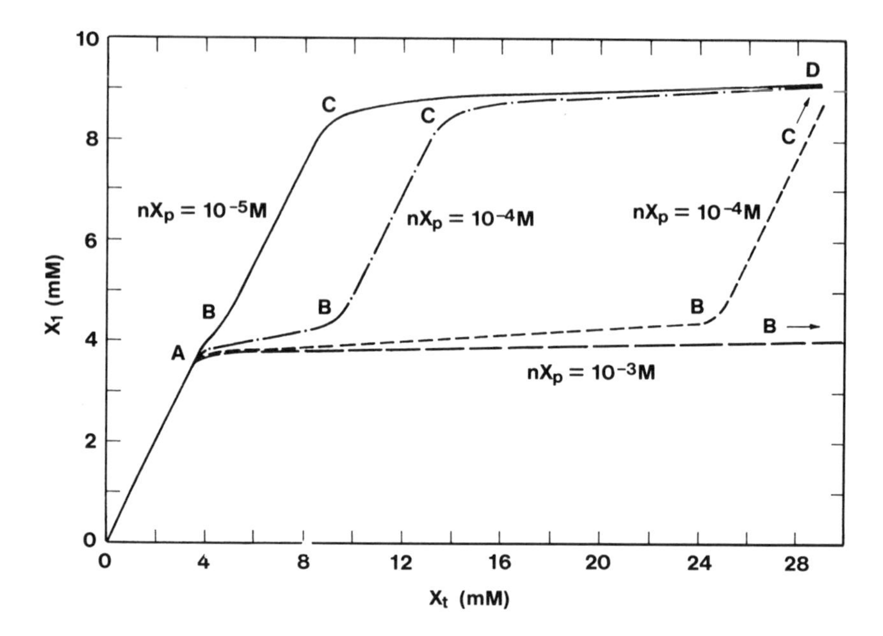

FIGURE 16. Calculated equilibrium concentrations of singly dispersed surfactant $(X)_1$ as a function of total SDS concentrations (X_t) for different mass fractions (nX_p) of PEO. See text for further explanation. (Reproduced from Nagarajan, R., *Colloids Surfaces,* 13, 1, 1985. With permission.)

Wolfram[54] on this system in the presence of 0.1 *M* $NaNO_3$ (Figure 17). The regions OA, AB, BC, and CD are identified on the experimental curves obtained for different concentrations of polymer in solution. One can see that at a low polymer concentration (0.25 g l^{-1}), all the regions are present, whereas at high polymer concentration (4 g l^{-1}), the surfactant is inadequate for saturating the polymer molecule. The constancy of X_1 in the regions AB and CD suggests that both the polymer-bound micelle and the free micelle are of substantial size. Agreement between predictions of the model and the experimental data is clearly very satisfactory.

Hall[100] has very recently presented a detailed thermodynamic treatment using a Donnan equilibrium approach, of the binding of ionic surfactants to polymers. In systems which can contain salt, monomeric surfactant, micellar surfactant, and adsorbed surfactant (as well as the original polymer), interesting possibilities are predicted for the conditions of increasing surfactant concentration, such as (1) the amount of the bound surfactant may decrease if micelles are present, and (2) the surfactant monomer concentration may exhibit two maxima. Hall avoids considering that the surfactant bound to the polymer exists as well-defined aggregated species (''micelles'') — a possibility that he considers unlikely (and difficult to treat theoretically). He introduces ''clustering'' as a situation where some polymer molecules have more bound surfactant than others, and then formally treats a situation (''all

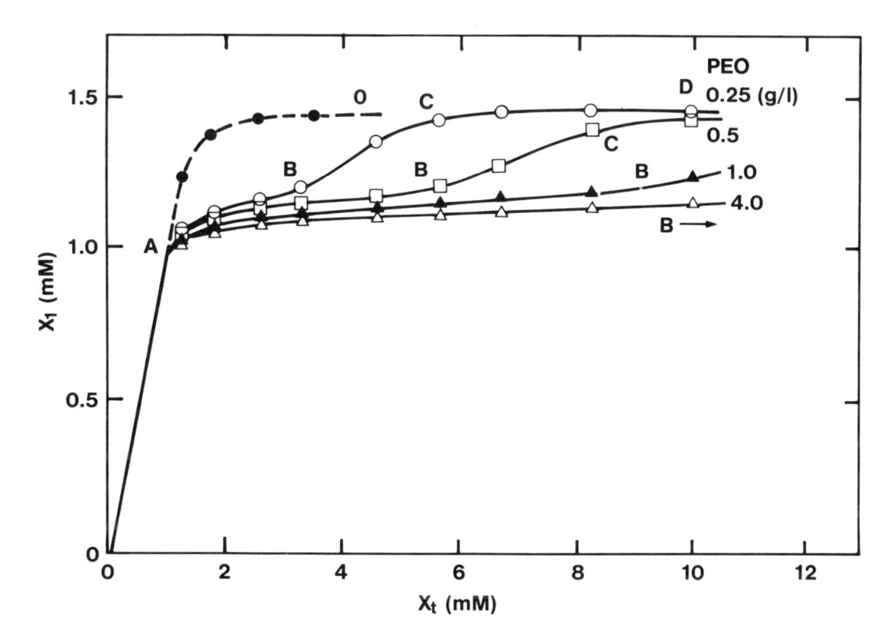

FIGURE 17. Experimental data of Gilanyi and Wolfram[54] for PEO/SDS solutions containing 0.1 *M* $NaNO_3$, for comparison with theoretical curves of Figure 16. (Reproduced from Nagarajan, R., *Colloids Surfaces,* 13, 1, 1985. With permission.)

or none'') in which polymer molecules saturated with clustered surfactant coexist with polymer containing nonclustered surfactant, and gives predictions as to the various conditions which can exist in the solution as values of the assigned equilibrium constants are varied. Although Hall states that there is strong evidence that the bound surfactant can exist in at least two forms, it must be stressed that there is also strong evidence, from SANS,[72] sedimentation,[17] and gel filtration[35] experiments, that the bound surfactant is divided equally (rather than ''all or none'') among the macromolecules. Hall concludes that, at present, published experimental data on binding, and especially ion activity, are insufficient or inadequate to allow systematic comparison of his theory with experiment and points out the areas where, and what, information is necessary. He notes, however, that the dialysis data of Jones et al.[101] for binding of SDS to lysozyme lend support to some of the predictions of his theory.

V. WHY DO POLYMER/SURFACTANT COMPLEXES FORM?

In viewing polymer/surfactant association, it is pertinent first to consider the self-association, or micellization, of a typical ionic surfactant, such as SDS, in water. The chief driving force for association is a reduction of the hydrocarbon/water contact area of the alkyl chains of the dissolved surfactant.

The process of micellization of the ionic surfactant represents a delicate balance between several forces favoring and resisting aggregation. One of the main forces resisting aggregation is the crowding together of the ionic headgroups at the periphery of the micelle. The consequently high charge density in this zone favors the tight binding of counterions which diminishes the electrical potential and headgroup repulsion. This is not a factor in the case of nonionic surfactants, and for these materials a much lower CMC results. One must bear in mind that in the well-accepted spherical, or Hartley, model of the micelle which describes the dilute solution properties of a typical anionic surfactant such as SDS, on a molecular scale there is a considerable distance between the headgroups at the periphery, if only for geometrical reasons of packing. Some of this space will accommodate counterions, but most will comprise areas of the hydrocarbon chain exposed to water — an obviously unfavorable situation. Early NMR data[102] indicated that the first three or four carbons (measured from the headgroup) of micellized molecules of SDS are in intimate contact with water. Accordingly, any change brought about to diminish either one of these unfavorable conditions will promote micellization. A well-known method is the addition of an intermediate chain length alcohol, geometrically compatible with the micellized surfactant assembly (i.e., still allowing a radial array of alkyl groups) and whose hydroxyl headgroups can act as an "interacting" insulator between the headgroups of the aggregated surfactant molecules. Addition of such an alcohol renders the surfactant aggregation process more favorable, as is manifested by a drop in the CMC, and is also accompanied by increased ionic dissociation of the micelle.[103] This insulating effect is, in fact, similar to that involved in mixed crystal formation in pairs of long-chain polar compounds and ionic surfactants reported previously by this author.[104,105] In this case, the alkyl groups of the two components align in the crystal and equal length alkyl groups facilitate the preservation of the characteristic bimolecular-layer structure which also exists in lamellar micelles and liquid crystals.

Let us now consider the effect of an additive with alternating hydrophilic and hydrophobic segments, in fact a typical synthetic water-soluble polymer. If the molecule has sufficient flexibility one can envisage a configuration allowing ion-dipole association between the dipole of the hydrophilic group, and the ionic headgroup of the surfactant, and contact between the hydrophobic segments of the polymer and the "exposed" hydrocarbon areas of the micelle — in effect resulting in screening of the electrical charges and diminution of the extent of these exposed areas. Remembering that for PEO/SDS and PVP/SDS saturated complexes there are approximately three monomer units of polymer per molecule of aggregated SDS, one can easily envisage a structure in which the aggregated SDS is surrounded by the macromolecule(s) in a "loopy" configuration, in a manner envisaged by Shirahama et al.,[32] Landoll,[106] and Cabane,[48] and depicted by Nagarajan in Figure 18. Consequences of the above would include several features already observed, such as (1) a more favorable free energy of association, as manifested in a lowered

FIGURE 18. Schematic diagram of polymer-surfactant complex. (Reproduced from Nagarajan, R. and Kalpakci, B., *Polym. Prepr. Am. Chem. Soc. Div. Polym. Chem.*, 23, 41, 1982. With permission.)

CMC (i.e., T_1 < CMC); (2) increased ionic dissociation of the aggregates; and (3) an altered environment of the CH_2 groups of the surfactant near the headgroup, as seen in the ^{13}C-NMR data of Cabane.[48]

Since the balance of forces leading to surfactant aggregation is changed in the presence of the polymer, it is not surprising that the aggregation number may change. It is also easy to accept the notion that certain polar groups in a water-soluble polymer are more favorable than others for interaction, that backbone flexibility is a favorable factor, that there is a minimum molecular weight required for reacting with the surfactant cluster, and that a high-molecular-weight polymer can loop around several surfactant clusters forming the necklace structure envisaged by Shirahama. The classical view that the complexes comprise ''site-bound'' aggregates of the surfactant along the polymer chain is certainly compatible with the above picture and really reflects a difference in interpretation regarding the initiation of the interaction rather than its final effect. Another factor expected to influence the interaction is the hydrophobicity of the polymer, as exemplified in the interaction of SDS with PPO of comparatively low molecular weight (<1200, which represents the limiting molecular weight for water solubility in this family of polymers). The higher hydrophobicity lowers the molecular weight limit required for interaction. In this case also, association has been found to increase the ionic dissociation of the SDS aggregates.[8]

There is another set of information regarding the influence on surfactant aggregation of adding molecules with alternating polar/nonpolar segments (as in oxyethylene chain units) to molecules such as SDS, but where the chains are short and attached chemically to the surfactant. We refer to the alkyl polyoxyethylene sulfate ($R(EO)_nSO_4^-Na^+$) family of surfactants.[80] The interpositioning of a hydrophilic segment in the surfactant molecule would be expected to increase the water solubility of the surfactant and this is indeed reflected in a lowering of the Krafft point of members of the polyoxyethylene-substituted dodecyl sulfate family. However, introduction of $(EO)_n$ groups (n = 1 to 4) results in a *progressive reduction* in the CMC (by a factor of about a half in the $C_{12}(EO)_nSO_x^-Na^+$ series with n = 0 to 4) and an increase in ionic dissociation of the micelle (by a factor of about two in the same series). In other words, micellization is favored as n increases, even though the molecules are intrinsically more hydrophilic than the parent SDS. Surfactant headgroup screening effects and reduction of micellar hydrocarbon/water exposed area would seem operative in this case, as postulated for the case where PEO is incorporated by simple addition. If these views are correct, much less interaction tendency would be found on the addition of PEO to solutions of $R(EO)_nSO_4^-Na^+$ surfactants, as has already been found[82] with PVP as the added polymer. Experiments are needed to confirm this. (We mention, in passing, that polymer-complexed ionic surfactants acquire some similarity to nonionic surfactants: this conclusion comes from the observation[10] that the chain-length dependence of the aggregation concentration $[T_1]$ of PVP-complexed alkyl sulfates lies in between that of the CMC of alkyl sulfate surfactants and alkyl polyglycol ether surfactants.)

In the discussion so far, emphasis has been placed on the role/fate of the surfactant during the association reaction. One must consider the role of the polymer also. An important clue comes from the polymer "reactivity" series of Breuer and Robb[3] where reactivity increases with polymer hydrophobicity. One infers that a driving force for association is a reduction of the polymer hydrophobic segment/water interfacial area by contacting of such areas with exposed hydrophobic areas of developing surfactant aggregates, as outlined above. In harmony with this picture, polymers with a very low level of hydrophobicity, e.g., dextran, HEC, and PAAm, would have very little interaction tendency with the surfactant, as has indeed been observed. (The subject of polymer hydrophobicity, as judged by ability to lower the surface tension of water and reactivity towards surfactants, will be the subject of a forthcoming paper from this laboratory.[107] A special case is one where the hydrophobic group interaction is reinforced by substantial hydrophilic group interaction. This occurs with PVP (and to some extent PEO) and anionic surfactants and can be considered, in turn, a special case of the much stronger interactions, involving oppositely charged ions, which are operative in the association systems treated in Part II.[108]

Finally, we address briefly the mechanistics of formation of the complexes. The question of how they form is certainly less easy to answer than why they form, but the fast kinetics measurements of Wyn-Jones make it clear that the timescale of the aggregation/deaggregation processes is of the same order as that involved in regular micelle formation. The facile nature of the process may seem surprising; however, co-occupancy of the total solution space by the polymer and the surfactant molecules (i.e., being "in place") facilitates fast response to, and participation in, any change in local conditions such as a local increase in surfactant concentration leading to aggregation. Indeed, the presence of the polymer may help to nucleate the aggregation of the surfactant. Whether this occurs by initial site binding of the surfactant on the polymer or by the polymer's responding to local aggregation of the surfactant, remains to be determined.

VI. ADDITIONAL READING

Just published is a book edited by Dubin[109] which contains some half-dozen papers on polymer surfactant systems, including full versions of References 21 and 27. Attention is also drawn to the article, in this compilation, by Gilanyi and Wolfram with results on the molecular weight of PVP/SDS complexes which show good agreement by the two techniques employed, viz., light scattering and dialysis.

REFERENCES

1. **Dervichian, D. G.,** in Lipo-proteins, *Discuss. Faraday Soc.,* 6, 1949.
2. **Saito, S.,** *Colloid Polym. Sci.,* 257, 266, 1979.
3. **Breuer, M. M. and Robb, I. D.,** *Chem. Ind.,* 530, 1972.
4. **Robb, I. D.,** in *Anionic Surfactants in Physical Chemistry of Surfactant Action,* Lucassen-Reynders, E. H., Ed., Marcel Dekker, New York, 1981, 109.
5. **Jones, M. N.,** *J. Colloid Interface Sci.,* 23, 36, 1967.
6. **Cockbain, E. G.,** *Trans. Faraday Soc.,* 49, 104, 1953.
7. **Schwuger, M. J.,** *J. Colloid Interface Sci.,* 43, 491, 1973.
8. **Pletnev, M. Yu. and Trapeznikov, A. A.,** *Kolloidn. Zh.,* 40, 948, 1978.
9. **Lange, H.,** *Kolloid Z. Z. Polym.,* 243, 101, 1971.
10. **Schwuger, M. J. and Lange, H.,** *Proc. 5th Int. Congr. of Surface Act. Agents,* Vol. 2, Barcelona, 1968, Ediociones Unidas SA, Barcelona, 1969, 955.
11. **Arai, H., Murata, M., and Shinoda, K.,** *J. Colloid Interface Sci.,* 37, 223, 1971.
12. **Murata, M. and Arai, H.,** *J. Colloid Interface Sci.,* 44, 475, 1973.
13. **Tadros, Th. F.,** *J. Colloid Interface Sci.,* 46, 528, 1974.
14. **Moroi, Y., Akisada, H., Saito, M., and Matuura, R.,** *J. Colloid Interface Sci.,* 61, 233, 1977.
15. **Gravsholt, S.,** *Proc. 5th Int. Congr. Surface Act. Agents, Barcelona, 1968,* Vol. 2, Ediociones Unidas SA, Barcelona, 1969, 993.
16. **Tokiwa, F. and Tsujii, K.,** *Bull. Chem. Soc. Jpn.,* 46, 2684, 1973.
17. **Francois, J., Dayantis, J., and Sabbadin, J.,** *Eur. Polym. J.,* 21, 165, 1985.
18. **Fadnavis, N. W. and Engberts, J. B. F. N.,** *J. Am. Chem. Soc.,* 106, 2636, 1984.
19. **Fadnavis, N. W., van den Berg, H. J., and Engberts, J. B. F. N.,** *J. Org. Chem.,* 50, 48, 1985.
20. **Muto, S., Ino, T., and Meguro, K.,** *J. Am. Oil Chem. Soc.,* 49, 437, 1972.
21. **Zana, R., Lang, J., and Lianos, P.,** *Polym. Prepr. Am. Chem. Soc. Div. Polym. Chem.,* 23, 39, 1982.
22. **Saito, S.,** *Kolloid Z.,* 154, 19, 1957.

23. **Isemura, T. and Imanishi, A.,** *J. Polym. Sci.,* 33, 337, 1958.
24. **Saito, S.,** *Kolloid Z.,* 137, 93, 1954.
25. **Saito, S. and Mizuta, Y.,** *J. Colloid Interface Sci.,* 23, 604, 1967.
26. **Takagi, T., Tsujii, K., and Shirahama, K.,** *J. Biochem. (Tokyo),* 77, 939, 1975.
27. **Nagarajan, R. and Kalpakci, B.,** *Polym. Prepr. Am. Chem. Soc. Div. Polym. Chem.,* 23, 41, 1982.
28. **Lewis, K. E. and Robinson, C. P.,** *J. Colloid Interface Sci.,* 32, 539, 1970.
29. **Prud'homme, R. K. and Uhl, J. T.,** *Soc. Pet. Eng. J.,* 24, 431, 1984.
30. **Bocquenet, Y. and Siffert, B.,** *Colloids Surfaces,* 9, 147, 1984.
31. **Kucher, R. V., L'vov, V. G., and Serdyuk, A. I.,** *Kolloidn. Zh.,* 40, 262, 1984.
32. **Shirahama, K., Tsujii, K., and Takagi, T.,** *J. Biochem. (Tokyo),* 75, 309, 1974.
33. **Shirahama, K.,** *Colloid Polym. Sci.,* 252, 978, 1974.
34. **Saito, S., Taniguchi, T., and Kitamura, K.,** *J. Colloid Interface Sci.,* 37, 154, 1971.
35. **Sasaki, T., Kushima, K., Matsuda, K., and Suzuki, H.,** *Bull. Chem. Soc. Jpn.,* 53, 1864, 1980.
36. **Sasaki, T., Tanaka, K., and Suzuki, H.,** *Proc. 6th Int. Congr. Surface Act. Agents,* Vol. 2, Zurich, 1972, Carl Hanser Verlag, Munich, 1973, 849.
37. **Szmerekova, V., Králik, P., and Berek, D.,** *J. Chromatogr.,* 285, 188, 1984.
38. **Chibowski, S.,** *J. Colloid Interface Sci.,* 76, 371, 1980.
39. **Chibowski, S. and Szczypa, J.,** *Pol. J. Chem.,* 56, 359, 1982.
40. **Saunders, F. L.,** *J. Colloid Interface Sci.,* 28, 475, 1968.
41. **Somasundaran, P.,** *J. Colloid Interface Sci.,* 31, 557, 1969.
42. **Ma, C.,** *Colloids Surfaces,* 16, 185, 1985.
43. **Gebhardt, J. E. and Fuerstenau, D. W.,** *ACS Symp. Ser.,* 253, 291, 1984.
44. **Hollander, A.,** M.S. thesis, Columbia University, New York, 1979.
45. **Moudgil, B. and Somasundaran, P.,** *Colloids Surfaces,* 13, 87, 1985.
46. **Fishman, M. L. and Eirich, F. R.,** *J. Phys. Chem.,* 75, 3135, 1971.
47. **Shirahama, K. and Ide, N.,** *J. Colloid Interface Sci.,* 54, 450, 1976.
48. **Cabane, B.,** *J. Phys. Chem.,* 81, 1639, 1977.
49. **Smith, M. L. and Muller, N.,** *J. Colloid Interface Sci.,* 52, 507, 1975.
50. **Birch, B. J., Clarke, D. E., Lee, R. S., and Oakes, J.,** *Anal. Chim. Acta,* 70, 417, 1974.
51. **Kresheck, G. C. and Constantinidis, I.,** *Anal. Chem.,* 56, 152, 1984.
52. **Botré, C., Crescenzi, V. L. and Mele, A.,** *J. Phys. Chem.,* 63, 650, 1959.
53. **Botré, C., DeMartis, F., and Solinas, M.,** *J. Phys. Chem.,* 68, 3624, 1964.
54. **Gilanyi, T. and Wolfram, E.,** *Colloid Surfaces,* 3, 181, 1981.
55. **Stigter, D., Williams, R. J., and Mysels, K. J.,** *J. Phys. Chem.,* 59, 330, 1955.
56. **Nunn, C. C., Schechter, R. S., and Wade, W. H.,** *J. Phys. Chem.,* 86, 3271, 1982.
57. **Saito, S.,** *Kolloid Z.,* 158, 120, 1958.
58. **Jones, M. N.,** *J. Colloid Interface Sci.,* 26, 532, 1968.
59. **Thomas, J. K.,** *Chem. Rev.,* 80, 283, 1980.
60. **Turro, N. J., Baretz, B. H., and Kuo, P.-L.,** *Macromolecules,* 17, 1321, 1984.
61. **Zana, R., Lianos, P., and Lang, J.,** *J. Phys. Chem.,* 89, 41, 1985.
62. **Bloor, D. M. and Wyn-Jones, E.,** *J. Chem. Soc. Faraday Trans. 2,* 78, 657, 1982.
63. **Zana, R., Yiv, S., Strazielle, C., and Lianos, P.,** *J. Colloid Interface Sci.,* 80, 208, 1981.
64. **Lissi, E. A. and Abuin, E.,** *J. Colloid Interface Sci.,* 105, 1, 1985.
65. **Rudd, P. J. and Jennings, B. R.,** *J. Colloid Interface Sci.,* 48, 302, 1974.
66. **Gettings, J., Gould, C., Hall, D. G., Jobling, P. L., Rassing, J., and Wyn-Jones, E.,** *J. Chem. Soc. Faraday Trans. 2,* 76, 1535, 1980.
67. **Bloor, D. M., Knoche, W., and Wyn-Jones, E.,** *Techniques and Applications of Fast Reactions in Solution,* NATO Adv. Study Inst. Ser., Ser. C (NASCDS), VC50, D. Reidel, Dordrecht, 1979, 265.
68. **Wyn-Jones, E.,** private communication.

69. **Tondre, C.,** *J. Phys. Chem.,* 89, 5101, 1985.
70. **Muller, N. and Johnson, T. W.,** *J. Phys. Chem.,* 73, 2042, 1969.
71. **Cabane, B. and Duplessix, R.,** *J. Phys. (Paris),* 43, 1529, 1982.
72. **Cabane, B. and Duplessix, R.,** *Colloids Surfaces,* 13, 19, 1985.
73. **Kresheck, G. C. and Hargraves, W. A.,** *J. Colloid Interface Sci.,* 83, 1, 1981.
74. **Goddard, E. D. and Benson, G. C.,** *Can. J. Chem.,* 35, 986, 1957.
75. **Shirahama, K., Tohdo, M., and Murahashi, M.,** *J. Colloid Interface Sci.,* 86, 283, 1982.
76. **Witte, F. M., Buwalda, P., and Engberts, J. B. F. N.,** *Colloid Polym. Sci.,* 265, 42, 1987.
77. **Shaginyan, A. A., Minasyants, M. Kh., Sarkisyan, A. Gh., and Khanamiryan, L. A.,** *Izv. Akad. Nauk Arm. SSR, Fiz.,* 15, 55, 1980.
78. **Singh, K., Deepak, S., Bahadur, P., and Bahadur, P.,** *Colloid Polym. Sci.,* 257, 57, 1979.
79. **Saito, S. and Kitamura, K.,** *J. Colloid Interface Sci.,* 35, 346, 1971.
80. **Saito, S.,** *J. Polym. Sci. A,* 7, 1789, 1969.
81. **Gravsholt, S.,** *Proc. Scand. Symp. Surface Chem.,* 2, 132, 1965.
82. **Saito, S.,** *J. Colloid Interface Sci.,* 15, 283, 1960.
83. **Schwuger, M. J.,** *ACS Symp. Ser.,* 253, 1, 1984.
84. **Saito, S.,** *Kolloid Z.,* 215, 16, 1967.
85. **Saito, S. and Yukawa, M.,** *J. Colloid Interface Sci.,* 30, 211, 1969.
86. **Saito, S. and Yukawa, M.,** *Kolloid Z.,* 234, 1015, 1969.
87. **Pal'mer, V. G. and Musabekov, K. B.,** *Fiz. Khim. Issled. Slozhnykh Sist.,* 147, 1981.
88. **Saito, S. and Taniguchi, T.,** *J. Colloid Interface Sci.,* 44, 114, 1973.
89. **Boscher, Y., Lafuma, F., and Ouivoron, C.,** *Polym. Bull. (Berlin),* 9, 533, 1983.
90. **Sabbadin, J., LeMoigne, J., and Francois, J.,** in *Surfactants in Solution,* Vol. 2, Mittal, K. L., Ed., Plenum Press, New York, 1984, 1377.
91. **Arai, H. and Horin, S.,** *J. Colloid Interface Sci.,* 30, 372, 1969.
92. **Wurzschmitt, B.,** *Fresenius Z. Anal. Chem.,* 130, 119, 1950.
93. **Murai, N., Makino, S., and Sugai, S.,** *J. Colloid Interface Sci.,* 41, 399, 1972.
94. **Takagi, T. and Isemura, T.,** *Bull. Chem. Soc. Jpn.,* 33, 437, 1960.
95. **Fishman, M. M. and Miller, R. S.,** *J. Colloid Sci.,* 15, 232, 1960.
96. **Palmstierna, H., Scott, J. E., and Gardell, S.,** *Acta Chem. Scand.,* 11, 1792, 1957.
97. **Barker, S. A., Stacey, M., and Zweifel, G.,** *Chem. Ind.,* 330, 1957.
98. **Nagarajan, R.,** *Colloids Surfaces,* 13, 1, 1985.
99. **Hill, A. V.,** *J. Physiol. (London),* 40, 190, 1910.
100. **Hall, D. G.,** *J. Chem. Soc. Faraday Trans. 1,* 81, 885, 1985.
101. **Jones, M. N., Manley, P., and Midgeley, P. J. W.,** *J. Colloid Interface Sci.,* 82, 257, 1981.
102. **Clifford, J. and Pethica, B. A.,** *Trans. Faraday Soc.,* 60, 1483, 1964.
103. **Abu-Hamdiyyah, M. and Rahman, I. A.,** *J. Phys. Chem.,* 89, 2377, 1985.
104. **Goddard, E. D., Goldwasser, S., Golikeri, G., and Kung, H. C.,** *Adv. Chem. Ser.,* 84, 67, 1968.
105. **Kung, H. C. and Goddard, E. D.,** *J. Phys. Chem.,* 68, 3465, 1964.
106. **Landoll, L. M.,** private communication.
107. **Ananthapadmanabhan, K. P. and Goddard, E. D.,** unpublished results.
108. **Goddard, E. D.,** *Colloids Surfaces,* 19, 301, 1986.
109. **Dubin, P., Ed.,** *Microdomains in Polymer Solutions,* Plenum Press, New York, 1985.

Chapter 4

POLYMER-SURFACTANT INTERACTION

PART II.

POLYMER AND SURFACTANT OF OPPOSITE CHARGE

E. D. Goddard

TABLE OF CONTENTS

I. INTRODUCTION

The case of polymer/surfactant pairs in which the polymer is a polyion and the surfactant is also ionic but bears the opposite charge is of special interest. It should be pointed out that, when the respective charges are of the same sign, association between the polymer and the surfactant can be expected to be feeble or absent, as has been reported between sodium carboxymethylcellulose (NaCMC)[1] or DNA[2] with sodium dodecyl sulfate (SDS). That association in the case of oppositely charged polyelectrolytes and surfactants is strong is not surprising, since very strong forces of electrical attraction are involved. In fact, the association which takes place can be considered an ion-exchange process where, as will be discussed later, the electrostatic forces of interaction are reinforced by a cooperative process involving aggregation of the alkyl chains of the bound surfactant molecules. An ion-exchange process would also involve pH shifts as — in addition to bound inorganic counterions, e.g., chloride (sodium) — bound hydroxyl (hydronium) ions are displaced by surfactant anions (cations). Such shifts have been found by van den Berg and Staverman,[3] and by Vanlerberghe et al.,[4] who also studied changes in the electrical conductance which accompany complex formation in such systems.

Historically, the case of mixed protein, ionic surfactant systems provides a useful starting point. Several investigators (e.g., Goddard and Pethica[5]) showed that when a protein, such as bovine serum albumin, is in solution on the acid side of its isoelectric point the addition of an "equivalent" amount of anionic surfactant results in stoichiometric precipitates, with equal numbers of positive charges (from the protein) and negative charges (from the surfactant), separating from the aqueous solution. A point of interest is that such precipitates could be resolubilized on addition of excess surfactant, giving rise to the concept that a second layer of bound surfactant ions, with their ionic groups pointing "outwards", was attached to the first bound layer through association with the hydrocarbon chains of the first layer. The excess negative charge conferred by the second layer transformed the "complex" into an anionic polyelectrolyte.

As may be imagined, such interaction frequently involves a substantial change in the conformation of the protein macromolecule. Indeed, this aspect is of recognized importance insofar as the properties of natural polymer/surfactant complexes, viz., lipoproteins, are concerned. Although protein systems will not be treated here in any depth, we will review later in this article work on the conformational changes undergone by protein analogs, viz., synthetic polypeptides, in mixtures with oppositely charged surfactants.

With the advent of a range of new polyelectrolytes, comprising both natural backbones, such as polysaccharides, and also completely synthetic materials, based on vinyl chemistry, interest in their properties has increased. Cationic polyelectrolytes, in particular, enjoy a high level of industrial importance in view of their ability to interact with, and condition, various negatively charged surfaces. Since such use may involve concurrent, or pre- or post-exposure to surfactants, understanding of the properties of mixed polymer/surfactant systems has acquired new importance. In fact, such systems have been investigated by a large number of methods and this review is largely structured along the lines of the different investigative techniques that have been employed for this purpose. In the author's laboratory a considerable amount of work has been done on the properties of a cationic cellulosic polymer derived from hydroxyethylcellulose, viz., Polymer JR (Union Carbide Corporation), in association with a number of surfactants. We will use some of the results obtained with mixtures of Polymer JR and SDS as an introduction to, and periodically an illustration of, this field of study.

II. SURFACE TENSION

As mentioned in Part I of this chapter, surface tension measurements afford a simple and informative method of studying mixtures of two components, of which one is highly active and the other relatively inactive at the air/water interface. The characteristics of the Polymer JR and SDS system are shown in Figure 1.[6] Similarities to the case of uncharged polymer/charged surfactant systems can be seen in the existence of an "interaction zone", but in the present case this zone is shifted to much lower surfactant concentration. It is of interest that no interaction was detectable between SDS and the uncharged parent polymer, viz., hydroxyethylcellulose.[6] Points to be noted in Figure 1 are (1) a synergistic lowering of surface tension at very low surfactant concentration, implying the formation of a highly surface-active complex; (2) the persistence of a low surface tension even in the zone of high precipitation where most of the originally added SDS is out of solution; and (3) eventual coincidence with the surface tension curve of the polymer-free surfactant system in the micellar region. The phenomena observed can be explained in terms of the simple diagrams shown in Figure 2, representing progressive uptake of surfactant by the polymer. A direct consequence of the increased surface activity of the polycation as it binds surfactant is its increased foaming power.[7] Interestingly, maximal foaming, as measured by simple cylinder shaking tests, occurred in the region of highest precipitation, i.e., maximal hydrophobization of the polymer. Stabilization of the foams may well be assisted by the presence of precipitate particles.

We also point out that broadly similar trends to the above can be inferred from Cockbain's[8] interfacial tension work on the bovine serum albumin/SDS system and the surface tension work of Knox and Parshall[9] on the gelatin/SDS system. Another point worth noting is that when these studies were extended by the author's group to include various vinyl polycations, it was found that those containing cationic charge centers essentially along the backbone led to much lower surface tension values in the precipitation zone with SDS; also, that in some instances it was not possible to solubilize the stoi-

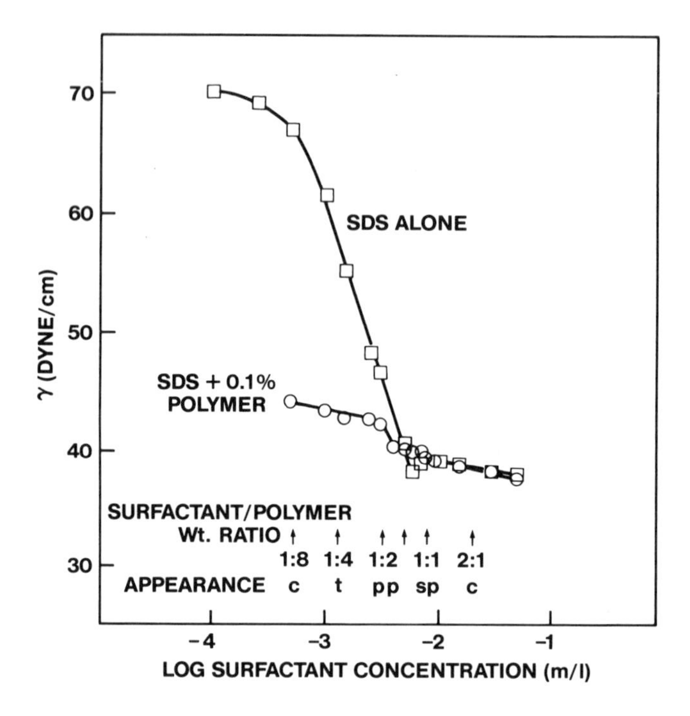

FIGURE 1. Surface tension/concentration curves of SDS with and without Polymer JR 400. The terms c, t, p, and sp refer to clear, turbid, precipitate, and slight precipitate, respectively. (Reproduced from Goddard, E. D. et al., *J. Soc. Cosmet. Chem.*, 26, 461, 1975. With permission.)

chiometric precipitation aggregates by addition of excess surfactant, especially if the charge density of the polyelectrolyte was high.[10] Two polymers exhibiting this behavior are Merquat 550 [copolymer (*N,N*-dimethyl-3,5 methylenepiperidinium chloride)/acrylamide, from Merck] and Carteretin F-4 (copolymer adipic acid/dimethylaminohydroxypropyldiethylenetriamine, from Sandoz). Similar trends were noted by Vanlerberghe et al.[4] for the polyethyleneimine (PEI)/SDS system. The interplay of polymer structure, charge density, surfactant type, and other relevant factors and the above phenomena seems worthy of further investigation.

We have implied that the formation of a highly surface-active macromolecular species with outwardly extended alkyl groups is responsible for the high surface activity noted in the precipitation region of the interaction diagram, as depicted in Figure 2. Another way to investigate this phenomenon makes use of the insoluble monolayer technique.[7] Specifically, the alkyl sulfate can be spread as a monolayer on a solution containing the polymer and compressed in the regular way. Sodium docosyl sulfate was used in order to attain water insolubility. An indication of the strength of the interaction comes from the observation that, even at the very low bulk polymer concen-

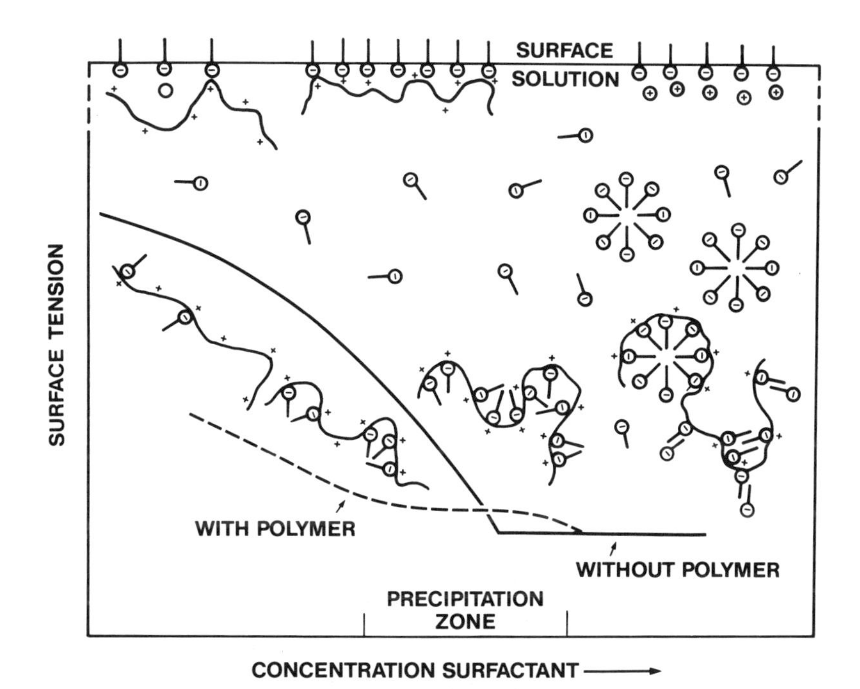

FIGURE 2. Conditions in bulk and surface of a solution containing a polycation (fixed concentration) and anionic surfactant. Full line is the hypothetical surface tension/concentration curve of the surfactant alone; dotted line is that of mixture with polycation. Simple countercations are depicted only in surface zone.

tration of 10 ppm, the surface pressure started to rise during the spreading process and some 15 min had to be allowed for attainment of a reasonably steady starting surface pressure. The qualitative difference between the force-area curves of docosyl sulfate, with and without polymer in the subsolution, together with the existence of a much higher monolayer collapse pressure in the former case, is consistent with the formation of a new surface-active species from the combination of the surfactant and the polymer, see Figure 3. It is significant that the difference in π-A curves corresponds to an expansion of the monolayer. This means that the polycation, though present at very low bulk concentration, is able to impose a packing condition on the monolayer molecules as it associates with them electrostatically. The "area per positive charge" in the adsorbed layer of this polycation, with its cellulosic backbone and flexible side chains where the cationic centers are located, will be sensitively dependent on its (unknown) configuration and hence any estimate of this area can only be qualitative and crude. Nonetheless, a completely flat configuration, parallel to the surface, could require an area greater than 100 Å.[2] Naturally, on compression, the configuration of the complex would be a compromise between the preferred packing arrangement of each component as the surface pressure is increased. (It should be noted that the

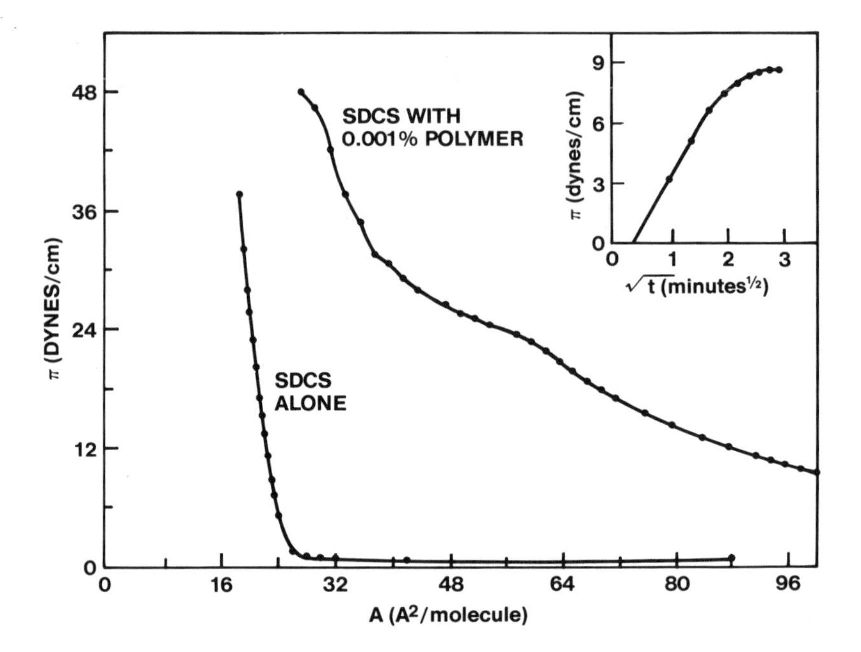

FIGURE 3. Surface pressure (π)-area (A) isotherms of sodium docosyl sulfate (SDCS) spread on 0.1 M NaCl and on 0.1 M NaCl plus 0.001% Polymer JR 400. Inset: π vs. square root of time after spreading monolayer at A = 100 Å^2 mol^{-1} on latter solution. (Reproduced from Goddard, E. D. and Hannan, R. B., *J. Colloid Interface Sci.*, 55, 73, 1976. With permission.)

polyvalent counterion need not be a macromolecule to observe this effect; indeed, Möbius and Grüniger[11] have found a similar expansion of a charged eicosylamine monolayer in the presence of a very small [10^{-6} M] bulk concentration of a tetrasulfonated porphin.)

A different interpretation of surface tension results of a mixed SDS/polycation (poly-L-lysine) system was offered by Buckingham et al.[12] They worked with solutions of varying SDS/polylysine ratios but at SDS levels sufficiently low that no precipitation- or surfactant-induced conformational changes of the polymer (by circular dichroism criteria) occurred. An interesting finding was that the equilibrium surface tension appeared to be solely dependent on the SDS concentration, i.e., was independent of that of polylysine, within the limits of the concentrations employed. These authors showed that, subject to certain assumptions, the Gibbs adsorption equation for the system reduces to the rather simple form:

$$-\,d\gamma = RT\,\Gamma_{DS^-} d\,\ln[C_{SDS} \cdot (C_{PBr_n})^{1/n}]$$

and with n, the degree of polymerization of polylysine (P^+), large (e.g., ≥20), further reduces to:

$$d\gamma \approx -RT\,\Gamma_{DS^-} d\,\ln C_{SDS}$$

One of the assumptions made in deriving the above equations was that the Gibbs excess values of the inorganic counterions were negligible, i.e., $\Gamma_{Br^-} = 0$ and $\Gamma_{Na^+} = 0$. The authors point out that unless the surface-adsorbed polymers were in the extended (β) configuration, it would be impossible for all the DS^- counterions to be close enough to the air/water interface to allow their aliphatic chains to be in the air. In other words, this would be tantamount to the surface "complex" proposed above (see Figure 2).

III. SOLUBILITY CHARACTERISTICS

As was found with solutions of proteins below their isoelectric point when mixed with anionic surfactants, the solubility behavior of cationic polyelectrolytes and anionic surfactants is complex. At low concentrations of added anionic surfactant the polymer solution remains clear, but surface tension and direct binding evidence (see above and below) indicates that binding of surfactant to the macromolecule starts to occur in this region. To reiterate, continued addition of surfactant results in the appearance of a precipitate, maximum precipitation in the case of the Polymer JR and SDS system corresponding to a stoichiometric 1:1 complex, based on charge neutralization.[6,7] Addition of excess surfactant normally results in complete clarification of the solution as the complex is resolubilized.

It has been found that many factors can affect the above phenomena. Already mentioned has been the difficulty of redissolution of the complex if the charge density of the polymer is very high. So far as the surfactant is concerned, "irregularities" introduced into the surfactant structure, such as chain branching, e.g., wing-shaped hydrophobic groups, or the introduction of polyethoxy chains into the hydrophilic headgroup, can render the dissolution process incomplete.[10]

Ohbu et al.,[13] working with a series of cationic hydroxyethylcelluloses of different degrees of cationic substitution, obtained solubility diagrams with added SDS in general conformity with the above. Direct surfactant binding studies by these authors will be referred to later.

Several other references can be found in the literature on the insolubility of pairs of polyions and oppositely charged surfactants in the range of molar equivalence. For example, Aleksandrovskaya et al.[14] reported this phenomenon for the polycation poly(1-butyl-2-methyl-5-vinylpyridinium bromide) and the three soaps, K caprate, K laurate, and K stearate, and Harada and Nozakura,[15] working with an "inverse" system, showed there was a sharp maximum in the turbidity of polyvinyl sulfate/cetyltrimethylammonium bromide (CTAB) mixtures at the 1:1 molar ratio. These authors also report an electron microscope study of this system, see below.

Often, as in ion binding studies, the formation of precipitates is regarded as a "nuisiance", but we will show that a study of the precipitation phenomenon itself can yield useful information. For example, in an investigation[10] of a cationic cellulosic system with alkyl sulfates of different chain length it

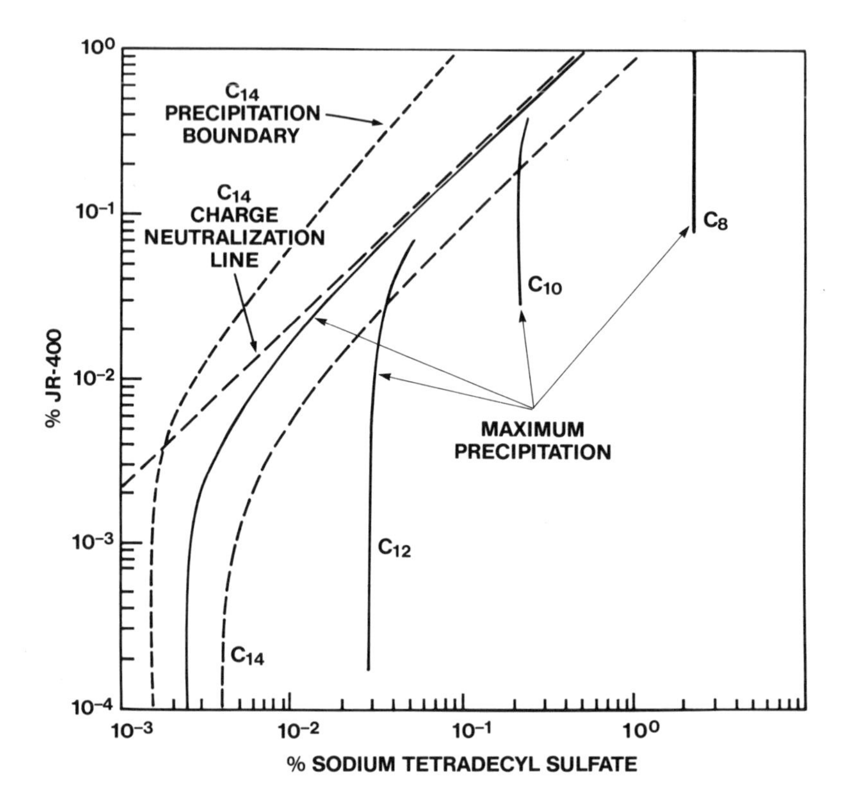

FIGURE 4. Solubility diagram of the Polymer JR 400/sodium tetradecyl sulfate system. Maximum precipitation lines for the corresponding C_{12}, C_{10}, and C_8 sulfates included. (Data from Reference 10.)

was found that the 45° stoichiometric association line in the solubility diagram — corresponding to maximum precipitation — terminated when the polymer concentration was lowered beyond a certain value which was determined by the chain length of the surfactant: the longer its alkyl chain the lower this critical polymer concentration, see Figure 4. At sufficiently low polymer concentration the line of maximum turbidity actually became independent of polymer concentration and registered as a vertical line on the polymer concentration surfactant concentration plot. This situation can be analyzed as follows: for conditions of maximum precipitation, the equilibrium is given by

$$C_t - C_e = \text{constant} \times [P]$$

where C_t and [P] are total concentrations, and C_e, obtained from the position of the vertical lines of the figure, represents the equilibrium concentration of surfactant required to maintain the existence of the stoichiometric, precipitated complex. Under conditions corresponding to high polymer concentration, i.e., in the 45° slope region, C_e represents a small fraction of C_t and hence the relation, C_t = constant × [P], holds. At lower concentrations of added polymer, especially when the surfactant is more weakly bound (as with octyl sulfate), the fraction C_e/C_t becomes progressively larger and finally approaches unity, when most of the added surfactant is required to maintain equilibrium with the precipitated stoichiometric complex. That is,

$$C_e \rightarrow C_t \text{ when } [P] \rightarrow 0$$

It is obvious from Figure 4 that the affinity of the association reaction increases strongly with increased chain length of the surfactant. In fact, the variation of C_e with alkyl chain length, which may be repesented as a linear semilog plot, allows a quantitative estimate of the interaction energy of the surfactants with the polymer to be made.

The equilibrium governing the adsorption of an anionic surfactant onto a positively charged polymer can be expressed as

$$C_{ads} = C_b \exp \{(e\psi_0 + n\phi)/kT\}$$

where C_b is the equilibrium concentration of surfactant in solution, C_{ads} is the concentration of adsorbed surfactant in undefined units, Ψ_0 is the electrical potential around the polymer, n is the number of CH_2 groups in the surfactant chain, Φ is the adsorption energy per CH_2 group, and k and T are the Boltzmann constant and absolute temperature, respectively. For conditions of maximum precipitation (i.e., charge neutralization), the equation simplifies to

$$C_{ads} = C_e \exp \{n\phi/kT\} = a \ constant$$

Hence, the slope of the $\log_{10}C_e$ vs. n plot is $\Phi/2.3\ kT$, and from the data a value of Φ is 1.1 kT was obtained, somewhat higher than the value for the free energy of micelle formation of this type of surfactant, but close to the value for hemimicelle formation on oppositely charged mineral solids.[16]

Verification of the correctness of these views was obtained from electrophoresis measurements[7] on fine particles of precipitate withdrawn from a series of aqueous compositions containing a constant concentration (0.1%) of Polymer JR and varying amounts of SDS. The electrophoretic profile vs. SDS concentration is shown in Figure 5. Below the stoichiometric SDS concentration the particles bear a positive charge, and above it a negative charge. Near the stoichiometric ratio, the particles in suspension have zero or little charge. Ohbu et al.[13] have reported similar findings.

Extensive data on the optical density of mixtures of a series of sodium alkyl sulfates and the cationic resin Hercosett 57 (a polyamide-polyamine-epichlorohydrin composition from Hercules) have recently been presented by Garcia Dominguez and co-workers.[17] Included are data from time studies on the precipitation of the polymer/surfactant complexes which form. Chain-

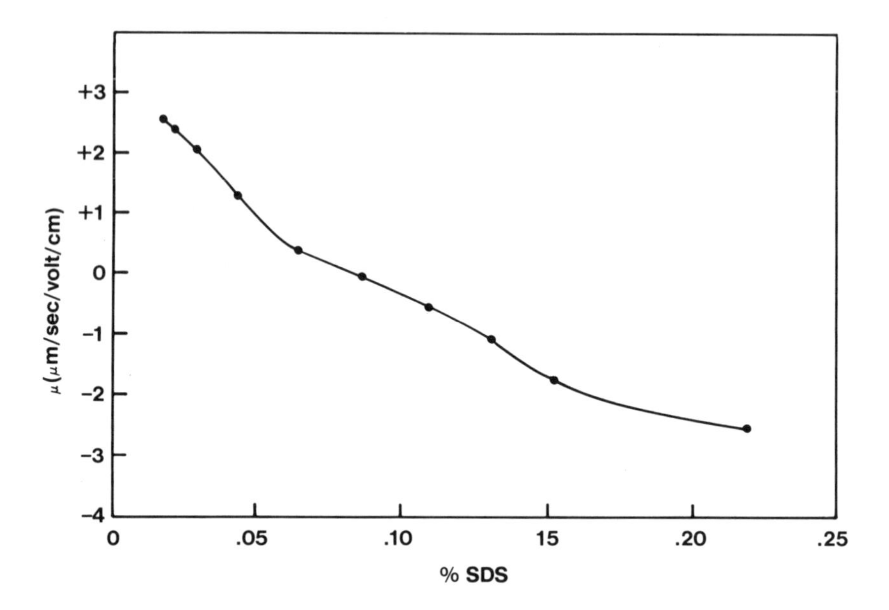

FIGURE 5. Electrophoretic mobility vs. percent SDS at constant 0.1% Polymer JR 400. (Reproduced from Goddard, E. D. and Hannan, R. B., *J. Colloid Interface Sci.*, 55, 73, 1976. With permission.)

length dependence of the location of the precipitation boundaries, in general, follow the patterns outlined above. It should also be mentioned here that some interesting work on the phase relationships of polycation/anionic surfactant mixtures has been reported recently by Dubin and co-workers.[18-20] They have examined the effect of adding a second surfactant (nonionic) and found that at a certain ionic strength and ratio of nonionic/ionic surfactant, precipitation can be inhibited.

IV. ADSORPTION ON SOLID SURFACES

Comparatively little work has been done on the adsorption properties of mixed surfactant/polyion systems. (Here, we do not include studies of "competitive" adsorption between surfactants and polymers bearing the same sign of charge,[21] nor like-charged oligomer/surfactant mixtures.[22]) Moudgil and Somasundaran[23] report differing results, depending on the order of addition, for the adsorption onto hematite of a cationic polyacrylamide in the presence of dodecanesulfonate. Adsorption of the polymer on quartz[23] was unaffected by the presence of dodecanesulfonate, but under these conditions[21] it did activate quartz for flotation by this surfactant. These results suggest[21] the presence of a primary layer of adsorbed polycation and a secondary layer of surfactant, anchored with the hydrocarbon tails pointing outwards. Moudgil and Somasundaran[24] found only small effects due to the presence of dodecylammonium chloride on the adsorption of an anionic polyacrylamide onto hematite, while Arnold and Breuer[25] reported variable results for mixtures of Polymer JR and SDS on alumina; depending on the conditions, each component could decrease the adsorption of the other. Approaches similar to those of Somasundaran have recently been taken by Musabekov et al.,[26] who have reported changes in the wetting and electrokinetic properties of quartz attending its exposure to a variety of anionic and cationic polyelectrolytes with ionic surfactants. The systems examined were SDS/poly-(2-methyl-5-vinylpyridine) (PMVP) and its copolymers with butylmethacrylate, and CTAB/polyacrylic acid (PAA) or polymethacrylic acid (PMMA). The influence of polymer concentration, polymer/surfactant ratio, and polymer molecular weight was investigated. The presence of SDS with PMVP led to increased hydrophobization of quartz, the effect being maximal when the SDS/base polymer mole ratio was unity and the polymer was present at the intermediate value (0.155%) of the three concentrations chosen. Likewise, the presence of SDS at this ratio was found to nullify the positive potential conferred on quartz by this polymer, higher ratios of SDS restoring the original negative sign of potential characteristic of uncoated quartz. The results were interpreted in terms of the hydrophobization and coiling of the polymer at the 1:1 ratio. An increase in water contact angle and potential sign reversal (as compared with the polymer alone case) was induced when CTAB was incorporated with PAA or PMMA but occurred at a lower surfactant/polymer ratio in this case.

In concluding this section we point out that extensive studies[27,28] have been carried out on the influence of SDS on the adsorption of radiotagged Polymer JR on keratin substrates which are negatively charged and somewhat water swellable. Small additions of surfactant progressively reduced the adsorption of the polymer owing to a reduction of its positive charge density. It is of interest that at high levels of SDS (in the postprecipitation zone) adsorption of this polymer was fully restored.[29]

V. SURFACTANT BINDING STUDIES

Direct binding studies of surfactants by a variety of polyelectrolytes have been carried out by a number of workers. In the early work on proteins, the dialysis equilibrium method was generally employed but in recent years surfactant ion-selective electrodes using surfactant-complex impregnated membranes have been used, since they allow much easier determination of binding. Of course, binding can be inferred from electrical conductance measurements and limited work has been done on polyelectrolyte/oppositely charged surfactant pairs by this technique, most recently by Garcia Dominguez et al.,[17] who report either three or four linear segments in the specific conductivity/concentration plot of aqueous solutions of dodecyl or decyl sulfate, respectively, in the presence of the cationic resin Hercosett 57. These authors analyzed the segments in terms of the various ionic exchange processes which can occur during the course of the "titration" with the anionic surfactant.

More direct measurements, however, involve analytical determination of the amount of bound surfactant. Ohbu et al.[13] used the dialysis method on cationic cellulose/SDS mixtures and their data, which revealed a number of important features, will be referred to first:

1. Binding was found to occur at very low concentrations of SDS: for example at 1/20th of the critical micelle concentration (CMC), the degree of binding β had already reached the value of 0.5 ($\beta = 1$ corresponding to 1 bound DS ion for each ammonium group of the polymer).
2. β vs. SDS equilibrium concentration curves were identical for those polymeric homologs with a degree of cationic substitution (CS) ≥ 0.23 (up to the limit tested which was CS $= 0.36$).
3. Precipitation was encountered when $\beta = 1$. The actual degree of binding increased in the precipitation zone, and in the resolubilization zone rose to a value of ~ 3, which was attained in the vicinity of the regular CMC of the surfactant.

Since binding in the case of the polymer with a CS value of 0.05 did not exceed a β value of ~ 1, the authors concluded that cooperativity in the binding process did not occur with this particular member, presumably since the initially bound DS^- anions were too far apart. Binding in the initial part of the isotherm was found to follow the Langmuir equation; the adsorption coefficient in this equation increased sharply at a CS of about 0.2, i.e., cooperativity became evident at this CS value.

Kwak and co-workers have made extensive studies[30-37] of the binding of cationic surfactants to a series of anionic polyelectrolytes under a variety of conditions, e.g., of added salt concentration, salt type, and temperature. The cationic surfactants comprised alkyl (C_{12} and C_{14}) trimethylammonium bromide (DTABr and TTABr) and alkyl (C_{11}, C_{12}, C_{13}, C_{14}) pyridinium bromides. The polyanion series comprised sodium dextran sulfate (SDexS), polystyrenesulfonate (PSS), DNA, PAA, alginate, pectate, and NaCMC. Their binding data are of high precision in view of the excellent performance of the surfactant-ion-selective electrodes which they employed.[31] In brief, the typical electrode employs a cell with a PVC membrane impregnated with a 1:1 complex of DTAB and SDS: it shows Nernstian response to DTAB over 3 decades of concentration.

A number of interesting points are found in Kwak's results.

1. Effect of polyanion — Binding affinity of DTA^+ to the polyanion varies considerably with the actual polyanion. One can infer from all of the Kwak group's work that the strength of binding sequence is PSS > DexS > PAA > DNA > alginate ≥ pectate > NaCMC. This effect can be seen in plots of the binding coefficient, β, and the concentration of free surfactant, C^f_D (or M^f_D), vs. total surfactant added, see Figure 6, or, more concisely, in

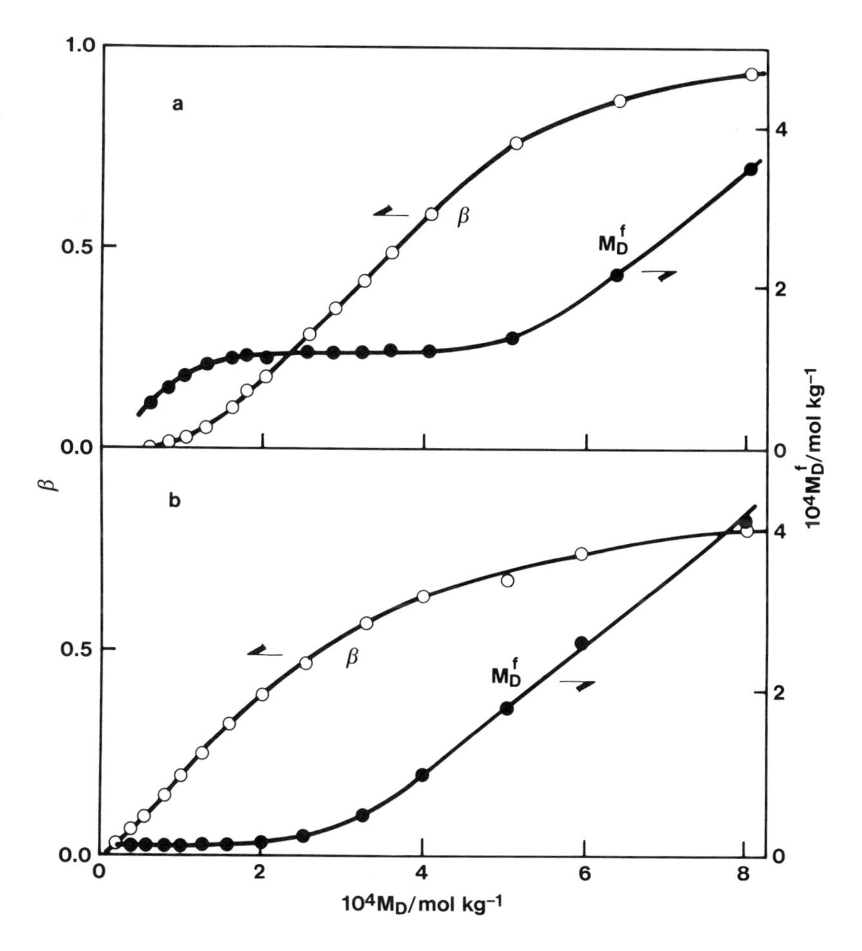

FIGURE 6. Dependence of degree of binding (β) and equilibrium concentration (M_D^f) of DTA^+ on added DTABr concentration in 0.04 *M* NaCl solutions containing 4.8×10^{-4} equivalents per kilogram of polyanion (P^-). P^- is (a) sodium dextran sulfate (SDexS); (b) sodium polystyrenesulfonate (PSS). (Reproduced from Hayakawa, K. and Kwak, J. C. T., *J. Phys. Chem.*, 86, 3866, 1982. With permission.)

tabulations of the binding constant Ku (see below). Affinity is high in the case of PSS: binding of DTA^+ occurs at concentrations of added DTABr as low as 1×10^{-4} *M*, leaving free levels of this surfactant, C_D^f, much less than this value in the early stages of binding.[30] Under these conditions, binding to SDexS occurs at tenfold higher concentrations. Once binding starts it is highly cooperative: β increases steeply over a relatively small concentration range of free DTABr concentration, and thereafter levels off at a value somewhat less than unity, i.e., based on saturation of the negative charges of the polyanion. The binding data of Table 1 show a range of about two orders of magnitude of the binding constant Ku and u, the cooperativity parameter. Quite clearly, specifics of the polymer structure are involved. For example,

TABLE 1
Cooperative Binding Constant *Ku* and Cooperative Parameter *u* at 30°C

		TTABr		DTABr	
Polymer	[NaCl] (mol^{-1} kg)	10^{-3} *Ku* (mol^{-1} kg)	*u*(±20%)	10^{-3} *Ku* (mol^{-1} kg)	*u* (±20%)
PAA		245	20	26.9	15
	0.01	30.5	600	2.69	500
Alginate		29.9	150	2.45	70
	0.01	7.5	2000		
Pectate		28.8	60	2.24	70
	0.01	7.2	2000		
NaCMC		25.1	7	1.78	4
	0.01	6.9	30		
DNA		107	20	9.3	6
	0.01	16.6	200	1.38	70
DexS	0.01			24.0	650
PSS[a]	0.01			(200)	(200)

[a] Interpolated from previous data from Hayakawa and Kwak.[30]

Data from Hayakawa, K. et al., *Biophys. Chem.*, 17, 175, 1983.

PSS, the most strongly interacting polyanion of the series, is also the most hydrophobic,[33,34] perhaps encouraging some contact of the bound surfactant with its backbone. Curiously, binding is less cooperative with this polyanion than with the more hydrophilic $DexS^-$. Flexibility of the backbone is evidently a factor and this perhaps explains the somewhat stronger binding observed with alginate over the less flexible pectate. This factor undoubtedly operates in the case of NaCMC, based on the well-known stiffness of the cellulose backbone, but here a longer distance of separation between neighboring anionic sites is invoked to explain the diminished cooperativity factor observed with this polymer. Hayakawa et al.[33] point out the importance of the so-called "linear charge density" parameter, ξ, but recognize there is no simple relation between this parameter and binding, i.e., insofar as it allows a ready differentiation of the binding characteristics of different polymers. The authors add "detailed local structure" of the polymer[33] to the three other factors specified above as being involved in determining the ultimate binding characteristics of the surfactant to the macromolecule.

2. Effect of salt — While adding salt increases the steepness, i.e., the "cooperativity", observed in the binding isotherm,[33] it substantially reduces the affinity of binding as seen by a steady increase in the concentration at which binding commences[30,36] (see Figure 7). Clearly, electrical shielding of charges is chiefly involved. This was verified[32] by increasing the valence of the cation of the added salt: $MgCl_2$, $ZnCl_2$, and $CaCl_2$ had a larger effect than NaCl and, even at the low level of 10^{-4} *M*, $LaCl_3$ caused an increase in the

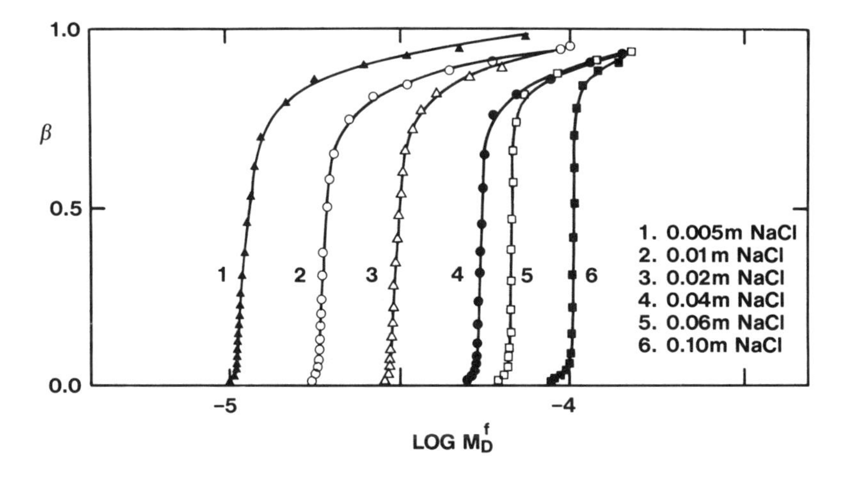

FIGURE 7. Binding isotherms for sodium dextran sulfate (5×10^{-4} *M*)-dodecylpyridinium chloride-NaCl systems. (Reproduced from Malovikova, A. et al., *J. Phys. Chem.*, 88, 1930, 1984. With permission.)

concentration for binding of DTABr onto SDexS and PSS. In the case of PSS, curves of β vs. logarithm of C^f_D, the free surfactant concentration, were indistinguishable for the three divalent cations; on the other hand, for SDexS, added $CaCl_2$ caused a larger increase in the value of the SDS concentration for binding than did $MgCl_2$ and $ZnCl_2$. In other words, some specificity in counterion effects can be encountered. We note that there are clear indications from conductance data[38] that $CaCl_2$ is more effective than $MgCl_2$ in screening the charges of a DS^- micelle. Analogies of the process of cooperative binding of the cationic surfactant to the polyanion in the presence of added salts to micellization of *anionic* surfactants in the presence of added salts have been drawn by the Kwak group.[30,32-37] Their data show that the slopes of the $(C^f_D)_{0.2}$ (or log *Ku*) vs. salt concentration plots have similar absolute values to those of corresponding CMC plots and the ratios of the slopes for added $MgCl_2$ vs. NaCl in the two systems are identical.[32] (Here $(C^f_D)_{0.2}$ represents the concentration of free surfactant when the degree of binding, β, equals 0.2.)

3. Effect of surfactant headgroup — Virtually all binding studies involving anionic surfactants onto polycations have utilized sodium alkyl sulfates. Studies on polyanions, in the main, have employed alkyltrimethylammonium (TMA^+) and alkylpyridinium bromides. Data of Malovikova et al.[35] on polyacrylate, alginate, and pectate polyanions allow a direct comparison of the binding of these two cationic surfactant series: at equivalent chain length the alkylpyridinium surfactants are bound slightly, but consistently, more strongly, see Figure 8. This is in line with the somewhat lower CMCs of these surfactants. The stronger binding is ascribed by these authors to the extra hydrophobicity of the pyridinium group and/or a steric hindrance effect as the TMA^+ group approaches a polyanion charge.

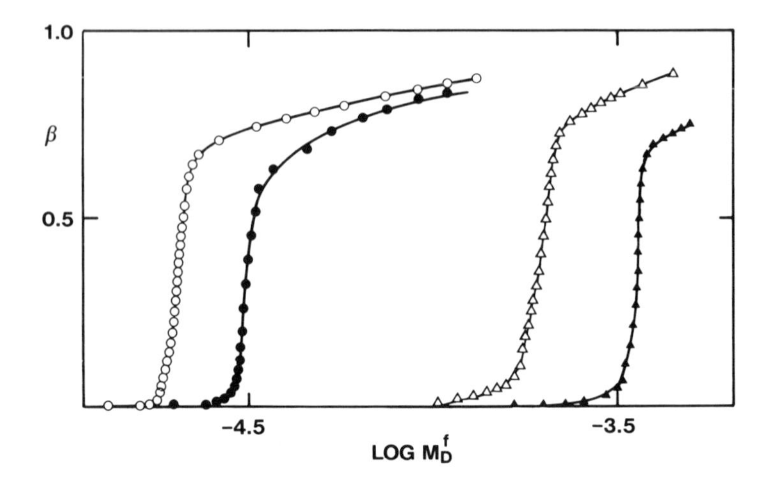

FIGURE 8. Binding isotherms of polyacrylate (5×10^{-4} *M*): comparison of alkylpyridinium and alkyltrimethylammonium ions. (△) DPyCl; (○) TPyBr; (▲) DTABr; (●) TTABr. (Reproduced from Malovikova, A. et al., *ACS Symp. Ser.*, 253, 225, 1984. With permission.)

4. Surfactant chain length effects — The extensive data of Kwak and coworkers on binding of families of cationic surfactants by a range of polyanions allow ready estimates of the free energy (ΔG) of binding, expressed, under the above heading, in units of *kT* per CH_2 group of the surfactant. Energy differences are obtained from the variation of the parameter, log $(C^f_D)_{1/2}$, which represents the free energy of binding per mole, with the chain length of the surfactant (see sections above and below on energetics of binding). For the binding of alkylpyridinium bromides (C_{11}, C_{12}, C_{13}, C_{14}) to SDexS,[36] ΔG had the value of 1.29 *kT* per CH_2, and to various polysaccharides and PAA,[35] 1.19 *kT* per CH_2. For the binding of DTA^+ and TTA^+ ions to DNA,[34] the value was 1.23 *kT* per CH_2, and 1.10 to 1.32 *kT* per CH_2 for the same cations to various carboxylated polyanions.[33] These values are again somewhat higher, but comparable to the free energy of micellization of these surfactants, suggesting some similarity of the two different aggregation processes, as was reported earlier for the binding of alkyl sulfates to a cationic cellulosic polymer.[7,10]

5. Effect of temperature — See next section.

In concluding this section, we point out that work on the binding of charged surfactants by (oppositely charged) polypeptides had been carried out in the mid-1970s by Satake and co-workers, who also employed membrane electrodes specific to the ionic surfactant. This work was a natural outgrowth of the earlier activity concerning the binding of ionic surfactants by proteins. The data by Satake and Yang[39] for binding of sodium decylsulfate (SDecS) by poly(L-lysine)HCl show the typical features of a polyion/oppositely charged surfactant system, viz., cooperative, high-affinity binding of the surfactant.

Interestingly, the binding isotherm is noticeably less steep for poly(L-ornithine) under the same conditions — a result showing the sensitivity of the binding process to polymer structure and connected with the fact that ornithine has one less methylene group than lysine in its amino side chains. Satake et al.[40] observed essentially the same binding characteristics for a polypeptide/surfactant system charged in the opposite sense, viz., polyglutamic acid/decylammonium chloride. Shirahama and co-workers also used liquid membrane electrodes in their studies of the cooperative binding of SDecS to a cationic copolymer composed of dimethyldiallylammonium chloride/sulfur dioxide,[41] and of decylpyridinium bromide to poly(vinyl sulfate) (PVS).[42] The influence of added salt was the same as that reported by Kwak, i.e., it resulted in a lower binding constant. Binding measurements were extended to higher surfactant levels (beyond the 1:1 stoichiometric value) in the latter case, and unusual and unexplained maxima and minima were observed in the binding isotherm. However, reference is made to Hall's prediction of such binding extrema under certain conditions.[43]

VI. BINDING MODELS

Shirahama and co-workers, and Satake and co-workers have developed theoretical treatments to describe the cooperative adsorption of surfactant molecules onto a polymer. A statistical mechanical treatment and Ising's model with nearest-neighbor interaction, as developed by Schwarz[44] for the binding of anionic dyes, were employed by Shirahama and co-workers.[41,42] The following relationships were derived:

$$K = [u(C^{\mathrm{f}}_{\mathrm{D}})_{1/2}]^{-1}$$

and

$$(\mathrm{d}/\beta/\mathrm{d}\ \ln C^{\mathrm{f}}_{\mathrm{D}})_{1/2} = \sqrt{u/4}$$

where K is the constant for binding to an isolated site on the polymer; u is a cooperativity parameter; $(C^{\mathrm{f}}_{\mathrm{D}})_{1/2}$ is the equilibrium surfactant concentration corresponding to the half saturation/binding point, i.e., $\beta = 0.5$.

Shirahama points out that the model applies to the region of the isotherm (i.e., for β values <0.5) where binding does not seriously alter the Ψ potential of the polymer by the ion-condensation effect, following Manning.[45]

Shirahama and Tashiro[42] make some interesting observations concerning the thermodynamic parameters derived for the decylpyridinium bromide/PVS system. The cooperativity parameter u is largely unchanged as salt is added to the system, whereas K drops as the salt concentration is increased, because of ion shielding. The reciprocal of K is a hypothetical concentration required for half-saturation of the binding sites ($\beta = 0.5$) in the absence of cooperativity. For the above system, K^{-1} has the very high value of 0.15 M whereas,

TABLE 2
The Average Cluster Size, *m*, of Bound Decyl Sulfate Ions to Polypeptides at Different Binding Ratios, *x*

	m		
x	Poly(L-ornithine) $u = 77$	Poly(D,L-ornithine) $u = 161$	Poly(L-lysine) $u = 10^4$
0.1	3.5	4.8	33.9
0.2	5.0	7.0	50.6
0.3	6.5	9.0	66.2
0.4	8.0	11.2	82.5
0.5	9.8	13.7	101.0

From Satake, I. and Yang, J. T., *Biopolymers*, 15, 2263, 1976. With permission.

experimentally, half-saturation occurs at $<0.5 \times 10^{-3}$ M, underlining the importance of the cooperativity contribution.

Precisely the same relationships as those above were derived by Satake and Yang,[39] who adapted the Zimm-Bragg theory[46] for coil-helix transitions to the cooperative binding process. [That any conformational transition makes an insignificant thermodynamic contribution to the binding process is an assumption, and this was borne out by data[39] for the two isomers poly(D,L-ornithine) and poly(L-ornithine).] The interaction or cooperativity parameter, u, is expressed by

$$u = \exp[(2E_{01} - E_{11} - E_{00})/kT]$$

where E_{ij} represents the interaction energy between neighboring ij pairs, and subscripts 0 and 1 refer to a free and to a surfactant bound site on the polymer. A knowledge of u and of β allows an estimate of the average cluster size, m, of bound surfactant to be made according to

$$m = 2\beta(u - 1)/[\{4\beta(1 - \beta)(u - 1) + 1\}^{1/2} - 1]$$

Satake and Yang[39] have compiled u and m values for the binding of decylsulfate to polyornithine and polylysine, and m values for binding of the decylammonium ion to poly(L-glutamate) are listed by Satake et al.[40] The data for the former systems are given in Table 2. It is seen that values of m increase steadily with β, i.e., as binding progresses, and that both the cooperativity parameter u, which is a measure of the interaction between alkyl groups of bound surfactant neighbors, and m, the cluster size, are considerably larger for poly(L-lysine) than for poly(L-ornithine) or poly(D,L-ornithine).

Very recently, Delville[47] has presented a model of the interaction of a charged surfactant with an oppositely charged polyion which differs in detail from the above. The model treats two additive effects, viz., (1) a contribution due to Poisson-Boltzmann ionic "condensation" of the surfactant ion (with due account of the important competitive effect of co-ions from added salt) and (2) a contribution due to cooperative specific site binding. Very satisfactory agreement between predictions of the model and actual measurements of the binding of DTABr to DNA, in the absence and presence of added NaCl (10^{-2} *M*), was found.

As stated previously, Kwak and co-workers have made extensive studies of the binding of cationic surfactants by polyanions, and have made correspondingly extensive compilations of the binding constant *Ku*, which can be determined quite accurately from the binding data, and of the less accurately determinable parameter, *u*. We have also drawn attention to the analogies drawn by Kwak between binding of the surfactant "counterion" and self-aggregation or micellization of the same surfactant ion. The other thermodynamic parameters of binding, viz. enthalpy and entropy, have until very recently (Shirahama,[48] Santerre et al.[37]) been little investigated. We mention that early data by Goddard and Benson[49] had shown a minimum in the CMC value of SDS with temperature at about 25 to 30°C which was interpreted by them in terms of structural effects of the water. Such a minimum has also been reported for cationic surfactants DTABr, and dodecylpyridinium bromide (DPyBr) by Schick,[50] and Adderson and Taylor.[51] Strengthening of the analogy between micellization and surfactant binding is provided by Kwak's work[37] on DexS/DTABr and Shirahama's work[48] on DNA/DPyBr where a change in sign of enthalpy, ΔH, from positive to negative occurred at about 30°C. Kwak's entropy, ΔS, values for binding are all positive (reflecting, at least in part, increased mobility of the surfactant chains on binding/aggregation) but decrease with temperature as the water structure is progressively broken down.

Finally, we recall Ohbu's[13] finding of adherence of binding data for SDS on a cationic cellulosic polymer to a Langmuir isotherm, at least in the early stages of binding. This bears some similarity to the finding by Kwak[33] of the relatively low level of cooperativity in the binding of a cationic surfactant by another cellulosic, viz. NaCMC, even though the charge density of the latter was higher than in Ohbu's polymers. Beyond a binding level of ~0.25, Ohbu et al.[13] noticed a progressive increase in binding coefficient which they ascribed to cooperativity.

VII. STRUCTURE OF THE (SOLUBLE) COMPLEXES

Binding of a charged surfactant to a polyion is expected to influence the microenvironment of both of these components — the surfactants through self-association and association with the macromolecule if it has hydrophobic areas, and the polyion through conformational changes and possible tendencies toward association with (or dissociation from) its neighbors. Those macromolecules, such as several of the polyamino acids previously referred to, exhibiting optical activity are amenable to investigation by such techniques

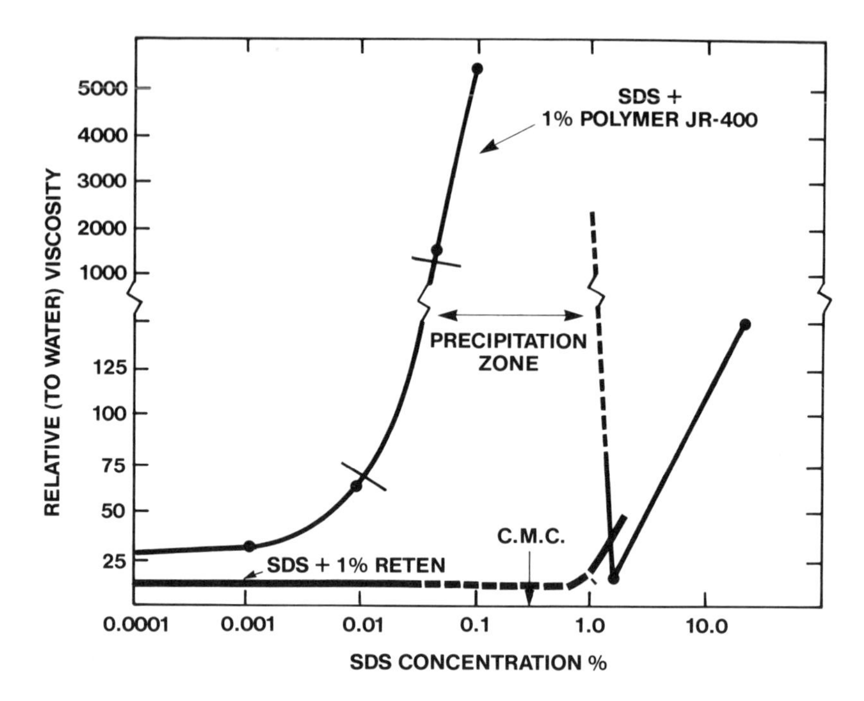

FIGURE 9. Relative viscosity of 1% Polymer JR 400 and of 1% Reten 220 as a function of SDS concentration. (Reproduced from Leung, P. S. et al., *Colloids Surfaces,* 13, 47, 1985. With permission.)

as optical rotatory dispersion (ORD) and circular dichroism (CD); they will be treated in a later part of this section. Other methods of obtaining relevant information include viscometric studies and measurement of solubilization of oil-soluble and fluorescent dyes. Together with results from a limited amount of work using neutron magnetic resonance (NMR) and small-angle neutron scattering (SANS), such information is presented below.

A. RHEOLOGICAL MEASUREMENTS

The author's group has reported viscometric[52-54] studies on two cationic polymers, viz., a cationic cellulosic (Polymer JR) and an acrylamide/β-methacryloxyethyltrimethylammonium chloride copolymer (Reten, Hercules) within a range of polymer and added SDS concentrations. Considerable differences in behavior between the two polymers were found. At the 1% Reten level, no change was detectable in viscosity at all levels of added SDS in the preprecipitation range but, by contrast, the viscosity of 1% Polymer JR solutions increased over 200-fold in this range, see Figure 9. Under the same conditions, the viscosity of 0.1% Polymer JR solutions decreased somewhat.[54] These results suggested little change in conformation accompanied the binding of SDS to Reten but, on the contrary, that intermolecular association was promoted in solutions of the cationic cellulose at higher concentration and intramolecular association at the low concentration. We will see from dye solubilization results that evidence favors an association mechanism involving the alkyl chains of bound surfactant molecules. This type of association would be expected to be shear sensitive. Indeed, rheological data on the 1% Polymer JR/0.1% SDS system show this to be the case. At low shear rates, up to about 25 s^{-1}, the system had a high viscosity and was Newtonian. Thereafter, the system was markedly shear thinning, until at about 400 s^{-1} it was again Newtonian with a comparatively low differential viscosity. The inference here is that the energy input at shear rates $\geq$400 s^{-1} is sufficient to break the intermolecular association bonds. Rheological results on this system in the postprecipitation (higher SDS concentration) range (1% Polymer JR, 1.2% SDS) were unusual in the sense of displaying rheopexy and a large hysteresis loop. We[53] offered the interpretation that the low-viscosity form is a somewhat tangled necklace structure of SDS micelles and polymer chains, and the higher-viscosity form, generated on shearing, a more open structure with a double layer of adsorbed SDS molecules.

Several references may be found in the literature to the compaction of polyelectrolytes on binding of oppositely charged surfactants.[26] Viscosity measurements have been carried out by Abuin and Scaiano[55] on mixtures of PSS of mol wt 130,000, and DTABr. When added to solutions of 0.36% PSS, DTABr in the preprecipitation zone brought about a progressive reduction in the viscosity of the solutions. For example, 5.8 m*M* DTABr was found to drop the reduced viscosity of PSS by a factor of ten, far exceeding the reduction effected by addition of the simple analog "surfactant", tetramethylammonium bromide. The explanation offered was that coiling of the flexible "vinyl" backboned polyelectrolyte occurs around small clusters of the surfactant which form under these conditions. Similar effects have been obtained by Bekturov et al.,[56] who studied the viscosity characteristics of a series of amphoteric (i.e., basic and acid group containing) polymers on addition of anionic (SDS) or cationic (CTABr) surfactants to their solutions. While a viscosity reduction was generally observed as surfactant was bound to the polyampholyte, the effects were rather complicated, being dependent on the point of zero charge of the polymer, the chemical groupings present, backbone hydrophobicity, and, of course, pH. The results are of some importance and interest as these polyampholytes can be considered as models for proteins.

B. DYE SOLUBILIZATION, FLUORESCENCE, AND PHOTOCHEMISTRY

A clear indication of the behavior of surfactant clusters in the preprecipitation zone of mixtures of Polymer JR and SDS was found[53,54] in the existence of a distinct solubilization area for the oil-soluble dye Orange OT at a SDS concentration one order of magnitude lower than its CMC. Such an area was not observed in the case of the vinyl copolymer Reten under the same con-

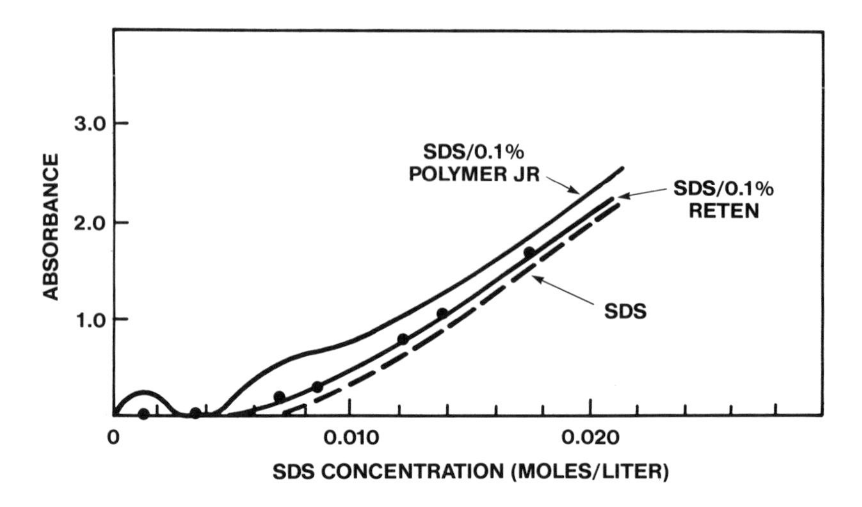

FIGURE 10. Solubilization of Orange OT by SDS alone and in the presence of 0.1% Polymer JR 400 and 0.1% Reten 220. (Reproduced from Leung, P. S. et al., *Colloids Surfaces,* 13, 47, 1985. With permission.)

ditions (0.1% polycation), Figure 10.[53] Beyond the precipitation zone, solubilization was again observed, the extent of solubilization exceeding that of micellar SDS at the same concentration of surfactant. Similar results were reported by Ohbu et al.[13] on the cationic cellulosic/SDS system for solubilization of the dye Yellow OB and by Garcia Dominguez et al. for solubilization of Oracet Red 3B by mixtures of the cationic resin Hercosett 57 and SDS or SDecS.[17] It should be mentioned here that the solubilization of oil-soluble dyes, hydrocarbons, and other organic materials by polyelectrolyte/surfactant complexes has been studied quite extensively in the case where the polyelectrolyte is a protein. Work on PEI/SDS systems has recently been published by Yui et al.,[57] who reported synergism of these components in the solubilization of benzene. In another paper,[58] information is provided on the effect of changing the molecular weight of the PEI on the solubilization of the dye Sudan III by this pair of compounds.

From studies of the solubilization of the dye Pinacyanol by clear supernatant solutions of mixtures in the precipitation zone of Polymer JR and SDS, Ananthapadmanabhan et al.[59] concluded that micelle-type structures are present in the solutions both in the initial (low SDS level) and the final (high SDS level, beyond maximum precipitation) part of the precipitation zone, but not in between. From the observation that the fluorescent dye pyrene is much more efficiently solubilized by the "low SDS" aggregates than the "high SDS" aggregates, these authors concluded that there are distinct structural differences between the two, the former possibly corresponding to hemimicelle-type structures, assemblies of low surfactant aggregation number, and identified as being quite surface active in the surface tension measurements, and the latter to more conventional micelles — the "beads" in an entangled necklace structure. Fluorescence decay measurements which, as shown by Zana et al.,[60] allow determination of the size of the aggregates (solubilizing pyrene), confirmed the latter picture. Measurements on the supernatant solution in a 0.1% Polymer JR, 3.5×10^{-2} *M* (''high'') SDS solution yielded a value of 70 as the aggregation number of SDS — a value typical of regular SDS micelles. One can infer from the data that the aggregates in the ''low'' SDS region are smaller, more efficient, solubilizing entities.

Abuin and Scaiano,[55] using the fluorescent probe 8-anilino-1-naphthalenesulfonic acid (ANS) as a sensor of ''free'' cationic surfactant in solution, obtained direct information on the extent of binding of DTABr and CTABr to PSS. This probe undergoes a fluorescent shift from ~520 nm, in polar media such as water, to ~480 nm when it associates with these surfactants. The presence of PSS, which effectively ''scavenges'' the first amounts of surfactant added, inhibited these shifts, and from their data these authors deduce that for up to 50% coverage of the polyelectrolyte (i.e., $\beta = 0.5$) less than 5 to 10% of the added surfactant remains free. Use of the probe in fact allows the determination of the binding characteristics of the surfactant which closely resemble those obtained by Kwak and co-workers who, in general, employed polymer concentrations some fiftyfold lower.

Other valuable information obtainable from the use of fluorescent probes is the polarity of the microenvironment in which they are dissolved. This information is obtained from the fluorescence characteristics at very low levels of the probe, viz., $\sim 10^{-6}$ *M*. For pyrene, for example, the sensitive parameter is the intensity ratio of the first and third emitted peak;[61] for pyrene aldehyde it is the position of the maximum in the fluorescence spectrum, according to the relationship

$$\lambda_{max} = 0.52D + 431.5$$

where D is the dielectric constant of the solubilizing medium.[62,63]

For SDS micelles, a D value of 50 has been deduced, indicating the probe experiences an environment comparable in polarity to ethanol. In the immediate pre- and post-precipitation zones of Polymer JR/SDS mixture with 0.01 and 0.1% polymer, pyrene aldehyde was shown to experience a much lower polarity medium, of dielectric constant down to as low as 30, indicating a less open and less hydrated palisade layer of the aggregated surfactant phase as compared to regular micelles.[59] With Reten 220, on the other hand, the corresponding solubilizing environments have a D value almost identical to that of regular micelles, suggesting a more open and hydrated structure in this case. Specific contributions of the polymer structure to the nature of the association complex are thus clearly established. Well into the post-precipitation zone the properties in both cases approximated those of SDS micellar solutions.

Abuin and Scaiano,[55] using pyrene as a probe, have reported similar results for mixtures of DTABr and the ''vinyl'' polyelectrolyte, PSS; in this case a slightly lower polarity of DTA^+ in the aggregate vs. that in DTABr micelles was found.

The last-mentioned authors have obtained other interesting information from their photochemical studies. From the lifetime of xanthone triplets present in mixed PSS/DTABr (28 m*M* PSS, 14 m*M* DTABr) solution vs. that in micellar DTABr solution (20 m*M*), viz. ~2.5 μsec and ~15 nsec, respectively, and the knowledge that the bromide ion is a very efficient triplet quencher, these authors conclude that in the complex the bromide ions of the surfactant are quantitatively replaced. Similar conclusions were reached from studies of the rate of formation of "light-absorbing transients" by irradiation of benzophenone in PSS/DTABr and DTABr solutions: it was shown that the rate was at least tenfold higher in the mixed solution than in simple DTABr micellar solution, again because of quenching by bromide ion in the latter case. Other quenching studies, in this case of xanthone triplets by 1-methylnaphthalene which is solubilized by aggregates of the surfactant, led to an estimate of the cluster size in the aggregate which was quite small, viz., 7 to 10 surfactant monomers.

C. SMALL-ANGLE NEUTRON SCATTERING (SANS)

SANS has a unique feature, which depends on the different scattering cross section of hydrogen and deuterium to neutrons, of being able to provide separate information on each member of a two-component mixture in a solvent such as water. Put simply, if one component (A) is in deuterated form and the other (B) in hydrogenous form, scattering information on A is obtainable when the solvent is H_2O, and on B when the solvent is D_2O — a technique known as the contrast variation method. Exact interpretation of the scattering information from mixtures of charged components such as surfactants and polymers is rather difficult; we report here the results of preliminary experiments[53] designed to highlight differences in behavior of a typical cellulosic and vinyl polycation in mixture with SDS, namely the Polymer JR and Reten 220 previously referred to. The results, obtained are summarized as follows:

1. The plot of scattering intensity I vs. wave vector Q (equal to $(4\pi/\lambda)\sin\theta/2$ where λ is wavelength and θ scattering angle) for 1% Polymer JR in D_2O is flat and featureless down to a Q value of 0.005 Å^{-1}, probably reflecting an extended rod configuration of this polymer. The addition of a small amount of SDS_D (0.05%) brought about a significant level of small-angle scattering, probably through local densification of the polymer chains through bound alkyl chain interaction. By contrast, a 1% Reten solution in D_2O has a clearly patterned small-angle scattering profile, which was unchanged on addition of a small amount of SDS, suggesting little or no change in shape on binding.
2. Addition of 1.5% Polymer JR to a 1.5% SDS_D solution in H_2O removed the strong intensity peaks ($Q \approx 0.05$ Å^{-1}) of the latter resulting from

structuring due to coulombic repulsions between the highly charged micelles. The polymer is apparently very effective in screening the micelles and/or disordering their array. Reten also eliminated the peak but led to a markedly different I/Q profile. In fact, the Guinier plot (ln I vs. Q^2) for the Reten/SDS system is linear, and this led to a derived radius of gyration much smaller than that observed for SDS micelles, suggesting the formation of much smaller (bound) surfactant aggregates under these conditions. On the other hand, the corresponding Guinier plot of Polymer JR/SDS is clearly not linear, again providing a clear indication of the difference in structure of the two polymer/surfactant systems; possibly a distribution of aggregate sizes is involved in the Polymer JR case. More work in the very low Q region is needed to clarify this point.

When SDS was present in large excess, as in 1% polymer/5% SDS_D in H_2O solution, a prominent peak was observed in the I vs. Q plots for both polymers, showing that micellar interaction effects again predominate.

One can conclude that SANS, like several of the other techniques referred to, shows clear differences in the properties of mixed polyelectrolyte/surfactant solutions depending on the particular polyelectrolyte employed.

D. OPTICAL MEASUREMENTS: CIRCULAR DICHROISM

The foregoing section has provided information, often indirect, concerning the structure of the association complexes which form in solution. For macromolecules possessing appropriate optical activity, the techniques of ORD and CD provide a powerful and direct method of measuring structural changes of the macromolecule during the association reaction. Polyamino acids offer such opportunities. Yang, Satake, and Shirahama and their co-workers have published extensively in this area in the last 10 years. Since we are concerned in this section with the review of charged polymers, the properties of surfactant/polyamino acid pairs will be discussed only when pH conditions are such that the latter component exists in solution alone as a polyelectrolyte. It is, however, appropriate to consider first the effect of ionization itself on the polyamino acids. Among these, the most widely investigated in pairs with surfactants is poly(L-lysine). The CD measurements of Tseng and Yang[64] show that at high pH poly(L-lysine) and its homologs exist in the ordered, helix form, while at neutral pH they adopt a disordered, coil form. Disordering is evidently brought about by charge repulsion effects as the polymer acquires polyelectrolytic character.

Intuitively, one can expect the binding of an oppositely charged surfactant to the ionized polyamino acid to restore an ordered structure by two possible mechanisms; (1) charge neutralization, and (2) a tendency of the adsorbed surfactants to adopt/impose some ordering through aggregation. Work by Satake and Yang[65] on poly(L-ornithine) and poly(L-lysine) with a series of even-numbered alkyl sulfates (C_8–C_{16}) confirmed these expectations. The

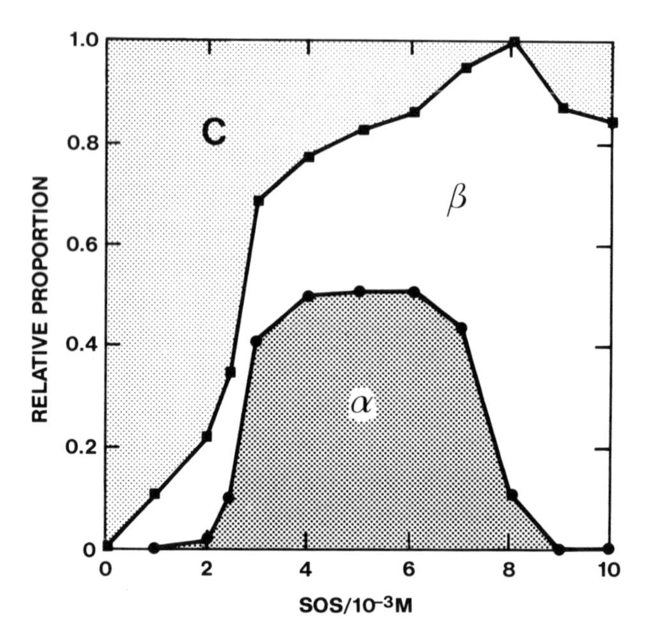

FIGURE 11. Relative proportions of α-helix (α), β-structure (β), and random structure (C) of poly(L-lysine) as a function of sodium octyl sulfate (SOS) concentration. (Reproduced from Takeda, K., *Bull. Chem. Soc. Jpn.*, 58, 1210, 1985. With permission.)

appearance of a CD spectrum with a characteristic double minimum at about 210 and 222 nm indicated a transition to the helical form of poly(L-ornithine) in solutions of all the alkyl sulfates, confirming earlier work of Grourke and Gibbs[66] with SDS. With added octyl sulfate, poly(L-lysine) also adopted a helical conformation but, interestingly, all the other alkyl sulfates induced another ordered form, viz., the extended (β) structure, as had been reported earlier by Li and Spector[67] with SDS, and characterized by a single minimum at about 217 nm. The work of Satake and Yang[65] covered a range of alkyl sulfate concentrations.

Recently, Takeda[68] has mapped the relative proportions of α-helix, β and random-coil structures in poly(L-lysine) solutions containing added surfactants (SDS, sodium octyl sulfate [SOS] and sodium octanesulfonate) by a computer simulation procedure using both the 190 to 250 and 200 to 250 nm regions, at 1-nm intervals. Simulation to the experimental spectra was carried out assuming additivity of the CD spectra of the various conformations of the polypeptide chains. The results are extremely informative and we show here the results of mapping of the SOS and SDS systems, Figures 11 and 12. Note the interesting appearance of the α-helix form of poly(L-lysine) in SOS solutions of intermediate concentration, and none in SDS solutions; Hayakawa et al.[69] report its appearance in octanesulfonate solutions only above 8 m*M* concentration.

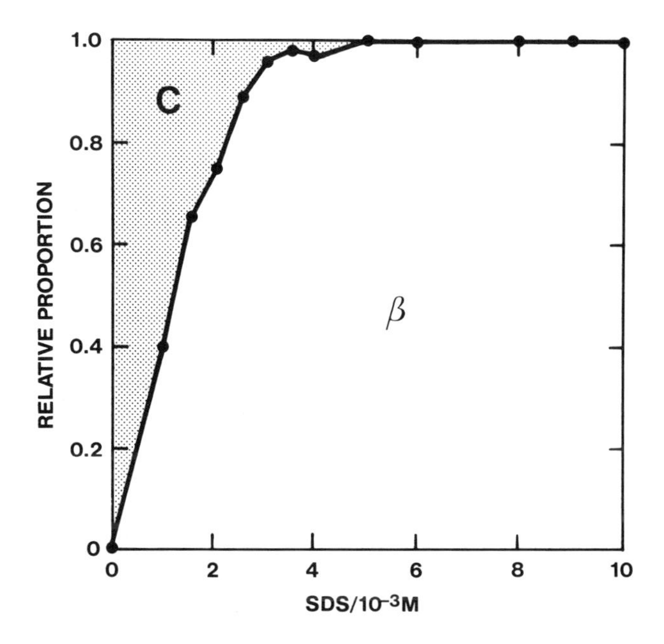

FIGURE 12. Relative proportions of α-helix (α) β-structure (β), and random structure (C) of poly(L-lysine) as a function of SDS concentration. (Reproduced from Takeda, K., *Bull. Chem. Soc. Jpn.*, 58, 1210, 1985. With permission.)

Tseng and Yang[64] studied the effect of SDS and SOS addition on a number of poly(L-lysine) homologs. If R = $-(CH_2)_nNH_2$ represents the side chain, n was varied from 3 through 7, $n = 4$ representing poly(L-lysine). All homologs, except poly(L-lysine) itself which favors the extended β-conformation, adopted the helical form in SDS solutions; in SOS solutions only the helix conformation was evident. It should be pointed out that the longer side chain ($n = 6, 7$) poly(L-lysine) homologs are partially helical themselves in neutral solution. Mattice and Harrison[70] compared the behavior of tri-, tetra-, and penta(L-lysine) with that of poly(L-lysine) in SDS solution. None of these oligomers, except the pentamer over a limited SDS concentration range, was found to undergo any conformational change.

The effects of SDS on the conformation of various copolypeptides of lysine (lys) and alanine (ala), and of lysine, and leucine, serine or glycine, have been reported by Kubota et al.[71,72] In the first series (all disordered in neutral solution) the conformation assumed depends on whether or not the copolymer is sequential, the length of the sequences [i.e., the value of x in $(\text{lys}^x\text{–ala}^x)_n$], and the ratio of the monomers. For example $(\text{lys}^1\text{–ala}^1)_n$ adopts a stable β-form in SDS solution, whereas $(\text{lys}^{50}\text{–ala}^{50})_n$ adopts the helical form. In the series $(\text{lys}^x\text{–ala}^y)_n$ where $x = 2$ or 3 and $y = 1$, a partial helix was seen at low SDS concentration but the polymer was disordered at high SDS concentration. Dependencies on monomer ratio, type, and sequence were also shown in the second series. These authors draw analogies between their results and the conformational changes of proteins, well known to be brought about by anionic surfactants.

Analogous data to the foregoing have been obtained with anionic polypeptide/cationic surfactant pairs, although such systems have been far less studied than the cationic polypeptide/anionic surfactant counterparts. First, we refer to studies by Satake et al.[40] on the binding and conformational behavior of poly(L-glutamic acid) in decylammonium chloride (Dec-$NH_3^+Cl^-$) solution. This polyelectrolyte was found to undergo a coil-to-helix transition. Initiation of the transition was related to the degree of binding (β) of $DecNH_3^+$ and was found to occur beyond a β value of 0.55. These authors attribute stabilization of the helical structure to the existence of surfactant clusters in the complex. A similar effect was proposed by Satake and Yang[39] for poly(L-ornithine) in SDecS solutions. Other studies have shown that the α-helix form of poly(L-glutamic acid) can also be induced by dodecyl-[73] and octadecyl-[74] ammonium chlorides. CD studies of the poly(L-glutamic acid)/dodecylammonium surfactant systems have recently been extended by Maeda et al.[75] to include $DNH_3^+Cl^-$, $DNH(CH_3)_2^+Cl^-$, and $DN(CH_3)_3^+Cl^-$. The maximum extent of helix formation occurred in each case when the mixing ratio of the surfactant was around unity. The individual surfactants showed differing, and rather complex, behavior when this ratio was exceeded.

A comprehensive review on the analysis and kinetics of conformational changes undergone by synthetic polypeptides (and proteins) in the presence of surfactants has been presented very recently by Takeda.[76]

VIII. MISCELLANEOUS: ELECTRON MICROSCOPY OF DRIED "COMPLEXES"

Harada and Nozakura[15] have investigated the structure of various polyelectrolyte-ionic surfactant pairs. Drying a 1:1 mole ratio (maximum turbidity) mixture of PVS and CTABr resulted in the appearance of long single strands, of diameter ~30 Å. With the 2 PVS:1 CTABr mixture, on the other hand, a multilayer, vesicular-type structure was observed with a layer spacing of about 25 Å. In the systems, SDS/ionene polymers $[N^+(CH_2)_3–N^+–(CH_2)_4]_n$, where the spacings between the charges along the backbone are greater, a multilayered, vesicular structure was observed at the stoichiometric 1:1 ratio.

IX. CONCLUDING REMARKS

Despite the possibility that precipitation can occur in their mixtures, the formation of association complexes between polyelectrolytes and oppositely charged surfactants is, at least in the soluble range, conceptually more straightforward than that involving an uncharged polymer, since in the former case, there are well-defined binding sites on the polymer. While the primary driving force for adsorption can be viewed as being electrostatic, the process is strongly reinforced by alkyl chain association of the adsorbing surfactant molecules. Indeed, the process can be considered a special case of surfactant aggregation in which the strong driving force for surfactants to reduce their hydrocarbon/water contact area is satisfied by alkyl chain association during adsorption. The aggregation process is again opposed by the electrical potential which develops in the assembly, but this is diminished by the array of opposite charges along the polyelectrolyte with which the surfactant ions associate. There are, in fact, strong analogies between the process of surfactant adsorption, leading to complex formation of surfactant with the polyelectrolyte, and adsorption, leading to hemimicelle formation of ionic surfactants on the surface of oppositely charged solids, such as minerals. In both cases an ion-exchange process is involved in which the counterion of the polyelectrolyte (or charged surface) is replaced by the surfactant ion and binding and formation of aggregates (hemimicelles) commence at concentrations orders of magnitude below the CMC of the surfactant. A differentiating feature in the case of the polyelectrolyte is the molecular flexibility of the charge-bearing substrate, meaning that its properties, such as conformation, can be substantially changed by the adsorption process.

We have seen that several factors such as chemical composition, charge density, location of charges, and backbone flexibility of the polymer influence (1) its reactivity toward the ionic surfactant; (2) the properties of the complexes, e.g., surface tension and configuration in solution; (3) solubility of the complex and the ability of the complex once precipitated, to be resolubilized; and (4) the size and configuration of the surfactant clusters. Factors (3) and (4) are obviously also affected by the detailed properties of the surfactant. The final conformation of the polymers and disposition of the surfactant molecules in the associated aggregates will be a compromise between the preferred structure of the polymer and the preferred packing and size of the surfactant assemblies: each of the components will have an effect on the other. Concentration and ratio of the two components also have important effects on the observed properties.

While the structure of the surfactant is of obvious importance, we have given several examples of how the properties of a surfactant's domains in the complexes are sensitively influenced by the particular polyion present. Included are cluster polarity, cluster size, ability to dissolve dyes, the "critical cluster concentration", and so on. Likewise the way in which the polyelectrolyte itself responds to addition of surfactant, as evidenced by solution rheology, optical dispersion, and neutron scattering, is again dependent on the specifics of the polyelectrolyte chosen.

Much work remains to be done to further understand the properties of these ionic polymer/surfactant systems and exploit them. In the latter regard, impetus is already provided in the use of the surfactant clusters as microenvironments in the study of photochemical reactions, and in the studies of polypeptide/surfactant systems as models for lipoproteins. It is confidently expected that more work along these, and other, lines will follow.

REFERENCES

1. **Schwuger, M. J. and Lange, H.,** *Tenside,* 5, 257, 1968.
2. **Chatterjee, R., Mitra, S. P., and Chattoraj, D. K.,** *Indian J. Biochem. Biophys.,* 16, 22, 1979.
3. **van den Berg, J. W. A. and Staverman, A. J.,** *Rec. Trav. Chim. Pays-bas,* 16, 1151, 1979.
4. **Vanlerberghe, G., Handjani-Vila, R. M., and Poubeau, M. C.,** *Proc. 7th Int. Congr. Surf. Act. Agents,* Moscow, 1976, V. M. Zimina, Moscow, 1978, 791.
5. **Goddard, E. D. and Pethica, B. A.,** *J. Chem. Soc.,* 2659, 1951.
6. **Goddard, E. D., Phillips, T. S., and Hannan, R. B.,** *J. Soc. Cosmet. Chem.,* 26, 461, 1975.
7. **Goddard, E. D. and Hannan, R. B.,** *J. Colloid Interface Sci.,* 55, 73, 1976.
8. **Cockbain, E. G.,** *Trans. Faraday Soc.,* 49, 104, 1953.
9. **Knox, W. J. and Parshall, T. O.,** *J. Colloid Interface Sci.,* 33, 16, 1970.
10. **Goddard, E. D. and Hannan, R. B.,** *J. Am. Oil Chem. Soc.,* 54, 561, 1977.
11. **Möbius, D. and Grüniger, H.,** *Bioelectrochem. Bioenerg.,* 12, 375, 1984.
12. **Buckingham, J. H., Lucassen, J., and Hollway, F.,** *J. Colloid Interface Sci.,* 67, 423, 1978.
13. **Ohbu, K., Hiraishi, O., and Kashiwa, I.,** *J. Am. Oil Chem. Soc.,* 59, 108, 1982.
14. **Aleksandrovskaya, S. A., Tret'yakova, A. Ya., and Barabanov, V. P.,** *Vysokomol. Soedin. Ser. B,* 26, 280, 1984.
15. **Harada, A. and Nozakura, S.,** *Polym. Bull.,* 11, 175, 1984.
16. **Somasundaran, P., Healy, T. W., and Fuerstenau, D. W.,** *J. Phys. Chem.,* 68, 3562, 1964.
17. **Garcia Dominguez, J., Erra, P., Julia, M. R., and Infante, M. R.,** *J. Disp. Sci. Tech.,* 6, 437, 1985.
18. **Dubin, P. L. and Oteri, R.,** *J. Colloid Interface Sci.,* 95, 453, 1983.
19. **Dubin, P. L. and Davis, D.,** *Colloids Surfaces,* 13, 113, 1985.
20. **Dubin, P. L., Rigsbee, D. R., and McQuigg, D. W.,** *J. Colloid Interface Sci.,* 105, 509, 1985.
21. **Somasundaran, P. and Cleverdon, J.,** *Colloids Surfaces,* 13, 73, 1985.
22. **Balzer, D. and Lange, H.,** *Colloid Polym. Sci.,* 257, 292, 1979.
23. **Moudgil, B. M. and Somasundaran, P.,** in *Fundam. Adsorp., Proc. Eng. Found. Conf., 1983,* Mayers, A. L. and Belfort, G., Eds., Eng. Found., New York, 1984, 355.
24. **Moudgil, B. M. and Somasundaran, P.,** *Colloids Surfaces,* 13, 87, 1985.
25. **Arnold, G. B. and Breuer, M. M.,** *Colloids Surfaces,* 13, 103, 1985.
26. **Musabekov, K. B., Omarova, K. I., and Izimov, A. I.,** *Acta Phys. Chem.,* 29, 89, 1983.
27. **Faucher, J. A. and Goddard, E. D.,** *J. Colloid Interface Sci.,* 55, 313, 1976.
28. **Faucher, J. A. and Goddard, E. D.,** *J. Soc. Cosmet. Chem.,* 27, 543, 1976.
29. **Goddard, E. D. and Leung, P. S.,** *Cosmet. Toiletries,* 97, 55, 1982.
30. **Hayakawa, K. and Kwak, J. C. T.,** *J. Phys. Chem.,* 86, 3866, 1982.
31. **Hayakawa, K., Ayub, A. L., and Kwak, J. C. T.,** *Colloids Surfaces,* 4, 389, 1982.
32. **Hayakawa, K. and Kwak, J. C. T.,** *J. Phys. Chem.,* 87, 506, 1983.
33. **Hayakawa, K., Santerre, J. P., and Kwak, J. C. T.,** *Macromolecules,* 16, 1642, 1983.
34. **Hayakawa, K., Santerre, J. P., and Kwak, J. C. T.,** *Biophys. Chem.,* 17, 175, 1983.
35. **Malovikova, A., Hayakawa, K., and Kwak, J. C. T.,** *ACS Symp. Ser.,* 253, 225, 1984.
36. **Malovikova, A., Hayakawa, K., and Kwak, J. C. T.,** *J. Phys. Chem.,* 88, 1930, 1984.
37. **Santerre, J. P., Hayakawa, K., and Kwak, J. C. T.,** *Colloids Surfaces,* 13, 35, 1985.
38. **Goddard, E. D.,** *Adv. Chem. Ser.,* 144, 67, 1975.

39. **Satake, I. and Yang, J. T.,** *Biopolymers,* 15, 2263, 1976.
40. **Satake, I., Gondo, T., and Kimizuka, H.,** *Bull. Chem. Soc. Jpn.,* 52, 361, 1979.
41. **Shirahama, K., Yuasa, H., and Sugimoto, S.,** *Bull. Chem. Soc. Jpn.,* 54, 375, 1981.
42. **Shirahama, K. and Tashiro, M.,** *Bull. Chem. Soc. Jpn.,* 57, 377, 1984.
43. **Hall, D. G.,** *J. Chem. Soc. Faraday Trans. 1,* 81, 885, 1985.
44. **Schwarz, G.,** *Eur. J. Biochem.,* 12, 442, 1970.
45. **Manning, G. S.,** *J. Chem. Phys.,* 51, 924, 1969.
46. **Zimm, B. H. and Bragg, J. K.,** *J. Chem. Phys.,* 31, 526, 1959.
47. **Delville, A.,** *Chem. Phys. Lett.,* 118, 617, 1985.
48. **Shirahama, K.,** paper presented to 35th Colloid Surface Chem. Symp., Kiryu, Japan, 1982.
49. **Goddard, E. D. and Benson, G. C.,** *Can. J. Chem.,* 35, 986, 1957.
50. **Schick, M. J.,** *J. Phys. Chem.,* 67, 1796, 1963.
51. **Adderson, J. E. and Taylor, H.,** *J. Colloid Sci.,* 19, 495, 1964.
52. **Goddard, E. D. and Leung, P. S.,** *Polym. Prepr. Am. Chem. Soc. Div. Polym.,* 23, 47, 1982.
53. **Leung, P. S., Goddard, E. D., Han, C., and Glinka, C. J.,** *Colloids Surfaces,* 13, 47, 1985.
54. **Goddard, E. D. and Hannan, R. B.,** in *Micellization, Solubilization and Microemulsions,* Vol. 2, Mittal, K. L., Ed., Plenum Press, New York, 1977, 835.
55. **Abuin, E. B. and Scaiano, J. C.,** *J. Am. Chem. Soc.,* 106, 6274, 1984.
56. **Bekturov, E. A., Kudaibergenov, S. E., and Kanapyanova, G. S.,** *Polym. Bull.,* 11, 551, 1984.
57. **Yui, T. S. T. I., Abilov, Zh. A., Pal'mer, V. G., and Musabekov, K. B.,** *Issled. Ravnovesnykh Sist.,* 78, 1982.
58. **Yui, T. S. T. I., Pal'mer, V. G., and Musabekov, K. B.,** *Izv. Akad. Nauk, Kaz. SSR Ser. Khim.,* 19, 1984.
59. **Ananthapadmanabhan, K. P., Leung, P. S., and Goddard, E. D.,** *Colloids Surfaces,* 13, 63, 1985.
60. **Zana, R., Yiv, S., Strazielle, C., and Lianos, P.,** *J. Colloid Interface Sci.,* 80, 208, 1981.
61. **Kalyanasundaram, K. and Thomas, J. K.,** *J. Am. Chem. Soc.,* 99, 2039, 1977.
62. **Kalyanasundaram, K. and Thomas, J. K.,** *J. Phys. Chem.,* 81, 2176, 1977.
63. **Turro, N. J. and Okubo, T.,** *J. Phys. Chem.,* 86, 159, 1982.
64. **Tseng, Y.-W. and Yang, J. T.,** *Biopolymers,* 16, 921, 1977.
65. **Satake, I. and Yang, J. T.,** *Biochem. Biophys. Res. Commun.,* 54, 930, 1973.
66. **Grourke, M. J. and Gibbs, J. H.,** *Biopolymers,* 5, 586, 1967.
67. **Li, L.-K. and Spector, A.,** *J. Am. Chem. Soc.,* 91, 220, 1969.
68. **Takeda, K.,** *Bull. Chem. Soc. Jpn.,* 58, 1210, 1985.
69. **Hayakawa, K., Ohara, K., and Satake, I.,** *Chem. Lett. Jpn.,* 647, 1980.
70. **Mattice, W. L. and Harrison, W. H.,** *Biopolymers,* 15, 559, 1976.
71. **Kubota, S., Ikeda, K., and Yang, J. T.,** *Biopolymers,* 22, 2219, 1983.
72. **Kubota, S., Ikeda, K., and Yang, J. T.,** *Biopolymers,* 22, 2237, 1983.
73. **Mattice, W. L., McCord, R. W., and Shippey, P. M.,** *Biopolymers,* 18, 723, 1979.
74. **Shirahama, K. and Yang, J. T.,** *Int. J. Pept. Protein Res.,* 13, 341, 1979.
75. **Maeda, H., Kato, H., and Ikeda, S.,** *Biopolymers,* 23, 1333, 1984.
76. **Takeda, K.,** *Hyomen,* 23, 351, 1985.

Chapter 5

POLYMER-SURFACTANT INTERACTIONS — RECENT DEVELOPMENTS

Björn Lindman and Kyrre Thalberg

TABLE OF CONTENTS

0-8493-6784-0/93/$0.00 + $.50

I. INTRODUCTION

An increasing interest is directed today towards polymer-surfactant systems, both for technical reasons (see Chapter 10), for the relevance of polymer-surfactant interactions in biology, and for the intriguing and fascinating features displayed by such systems, which continuously provide new challenges to the scientific community. Several review articles have appeared over the years,[1-4] of which one, written by Goddard in 1986,[2] is included as Chapter 4 in this volume. The present article will focus on the developments in the field during the last 5 years, but will also consider a number of general aspects.

There are quite different types of aqueous polymer-surfactant systems. Of major significance is the presence of charged groups on the polymer and/or on the surfactant, and the sign of these charges. A division of system types according to pattern of charges is given in Table 1, where P stands for polymer, S for surfactant, and the superscript indicates the charge. As is seen, nine different types are possible. Most studies have been carried out for systems of a water-soluble uncharged polymer (P^o) and an ionic surfactant (S^+ or S^-) and for systems of a polyelectrolyte ($P^{\pm}$) and an oppositely charged surfactant ($S^{\mp}$). Since the two kinds of systems display rather different features, a division of the area is clearly motivated and Goddard accordingly divides his review into two parts. Recently, a review article by Hayakawa and Kwak, dealing with the interaction between polymers and cationic surfactants (i.e., the systems P^o-S^+ and P^--S^+) has appeared.[4]

Other combinations of polymers and surfactants also have been subject to study, and the interactions of nonionic surfactants (S^o) with both uncharged polymers (P^o) and with polyelectrolytes ($P^{\pm}$) have been reviewed by Saito.[3] A few studies have been performed for systems containing a polyelectrolyte and a surfactant of the same charge. Besides, several reports on the interaction of surfactants with hydrophobically modified polymers have appeared in the last few years.

A second way of classifying polymer-surfactant investigations is according to the concentration regime in which they are carried out. This approach is pursued here. The majority of polymer-surfactant studies has been directed to relatively dilute solutions, and focuses on the thermodynamics of the interactions in terms of binding isotherms etc., and the structure of the formed polymer-surfactant complexes. The progress on these topics is discussed in Section III. The scope of polymer-surfactant investigations has been significantly broadened in the last few years, and most notable is a renewed interest in the phase behavior, which has led to an improved knowledge concerning phase separation phenomena in the water-rich parts of the phase diagrams. These findings will be presented in Section IV. Because of the broad range of systems studied, Sections III and IV are further divided into subsections according to the presence and sign of charges.

TABLE 1
Division of Polymer-Surfactant Types According to Charges

Uncharged polymer-anionic surfactant	P^o-S^-
Uncharged polymer-cationic surfactant	P^o-S^+
Uncharged polymer-nonionic surfactant	P^o-S^o
Anionic polymer-cationic surfactant	P^--S^+
Anionic polymer-anionic surfactant	P^--S^-
Anionic polymer-nonionic surfactant	P^--S^o
Cationic polymer-anionic surfactant	P^+-S^-
Cationic polymer-cationic surfactant	P^+-S^+
Cationic polymer-nonionic surfactant	P^+-S^o

Polymer-surfactant systems show a rich rheological pattern, which is one major reason for their many uses; novel systems showing highly viscous solutions or gels are discussed in Section V. Efforts made to investigate the influence of added polymer on microemulsions and liquid crystalline phases are described in Section VI. Finally, some studies on the adsorption from polyelectrolyte-surfactant solutions to surfaces are briefly described in Section VII. Unless novel techniques have been applied, experimental details and principles will generally be omitted; the reader is referred to the original work or to the review by Goddard (Chapter 4). However, brief mention is appropriate as regards a few newly introduced techniques, which have been extensively used and/or are expected to have particular prospects in future work.

A wide range of methods are available for the study of polymer-surfactant interactions. The suitability of a method may, however, differ largely between different systems. As regards the determination of binding isotherms, *surfactant selective electrodes* continue to be very important and development in this area will continue as they provide very basic information. As a complement, which also has a broader scope, *surfactant self-diffusion* studies are very relevant. From the surfactant self-diffusion coefficient, the concentrations of free and polymer-bound surfactant molecules may be directly deduced according to

$$D = c_f/c_tD_f + c_b/c_tD_b \tag{1}$$

where subscripts f and b refer to free and bound surfactant, respectively, and c_t is the total surfactant concentration. D_f is obtained from measurements in the absence of polymer. D_b is taken as the D value of the polymer-surfactant aggregate and may be obtained from studies of the polymer diffusion; except for low-molecular-weight polymers, the second term of Equation 1 may be neglected without introducing a significant error. Diffusion studies can then be used to obtain the distribution of surfactant between different environments. We do not expect any difference between the concentrations obtained here and the activities obtained in electrode work. A complication in both these approaches, as well as in others, is to account for a coexistence of polymer-bound and free surfactant micelles.

Self-diffusion coefficients can be obtained experimentally in several different ways — in previous work, mainly by using a technique based on radioactive labeling. However, the increased popularity of self-diffusion, in many areas of surfactant science, can be referred to the development of the Fourier transform nuclear magnetic resonance (NMR) pulsed gradient spin echo technique which allows the simultaneous determination of the D values of several (typically all) components of a complex mixture without labeling of the constituents.[5,6] If feasible, determination of the self-diffusion coefficients of the surfactant, water, and polymer molecules as well as of, if relevant, surfactant and/or polyelectrolyte counterions gives a quite complete description of the solutions: surfactant binding, counterion binding, and hydration. The use of self-diffusion is, however, not limited to these aspects; for concentrated solutions or gels it can provide unique structural information. (It has become a standard approach to establish the microstructure of microemulsions.)

More "genuine" NMR methods also continue to be developed to provide insight into molecular features of polymer/surfactant (PS) complexes. In general these are adopted from the field of NMR of simpler surfactant systems,[7] and this applies for two recent techniques based on *NMR relaxation*. In one, the effect of paramagnetic molecules on the relaxation of different parts of a surfactant molecule provides unique insight into geometrical features of a PS complex.[8] In another method, the relaxation of different parts of the chain (generally ^{2}H and/or ^{13}C nuclei) is studied at different magnetic fields.[9,10] Such data, analyzed by a theory assuming a time separation of the different motional processes responsible for relaxation,[11] provide (for positions along an alkyl chain) the following parameters:

1. A slow motion correlation time, corresponding, for example, to the time constant of the reorientation of the PS complex
2. A fast motion correlation time characterizing the local motions in the alkyl chain, for example gauche-trans isomerism
3. One order parameter characterizing the degree of orientation of the surfactant alkyl chains with respect to the micelle surface.

For the adsorption properties of surfactant-polymer complexes novel techniques are (*in situ*) ellipsometry[12] and direct force measurements;[13] so far applied to a very limited extent, but capable of providing detailed information on PS aggregation at solid surfaces and the forces between PS covered surfaces down to contact. Such information is vital in connection with the design of stabilizers for various dispersions.

Finally, it should be mentioned that the *fluorescence quenching technique*[14] manifests itself as a very important source of information for the aggregation numbers in a wide range of PS systems, both solutions and gels. An extension of this method is to covalently link the fluorescent probe to the polymer. By this technique, it is possible to obtain information about the polymer behavior in the PS interaction.

II. GENERAL ASPECTS OF INTERACTIONS AND MODELING

A. DRIVING FORCES

The interactions responsible for association phenomena in polymer-surfactant systems are mainly:

1. Hydrophobic interactions between polymer and surfactant molecules; this interaction will be particularly important for block copolymers with hydrophobic segments or for so-called hydrophobe-modified polymers
2. Hydrophobic interactions between surfactant molecules
3. Hydrophobic interactions between polymer molecules
4. Electrostatic interactions between polymer molecules
5. Electrostatic interactions between polymer and surfactant molecules; these may be attractive or repulsive, depending on whether the molecules have similar or opposite charges
6. Electrostatic interactions between surfactant molecules; these (repulsive) interactions are strongly unfavorable for surfactant micellization and a modification of them due either to amphiphilic portions of a polymer chain (''dilution effect'') or to the net charges of a polyion (neutralization) can dramatically facilitate surfactant self-assembly.

The different electrostatic interactions can generally be well understood and accounted for in terms of the entropy in the counterion distribution.

Of the different interactions mentioned, the main driving force for association in polymer-surfactant systems in general comes from the hydrophobic interactions between surfactant molecules as can be established in comparing surfactants with different alkyl chain lengths. Because of delicate energetical balances, even quite small modifications of the free energy of normal micellization and small contributions from the other interactions can have dramatic influences on the self-assembly and induce important surfactant-polymer interactions. Ionic surfactant micellization will, because of the electrostatic interactions, be particularly susceptible to changes. This is the reason why ionic surfactants in general interact strongly with polymers, while nonionic surfactants interact quite weakly or insignificantly.

B. SURFACTANT BINDING AND SELF-ASSEMBLY

There are different ways of approaching the problem of describing the equilibria in polymer-surfactant systems. First, we can discuss these systems in terms of an interaction of a polymer with either single surfactant molecules or with micelles; an interaction will, of course, change the distribution between the two states of the surfactant molecules. More basically, we may consider the interaction either in terms of the binding of surfactant molecules to a polymer molecule or in terms of the effect of the polymer chains on surfactant self-assembly.

In the first type of treatment, binding has generally to be considered to involve a certain degree of *cooperativity;* typically the inferred cooperativity is very high. The interaction is discussed in terms of the binding of the surfactant molecules to the polymer chains and binding is characterized by an equilibrium constant for binding to specific sites. To account for the cooperativity in the binding, lateral interactions are considered and different equilibrium constants introduced to account for a binding affinity, which depends on whether the next-neighbor position is occupied or not. While approaches like the Zimm-Bragg model have been very successful for the description of a number of phenomena displayed by polymer solutions,[15] they may be misleading and inconsistent with the main driving force for the case of polymer-surfactant interactions. Such models cannot explain some basic observations for polymer-surfactant solutions, like that of limited rather well-defined surfactant clusters.

As mentioned above, the dominating force in polymer-surfactant systems is the hydrophobic interaction among the surfactant chains. Therefore, a natural starting point of discussions of polymer-surfactant interactions appears to be to consider the effect of polymer molecules on surfactant self-assembly, notably micelle formation. Surfactant micellization is well understood,[16,17] which makes this approach particularly relevant and fruitful. We are led then to study the interactions in surfactant aggregates and it is known that by accounting for electrostatic interactions and simple geometrical packing effects, a good description of surfactant aggregates is obtained.[17] For example, the distribution of ionic micelles relative to a polyelectrolyte can be described by electrostatic theory, while the distribution relative to nonionic polymers has analogies with the effects of cosurfactants on micelles. The main effect of the polymer would, in general, be in its influence on the interaction between surfactant headgroups. The direct interaction of the surfactant with the polymer is, on the other hand, of secondary importance. Exceptions are evidently polymers, which from their structure can be judged to show specific interactions with a surfactant, such as polymers having alkyl chains, or other nonpolar groups, or polymers with a block character of nonpolar and polar blocks.

For cases where a "binding" approach is appropriate, it is instructive to obtain binding isotherms presenting the concentration of polymer-bound surfactant as a function of the surfactant activity (= free surfactant unimer concentration) or the total surfactant concentration. Binding isotherms are particularly appropriate for the case where the polymer affects surfactant self-assembly through short-range interactions. Then a plateau value and a saturation of binding result, and the formation of free micelles starts at a certain concentration above the saturation concentration. Coexistence of free and polymer-bound micelles then offers an appropriate description of the solutions. On the other hand, this will not necessarily be so for long-range interactions, like for polyion-ionic surfactant systems. Here it would typically be uncertain to assign a saturation in binding, or to distinguish between free and bound micelles.

The viewpoint of considering (cooperative) binding of surfactant molecules to a polymer has also other conceptual problems. While it is a natural approach for dilute solutions in the absence of long-range interactions, it needs to be modified for semidilute or concentrated polymer solutions. Here, a surfactant cluster can potentially interact with two or more polymer molecules. Surfactant self-assembly encompassing two polymer chains (or different parts of the same chain), which can be envisaged to have major influences on the properties, notably the rheological ones, depends strongly on the polymer-surfactant ratio.

For many of the polymers which have been investigated, and in particular the nonionic ones, polymer-polymer association phenomena are significant, which also causes problems with a simple model of surfactant molecules binding to single polymer chains. For many nonionic polymer solutions, the solvent quality is not good, as is evident from the closeness of a phase separation into one polymer-rich and one polymer-poor phase; for example, poly(ethylene oxide), nonionic block copolymers, and nonionic cellulose ethers phase separate at higher temperature. Here it is natural to consider the effect of surfactant on the clouding phenomenon and thus on the polymer-polymer association. Studies of cloud points offer a very facile and informative way of investigating the polymer-surfactant interactions in such systems.

The situation will be somewhat special for the case where the polymer chain either contains sites which interact specifically with surfactant molecules or when a polymer chain contains ''nucleation sites'' for micelle formation. The latter refers in particular to the so-called hydrophobe-modified polymers. Here special interactions can be envisaged. First, it is clear *inter alia,* from rheological investigations that these polymers self-associate in the absence of surfactant. This self-association may either be promoted or weakened by surfactant molecules. In particular, there will be ''mixed micelle'' formation between the polymer and the surfactant alkyl chains.

As regards the thermodynamic treatment of polymer-surfactant interactions, these systems thus span a broad range of cases from specific complex formation, where the driving force lies mainly in the direct polymer-surfactant interaction, to the situation where the polymer mainly influences the surfactant micellization. The main driving force of surfactant-polymer interactions is, however, in the majority of cases, the hydrophobic interaction between the surfactant alkyl chains; for hydrophobe-modified polymers, the hydrophobic interaction between polymer and surfactant alkyl chains will contribute more or less strongly, depending on the system. There is obviously an interrelation between specificity and degree of cooperativity such that for a higher degree of specificity, cooperativity will be less important and smaller surfactant aggregates may form (at least initially).

The variation in driving forces will have its counterpart in the structural description. For a specific complex formation, a considerable surfactant-polymer contact is expected, while in the case of surfactant-induced or -promoted micellization, the structure may be highly variable. In certain cases, like for polymers with amphiphilic segments or with nonpolar groups, the polymer chains may penetrate more or less deeply into the micelles, in others we expect the polymer chains to be mainly external to the micelle hydrocarbon core. For many polyelectrolyte-ionic surfactant systems, there will be just a long-range effect on micellization and the polymer chains and the surfactant micelles may remain at some distance.

C. EQUILIBRIA INVOLVED AND MODELING

These considerations emphasize the complexity of polymer-surfactant-solvent systems and the need to account for a large number of different interactions in any modeling and it is obvious that any understanding must be based on a very solid understanding of the different two-component systems, notably the polymer-solvent and surfactant-solvent systems. While these are treated in separate chapters in this volume, it is convenient to offer some basic comments[18] as a background for our discussion. As regards the polymer-solvent system, we have noted in particular that for many nonionic polymers, like those containing ethylene oxide groups, the solubility of the polymers (in water and formamide) decreases with increasing temperature. This must mean that the effective interaction between the polymer segments gets more attractive at higher temperatures. In the model of Karlström,[19] this is a consequence of the equilibrium between the polar and nonpolar conformations of the ethylene oxide groups. When temperature is increased, entropy populates the nonpolar conformations, since they can be realized in more ways than the polar ones, and thus the effective polarity of the polymer decreases with increasing temperature. In an alternative picture one may regard the phase separation as a consequence of a series of equilibria similar to those responsible for micelle formation (see below).

$$P + P \rightleftarrows P_2 \tag{2}$$

$$P + P_2 \rightleftarrows P_3 \tag{3}$$

$$P + P_n \rightleftarrows P_{n+1} \tag{4}$$

Note that we have regarded the phase separation process occurring in a polymer-solvent solution as an aggregate formation. This is only correct for infinite n values. A problem with these equations is the meaning of the complexes — P_2, P_3, etc. — in the semidilute concentration regime, but once the definition of these concepts has been agreed on it is obvious that an increased effective attraction between the polymer segments will favor the right-hand side of the equations.

Similar effects to those observed for the nonionic polymers on temperature increase can be observed for ionic polymers when salt is added to the system. Salt will screen the electrostatic repulsion between the polymer molecules, and if the backbone of the polymer system is hydrophobic enough, phase separation may occur.

The micellization in water is a highly cooperative process.[17,20] It can be regarded as a result of a set of equilibria

$$S + S \rightleftarrows S_2 \tag{5}$$

$$S + S_2 \rightleftarrows S_3 \tag{6}$$

$$S + S_n \rightleftarrows S_{n+1} \tag{7}$$

In systems which are coupled in this way it is well known that for a suitable choice of equilibrium constants, the concentration of S_n for n roughly 100 may increase dramatically when the free S concentration is only slightly increased.[17] Micelle formation for ionic surfactants can be understood from the interplay between hydrophobic forces, which try to minimize the water-hydrocarbon contact, and electrostatic forces, which try to maximize the ion-water contact. If the effect of the counterions on the micellization process is considered, an observation, which at first may seem somewhat confusing, can be made.[21] It is then possible to write the equilibrium conditions for micelle formation

$$nS + \alpha nI \rightleftarrows S_nI_{\alpha n} \tag{8}$$

$$[S_nI_{\alpha n}]/[S]^n[I]^{\alpha n} = K \tag{9}$$

where I denotes the counterion and α is the degree of counterion binding. However, such a treatment does not predict some striking features of ionic surfactant solutions. This is due to not considering the long-range electrostatic interactions. One important observation is that of a closely concentration-invariant degree of counterion association, a phenomenon also encountered for polyelectrolyte systems and usually denoted *counterion condensation.* Another is that of a decreased surfactant monomer concentration and activity at increasing the surfactant concentration above the critical micelle concentration (CMC). This can be understood from the unfavorable contribution to micellization, resulting from the negative entropy term arising from the inhomogeneous counterion distribution, and its partial elimination with increasing surfactant concentration due to the increased counterion concentration in the bulk.[21]

The starting point for the discussion of the interactions between a nonionic polymer and an ionic surfactant may be that both these species contain fairly hydrophobic parts, and that they may take part in cooperative processes, which are induced by the interaction between these hydrophobic domains. Since the origin for the cooperativity is the same (hydrophobicity) it seems reasonable to assume that there may be an interplay between the processes behind the cooperativity. Thus, it seems plausible to write, in accordance with Equations 5 to 7:

$$S + P \rightleftarrows SP \tag{10}$$

$$S + SP \rightleftarrows S_2P \tag{11}$$

$$S + S_n \rightleftarrows S_{n+1}P \tag{12}$$

The process illustrated in these equations is the formation of a micelle bound to a polymer. Often it is assumed that the aggregation number for these micelles is somewhat smaller than that for free micelles. It is also standard to assume that the amount of aggregates of type S_2P, S_3P, and so on, where no real micelles are formed, is so small that their contribution to the amount of bound surfactant can be neglected. On the other hand, we may also write, in accordance with Equations 2 to 4:

$$P + S \rightleftarrows PS \tag{13}$$

$$P + PS \rightleftarrows P_2S \tag{14}$$

$$P + P_nS \rightleftarrows P_{n+1}S \tag{15}$$

Here we have considered the effect of the surfactant on the self-association (underlying clouding phenomena) for the polymer-solvent system. If the ionic part of the surfactant is ignored, one would thus expect that addition of surfactant molecules would lower the cloud point temperature. If an ionic surfactant in very low concentrations is added, it seems reasonable to assume that it induces the same type of aggregation, but that the electrostatic repulsion between the aggregates prevents phase separation. Salt may, however, screen this long-range repulsion. From this discussion it seems necessary to consider the cooperativities in both the polymer system and the surfactant system and to write for the general complex formed:

$$nS + mP \rightleftarrows S_nP_m \tag{16}$$

Complexes with large m values and small n values are apparently important close to the clouding temperature and at low surfactant concentrations. One may also expect that complexes with high n values and low m values are most important at low temperatures and high surfactant concentrations.

One of the standard ways to model a polymer solution is to use Flory-Huggins (F-H) theory.[22] Within this model the mixing enthalpy can be written

$$U = N\Phi_1\Phi_2 w_{12} \tag{17}$$

where w_{12} is the effective interaction parameter between a polymer segment and a neighboring solvent molecule. Note that the w used here differs from the ordinary χ parameters used in F-H theory, in that the trivial temperature dependence has been removed. N is the total number of cells (solvent molecules + polymer segments in the system) and Φ_1 and Φ_2 the volume fractions of solvent and solute. The effective interaction parameter w_{12} is related to the direct interaction parameters w'_{ab} according to

$$w_{12} = w'_{12} - (w'_{11} + w'_{22})/2 \tag{18}$$

The corresponding entropy expression is

$$S = -kN\,(\Phi_1 \ln\Phi_1 + \Phi_2/M_2 \ln \Phi_2) \tag{19}$$

where k is Boltzmann's constant and M_2 the degree of polymerization of the polymer. As described in Reference 23, Equations 17 and 19 can be modified easily to include a two-conformational description of the polymer segments. Equations 17 and 19 can also be generalized to include the effect of a second polymer. One then obtains

$$U = U' + \Phi_1\Phi_3 w_{13} + \Phi_2\Phi_3 w_{23} \tag{20}$$

$$S = S' - kN\Phi_3/M_3 \ln \Phi_3 \tag{21}$$

U' and S' in these equations are defined as U and S according to Equations 17 and 19.

One possible starting point in modeling polymer-surfactant systems is to use the F-H theory and it is then found that it is necessary to treat the surfactant as a second polymer. However, the model is extremely crude in its treatment of the interactions in the system, and the limitations must be borne in mind, not the least for systems with long-range electrostatic (or other) interactions.

III. SOLUTIONS

A. GENERAL ASPECTS

1. Binding Isotherms

As discussed by Goddard in Chapter 4, it was recognized early that ionic surfactants "bind" to uncharged water-soluble polymers. Although no spe-

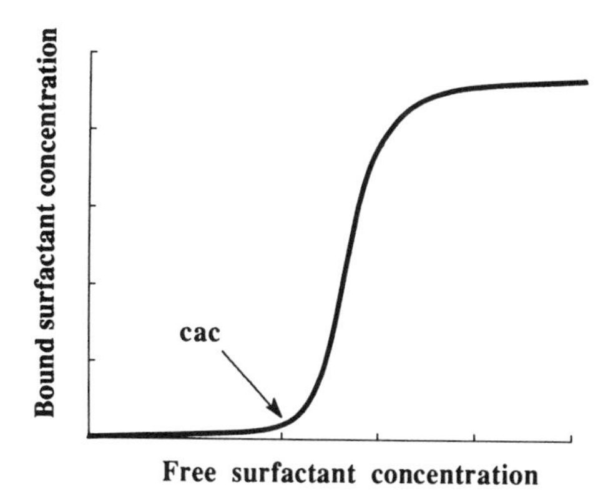

FIGURE 1. A typical binding isotherm, for the binding of a surfactant to a polymer. The critical aggregation concentration, cac, is indicated.

cific binding sites for the surfactant molecules have been identified on these polymers (as compared to proteins), the notion of surfactant "binding" to the polymer, formerly used for protein-surfactant interactions, has been adopted also for polymer-surfactant systems. Methods that can monitor the amount of polymer-bound surfactant directly as a function of the total surfactant concentration have significantly contributed to our present understanding of polymer-surfactant interactions. In this way, a characterization of the surfactant binding to the polymer (at a fixed polymer concentration) can be obtained and is usually represented as a "binding isotherm". The development of surfactant-specific electrodes has made possible rapid progress in this field (see Chapter 4). A typical binding isotherm is shown in Figure 1. The isotherm has a marked sigmoidal shape, which is an indication of cooperative binding, and the onset of surfactant binding often occurs at a certain, rather well-defined, surfactant concentration, which we will call the critical aggregation concentration, denoted *cac*. This notion, which was introduced by Chu and Thomas,[24] making an analogy with surfactant micellization, indicates that the surfactant molecules form aggregates upon interacting with the polymer chains. (The symbols T_1, c_1, and x_1 have also been used to denote the onset of binding.) In general, a steep binding isotherm (a large cooperativity) is expected if the polymer-adsorbed surfactant molecules form micelles which are similar to normal free micelles. Sometimes, the free surfactant concentration at the midpoint of the binding isotherm is used instead of the cac to characterize the interaction. (This value may be obtained with great accuracy from a statistical mechanical analysis of the binding isotherm.[4]) For a highly cooperative binding there will be little difference between this value and the cac.

The leveling out of the binding isotherm at higher surfactant concentrations is due to saturation of the polymer with surfactant and indicates the maximum amount of surfactant that can be bound per polymer unit. For many systems, this level is not reached due to phase separation, or is indistinct or obscured by the formation of free micelles.

2. Strength of Interaction

A description in terms of a cac value clearly presupposes a driving force similar to that of normal surfactant micellization and a strongly cooperative binding. The degree to which the cac is lower than the CMC depends on the magnitude of the interaction, but both local and long-range interactions may cause a major lowering.

To a first approximation, adopting the phase separation model for micelle formation,[17] the (molar) free energy of micellization can be written

$$\Delta G^o_m = RT \ln \text{cmc} \tag{22}$$

and the free energy of surfactant binding to a polymer is analogously expressed as

$$\Delta G^o_b = RT \ln \text{cac} \tag{23}$$

Therefore, we can derive the free energy per mole surfactant for the reaction: free micelle ⇄ polymer-bound micelle, as

$$\Delta G^o_{PS} = RT \ln (\text{cac}/\text{cmc}) \tag{24}$$

This quantity is a convenient measure of the strength of the interaction between the surfactant and the polymer. In particular, it is pointed out that it is impossible to have a cac which is higher than the CMC of the surfactant — in that case the surfactant molecules would prefer to form free micelles instead of polymer-bound ones and surfactant binding to the polymer will not occur.

The question of whether polymer-surfactant interactions need to be accompanied by a decreased CMC has been considered by different authors, most recently by Brackman and co-workers.[25,26] These authors present results which suggest a significant interaction with a cac equalling the CMC of the surfactant. In our opinion, such a situation would only apply for the, somewhat unlikely, situation where the surfactant would bind in clusters and as single molecules with equal affinity. If the polymer preferentially interacts with a micelle, the free energy of the micelle, and thus the CMC, should be lowered. The interesting work of Brackman et al. should clearly be followed up both by further experimental work and by a closer analysis of the general problem.

On the other hand, while a lowering of the CMC demonstrates an interaction, it is not necessarily an indication of a complex formation between the polymer and the surfactant. For example, addition of salt also reduces the CMC, and addition of polyelectrolyte of the same charge as the surfactant has a similar effect (see below).

The strength of the interaction varies considerably between the different types of polymer-surfactant systems. In particular, the interaction between a polyelectrolyte and an oppositely charged surfactant is strongly favored, and a decrease of the cac relative to the CMC by several orders of magnitude is commonly observed. It is pointed out, however, that also in the case of such strong interactions, the surfactant molecules retain a large motional freedom within the aggregate, as demonstrated by NMR relaxation measurements (see Section III.C).

B. INTERACTION BETWEEN NONIONIC POLYMERS AND IONIC SURFACTANTS

Systems of uncharged polymer and ionic surfactant continue to be the main type of polymer-surfactant systems studied. Many (but not all) of the neutral polymers investigated have been reported to interact with ionic surfactants. In Chapter 4, Part I, the different techniques employed are presented and the influence of various parameters, such as the properties of the surfactant and the polymer, the temperature, and addition of simple salt, is described. Noteworthy is the finding that cationic surfactants interact significantly more weakly than their anionic analogs. Before discussing the recent findings, we will briefly summarize the general features of the interaction for this kind of system.

1. Effect of the Polymer

The binding isotherm for an ionic surfactant interacting with an uncharged water-soluble polymer is typically of the kind shown in Figure 1, with cac values normally up to one order of magnitude lower than the CMC of the surfactant. As an example we take the most studied polymer-surfactant system, that of poly(ethylene oxide), PEO, and sodium dodecyl sulfate, SDS, where the cac is reduced from 8 m*M* (the CMC) to approximately 4 m*M*,[27,28] corresponding to a free energy of -1.7 kJ/mol of surfactant (at room temperature). It should be noted that the cac is largely independent of polymer concentration. The interaction is enhanced by an increased hydrophobicity of the polymer. Thus, the cac for SDS with poly(vinylpyrrolidone), PVP, is about 2.5 m*M*,[29] corresponding to a free energy of -2.8 kJ/mol. (In Table 2 the commonly used uncharged polymers are listed and their abbreviations are given.)

The influence of added salt on the interaction is relatively small. At 0.1 *M* of NaCl, the cac of the PEO-SDS system is about two thirds of the CMC (at this salt concentration),[30] and also at 0.4 *M* of NaBr, the interaction seems to be quite similar to the salt-free case.[31] As will be seen, this is in sharp contrast to the influence of salt on the interaction between polyelectrolyte and oppositely charged surfactant (Section IV.B).

The molecular weight of the polymer has little influence on the interaction, as long as it is above a certain minimum value. For the PEO-SDS system, all PEO species with a molecular weight above 4000 (about 90 EO units) have approximately the same free energy of interaction, while for shorter chains, a reduction in the interaction is observed. The lower limit of the polymer molecular weight required for a strong interaction may, however, differ between different systems.[25,32]

TABLE 2
Uncharged Polymers Used in the Study of Polymer-Surfactant Interactions

Poly(ethyleneoxide)	PEO
= Poly(ethyleneglycol)	(PEG)
Poly(propyleneoxide)	PPO
Poly(vinylmethylether)	PVME
Poly(vinylpyrrolidone)	PVP
Poly(vinylalcohol)	PVA
Poly(vinylacetate)	PVAc
Poly(*N*-isopropylacrylamide)	PNIPAAM
Hydroxypropylcellulose	HPC
Ethylhydroxyethylcellulose	EHEC

2. Effect of the Surfactant

While alkyl sulfates, and in particular SDS, were almost exclusively used in the past, a comparison of these with alkyl phosphates has recently been made.[32] Sodium alkylphosphates were seen to interact more weakly than their sulfate analogs. In the same work, the charge of the alkyl phosphate headgroup was varied. An increase in charge from one to two results in a marked reduction of the interaction with PEO.

Chari and Lenhart compared the interaction of SDS and the double-chained surfactant sodium bis(-ethylhexyl)sulfosuccinate (Aerosol OT) with PVP.[33] Both surfactants form micelles adsorbed to the polymer, but the free energy of binding of Aerosol OT to PVP was found to be only half of that of SDS.

By ^{1}H NMR, Gao et al. studied the aggregation behavior of ω-phenyldecanoate and its interaction with PEO.[34] The free energy of binding is similar to that of SDS. Chemical shifts for the PEO protons indicate that the adsorbed polymer segments are in close proximity to the phenyl groups.

The counterion of the ionic surfactant is known to be of great importance. However, no systematic study of this effect has appeared since the work of Saito and Yukawa.[35] Treiner and Nguyen investigated the interaction of $Cu(DS)_2$ with PEO and PVP using a Cu^{2+}-specific electrode.[36] The surfactant forms micelles adsorbed to PEO and the degree of counterion binding decreases with an increasing polymer concentration. A second transition in the binding behavior was attributed to the formation of free micelles, in equilibrium with the polymer-bound ones. Surprisingly, the surfactant concentration at the onset of the second transition was found to decrease with an increasing polymer concentration.

3. Anionic vs. Cationic Surfactants

Several reports have dealt with the difference between anionic and cationic surfactants in their binding to uncharged polymers. With few exceptions (see Chapter 4), anionic surfactants exhibit significantly stronger interactions than cationic ones of a similar chain length. As an example, hexadecyltrimethylammonium bromide, $C_{16}TAB$, does not interact appreciably with PVP and only very weakly with PEO, while for the more hydrophobic polymers PVME and PPO, a somewhat stronger interaction is observed. Several recent studies have been directed to the interaction of cationic surfactants with these polymers, and various techniques such as surfactant-specific electrodes,[37,38] molar volumes and heat capacities,[39] electric birefringence,[40] conductivity, electron spin resonance and luminescence quenching of $Ru(bipy)_3$,[25,41,42] have been used. All these investigations confirm the picture, indicating that the weak interaction between polymers and cationic surfactant is a quite general phenomenon. This is an intriguing finding and several explanations have been advanced for the origin of this difference. The first and most popular one is that the large headgroups of the cationic surfactants studied (e.g., the trimethylammonium group) will reduce the access of the polymer to the surface of the cationic micelles, and this mechanism has been modeled theoretically by Nagarajan[43,44] and by Ruckenstein et al.[45] (These theories are presented in Section III.G.) It can, however, be noted that also (primary) alkylammonium halides interact more weakly with polymers than do comparable anionic surfactants, in spite of their small headgroups. Shirahama et al.[38] report that dodecylammonium chloride ($C_{12}AC$) interacts more strongly than dodecylpyridinium chloride, $C_{12}PC$, with PVA90 (i.e., PVA with 10% of the OH groups acetylated), and this was ascribed to the difference in headgroup size (these surfactants have quite similar CMC values[46]). The size of the surfactant headgroup may have a certain influence on the interaction with polymers, but is only part of the explanation as to why anionic surfactants interact more strongly than cationic ones with uncharged polymers. A more systematic variation of the headgroup size for both cationic and anionic surfactants would be needed to enlighten this question. In cloud point (cp) measurements it is found that anionic surfactants interact much more strongly with EHEC than cationic ones in water[47] while the reverse is true with formamide as a solvent.[48] It was also found that in water, $C_{12}NH_3^+$ interacts similarly to $C_{12}N(CH_3)_3^+$ but much more weakly than analogous anionic surfactants, even those having larger headgroups.

In another explanation of the difference between anionic and cationic surfactants, a fractional positive charge on the polymer molecules has been invoked, which is thought to originate from protonation of the ether oxygens in the case of polyethers and the amide moiety in the case of PVP. The very low pK_a value of an ether or amide, however, raises serious doubts about the importance of protonation at neutral pH. A third explanation, recently proposed by Witte and Engberts,[42] suggests that there is a difference in the interactions of cations and anions with the hydration shell of the polymer, which would favor the interaction of anionic surfactants.

We believe that a fundamental approach is necessary in order to obtain a better understanding of the difference between anionic and cationic surfactants in the interaction with uncharged polymers. In particular, the discussion should be based on the general self-assembly of the two types of surfactants and comparing the interactions in different solvents. Factors to consider are the slightly higher CMC for a cationic surfactant and the higher degree of counterion binding for a simple cationic micelle. The latter can be referred to as a relatively weak hydration of the counterions. On the basis of differences between solvents, it has been suggested that the solvent molecules orient at the polymer surface to create a weak surface charge density around the polymer, which would favor binding of an oppositely charged surfactant.[48]

4. Surfactant Aggregate Structure

A structure of micelle-like surfactant clusters bound to the polymer chains is the general picture emerging from the large body of work on polymer-surfactant systems (see Chapter 4). The majority of investigations directed to the structure of the polyelectrolyte-surfactant complexes have focused on the properties of the surfactant aggregates; notably their aggregation numbers, but also the micropolarity in the adsorbed micelles and the dynamics of the surfactant molecules. Recently, however, a number of studies which have been concerned also with the behavior of the polymer have appeared, as will be discussed below.

A change in surfactant aggregation number compared to the value in the absence of polymer is an important indication of a polymer-surfactant interaction. Several techniques have been applied for the determination of aggregation numbers, and, in particular, techniques which make use of the solubilization of fluorescent probes in the micelles have become increasingly popular. Certainly, the addition of a probe may alter the behavior of the system as well as the structure, but comparisons with other methods have shown that, in most cases, the influence of an added probe (in submillimolar concentration) on the aggregates is small.

Generally, a decrease in the aggregation number, relative to the surfactant-water system, is reported, but it is important to consider that the polymer-adsorbed surfactant aggregates will undergo changes as the surfactant concentration in the system is increased. In the SDS-PEO system, the aggregation number changes with concentration from about 20 at the cac to a value close to the aggregation number of normal micelles (about 70), as has recently been demonstrated by the use of different probes.[42,49]

Of particular interest concerning the structure of the surfactant aggregates are the investigations by Brackman and Engberts[50,51] of the influence of polymers on rod-like micelles. Both cationic (C_{16}TASal)[50] and anionic[51] rod-like micelles have been investigated, and in both cases addition of polymer (PVME, PPO, PEO) induces a breakdown of the rods and formation of small, almost spherical, micelles.

The steady-state spectra of the fluorescent probes can give information about the polarity of the local environment. For pyrene, it is the ratio between the first and the third vibronic peaks, denoted I_1/I_3 (or its inverse), which is used. This ratio is quite sensitive to the local polarity and changes from about

1.8 in water to 1.25 to 1.33 in micelles. This is, per se, a simple and versatile method for the determination of cac values, as is further described in Chapter 4. By this method, Winnik et al. could establish complex formation between HPC and C_{16}TAC.[52] Witte and Engberts adopted an electron spin resonance (ESR) probe method to systems with PEO or PPO and SDS or C_{16}TAB.[41] It was concluded that the spin probes experience a more polar microenvironment and reorient faster in the polymer-adsorbed micelles as compared to the ordinary free micelles, i.e., the polymer-surfactant complexes have a more "open" structure.

The possibility for binding of surfactant monomers or very small aggregates to the polymer at concentrations below the CMC has been indicated in several papers.[38,47,53,54] Clearly, there is always a distribution equilibrium for the surfactant monomers between the polymer and the bulk. The structural details of such a "preassociation" are difficult to measure due to the extremely low concentrations of preassociated molecules. An approach to overcome this difficulty may be to use surfactant molecules which possess a fluorophore, as has been demonstrated by Schild and Tirrell.[53] Preassociation of 2-(*N*-dodecylamino)naphthalene-6-sulfonate to PNIPAAM could thus be observed, and it was inferred that the micropolarity of the preassociated monomers is lower than for the fluorophore in SDS micelles adsorbed to the polymer. For the system of amylose and sodium myristate, polymer-associated surfactant monomers have been proposed.[54] From surface tension and optical activity measurements, a first binding step including helix formation of the polymer around the surfactant molecules is suggested, which at higher surfactant concentrations is followed by an ordinary second binding step, leading to the adsorption of surfactant micelles to this complex.

As regards the size and size distribution of surfactant aggregates, it was suggested above that they should depend strongly on polymer hydrophobicity, and in particular on the presence of strong hydrophobic groups, but it seems that this aspect has not been systematically investigated.

5. Polymer Conformation and Association

Information about the conformational changes which the polymer molecules undergo on interacting with a surfactant has been obtained mainly from viscosity measurements. In general, a marked increase in the viscosity for a dilute solution, corresponding to a more extended conformation of the polymer chains, occurs as a result of the binding of mutually repelling charged micelles. However, a decrease in the viscosity may also occur (see also Chapter 4, Part I) as found recently for the system of EHEC-SDS by Holmberg et al.[55] It seems natural to suggest that a decreased viscosity on surfactant binding in dilute polymer solutions can be referred to surfactant-induced (or enhanced) intrachain attractions. This is in line with observations for semidilute solutions of a very pronounced maximum in viscosity as a function of surfactant concentration, and even of the formation of gels (to be described in Section V).

The clouding behavior of certain polymers is likely to be altered by the interaction with surfactants. PPO and many cellulose ethers display this kind of behavior, i.e., the polymer solutions turn turbid above a certain temperature, denoted the cloud point, cp (which depends on the polymer concentration). Binding of ionic surfactant to such polymers generally leads to an increase in cp, due to the charging up of the polymer chains. Schild and Tirrell have thoroughly investigated the cp behavior in systems of PNIPAAM and alkyl sulfates of varying surfactant chain length.[56] Cloud point measurements constitute a facile approach to the interactions in polymer-surfactant systems and can be used to obtain phase diagrams. They are considered further in Section IV.A.

Ricka et al.[57] studied the PNIPAAM-alkyl sulfate system using dynamic light scattering. The collapse of the polymer chains at the clouding temperature could be easily observed by this method, and it was concluded that surfactant binding starts at a well-defined surfactant concentration and entails a drastic increase in the dimensions of the polymer chains.

Covalent labeling of the polymer with fluorescent probes is another method for the study of conformational changes in the polymer, which has become increasingly popular.[52,58] For the most used probe, i.e., pyrene, it is the ratio between the excimer and monomer emission peaks of the fluorescence spectra, I_E/I_M, that is considered. In the absence of surfactant, the pyrene moieties from the same or different chains interact and form excimers. This tendency is normally reduced on addition of surfactant, indicating solubilization of the pyrene groups in surfactant aggregates and suggesting an uncoiling of the polymer chains. It should be noted that the surfactant binding capacity of the polymer will be somewhat altered by the incorporation of hydrophobic probes, but for the present example, parallel studies were carried out with the unlabeled polymers and unbound pyrene,[59] and the perturbation due to pyrene labeling was seen to be negligible.

An interesting investigation by Gao et al.[8] on the behavior of the polymer in the PEO-SDS system makes use of paramagnetic NMR relaxation. A paramagnetic anion is dissolved into the aqueous phase, and is repelled from the SDS-PEO micelles. The PEO segments which are associated with the micelles are much less affected than the PEO segments in the bulk water by the paramagnetic probe, and the contributions from the two states to the observed relaxation can be resolved. The method thus provides the fractions of micelle-bound and unbound PEO segments, and by varying the ratio between SDS and PEO, the distribution of PEO segments as a function of composition can be obtained. At low polymer segment-to-surfactant ratios, between 80 and 90% of the PEO segments are directly bound to the SDS micelles, but above a PEO/SDS ratio of 1, this fraction starts to decrease. The average number of bound PEO segments per micellar surfactant molecule increases with increasing PEO/SDS ratio up to a plateau value of about 1.9. The interaction is reduced when the PEO molecular weight falls below 4000, and for tetraethylene glycol, TEG, no interaction was found. If the hydroxyl endgroups of TEG are replaced by methyl ethers, however, a weak interaction could be observed. In general, the results obtained by Gao et al. support a picture of SDS micelles with PEO segments on the surface, and gives further insight into the detailed organization of the complexes. The degree of direct binding of the PEO segments to the micelles is significantly larger than reported earlier; at a PEO/SDS ratio of 3.3, 50% of the PEO segments were found to be adsorbed to the micelles, while Cabane at this PEO/SDS ratio inferred a structure with only about 10% of the PEO monomers directly adsorbed to the micelle, the remainder forming loops in the surrounding aqueous phase.[60]

C. POLYELECTROLYTES AND OPPOSITELY CHARGED SURFACTANT

1. Introduction

Systems of a polyelectrolyte and an oppositely charged surfactant have been extensively studied in dilute solution, and due to the strong attraction between the two species, the interaction starts at very low surfactant concentrations and is difficult to study by most conventional techniques. The development of surfactant-sensitive electrodes was the remedy to this problem, and by this technique, Kwak and co-workers have systematically investigated a great number of systems containing cationic surfactant and anionic polymer, of both biological and synthetic origins. Their earlier results are discussed in Chapter 4, Part II, together with other work on this type of systems. Recently, a review article by Hayakawa and Kwak has appeared,[4] in which, among other things, the details of the surfactant electrode technique are given.

Often, binding of a cationic surfactant to polyanions starts at a concentration which is several orders of magnitude lower than the CMC in polymer-free solution. We note that the surfactant binding is highly cooperative also in these systems, pointing to the contribution from interactions among the adsorbed surfactant molecules and the formation of polymer-adsorbed micelles or micelle-like clusters. A popular approach for this kind of system is to apply the Zimm-Bragg model[61] or a similar treatment, in which the cooperativity in surfactant binding can be evaluated. Such treatment gives a good characterization of part of the binding behavior. (It should not, however, be used as a picture for the structure of the formed complexes, as will be discussed in Section III.H).

Before going into the details of particular systems, we want to point out some general features of relevance for systems of polyelectrolyte and oppositely charged surfactant. First, it should be noted that, due to general electrostatic interactions, polyelectrolytes give rise to an uneven distribution of ions in a solution. The concentration of counterions is strongly enhanced close to the polyelectrolyte and decays rapidly with an increased distance from it. The uneven distribution of counterions also applies to monomeric surfactant counterions, present in the polyelectrolyte solution. Thus, the concentration of surfactant ions is enhanced close to the polyelectrolyte in systems of polyelectrolyte and oppositely charged surfactant. This general electrostatic effect is not related to the amphiphilic nature of the surfactant ions and is, therefore, of little interest here. Knowledge of its magnitude, for instance using the Poisson-Boltzmann theory, can be quite important for a correct interpretation of data, as the electrostatic interactions influence the activity of the surfactant ions.

The major reason for cooperative binding of surfactant molecules to an oppositely charged polyelectrolyte is the electrostatic stabilization of the surfactant micelles. We may thus picture surfactant ''binding'' to polyelectrolytes to a substantial degree as counterion binding of the polyelectrolyte charges to the surfactant micelle. For hydrophilic polyelectrolytes, such as polysaccharides, there is no driving force for the polymer segments to penetrate into the hydrophobic interior of the micelle. The situation is thus very different from systems of an uncharged polymer, where forces other than pure electrostatic ones play a major role. The situation will be altered if the polyelectrolyte also contains hydrophobic moieties, as is discussed below.

2. Influence of the Surfactant

Mainly two types of cationic surfactants have been used, i.e., alkyltrimethylammonium halides (abbreviated C_nTAX, where n indicates the number of carbons in the alkyl chain and X is C for chloride and B for bromide) and alkylpyridinium halides (abbreviated C_nPX). The latter kind shows slightly stronger interaction with polyanions.[4] The length of the hydrocarbon tail of the surfactant is a crucial parameter for the interaction with a polyelectrolyte. This is illustrated in Figure 2, where cac values are given for systems of C_nTAB and the anionic polysaccharide hyaluronan, Hy, used as a sodium salt[62] (i.e., sodium hyaluronate; for abbreviations of commonly used polyelectrolytes, see Table 3). An increasing difference is seen between log(CMC) and log(cac) when the surfactant chain length is increased, indicating that the interaction is enhanced for a surfactant of longer hydrocarbon tail. This can be attributed to the uneven distribution of counterions between the bulk and the micellar surface, which is unfavorable for the formation of normal micelles and which is more pronounced for a longer surfactant; a lower CMC gives a lower intermicellar electrolyte concentration. The binding of an oppositely charged polyelectrolyte to the micelles entails a release of the ordinary counterions, and therefore, leads to a larger increase in the entropy of the counterions for a surfactant of longer chain length.[62] It can be noted that for surfactants with less than a certain number of carbons in the alkyl chain (10 for the Hy-C_nTAB system, Figure 2), there will be no binding to the polyelectrolyte. In this case normal free micelles are favored relative to the polyion-adsorbed ones.

A second way of conceiving the interaction is to consider the formation of polyelectrolyte-surfactant aggregates with a relatively low charge (as compared to normal free micelles). The surfactant chain length dependence for the formation of these aggregates will therefore be similar to the alkyl chain

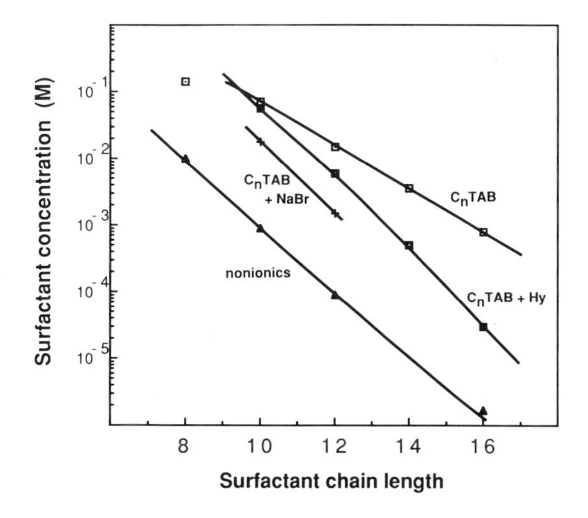

FIGURE 2. cac values for C_nTAB and Hy at 25°C (filled squares) compared with the CMC values of the surfactants (open squares). Also included are CMC values for these surfactants in the presence of 0.5 *M* NaBr (+) and CMC values for nonionic surfactants of the $C_n(EO)_6$ type (filled triangles). (From Thalberg, K. and Lindman, B., *J. Phys. Chem.*, 93, 1478, 1989. Reproduced with permission from the American Chemical Society.)

TABLE 3
Polyelectrolytes Used in the Study of Polymer-Surfactant Interactions

Poly(acrylate)	PA
Poly(methylacrylate)	PMA
Poly(styrenesulfonate)	PSS
Poly(vinylsulfate)	PVS
Dextran sulfate	DxS
Hyaluronan	Hy
Poly(diallyldimethylammonium chloride)	PDADMAC
Copolymers of maleic acid and	
Ethylene	PMAE
Butylvinylether	PMABVE
Styrene	PMASt
Indene	PMAIn

length dependence for the formation of nonionic surfactant micelles, as is clearly seen in Figure 2.

Little attention has been paid to the significance of the surfactant counterion. Some interesting results were obtained by Binana-Limbelé and Zana,[63] who, in their investigation, included the chloride, bromide, and iodide salts of the decyltrimethylammonium ion. From pyrene solubilization results, the cac value for the interaction with PMABVE is inferred to be lowest for the chloride specimen, followed by the bromide and the iodide. This is exactly opposite to the order between the CMC values for these surfactants.

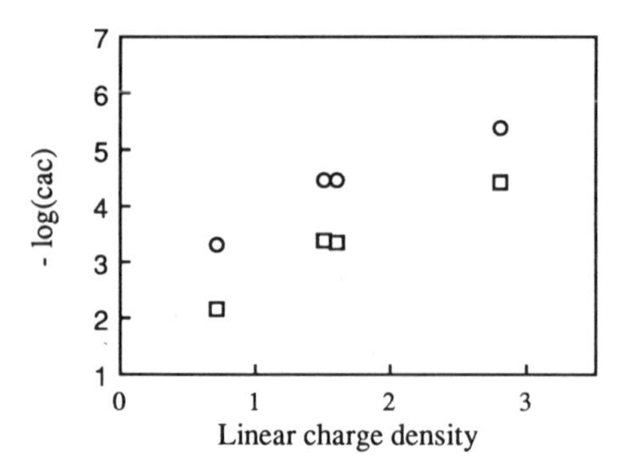

FIGURE 3. −Log (cac) for $C_{12}TAB$ (squares) and $C_{14}TAB$ (circles) and polyelectrolytes of varying the reduced linear charge density, ζ. The polyelectrolytes are PA (ζ = 2.8), alginate (ζ = 1.5), and pectate (ζ = 1.6) (data from Reference 4) and Hy (ζ = 0.7) (data from Reference 62).

3. Influence of the Polyelectrolyte

The properties of the polyelectrolyte are also of major importance for the interaction with surfactants. A variety of both synthetic polyions and polyelectrolytes of biological origin have been investigated, as is seen in Table 3. One important parameter is the reduced linear charge density of the polyelectrolyte, ζ, which is related to the distance, b, between adjacent charges along the polymer backbone by the relation

$$\zeta = e^2/4\pi\epsilon bkT \tag{25}$$

cac data for different systems as a function of ζ are given in Figure 3. The trend is clearly that an increase in linear charge density gives rise to a stronger interaction. It is, however, apparent that for polyelectrolytes of a given charge density, cac can still vary significantly. For example, poly(styrenesulfonate), PSS, shows a stronger interaction than dextran sulfate, DxS, which in turn interacts more strongly than poly(acrylate), PA, although these polyelectrolytes all have a similar linear charge density.[4] Apparently, other forces also play a role besides the pure electrostatic ones, and modulate the interaction between the polyelectrolyte and the surfactant micelles. Such contributions may arise from different types of charged groups or from the presence of hydrophobic moieties on the polyelectrolyte. In addition, the flexibility of the polymer backbone and the type of counterions present may influence the interaction.

The cooperativity in surfactant binding is also influenced by the charge density of the polyelectrolyte. Recently, Satake et al.,[64] who studied the interaction between $C_{12}TAB$ and partially hydrolyzed PVS, reported that the cooperativity decreases rapidly with a reduced charge density of the polyelectrolyte. In contrast to this, a marked cooperativity is observed for the binding of $C_{10}TAB$ to Hy, which has a relatively low charge density.[62,65] In this case the cac is quite close to the normal CMC, and the polymer-adsorbed surfactant aggregates are quite similar to normal free micelles. Further investigations would be welcome in order to clarify the relation between the cooperativity and the strength of interaction. The cooperativity also depends on other features of the polyelectrolyte, such as the flexibility of the chains and hydrophobic character of the repeating units.

A study directed to the influence of hydrophobic groups in the polyelectrolyte chain was carried out by Shimizu et al.,[66] using copolymers of maleic acid and different vinyl ethers (for abbreviations, see Table 3). It was found that the interaction is enhanced by a larger size of the hydrophobic group (Figure 4). An interesting observation is that for the more hydrophobic polyelectrolytes, two-step binding isotherms result (Figure 4), suggesting two different states for the bound surfactant molecules to these polymers.

Another observation made by Shimizu et al.[66] was that the interaction is somewhat stronger at a lower degree of neutralization of the polyacids. Apparently, the interaction is weaker for the more charged polymer, in contradiction to the trend in Figure 2. Similar findings were made by Chandar et al.,[67] who reported that the cac for binding of $C_{12}TAB$ to polyacrylate is largely independent of the pH of the solution, and thus of the degree of neutralization of the polymer, in the absence of added salt. This observation is indeed remarkable, in view of the much stronger interactions normally seen in P^--S^+ systems as compared to P^0-S^+ systems. A possible explanation of these observations, which should be further examined, may be that the polyacids titrate during the association with the oppositely charged surfactants and adopt a local charge density at the site of surfactant binding which is different from the overall charge density.

4. Effect of Salt

Several investigators have studied the effect of salt in dilute systems of polyelectrolyte and oppositely charged surfactant.[2,4,63,68,69] Almost all results show that the critical aggregation concentration increases when simple salt is added. This means that the interaction between polyelectrolyte and surfactant is reduced by addition of salt. The valency of the salt is of importance, i.e., a higher valency gives a larger increase in cac.[69] Intuitively, we may refer these observations to the screening of the electrostatic interactions between polyion and surfactant micelle.

The effect of salt on the polyelectrolyte-surfactant complexes is thus opposite to the influence of salt in micellar systems, where stabilization occurs, manifested by a lowering of the CMC.[17] At high concentrations of added salt this effect will also dominate in polyelectrolyte-surfactant systems. The effect of salt is thus twofold: (1) reduction of the electrostatic interaction between the polyelectrolyte and the surfactant, and (2) stabilization of the surfactant aggregates. The first mechanism will dominate at low ionic strength, while at high ionic strength, the second mechanism will take over. A decrease in the cac at high salt concentrations, similar to the CMC behavior, can

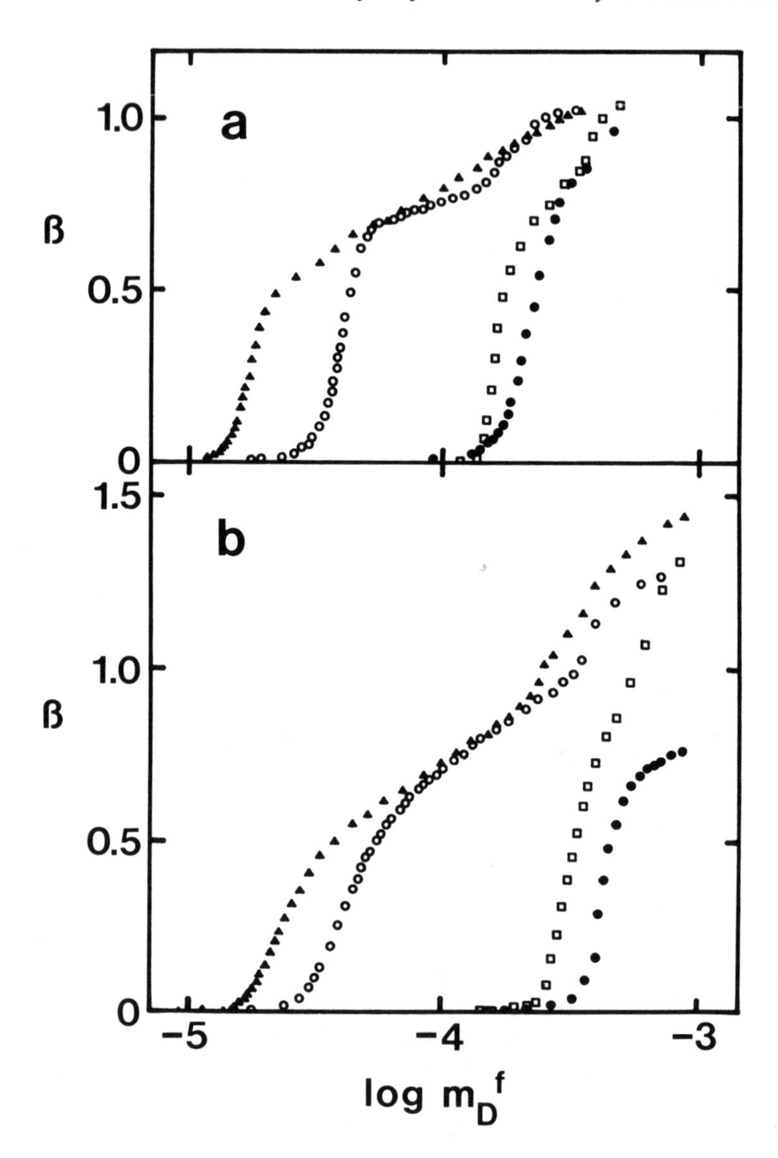

FIGURE 4. Binding isotherms for $C_{12}P^+$ to PMAE (filled circles), PMAEVE (open squares), PMASt (open circles), and PMAIn (triangles) in the presence of 0.025 *M* NaCl. The degree of neutralization was (a) 0.5 and (b) 1.0. (Reproduced from Shimizu, T. et al., *Colloids Surf.*, 20, 289, 1986. With permission from Elsevier Science Publishers B.V.)

therefore be expected. Further experimental investigations as well as theoretical approaches will be needed in order to fully disclose the effect of salt on the cac.

The effect of added salt can also be discussed in terms of the cooperativity of the surfactant binding.[4] Thus, an increase in the cooperativity parameter u results when salt is added to the system. This is due to (1) the screening of the repulsion between the adsorbed micelles, and (2) the adsorbed micelles being more akin to ordinary surfactant micelles on addition of salt. (It is pointed out that for normal micelles of an ionic surfactant, the free surfactant monomer concentration decreases above the CMC,[70] which would correspond to an infinite cooperativity parameter. A decrease in the concentration of free monomers has also been inferred for a system of cationic surfactant and oppositely charged polyelectrolyte.[65])

5. Influence of the Polyelectrolyte Concentration

The cac data given in Figures 2 and 3 refer to dilute polyelectrolyte solutions. However, there is also a dependence of cac on the polyelectrolyte concentration. Although additional studies of this phenomenon are needed, its main features have been deduced.[71,72] The first added polyelectrolyte chains give rise to a drop in the cac from the initial CMC value. When more polyelectrolyte is added, an increase in the cac occurs, followed again by a decrease at high polyelectrolyte concentrations. In fact, the behavior is similar to that discussed above concerning the addition of salt. The reason for this is that the polyelectrolyte and its counterions contribute to the ionic strength in the system and a cac behavior related to that effected by addition of salt is therefore expected.

6. Aggregation Numbers

Also for systems of polyelectrolyte and oppositely charged surfactant, a structure of micelle-like aggregates adsorbed to the polymer chains has thus been inferred in dilute solutions, and this structure seems to apply also in the concentrated phase of phase-separating mixtures, as will be discussed in Section IV.B. In particular, the use of fluorescence techniques has contributed to the establishment of this structure, but also different NMR techniques have been applied. However, no clear understanding of how the polyelectrolyte influences the micellar structure has yet been obtained; both an increase[24] and a decrease[73] in aggregation number of the polyelectrolyte-bound micelles, as compared to ordinary ones, have been reported.

The first thorough time-resolved fluorescence investigation was reported by Abuin and Scaiano in 1984,[73] dealing with the system of sodium poly(styrenesulfonate), PSS, and $C_{12}TAB$. The viscosity of the PSS solution is markedly reduced on addition of the surfactant, indicating a coiling up of the polymer chains. The surfactant molecules were inferred to form small clusters of only 7 to 10 monomers adsorbed to the polyelectrolyte chains. Furthermore, it was found that probes located in the clusters were protected from quenching by negatively charged quenchers, while the same surfactant in water protects from quenching by positively charged quenchers. It was concluded that the negatively charged PSS groups determine the charge of the micellar cluster. Recently, Almgren et al.[74] confirmed the small size of $C_{12}TAB$ clusters adsorbed to PSS, using time-resolved fluorescence quenching with pyrene as a probe. They obtained aggregation numbers around 20, slightly increasing with an increased surfactant-to-polyelectrolyte ratio. For $C_{16}TAB$, the aggregation number in the presence of PSS ranges from 36 to 55, as compared to 147 for free micelles. No change in aggregation number or pyrene lifetime was observed when the surfactant counterion was changed to chloride, further supporting the picture that the counterions are expelled from the surface of the aggregates.

Thermodynamic investigations of the systems with PSS and $C_{12}PC$ or $C_{16}PC$,[75,76] are also consistent with a picture of quite small, discrete micelles, adsorbed to the polyelectrolyte chains.

Chu and Thomas[24] investigated solutions of $C_{10}TAB$ and sodium poly(methylacrylate), PMA. Interestingly, a significant increase in aggregation number was found, indicating that the presence of the polyelectrolyte induces a considerable growth of the micelles. Thalberg et al.[65] determined aggregation numbers for $C_{10}TAB$ and $C_{12}TAB$ in the presence of Hy. The Hy-bound micelles were seen to have the same aggregation numbers as normal free micelles at the corresponding surfactant concentration.

In conclusion, quite different aggregation behavior has been reported for different polyelectrolyte-surfactant systems, and at present, no consistent picture has been obtained. Further systematic studies will be required, concerning parameters such as the surfactant chain length, the polyelectrolyte charge density and flexibility, and the presence of hydrophobic groups on the polyelectrolyte, and also concerning the relation between the degree of binding and the aggregation number.

7. Organization of the Surfactant-Polyelectrolyte Complex

The pyrene lifetime in the absence of quencher, τ_o, may give information about the local structure around the micellar aggregate. As bromide ions quench the fluorescence of pyrene, an increase in τ_o in the presence of polymer indicates displacement of the bromide ions from the micellar surface. This has been reported by Binana-Limbelé and Zana for a system of a copolymer of maleic acid and butylvinylether, PMABVE, and $C_{10}TAB$,[63] and by Thalberg et al. for systems of Hy and C_nTAB.[65] In the former study, complete displacement of the bromide ions from the micellar surface was inferred, while in the latter study the bromide ions were only partially displaced. The effect was larger for $C_{12}TAB$ than for $C_{10}TAB$, and furthermore, a relation between τ_o and the polymer-to-surfactant ratio was obtained. τ_o first increases with increasing polymer-to-surfactant ratio, but above a certain ratio, a fairly constant plateau level is attained, suggesting saturation of the micellar surface with polymer segments.

By means of NMR, both structural and dynamic aspects of the polyelectrolyte-surfactant complexes can be studied. Surfactant ^1H chemical shifts for $C_{12}TAB$ induced by polyelectrolytes containing aromatic groups (PMASt and PSS) were investigated by Gao et al.[77] Such shifts, due to "ring current" effects, are well known from the dissolution of, e.g., benzene in micelles and can give information about the location of the aromatic rings in the micelles. For both PMASt and PSS, large ring current effects were observed, as compared to normal $C_{12}TAB$ micelles, while for the corresponding polyelectrolytes without aromatic groups (PMAE and PVS), no chemical shifts were observed. The shifts are upfield for the *N*-methyl and α-methylene protons and downfield for the *n*-methylenes and the ω-methyl group. The results clearly indicate that the aromatic groups of the polymers take part in the aggregate formation, and are, furthermore, interpreted to suggest that the *n*-methylene and ω-methyl protons are, on average, in the plane of the ring while the *N*-methyl and α-methylene protons are above the phenyl groups, i.e., the phenyl groups are rather deeply buried into the interior of the micelle. The solubilization of benzene and some other aromatic compounds in the system of PMABVE and $C_{12}TAB$ was studied by the same group.[78] The chemical shifts observed were much smaller than in the case of phenyl-containing polyelectrolytes, but downfield shifts were present, notably for the α- and β-methylene protons. The results indicate that the aromatic compounds are primarily solubilized near the surfactant headgroups. Very small ring current shifts were observed for the protons in the butyl groups of the polymer.

A ^{13}C relaxation study of the system PSS-$C_{12}TAB$ was performed by the same group.[9] The spin-lattice relaxation time, T_1, and the Nuclear Overhauser Effect, NOE, were analyzed by use of the ''two-step'' model.[11] Slightly larger order parameters and a longer correlation time for the fast motion, τ_c^s, were obtained as compared to ordinary $C_{12}TAB$ micelles, especially for the carbon atoms close to the surfactant headgroup, but in general, the profiles were similar to those of ordinary micelles, pointing to a close structural similarity.

8. Polyelectrolyte Conformation

Few studies have touched upon the question of how the polyelectrolyte conformation is influenced by the interaction with surfactants. Information on this topic is mainly obtained indirectly from viscosity measurements on dilute polyelectrolyte solutions. Abuin and Scaiano[73] noted a dramatic decrease in the viscosity of a PSS solution on addition of $C_{12}TAB$, which was ascribed to coiling up of the chains accompanying the surfactant binding. A decrease in the dimensions of the polyelectrolyte coils was also observed by Thalberg et al. for NaPA-C_nTAB solutions,[72] but the study was limited by the onset of phase separation. Certainly, coiling of the chains can be expected both for electrostatic reasons, as polyelectrolyte chains in pure water are extended, and for topological reasons, i.e., in order to create a large contact area between the polymer chain and the (compact) micelle. Herslöf and Sundelöf[79] studied the viscosity in dilute solution for the NaHy-$C_{16}TAB$ system on addition of salt (NaBr). A reduction in the viscosity was found as compared to the surfactant-free system, but only below a certain salt concentration. Apparently, the polyelectrolyte-surfactant interaction is fully screened out at this salt concentration. Below this concentration, the viscosity decreased rapidly with a reduced salt concentration until the phase separation limit was reached.

Circular dichroism or optical activity measurements is another technique available for the study of polymer conformational changes on association with surfactants. Satake, Yang, and their co-workers have pursued this line of research since the 1970s focusing especially on polypeptides.[61,80-82] Oppositely charged surfactants have been found to induce both helix and β-types of conformation, depending on the particular polypeptide and surfactant, the temperature, and the pH of the solution.[80] The results are often unexpected. For example, poly(L-lysine) adopts a β-form in neutral SDS solution at 25°C, while its homologs with a number, m, of methylene groups in the side chain ranging from 2 to 7, all give right-handed helices. The $m = 1$ homolog is also exceptional, as it gives a left-handed helix. Recently, a model for the influence of surfactant binding on polypeptide coil-helix transitions was presented.[83] However, no explanation for the striking differences between different polypeptides has yet been given. In our view, the interaction between the polypeptide and the surfactant clusters is the dominant one, while the conformational change of the polypeptide can be regarded as an adaptation to the new environment; therefore, predictions of how a specific polypeptide will behave are extremely difficult to make.

It can be expected that the use of scattering techniques will provide further information and contribute to a more detailed picture of the polyelectrolyte behavior upon association with oppositely charged surfactant micelles. In particular, the importance of the flexibility of the polymer backbone for the interaction and the structure of the formed complexes is at present poorly understood.

9. Solubilization Properties of Polyelectrolyte-Surfactant Systems

The solubilization capacity of polyelectrolyte-surfactant complexes has been well demonstrated, as is reviewed in Chapter 4, Part II. Hayakawa et al. recently presented a study of the solubilization of the dye yellow OB in complexes of PVS and C_nTAB with $n = 10$ and 12, which is more thorough than previous work.[84] Solubilization of the water-insoluble dye correlated well with the surfactant binding isotherm which was obtained simultaneously by a surfactant-sensitive electrode, and the cac value was decreased by the presence of the dye. A theoretical treatment of the solubilization was given in terms of a concerted binding of both dye and surfactant ion.

10. Systems with Mixed Surfactants

By ''diluting'' an ionic surfactant with a nonionic one, the interaction with oppositely charged polyelectrolyte can be moderated. This has been clearly shown by Dubin and co-workers in a series of articles.[85-91] Systems of both anionic and cationic polyelectrolytes have been investigated, and for both types, the interaction (and the phase behavior) can be controlled by two parameters, i.e., the fraction of ionic surfactant molecules, Y, and the ionic strength of the system, I. Complex formation takes place only above a critical Y value, Y_c, and there is a linear relation between Y_c and $I^{1/2}$, largely independent of polymer and surfactant concentration. A difficulty in the analysis is that the fraction of ionic surfactant in the polyelectrolyte-adsorbed micelles is not the same as the total fraction Y, and it was shown that polymer-adsorbed and free micelles coexist.[85] In several systems, phase separation occurs immediately when Y exceeds Y_c, but in some systems, the formed polyelectrolyte-surfactant complexes may be soluble in the interval $Y_c < Y < Y_p$, where Y_p denotes the fraction of ionic surfactant at the onset of precipitation. There are, furthermore, two different types of soluble aggregates, those involving only one polyelectrolyte chain, and higher-order aggregates involving several chains.[90,91] The latter type is formed at higher polymer concentrations and when the molecular weight of the polymer is very high.

The surface potentials of the mixed micelles at Y_c were found to be rather constant for a certain system (at different I) and it was concluded that complex formation takes place when a critical potential is reached at the locus of closest approach of the polymer chains to the micellar surface.[89]

11. Special Types of Polyelectrolytes

Yomota et al.[92] compared the binding of $C_{12}TAB$ to chondroitin sulfate (linear) and to arabate, which is branched. In spite of similar distances between adjacent charges, the interaction with arabate was weaker and of very low cooperativity. It may be suggested that the degree of branching, probably through the loss of flexibility, markedly influences the interaction.

A different kind of polyelectrolyte is the "starburst dendrimers" which have been developed by Tomalia et al.[93] These consist of short polymer segments, connected in the center, and with an increased degree of branching outwards. All branches are terminated by charged groups, so in general they may be considered as charged microspheres. By the use of photochemical probes, Caminati et al.[94] could demonstrate the formation of complexes with oppositely charged surfactants consisting of micelle-like surfactant aggregates adsorbed at the surface of the starburst dendrimer.

D. POLYELECTROLYTE AND SURFACTANT OF THE SAME CHARGE

For P^-S^- or P^+S^+ systems, we do not, in general, expect any appreciable association unless the polymer is markedly hydrophobic and indeed, as will be discussed below, the most interesting systems in this category are those of hydrophobe-modified polymers. For other systems, the effects are small. For example, Binana-Limbelé and Zana[95] noted that the addition of polyacrylate to solutions of SDS results in a slight decrease in the CMC and in somewhat higher aggregation numbers. The effect was compared to that induced by addition of simple salts and was found to be of the same type, albeit weaker. The authors therefore concluded that no binding of the surfactant to the polyelectrolyte occurs. This is in line with previous findings for SDS interacting with carboxymethylcellulose[96] and for SDS or sodium dodecylbenzenesulfonate interacting with partially hydrolyzed polyacrylamide.[97] Another observation for this type of system is that the precipitation of SDS by Ca^{2+} is inhibited by the addition of polyelectrolyte.[95]

E. UNCHARGED POLYMER AND NONIONIC SURFACTANT

The interaction between an uncharged water-soluble polymer and an uncharged surfactant is generally weak and hydrophilic polymers such as PVA, PEO, and PVP show no sign of interaction with polyethoxylated nonionic surfactants.[3] EO-type nonionic surfactants are, however, known to interact with polycarboxylic acids. This interaction is analogous to the interaction between PEO and polycarboxylic acids,[98] and is suggested to proceed by a combination of cooperative hydrogen bonding and hydrophobic interaction. Such systems are thoroughly described in the review by Saito.[3]

For more hydrophobic polymers such as PPO and partially hydrolyzed polyvinylacetate, PVA-Ac, the addition of polyethoxylated nonionic surfactants induces a rise in the cloud point (cp).[3] The experiments were carried out at temperatures well below the cp of the surfactants used, and the increased cp can be attributed to the binding of the surfactants to these polymers.

Brackman et al. have recently investigated systems of uncharged polymers (PPO, PEO, and HPC) and a nonionic surfactant where the hydrophilic group is a sugar moiety (i.e., octyl-β-thioglucoside, OTG).[26] For all the polymers studied, the CMC in the presence of polymer is essentially equal to the CMC of the pure surfactant, indicating no, or only insignificant, interaction. For PPO, however, other techniques indicate that the surfactant micelles are bound to the polymer chains. Thus, a PPO solution, which is hazy at 25°C, turns transparent in the presence of OTG at concentrations above the CMC, which is attributed to a surfactant-induced uncoiling of the compact disc-like conformations of the PPO chains. Furthermore, the Krafft temperature of the surfactant is lowered from 30°C to below 20°C (in D_2O) in the presence of the polymer. A microcalorimetric titration of the polymer with OTG provided further indication for an interaction, and showed that the enthalpy of the polymer-surfactant interaction is endothermic.

Winnik recently reported on the interaction between HPC and the same sugar-containing nonionic surfactant (OTG), using a pyrene-labeled HPC.[99] The excimer formation of pyrene abruptly decreases at the CMC of the surfactant, implying an interaction between the modified HPC and the surfactant micelles. Again, no decrease in CMC was observed.

Also for the uncharged semipolar surfactant *n*-dodecyldimethylamine oxide, Brackman[25] reported binding to PPO, in particular inferred from a marked increase in cp. The isomeric polymer PVME, on the other hand, showed no significant change in cp when mixed with this surfactant. Both polymers, however, induce a slight decrease in the aggregation number of the micelles. Again, the CMC of the surfactant was unaltered by the presence of the polymers.

As also seen in cp studies of solutions of EHEC and some nonionic surfactants,[47] the interactions in P^oS^o systems are, thus, typically quite weak compared to other classes of PS systems. The cac values are often close to the CMC, while a number of changes in the physicochemical behavior suggest a significant, although weak, interaction. It seems that further progress should benefit from direct studies of binding (binding isotherms) or of interactions. For certain systems, attractive interactions can be inferred from the phase behavior (Section IV.D).

F. HYDROPHOBICALLY MODIFIED POLYMERS

Polymers in which, usually in low amounts, strongly hydrophobic groups have been introduced are of great interest for quite a number of products and this, as well as a fundamental interest, has initiated quite extensive research related to the structures, dynamics, and interactions in aqueous solutions of these hydrophobically modified or HM polymers. The main significance with respect to applications arises from a more or less extensive self-aggregation. The self-aggregation results from hydrophobic interactions and leads to an interesting rheological behavior, in particular a strongly shear-rate-dependent viscosity. These compounds, which may be ionic or nonionic, are often referred to as associative thickeners and are extensively used, for example in paints, to produce a suitable consistency.

The HM polymers are, of course, expected to display a rich, and also distinctly specific, pattern of interaction with surfactants. This provides a further possibility of controlling rheology and designing systems for various applications. However, this principle of devising organized systems is also of considerable fundamental interest. Some features that may be expected are

1. Binding of surfactant molecules to HM polymers will start at a lower surfactant concentration than for the corresponding parent polymer and binding will preferentially take place at the hydrophobic groups
2. Because of the stronger and more specific interaction, the cooperativity in surfactant binding will be lowered, leading to smaller aggregates (at least in the concentration region just above the cac)
3. Because of the hydrophobic interaction, surfactant binding can take place in cases where the polymer-surfactant interaction without the modification is strongly repulsive, like in polymer-surfactant systems where the two solutes have like charges; systems of a nonionic polymer and a nonionic surfactant provide another example where hydrophobe modification is expected to introduce attractive interactions
4. Surfactant binding will strongly modify the polymer-polymer association and, depending on the conditions (mainly the stoichiometry), an enforced or weakened polymer-polymer interaction can result
5. The presence of the hydrophobic groups will facilitate a crosslinking between different polymer chains as induced by surfactant binding

6. Rheology is, for these polymers, predicted to be highly sensitive to surfactant binding and the pattern will be complex due to a competition between crosslinking effects and an elimination of polymer-polymer association.

In recently started research by several groups these different features are illustrated, and here a few examples will be given. The HM polymers can, except for the general classification used above in distinguishing between nonionic and ionic polymers, also be classified according to the distribution of the hydrophobic groups. In much work, and especially fluorescence work, using pyrene as the hydrophobic group, hydrophobic groups have been attached to the two ends of a polymer. In another type of HM polymer, of the "comb" type, the hydrophobic groups are distributed more or less randomly along the polymer chain and the degree of hydrophobe modification has been varied.

In the latter group falls recent work by Iliopoulos et al.[100] and Wang and co-workers,[101-103] who have initiated a program of synthesizing mainly HM polyelectrolytes and performing physicochemical characterization of their solutions in the presence of ionic surfactants. Their findings provide nice illustrations of the field. The polymers used here are HM polyacrylates obtained by reacting a small fraction of the carboxylic acid groups with alkylamines. The alkyl chains have 8, 14, or 18 carbon atoms and the degree of HM substitution is varied in the range up to 10%. These polymers interact strongly with various surfactants, also with those of similar charge. The rheology displays a complex pattern with dramatic changes as a function of polymer concentration, electrolyte addition, degree and type of substitution, pH, and surfactant concentration. Above a certain concentration of polymer the viscosity increases dramatically and can be several orders of magnitude above that of regular polyacrylate. Electrolyte or surfactant addition gives a further increase of the viscosity. As a function of surfactant concentration the viscosity passes through a high maximum and then decreases strongly. The increase can be attributed to the formation of mixed aggregates through the interaction of the surfactant molecules with the alkyl chains. These aggregates will enforce polymer-polymer crosslinks as well as give new ones. The decrease at higher concentrations can be ascribed to an excess of surfactant micelles relative to alkyl groups of the polymer giving surfactant clusters at individual polymer alkyl chains rather than clusters involving more than one, and is thus a simple stoichiometric effect.

As mentioned above, anionic polymers do not, in general, associate with anionic surfactants and it is only for quite hydrophobic polymers that a significant interaction is noted.[100] Studies of HM polyacrylate, as illustrated in Figure 5, show, on the contrary, a very strong interaction. These solutions are in the semidilute regime and the rheological observations can be attributed to the formation of mixed hydrophobic clusters of polymer and surfactant alkyl chains. These clusters, which act as crosslinks, change in composition

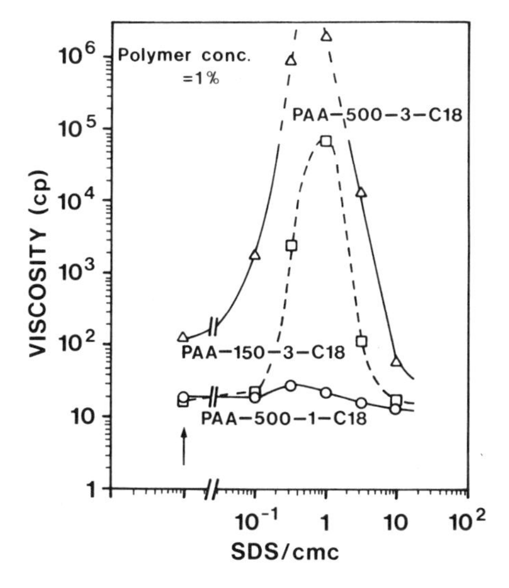

FIGURE 5. Plot of the viscosity of aqueous solutions of HM polyacrylates as a function of SDS concentration (expressed in CMC). The polymers are: (circles) PA-500-1-C18; (triangles) PA-500-3-C18; (squares) PA-150-3-C18, where the first number refers to the PAA molecular weight, the second to the percentage of hydrophobic groups, and the third to the number of carbons of the alkyl chains. Arrow indicates polymer solutions in pure water. (From Iliopoulos, I. et al., *Langmuir,* 7, 617, 1991. Reproduced with permission from the American Chemical Society.)

as a function of the relative amounts of surfactant and polymer alkyl chains. At higher surfactant concentrations, there is a coexistence of polymer-free micelles and micelles containing only one polymer alkyl chain and the cross-linking effect is lost. The general model is well supported by the observed changes on varying the degree of substitution.

This and similar association models have been proposed also for several other types of systems.[58,59,104-111] The field can be exemplified by studies of HM hydroxyethylcellulose interacting with sodium oleate,[104] SDS,[105,112] and nonionic surfactants.[113] In order to investigate especially the local dynamics of the crosslinks, Tanaka et al. performed ESR studies on a spin probe covalently linked to the polymer chain.[105]

Besides alkyl chains, hydrophobe modification has been achieved by introducing pyrene groups, either at the two ends of the polymer chains or at several locations along the chain. Pyrene substitution is generally made to allow fluorescence studies which can provide a good insight into important aspects of the surfactant clusters. Hu et al.[111] have, *inter alia,* studied polyethylene oxide endcapped with pyrene groups and found an excimer emission which is first enhanced and then suppressed with increasing concentration of SDS. These results give a nice description of the pyrene-pyrene, pyrene-surfactant, and surfactant-surfactant interactions, as is illustrated in Figure 6, and it is shown that the surfactant at low concentration can give an intrapo-

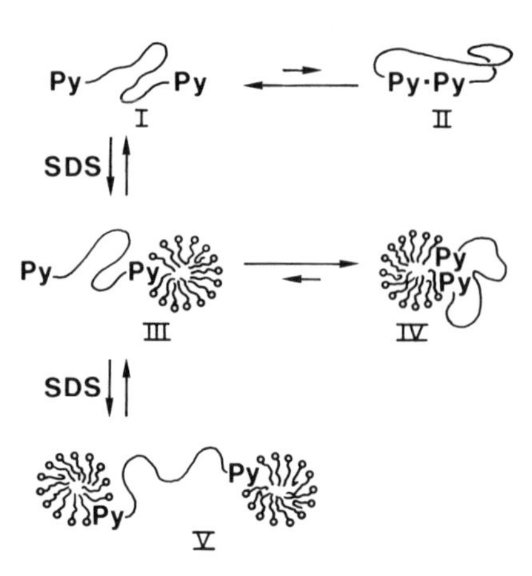

FIGURE 6. Schematic model of the interaction of pyrene-endcapped PEO with SDS in aqueous solution with increasing SDS concentration. (From Hu, Y.-Z. et al., *Langmuir,* 6, 880, 1990. Reproduced with permission from the American Chemical Society.)

lymer pyrene-pyrene association which is eliminated at excess SDS concentration. Winnik and co-workers[52,99,111] made extensive studies of the interaction of hydroxypropylcellulose labeled with pyrene groups along the chain with different surfactants like *n*-octyl β-D-thioglucopyranoside, *n*-octyl β-D-glucopyranoside, SDS, and C_{16}TAC. Significant interaction was noted and a picture similar to that noted above emerged. Recently, Winnik et al.[58,59] studied the interaction between the same surfactants and copolymers of *N*-isopropylacrylamide and (small amounts of) *N*-*n*-alkylacrylamides, *inter alia,* by using pyrene as a probe in fluorescence quenching studies. The surfactant forms clusters at the polymer alkyl groups and these clusters contain about 30 surfactant molecules for the C_8 or C_{12} surfactants but are smaller for the C_{16} one. The PS clusters appear to be more rigid than simple surfactant micelles.

Another interesting study concerns the interaction between pyrene-labeled polyacrylate and dodecyltrimethylammonium bromide,[67] using pyrene labeling for probe purposes. A detailed picture of surfactant-pyrene association and its implication for polymer chain conformations at different pHs emerges. Schild and Tirrell[114] have used an amphiphilic probe in studies of polymer-surfactant interactions for HM *N*-isopropylacrylamide and obtained a detailed description of association structures from fluorescence work.

G. MODELING OF POLYMER-SURFACTANT INTERACTIONS

Several approaches to model the interactions between a polymer and a surfactant in solution have been put forward during the last decade, and the majority deal with P^oS^+ and P^oS^- systems. The first quantitative model which could account for the strongly cooperative binding isotherms observed in such systems was given by Gilanyi and Wolfram in 1981.[115] The authors correctly realized that this was due to the formation of micelles adsorbed to the polymer chains, and their model is able to reproduce experimentally obtained activity data for both the surfactant ion and its counterion at various polymer concentrations. In 1985, Hall presented a general, model-independent theory for the thermodynamics of polymer-surfactant interactions.[116] Interesting predictions were made, but the wide generality of the theory makes its predictive value low, and the theory has not been thoroughly tested against experimental data.

1. The Theories of Nagarajan and Ruckenstein

Elaborate developments of the treatment of Gilanyi and Wolfram have been presented by Nagarajan[43,44] and by Ruckenstein and co-workers,[45] who try to relate the cac values and the aggregation numbers to particular properties of the surfactant and the polymer molecules. Both these theories consider the interactions from the surfactant micelle point of view, and as the theories are quite similar, they will be discussed together.

Assuming only one (optimal) aggregation number and adopting the pseudophase separation theory for micelle formation, we have the relation

$$\ln \text{cmc} = \Delta\mu_n^o/kT \tag{26}$$

where n is the aggregation number and $\Delta\mu_n{}^o$ is the change in chemical potential relative to the unassociated surfactant molecules. An analogous expression is used for the polymer-bound micelles, i.e.,

$$\ln \text{cac} = -\Delta\mu_n^o/kT \tag{27}$$

By providing expressions for $\Delta\mu_n{}^o$ in the presence and in the absence of polymer, it would, in principle, be possible to predict polymer-micelle association and to evaluate its strength and the optimal aggregation numbers. Nagarajan gives the following expression for $\Delta\mu_n{}^o/kT$ in the presence of polymer (the case without polymer should be obvious):

$$\begin{aligned}\Delta\mu_n^o/kT = {} & n\Delta\mu_{tr}^o/kT + n\Delta\mu_{ex}^o/kT + n\sigma(a - a_o - a_{pol})/kT \\ & - n(\ln(1 - a_p/a - a_{pol}/a) + \Delta\mu_{elec}^o\end{aligned} \tag{28}$$

The first term on the right-hand side refers to the free energy change associated with the transfer of the alkyl chain from water to a liquid hydrocarbon phase, and the second term corrects for the ordering of the alkyl chains in the micellar interior. The third term represents the free energy of formation of the micellar core-water interface, where the interfacial tension, σ, is taken to be the same as the macroscopic interfacial tension between liquid hydrocarbons and water. The parameter a represents the surface area per surfactant molecule in the micelle and a_o is the area shielded from contact with water by the polar headgroup of the surfactant. The parameter a_{pol} refers to the (mean) area per surfactant covered with polymer, and serves as a quantitative measure of the effectiveness of micellar binding to the nonionic polymer. The fourth term refers to steric repulsions between surfactant headgroups and between headgroups and the polymer segments. The area a_p is the cross-sectional area of the polar headgroup. The last term accounts for the electrostatic interactions and is calculated by means of the linearized Poisson-Boltzmann equation. The degree of counterion binding, β, is assumed to be the same for free and polymer-bound micelles.

Model calculations showed that the interaction increases with an increase in a_{pol}, accompanied by a decrease in the aggregation number. If, however, a_p is large, the steric repulsions will be large already at small values of a_{pol}, and normal free micelles will be favored to polymer-bound ones. This was obtained using parameters chosen to correspond to Triton X-100.

Ruckenstein's treatment differs from that of Nagarajan mainly in the third term of Equation 28, which describes the formation of the core-water interface. Ruckenstein et al.[45] use the expression

$$\text{interfacial term} = (\sigma - \Delta\sigma)(a - a_p) + a_p\Delta\sigma_p \text{ if } a_p < a_h \qquad (29)$$

where a_h denotes the cross-sectional area of the hydrocarbon chain and $\Delta\sigma$ and $\Delta\sigma_p$ are the changes in interfacial tensions between the hydrocarbon core and water, and between the headgroups and water, respectively, caused by the presence of the polymer. A division is thus made between the influence of the polymer on the hydrocarbon-water interface and on the headgroup-water interface, but in the calculations presented, $\Delta\sigma_p$ was set equal to $\Delta\sigma$. The point made by Ruckenstein et al. is that the parameter $\Delta\sigma$ can be obtained experimentally as the change in interfacial tension between water and *n*-octane, when the polymer is added to the aqueous phase. To a large extent, however, the results obtained by Ruckenstein et al. are analogous to those of Nagarajan.

A problem is that both theories contain several geometrical parameters, characterizing the formed complexes, which may be difficult to evaluate but are crucial for the interaction. Furthermore, the treatment of the electrostatic interactions is quite crude.

Brackman, who determined cac values and aggregation numbers for a large number of systems,[25] tried to adapt the theories of Nagarajan and Ruckenstein to her results. Experimental cac and aggregation numbers were used as input data, and the geometrical parameters of the models were estimated. It turned out that, although several values for both a_{pol} and $\Delta\sigma$ of the respective theories were tried, the data for the SDS-PPO system could not be accounted for by the models.[25] Instead, the "dressed micelle" model of Evans et al.[117] was employed for a rationalization of the data (see below).

It should be mentioned that Nagarajan has further developed his theory,[44] by adding an additional term to account for the changed environment of the polymer segments, and furthermore, the theory is extended to include other types of surfactant structures such as rod-like micelles, bilayers, and microemulsions.

2. The "Dressed Micelle" Theory

The "dressed micelle" theory by Evans et al.[117] may be used with the cac and the aggregation number as input data to give information about the different energy contributions of the polymer-surfactant interaction, i.e., it may reveal the much-debated driving force for the interaction. The basic equation is still Equation 26, and the free energy difference is divided into two contributions, (1) the hydrophobic free energy of transfer of hydrocarbon tails from water to the interior of a micelle (g_{tr}), and (2) surface contributions (g_s). Thus:

$$\Delta\mu_n^o = g_{tr} + g_s \quad (30)$$

To proceed, g_s is further divided into an electrostatic part and an attractive hydrophobic term, according to:

$$g_s = \gamma_o a + g_{el} \quad (31)$$

where g_{el} is calculated using the nonlinear Poisson-Boltzmann equation (the aggregation number is used to obtain the radius of the micelle). γ_o is called the effective interfacial tension and contains a number of unknowns lumped together. The "dressed micelle" theory has, however, been shown to be very useful for micelle formation and is well based in statistical mechanics.[117] At equilibrium, the attractive and repulsive forces must exactly balance, which gives

$$\frac{\partial g_s}{\partial a} = \gamma_o - \frac{\partial g_{el}}{\partial a} = 0 \quad (32)$$

Since g_{el} can be calculated explicitly, the effective interfacial tension γ_o can be obtained, and g_s can thus be calculated using Equation 31. Finally, by combining Equation 26 with Equation 30, the quantity g_{tr} is obtained.

In adopting the "dressed micelle" model to her data, Brackman obtained reasonable and consistent values for the parameters of interest.[25] In particular, it was found that the effective interfacial tension is lower in polymer-micelle aggregates than in simple micellar solutions, but due to the increase in surface area per molecule, the term $\gamma_o a$ is consistently higher, i.e., less favorable. The electrostatic energy, g_{el}, on the other hand, in most cases shows a slight reduction. Therefore, the value of g_s is found to be equal or slightly higher in the presence of a polymer, and cannot be a driving force for the interaction. It was concluded that "the reduction in CMC mediated by the polymer is due to the increased negative contribution from the energy of transfer of hydrocarbon chains from water to the micellar core combined with that of the transfer of polymer segments to the micellar surface''. One parameter, however, does not agree satisfactorily between the theory and experiments, namely the counterion binding, where the experimentally obtained reduction is much larger than that given by the ''dressed micelle'' theory. This point is in fact rather worrying, as the calculation of the electrostatic energy is the starting point of the analysis.

In conclusion, it seems that quantitative predictions are still not possible from the existing theories but that further refinements are needed. The ''dressed micelle'' theory may, however, give insight into the driving force for the association, when the cac value and the aggregation number are known.

3. Computer Simulations

A computer simulation model for polymer-surfactant interactions has been developed by Balazs and Hu.[118,119] In brief, polymer chains containing ''stickers'' at the ends and surfactant chains which have stickers at one end only undergo a random walk in a three-dimensional lattice. If a sticker jumps into a position parallel and adjacent to another sticker, an irreversible bond is formed, and by introducing the chains one by one, the growth of the aggregate can be followed. The size of the formed aggregates is studied as a function of the ratio between polymer and surfactant chains and the length of the surfactant chains. Certainly, the model molecules are very crude representatives of real polymers and surfactants. Nevertheless, interesting relations between the polymer-to-surfactant ratio and the aggregate size are obtained, which are claimed to be of relevance for the modeling of the enhancement of the viscosity of polymer-surfactant solutions. Computer simulations of polymer-surfactant systems are highly appropriate, and increased activity in this area, using different approaches, can be envisaged.

4. Polyelectrolyte-Surfactant Interaction Models

As for the interaction between a polyelectrolyte and an oppositely charged surfactant, several major questions still remain to be settled, as for example, the aggregation number and the influence of the polyelectrolyte and the surfactant charges on the formed complexes. Consequently, the modeling of these systems is not so well developed. In order to account for the cooperativity in binding, the one-dimensional Ising model,[120] sometimes referred to as the Hill equation, or its analog, the Zimm-Bragg model,[15] has been commonly adopted. The polymer is then pictured as a linear array of (monovalently charged) binding sites for single surfactant ions with a binding constant K, and lateral interactions between adjacently bound surfactant molecules are taken into account by a cooperativity parameter, u. This treatment often gives a good fit to the experimentally obtained isotherms, and is an excellent way to characterize the binding strength (1/Ku equals the free surfactant concentration at 50% degree of binding) and the cooperativity in the interaction. (If the cooperativity is high, 1/Ku is equal to the cac.) However, the structural picture of an array of bound surfactant molecules along the polymer chain is clearly unphysical, in particular that of the surfactant tails pointing out into the bulk water, and has long since been abandoned for the uncharged polymer-surfactant systems and also by most groups working in the field of polyelectrolyte-surfactant interactions.

Skerjanc et al.[75] have presented a straightforward way to estimate the cac value in a system of an ionic surfactant and a polyelectrolyte. It is presumed that the surfactant ion concentration is enhanced close to the polyelectrolyte for pure electrostatic reasons, which in turn entails an accumulation of the nonpolar tails, and as a consequence aggregation will take place in these regions at a total surfactant concentration that is considerably lower than the CMC. By using the Poisson-Boltzmann theory in the cylindrical cell geometry, the counterion concentration profile outside the polymer chain is calculated. The difference in size between the surfactant ions and the ordinary polyelectrolyte counterions is taken into account by dividing the radial distance into one zone accessible only to the smaller counterions, and one zone accessible to both types of ions. From an estimate of the size of the surfactant ion, the total surfactant concentration at which the local concentration close to the polymer equals the CMC can be calculated. This total concentration is thus the cac. From model calculations on the systems of $C_{12}PC$ or $C_{16}PC$ and PSS, it was concluded that the cac values obtained are of the right order of magnitude. The authors also use the model in a qualitative way to predict that resulting small aggregates behave as multivalent minimicelles trapped in the regions of high electrostatic potential close to the macroion. This approach seems most appropriate, especially for systems where the polyelectrolyte has a high charge density and thus may dominate the interaction.

IV. PHASE BEHAVIOR

Phase diagrams of surfactant-water and of polymer-water systems have been determined for a large number of cases and are rich in information on solute-solvent and solute-solute interactions as well as on self-assembly, interaggregate interactions, and structural progression. Also, studies of systems of water and two surfactants or of two polymers are rather abundant in the literature and provide useful information on the interactions between the two solutes. The two types of ternary-phase diagrams are different in nature since two surfactants easily mix and form mixed aggregates while for polymer molecules the entropic drive to mixing is small. Therefore, unless strongly attractive interactions occur — as between two oppositely charged polyelectrolytes — two polymers have the tendency not to mix extensively. Typically, we observe in such a ternary system two solution phases, each rich in one of the polymers (polymer incompatibility).[121]

Somewhat surprisingly, phase diagrams have not been extensively studied, until very recently, for polymer-surfactant-solvent systems. We can expect that such studies should provide quite detailed information on the interactions in the systems and depending on the types of polymer and surfactant, very different phase diagrams may be expected, displaying, *inter alia,* pronounced aggregation and segregation phenomena.

The behavior of single-phase systems of polymer and surfactant is, on the other hand, well understood. These systems can be fairly well described by only two parameters, namely the cac and the maximum amount of surfactant that can be bound to the surfactant per polymer unit.[60,122] (The systems are further described in Chapter 4, Part I.) A more recent small-angle neutron scattering (SANS) study by Cabane and Duplessix in semidilute PEO solutions further confirms the picture.[123] From the scattering behavior at addition of SDS, two additional critical lines are found, (1) the chain overlap concentration, c*, for the micelle-containing PEO chains and (2) a critical line at higher PEO concentration, above which the spatial distribution of the (polymer-adsorbed) micelles is the same as for normal free micelles.

A. UNCHARGED POLYMER AND IONIC SURFACTANT

Systems of a nonionic polymer and an ionic surfactant are characterized by repulsive surfactant-surfactant interactions while the polymer-polymer interaction is typically more or less attractive. In many cases studied, the attraction is strong and the solution of the polymer alone is not too far from a lower consolute curve (cp curve). These systems are particularly well suited for phase diagram work and deductions of the polymer-surfactant interaction from the appearance of the phase diagram. Thus changes in the cp (which are usually easy to measure) of a polymer solution on addition of a surfactant first give direct information on the polymer-surfactant interaction and second, make it easy to map the phase diagram.

The general effect expected is that a surfactant will raise the cp since on binding to the polymer it gives charged PS complexes which repel each other by long-range electrostatic interactions. This is indeed what is observed for many systems with a long-chain surfactant,[47,124,125] as is illustrated in Figure 7a. However, for a short- or medium-chain surfactant, like SDS (see Figure 7a), in addition to an increase in cp at higher concentrations, there is an initial decrease and a minimum. As discussed below, studies of cp curves in the presence of an electrolyte indicate that the minimum is related to the presence of charged species in the bulk solution; thus, the combination of an ionic surfactant and an electrolyte may give a dramatic lowering of the cp (Figure 7b).

The clouding phenomenon is a most conspicuous feature of nonionic systems and from a technical point of view, the influence of different cosolutes on the phase separation is highly significant. In order to understand the effects of surfactants we need also to consider simple cosolutes.[47,126] Simple electrolytes affect the cp only slightly; 100 m*M* concentrations shift the cp by a few degrees centigrade. The cp may either increase or decrease and is generally mainly determined by the anion. The cp changes effected by the salts follow the classical Hofmeister or lyotropic series found to apply for a very broad range of systems: for example, I^- and SCN^- give salting-in effects and

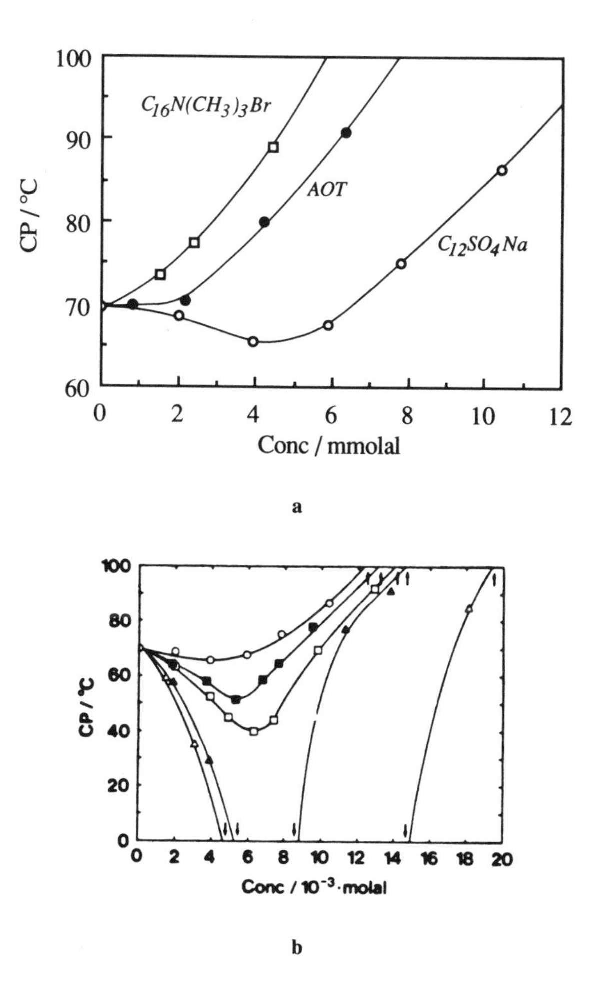

FIGURE 7. Cloud point of 0.9 wt% EHEC in water vs. the concentration of surfactant. (a) SDS, Aerosol OT, and C_{16}TAB. (From Karlström, G. et al., *J. Phys. Chem.*, 94, 5005, 1990. With permission from the American Chemical Society.) (b) SDS at different amounts of NaCl added: 0% (open circles), 0.009% (filled squares), 0.019% (open squares), 0.046% (filled triangles), and 0.11% (open triangles). Arrows represent clouding occurring below 0°C and above 100°C. (Reproduced from Carlsson, A. et al., *Langmuir*, 2, 536, 1986. With permission from The American Chemical Society.)

SO_4^{2-} gives a salting-out effect. Another example of effects of cosolutes on cp is provided by alcohols. Short-chain alcohols give a slight increase in cp while long-chain ones give a major decrease.

Cosolute effects on the cp and solubility may be discussed in terms of the distribution of the cosolute in the system and the relative polarity of the polymer, the solvent, and the cosolute. It is important to consider all pair interactions in the system and for charged polymers or cosolutes the effect of long-range electrostatic interactions must also be taken into consideration.

For salting-in ions it has been observed that there may be a partitioning in favor of the polymer (or a macroscopic interface) while salting-out ions are depleted in the vicinity of the polymer. Which effect is dominating in an electrolyte will determine the effect.

However, it may also be necessary to consider the interpolymer electrostatic repulsion due to the inhomogeneous ion distribution. This will be particularly significant for amphiphilic cosolutes. For alcohols, the short homologs will have a weak favorable interaction with the polymer and increased solubility results, while less hydrophilic alcohols will effectively interact more strongly with the polymer, thus causing phase separation.

A situation with a strong interaction between polymer and cosolute, involving strong adsorption of the cosolute to the polymer, rendering the polymer more polar and increasing its solubility, is thought to be significant mainly with amphiphilic cosolutes. In particular ionic surfactants can be expected to show complex behavior. Thus, these often bind strongly to nonionic polymers and are then expected to have dual effects: (1) by the alkyl chains they introduce hydrophobic groups; and (2) by the introduction of the charged groups they increase polymer polarity and, in particular (at low electrolyte and polymer contents) a strong interpolymer repulsion is created.

The change in cp (as well as many other properties) will depend on the balance between these two mechanisms and will, hence, be sensitive to counterion and the presence of (simple) electrolyte. In the absence of added salt an increase in solubility is expected because the electrostatic effect dominates.

At first sight, the strong synergistic surfactant-electrolyte effect is rather striking in that the simultaneous presence of two cosolutes, which both separately induce a cp increase (or only a slight decrease for many electrolytes), causes a dramatic decrease in cp at quite low concentration of the two cosolutes (Figure 7b). The effect has been found to be very general:[47,124,127] it has been documented for a large number of electrolytes, for a large number of ionic surfactants, and quite a number of polymers (including poly[ethylene oxide], methylcellulose, hydroxypropylcellulose, methyl[hydroxypropyl]cellulose and methyl[hydroxyethyl]cellulose). While the effect is qualitatively the same for different systems, there are important quantitative differences:

1. The effect is more marked, the longer the surfactant alkyl chain
2. Cationic surfactants give a smaller effect than anionics
3. Divalent counterions give a two-phase region extending over a wider surfactant concentration range than monovalent ones
4. Quantitatively, the effect is very different for different polymers
5. The effect is observed also with formamide as solvent but here cationic surfactants give a stronger effect[128]

We note that the balance between effects mentioned above would predict that, at high salt relative to surfactant concentration, the hydrophobic effect should dominate (electrostatic repulsion screened). At low salt relative to surfactant concentration the electrostatic effect should dominate, as observed in the increase in cp. However, the situation is complex since binding of surfactant is expected to affect polymer conformation and since addition of (especially strongly hydrated) electrolytes is expected to screen intra- and interpolymer repulsions differently.

In recent work, a model based on temperature-dependent solute-solute interactions has been advanced to rationalize the behavior of aqueous solutions of PEO and some other nonionic polymers.[129] There can be several mechanisms behind such variations, but the most apposite one is based on the assertion that the polymer conformation may vary with temperature. This led to analyses of the energies of different conformations in a polar medium.[130] In quantum mechanical calculations of the energies of the conformations of the –OCCO– segment it was found that the conformation that is gauche around the C–C bond and anti around the C–O bond has the lowest energy but a low statistical weight. It is interesting to note that this conformation, which will be the dominating one at low temperature, has the highest dipole moment of all conformers. At higher temperature, other conformers with smaller or no dipole moment will become increasingly populated. According to these calculations, the temperature-dependent conformational changes will make the EO chains progressively less polar as temperature is increased. As the chains become less polar, a solute-polar solvent interaction will become less favorable while the solute-solute interaction will become effectively more attractive.

In a recent study, the phase behavior of the EHEC-water-ionic surfactant-salt system was modeled.[47] The purpose was to gain insight into the molecular interactions causing the observed phase behavior. The basic concept is that at elevated temperature there are relatively few polar conformers, which prefer to interact with water, and more of the nonpolar ones which prefer to interact with themselves. From these types of assumptions, it is possible to reproduce the phase diagram of the PEO-water system reasonably well. If a third component is introduced into the system and this component has a slight preference for the solvent (short-chained alcohols), the solubility (i.e., the cp) increases. If the preference for the solute is fairly strong then the cp will decrease. However, if the additive prefers to interact with the solute (the polymer) and mostly with its polar conformer, and furthermore does not like to interact with itself, then a phase diagram like the one shown in Figure 8a can be obtained.[126] The agreement between theory and experiment is even better if the cosolute is modeled as a polymer, which is done in Figure 8b.

If the interaction between the cosolute and the polymer is made slightly more attractive, or if the attraction between the cosolute and water is made less attractive, the cp depression is increased. We can thus see that changes in the polymer or changes in the solvent (e.g., induced by addition of salt) may drastically change the cp curve.

The studies of the phase behavior of these ternary systems have recently been extended and more complete ternary-phase diagrams have been obtained at different temperatures.[131] As is illustrated in Figure 9, the phase diagrams

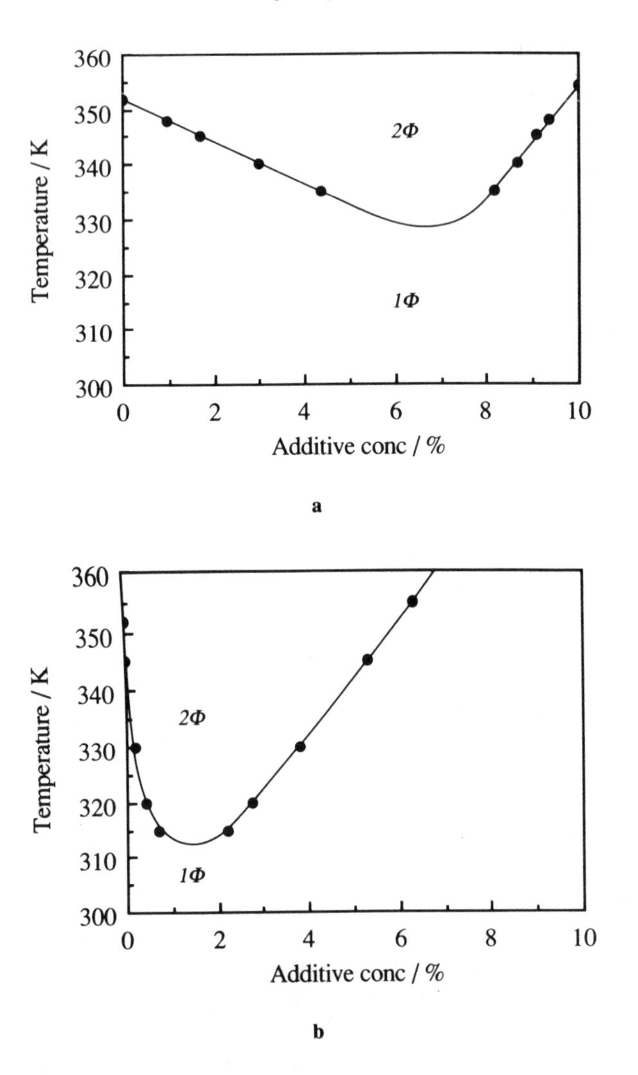

FIGURE 8. Calculated phase diagrams for 1% polymer in water where the additive is treated as a monomer (a) and a polymer (b). (For the details of the calculations, see Reference 47.) (Reproduced from Reference 124.)

are dominated by a two-phase region which extends to higher surfactant contents as the temperature is increased. The tie-lines are almost parallel to the water-polymer axis. Model calculations, using the F-H theory for a system of a solvent and two polymers, show that the surfactant component must be treated as a second polymer and that there is a relatively strong attraction between the polymer and the surfactant molecules.

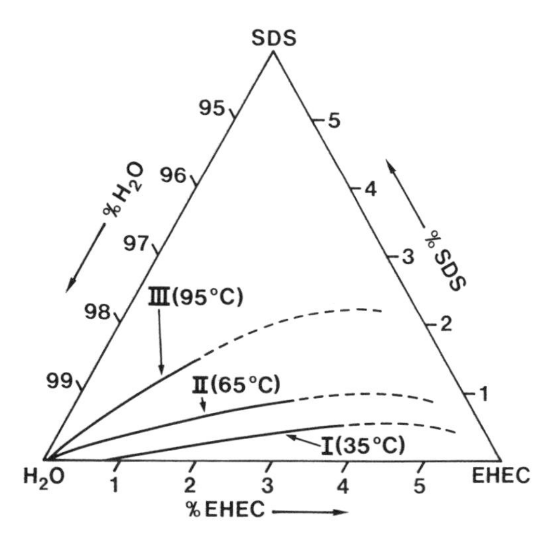

FIGURE 9. Phase diagrams at 35, 65, and 95°C for the water-EHEC-SDS system. Dashed part of the curve is uncertain due to slow equilibration. Above the curve there is a one-phase solution, while below there is a separation into two phases. (From Zhang, K., Karlström, G., and Lindman, B., *Prog. Colloid Polymer Sci.*, 88, in press.)

B. POLYELECTROLYTE AND OPPOSITELY CHARGED SURFACTANT

1. Phase Separation

Phase separation is a quite general feature for systems of a polyelectrolyte and an oppositely charged surfactant. While for an uncharged polymer, binding of surfactant leads to a charging up of the polymer, surfactant binding to an oppositely charged polyelectrolyte leads to a reduction of the overall charge of the resulting complex, and phase separation may be expected. In early studies in this field,[132-134] so-called "solubility diagrams" for various polyelectrolyte-surfactant systems were established, while recent studies have attempted to investigate the phase behavior more fully.[71,72,135-138]

Phase separation may be expected at a binding ratio close to one surfactant ion per charged polyelectrolyte segment. Interestingly, phase separation was observed at very low degrees of surfactant binding in the Hy-C_nTAB systems.[62,71] For $C_{12}TAB$ in particular, surfactant binding could not be detected prior to phase separation (at several Hy concentrations). This may seem surprising, as many polyelectrolytes are reported to bind at least 50% of surfactant (calculated as bound surfactant molecules per polyelectrolyte charge), with no phase separation (see Chapter 4, Part II). Moreover, phase separation at a very low degree of surfactant binding was inferred also in systems of poly(acrylate) and a cationic surfactant.[72] The same polymer is reported to bind 60% of $C_{12}TAB$ in dilute solution.[68] This suggests that phase separation is highly dependent on the polyelectrolyte concentration. The situation is schematically outlined in Figure 10, where the cac line is also included. At

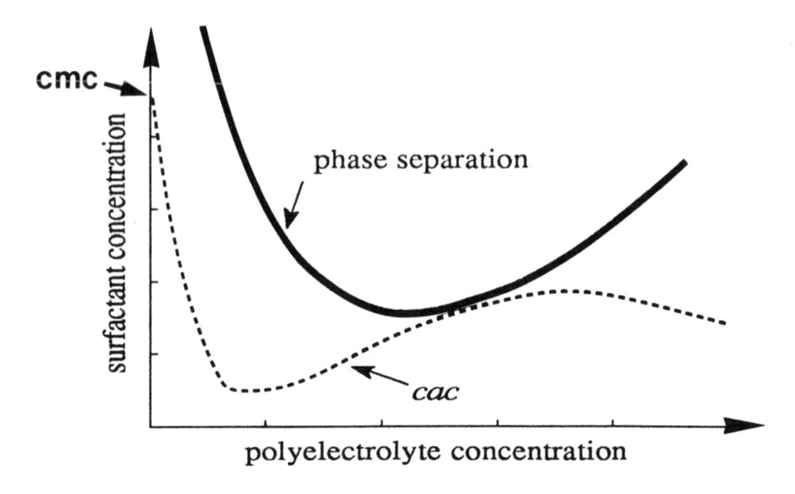

FIGURE 10. Generalized cac and phase separation behavior in a system of polyelectrolyte and oppositely charged surfactant. (Reproduced from Thalberg, K. and Lindman, B., *Surfactants in Solutions,* Mittal, K. and Shah, O. O., Eds., Plenum Press, New York, 11, 243, 1991. With permission.)

intermediate polyelectrolyte concentrations, the phase separation line may come very close to the cac line and phase separation will start at very low degrees of surfactant binding, while at both lower and higher polyelectrolyte concentrations, polyelectrolyte-surfactant complexes are stable up to high degrees of surfactant binding. Apparently, phase separation is not determined solely by charge neutralization of the polyelectrolyte-surfactant complexes, but has to be understood in the light of the complete phase behavior.

2. Critical Electrolyte Concentration

If a large amount of salt is added, phase separation may be completely suppressed. The concentration of salt needed is called the critical electrolyte concentration, abbreviated cec.[62] For a given salt, cec depends on the chain length of the surfactant and on the concentrations of both the surfactant and the polyelectrolyte. cec for NaBr was investigated in systems of 1.0 m*M* NaHy and C_nTAB,[62] and it was concluded that the cec increases with increased surfactant chain length. An increased linear charge density of the polyelectrolyte also increases the cec.[72]

Addition of excess surfactant to the system may also entail redissolution of the polyelectrolyte-surfactant dispersion. In the NaHy-C_nTAB systems, redissolution was attained at a relatively high surfactant concentration (far above the CMC).[62,71] Redissolution (sometimes called resolubilization) for some systems is reported to occur at very low surfactant concentrations. For instance, a sample of 0.1 wt% of the cationic cellulose derivative Polymer JR phase separates at 1.0 m*M* of SDS, but at 6 m*M* of SDS, the sample is again a clear one-phase solution.[139] This behavior was attributed to a charge reversal of the polyelectrolyte-surfactant complex, excess surfactant binding giving soluble polymer-surfactant complexes of a reversed charge. It is, furthermore, reported that for cationic polyelectrolytes, those of a relatively low charge density display this behavior, while for more highly charged polyelectrolytes, a very high surfactant concentration ($\gg$ CMC) is needed in order to bring about redissolution.[134] Apparently, redissolution by excess surfactant may proceed by different mechanisms and our understanding of the redissolution phenomenon is not complete. Obviously, a degree of binding above one (i.e., bound surfactant charges divided by polymer charges) is not sufficient for redissolution to occur. If redissolution proceeds by a charge reversal mechanism, the free surfactant monomer concentration at redissolution must be lower than the CMC; if free micelles are formed in the system, the formation of these will be favored at further addition of surfactant (the free monomer concentration decreases above the CMC for ionic surfactants) and the degree of binding to the polyelectrolyte will consequently decrease.

3. Ternary-Phase Diagrams

If phase-separating samples of Hy and C_nTAB are left for some days, macroscopic separation into two phases will occur. Interestingly, two transparent and apparently isotropic phases result, of which one is dilute and one concentrated with respect to both Hy and surfactant. The concentrated phase is often of marked viscosity, and is also called the ''gel'' phase.

The two-phase regions for systems of Hy and C_nTAB were mapped out by a procedure including thorough equilibration of phase-separating samples, followed by physical separation of the two phases and analyses of the contents of all ionic species in the dilute phase.[135-138] Systems of a polyelectrolyte and an oppositely charged surfactant in aqueous solution are four-component systems. (There are four different ionic species and the solvent; the requirement of charge neutrality reduces the number of independent components to four.) We thus need a three-dimensional phase diagram in order to fully represent the system. Such a phase diagram has been developed, and is briefly discussed below. For many purposes, however, it is convenient to represent the systems as pseudo-three-component ones, using the more conventional triangular type of phase diagram. The polyanion, the surfactant cation, and the water are then considered as the main species of the system, i.e., the distribution of Na^+ and Br^- ions between the two phases is disregarded.

Such a phase diagram is given in Figure 11. A droplet-shaped two-phase region results, and is anchored in the water corner. The tie lines are directed from this corner and the water-surfactant side towards the NaHy-$C_{14}TAB$ side of the diagram. The two-phase region displays a marked dissymmetry with respect to the bisector of the water corner. Addition of small amounts of Hy to a micellar solution immediately causes phase separation, while some surfactant can be contained in a (more concentrated) Hy solution.

Systems of sodium poly(acrylate), NaPA, and the same cationic surfactants show substantial similarities to the NaHy systems.[72] The two-phase region is somewhat larger, however, when the same cationic surfactant is

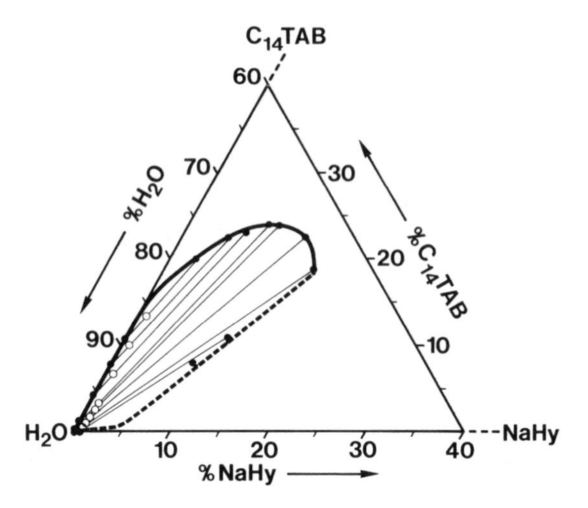

FIGURE 11. Pseudo three-component phase diagram for the system NaHy-C_{14}TAB-H_2O. Open circles refer to initial sample compositions and filled circles connected by tie lines refer to the two phases in equilibrium. (From Thalberg, K., Lindman, B., and Karlström, G., *J. Phys. Chem.*, 94, 4289, 1990. With permission from the American Chemical Society.)

used, which can be ascribed to the higher linear charge density of PA, and is in agreement with the stronger interaction in dilute solution for this polyelectrolyte. Another difference is that the two-phase region extends in the direction of charge neutrality between surfactant and polyelectrolyte in the NaPA system. This is not the case in the systems with NaHy, where the surfactant is in excess in the concentrated phase.

Preliminary investigations in a system of a polycation (PDADMAC) and anionic surfactants indicate that the interactions are stronger in such systems than in the polyanion-cationic surfactant systems.[72] This would be analogous to observations for systems of uncharged polymers, where anionic surfactants display a stronger interaction than cationic ones. The strong interactions often lead to the formation of dense opaque precipitates and make it difficult to obtain an equilibrium between the separating phases. The systems also show a weaker salt dependence than those of a polyanion and a cationic surfactant.

It is worth noting that this phase behavior is of the same type as that of the system gelatin-gum arabic, described by Bungenberg de Jong[140] in the first part of this century and named "complex coacervation". This suggests that the present system can be regarded as belonging to the type "colloid anion + colloid cation". More recent studies of systems of two oppositely charged linear polymers also give phase diagrams of this type.[141] Theoretical modeling of the observed phase behavior, using a mean-field theory for two polymers in a common solvent, further emphasizes the similarity between the present systems and systems of two favorably interacting polymers (see below).

4. Effect of Surfactant Chain Length and Polyelectrolyte Molecular Weight on Phase Behavior

When a cationic surfactant of shorter hydrocarbon chain is used, the area of the two-phase region decreases.[136] This is due to the weaker attraction between the polyelectrolyte and the surfactant, and to the smaller micelles formed by a shorter surfactant. The shape and position of the two-phase region are largely retained, as well as the slope of the tie lines. For $C_{10}TAB$, the area is considerably smaller than for the other surfactants. This can be explained by the relatively high concentration of free surfactant monomers in this system (cac $\approx$ 50 to 60 m*M*). The surfactant monomers with their counterions act as simple salt in the system and screen the interaction between the surfactant aggregates and the polyelectrolyte. This effect is still more pronounced in the C_9TAB system, where the two-phase region is totally absent. For C_9TAB, the CMC is about 140 m*M*;[46] this concentration of electrolyte is alone sufficient to prevent phase separation.

The polyelectrolyte molecular weight is of minor importance for the phase behavior.[136] A slightly increased polyelectrolyte concentration in the concentrated phase is observed when Hy of a lower molecular weight is used.

5. Effect of Salt on the Phase Behavior

Addition of salt leads to a reduction of the area of the two-phase region. Addition of 75 m*M* NaBr to the NaHy-$C_{14}TAB$-water system causes a reduction of the area by more than 50%.[137] From a closer examination, it was concluded that the polyelectrolyte binds more surfactant per repeating unit when salt is added. Further addition of salt to the system will make the two-phase region totally disappear. On addition of 250 m*M* NaBr to the NaHy-$C_{14}TAB$ system, phase separation is no longer observed.[137] (Again, we note a behavior parallel to the "complex coacervation", reported by Bungenberg de Jong.[140])

A striking observation is that, at relatively high salt concentrations, phase separation again occurs in the NaHy-$C_{14}TAB$ system with two clear and isotropic phases in equilibrium.[137] The phase behavior at high salt concentrations is, however, quite different from that observed at no or low salt concentrations. Without added salt, a phase concentrated in both polyelectrolyte and surfactant separates out, while at high salt concentrations, the supernatant phase is enriched in surfactant and the bottom phase in polyelectrolyte. Phase diagrams at 0.5 and 1.0 *M* of added NaBr are shown in Figure 12, together with phase diagrams at 0, 75, and 250 m*M* of NaBr. It is clear that the phase behavior of the NaHy-$C_{14}TAB$ system is extremely salt dependent, indicating that it is mainly governed by electrostatic interactions. At high salt concentrations, where the electrostatic interactions are highly screened, the phase behavior is related to that of two uncharged polymers in a common solvent, normally referred to as "polymer incompatibility."[121]

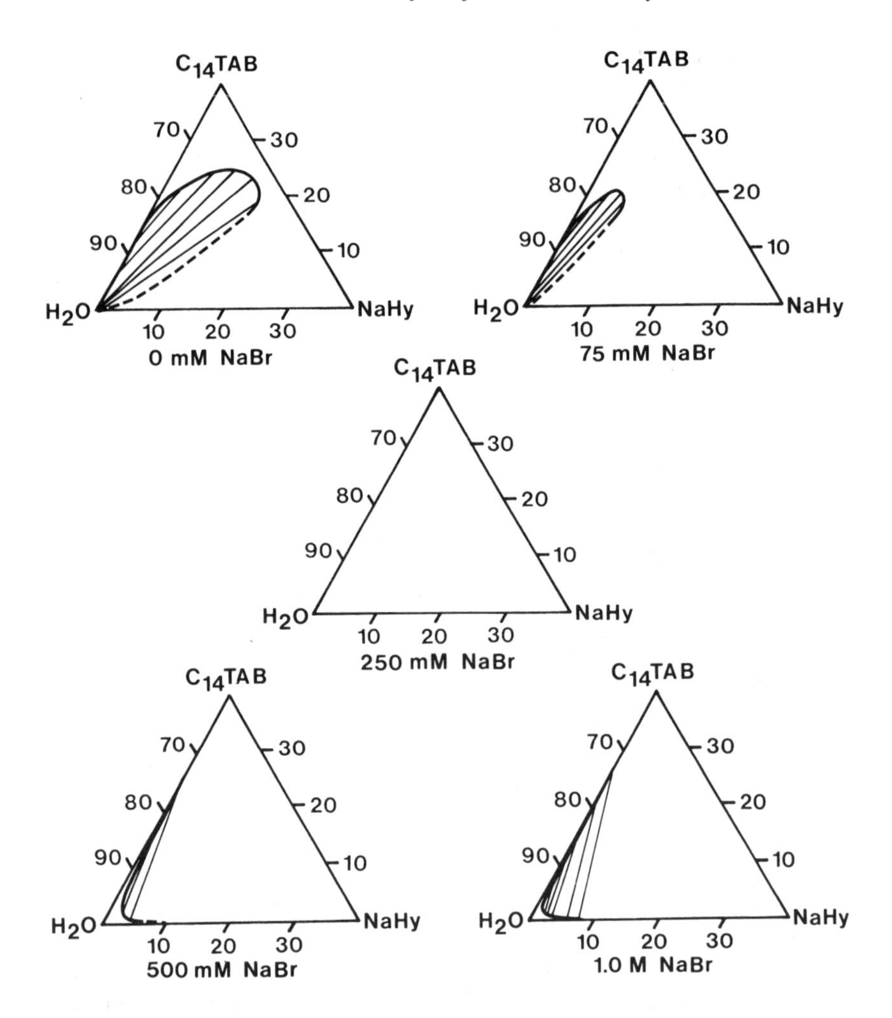

FIGURE 12. Pseudo three-component phase diagrams for the system NaHy-C_{14}TAB-H_2O at different concentrations of added salt. (From Thalberg, K. and Lindman, B., *Surfactants in Solution,* Mittal, K. and Shah, O. O., Eds., Plenum Press, New York, 11, 243, 1991. With permission.)

6. Pyramid-Shaped Phase Diagrams

A three-dimensional representation, shaped as a pyramid, is shown in Figure 13.[137] Water is placed at the top and the four sides of the pyramid base are assigned to the four ionic components of the system, with ions of the same charge located at opposite sides. In this way, the corners of the base will correspond to the four possible salts; in the present case Na^+Hy^-, Na^+Br^-, $C_{14}TA^+Br^-$, and $C_{14}TA^+Hy^-$. It is suitable to use molar concentrations of the ionic species in the pyramid representation; a horizontal plane through the pyramid then includes all compositions which have the same total ionic concentration (e.g., 1.0 *M* of NaHy, 1.0 *M* of C_{14}TAB, or 0.5 *M* of NaHy

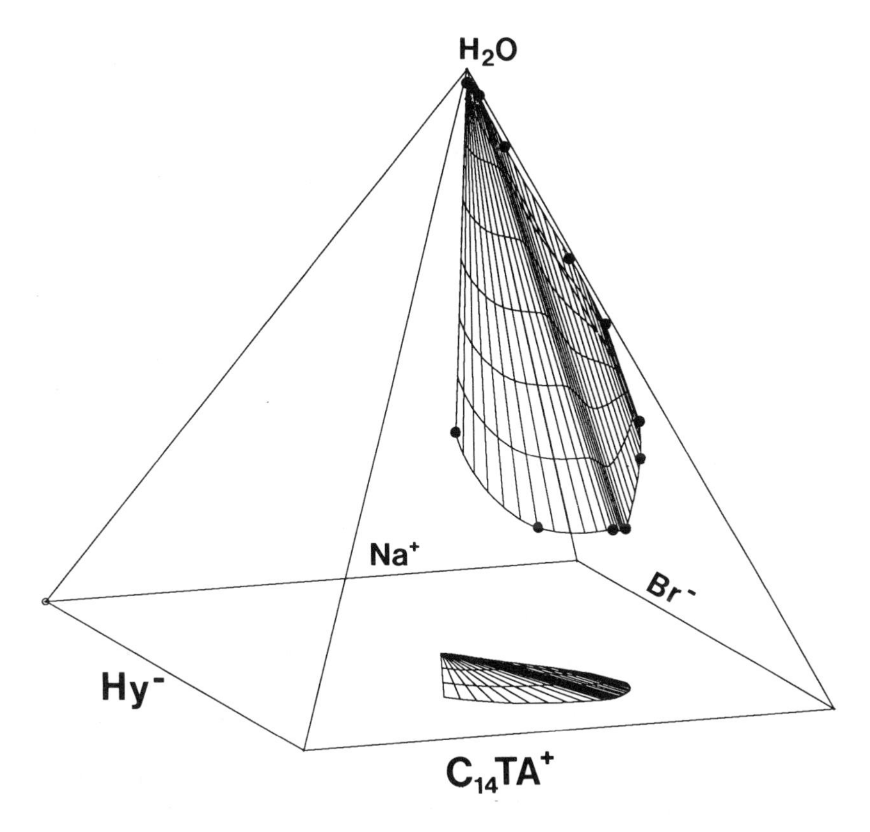

FIGURE 13. Pyramid representation for the system of Na^+, Hy^-, $C_{14}TA^+$, Br^-, and H_2O. The "sail" hanging from the pyramid top corresponds to the two-phase region of Figure 11. The projection of this sail on the pyramid base is also shown. The indicated base refers to a total concentration of ionic species of 1.0 *M*. (From Thalberg, K. et al., *J. Phys. Chem.*, 95, 6004, 1991. With permission from the American Chemical Society.)

and 0.5 *M* of $C_{14}TAB$). The distance of this plane from the pyramid base is roughly proportional to the water content.

In the pyramid of Figure 13, the phase-separating samples shown in Figure 11 are included. The samples form a "sail", hanging from the top of the pyramid. The dilute phase is located to the upper right side, and the concentrated phase to the lower side of this sail. In order to better visualize the position of the sail, its projection on the pyramid base is also shown.

The pyramid representation may contribute to a conceptual progress for this type of systems. Obviously, a two-phase region translates into a three-dimensional body in the pyramid. The sail shown in Figure 13 corresponds to a cut through this body. Samples with 75 m*M* of added NaBr make another cut through the body. The absence of phase separation at 250 m*M* of added NaBr indicates that the body is limited in extension towards the NaBr corner. The different type of phase separation, which occurs at high NaBr concentrations, corresponds to another "two-phase body", located in the same pyramid. The pyramid representation can be applied to all systems containing four different ionic species, two of each charge.

7. Theoretical Modeling of the Phase Behavior

In order to rationalize the observed phase behavior of NaHy-C_nTAB systems, a simple theoretical model was introduced, which is based on the F-H theory for polymer solutions (see Section II).[135] The systems are treated as three-component systems of solvent (index 1), polymer A (index 2, representing the polyelectrolyte), and polymer B (index 3, representing the surfactant). The phase behavior is then fully determined by the three interaction parameters, w_{ij}, and the polymerization numbers for polymer A and B, denoted L_2 and L_3. (It is impossible, even qualitatively, to obtain the experimentally observed kind of phase behavior without treating solute B as a second polymer component.) Calculations show that a phase diagram with a closed two-phase region, located in the water-rich part of the system, results if the interaction between A and B is more favorable than the interaction of the polymers with the solvent. If $L_2 = L_3$ and $w_{12} = w_{13}$, the two-phase region is symmetric with respect to the bisector of the water corner. By decreasing w_{12} relative to w_{13} (i.e., making polymer A more hydrophilic than polymer B), and increasing L_2 relative to L_3, a phase diagram similar to the experimentally observed one for the NaHy-$C_{14}TAB$ system (without added salt) results.

The reduced area of the two-phase region, resulting from a decrease in surfactant chain length, can be accounted for both by a reduction of the polymerization number for polymer B (since a surfactant of shorter chain length forms smaller micelles), and by a reduction of the interaction between polymers A and B.[136] A reduced polyelectrolyte molecular weight corresponds to a lowering of L_2 in the model. Changes in L_2 cannot, however, fully explain the experimentally observed phase behavior.[136] Obviously, more sophisticated models are required in order to give an explanation of the effect of polyelectrolyte molecular weight on the phase behavior.

The salt dependence of the phase behavior introduces a new challenge to the model. In order to circumvent the introduction of a fourth component in the model system, the added salt can be treated as included into the water component. By this approach, a phase behavior of the same kind as experimentally observed in the presence of 1 *M* NaBr (Figure 12), can be obtained.[137] The crucial condition for this kind of behavior is that w_{12} is lower than w_{23}, which, translated to the real system, means that the polyelectrolyte interacts more favorably with the solvent than with the surfactant micelles. It is pointed out that the F-H model is a crude tool for the modeling of polymer-surfactant systems. Apart from (well-known) general limitations of the model, such as only including nearest-neighbor interactions and assuming the same size of all the component monomers, it is also clear that the model calculations assume only one state for each component (e.g., the free surfactant monomers are not taken into account) and, furthermore, that the properties of a component are constant over the entire concentration range. This implies that the micelles

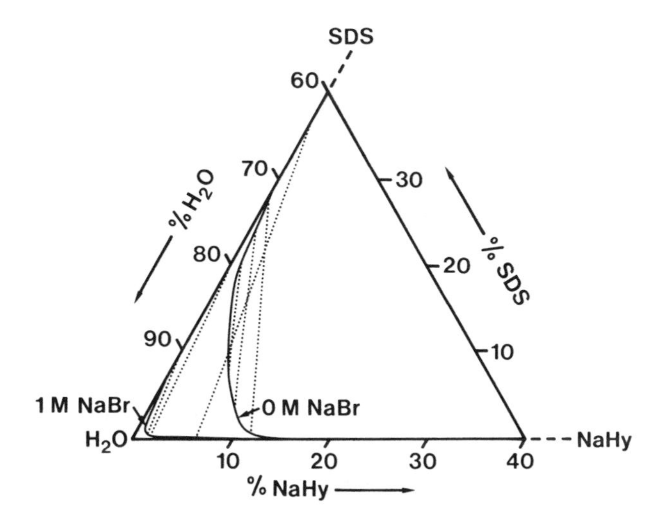

FIGURE 14. Phase diagram for the system NaHy-SDS-water at 40°C without added salt and at addition of 1.0 *M* NaBr.

are assumed to be of the same size whether they are polymer-bound or not. Above all, it will be important in further modeling to account for the long-range electrostatic interactions.

C. POLYELECTROLYTE AND SIMILARLY CHARGED SURFACTANT

As systems of polyelectrolyte and oppositely charged surfactant show phase separation of the "polymer incompatibility" type when the electrostatic interactions are screened out, one may expect to observe phase separation of this type in systems of polyelectrolyte and surfactant of the same charge, and also in systems of uncharged polymer and ionic surfactant in the presence of salt. Phase diagrams have recently been established for two similarly charged polymers in a common solvent[142] and compared to the diagrams for one charged and one uncharged polymer.[143] These results may give a clue to the behavior of the corresponding polymer-surfactant systems. In general, it was found that phase separation occurs more readily when both polymers are (similarly) charged, and that the sensitivity to added salt, which is very strong in the charged-uncharged polymer mixture, is weakened considerably.

Recent results in our laboratory have shown that phase separation of the "polymer incompatibility" type occurs in a system of NaHy and SDS. The phase diagram for this system is presented in Figure 14. It can be noted that, as the polyelectrolyte and the surfactant have the same counterion, this is a true three-component system. When 1.0 *M* of salt (NaBr) is added, phase separation is seen to be markedly facilitated.

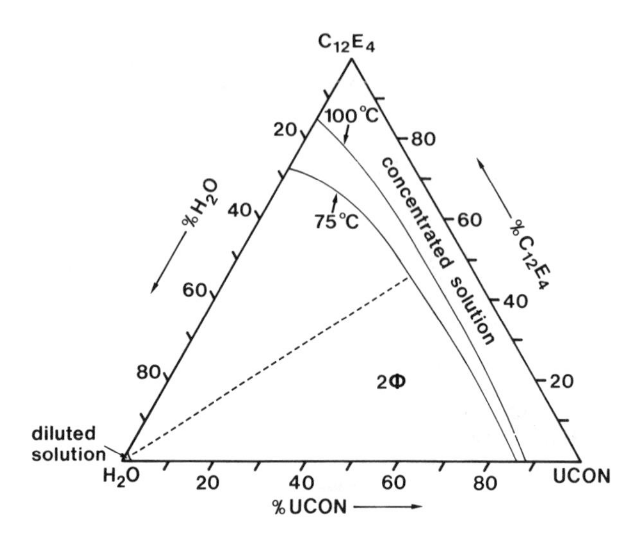

FIGURE 15. Phase diagrams for the system UCON-$C_{12}EO_4$-water at 75 and 100°C. (From Reference 141.)

D. UNCHARGED POLYMER AND NONIONIC SURFACTANT

A number of ternary-phase diagrams of systems of a nonionic polymer and a nonionic surfactant at different temperatures have recently been determined by two groups.[144,145] Surfactants studied by our group are $C_{12}E_4$ and $C_{12}E_8$ and the polymers are EHEC and UCON, a random copolymer of EO and PO. Depending on the polarity of the two cosolutes, different types of phase diagrams are obtained. For example for $C_{12}E_4$ (or $C_{12}E_8$) and UCON at higher temperature (Figure 15) there is a separation into one dilute solution phase and one phase concentrated in both surfactant and polymer, pointing to an appreciable attractive interaction.

Wormuth[145] studied systems of ethoxylated alcohols and PEO and found that miscibility of the two cosolutes decreases strongly upon increasing the molecular weight of either the surfactant or the polymer. This work also emphasizes the significance of micelle formation for phase separation and finds phase behavior patterns that resemble those of two polymers in a common solvent.

These studies demonstrate that, depending on the conditions, both aggregation and segregation phenomena are possible. Most probably the hydrophobicity of the polymer will be the most important factor in determining the type of phase separation.

V. NOVEL GELLING STRUCTURES

One of the most significant aspects of polymer-surfactant systems from a practical point of view is that of rheology control and viscosity enhancement, and several attempts have been made in recent years to design novel gelling structures. In addition to a study by Brackman[146] on the development of viscoelasticity in very high molecular weight polymer (PEO) solutions by introduction of an ionic surfactant, work can be broadly grouped into three categories: (1) crosslinking of hydrophobe-modified polymers by surfactant; (2) polyelectrolytes interacting with ionic surfactants; and (3) using the temperature-dependent solvation of certain nonionic polymers to obtain a thermal gelation with ionic surfactants.

One example of the dramatic increases in viscosity that a surfactant can produce in a solution of a HM polymer was given above (Figure 5). The rheology is very sensitive to the surfactant-to-polymer ratio. (See References in Section III.F.) The viscosity enhancement is due to crosslinking of the polymer chains by surfactant micelles, with the formation of these crosslinks being strongly facilitated by the presence of pendant alkyl chains or other hydrophobic centers. At higher surfactant content the number of polymer groups per micelle decreases and the viscosity enhancement is lost. These phenomena are quite general and different types of HM polymers interacting with different types of surfactants have been studied. For polyelectrolytes, electrolyte addition has a strong effect and can be utilized to control rheology. HM polymers constitute a very intense field of research and some other aspects have been discussed above (Section III.F).

Very interesting novel gels in quite dilute polymer-surfactant systems were recently described by Goddard et al.[147] and Leung and Goddard.[148] This group previously reported on the dramatic increases in viscosity which occur on adding small amounts of an anionic surfactant to a solution of a cationic cellulose derivative (see Chapter 4, Part II). In their recent work they have shown that by using a higher molecular weight polymer, even more dramatic increases in viscosity are encountered in the presence of SDS, resulting in a formation of transparent and strong gels. A careful rheological characterization was reported for different compositions, and as exemplified in Figure 16, the elastic properties dominate over the viscous ones even down to very low frequencies in an oscillatory measurement.

These gels, which increase in strength with the degree of cationic substitution, appear in quite dilute systems (1% and below of polymer), i.e., in the preprecipitation range of the phase diagram. In our characterization of phase separation of quite analogous systems, gels were identified in another range of compositions. Thus it is found, with hyaluronan and polyacrylate as examples, that on exceeding the precipitation limit of the surfactant concentration, there is a separation into two clear phases, one low viscous and dilute in both solutes and one which is typically quite stiff and is concentrated in both polymer and surfactant. This gel is transparent, very stable (and coexists with a dilute aqueous solution), contains hydrophobic domains, and is characterized by an extremely slow surfactant diffusion.[149] Rheological characterization (Figure 17) shows that the loss modulus dominates at low frequencies and the elastic modulus at high, but indicates no fundamental difference to concentrated solutions of the polymer alone.[71]

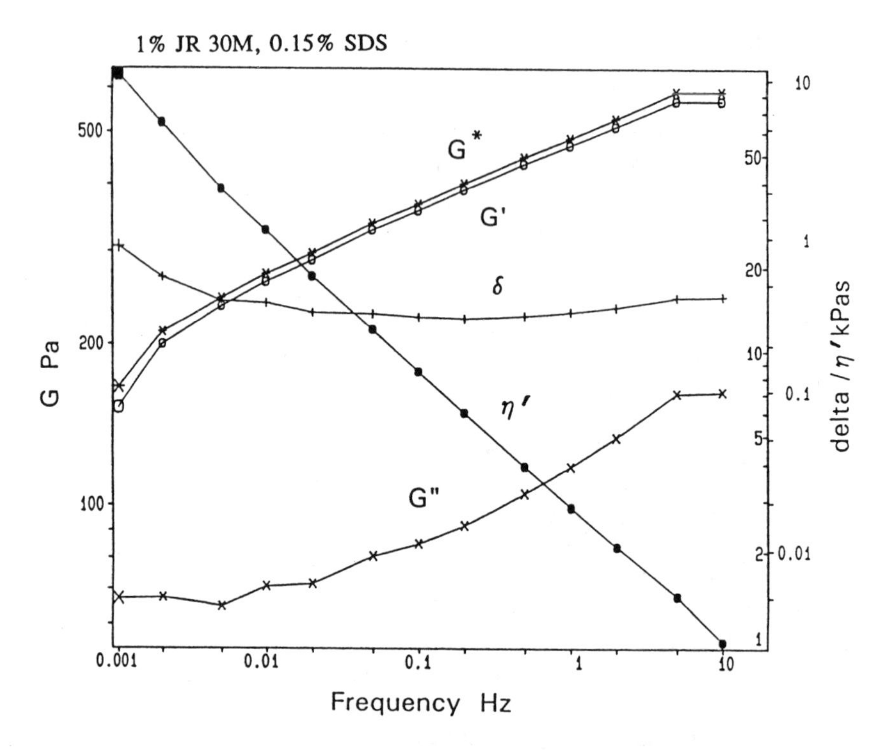

FIGURE 16. Viscoelastic parameters vs. oscillation frequency for 1% Polymer JR 30M (a high molecular weight cationic cellulose ether) — 0.15% SDS mixture. The parameters are elastic modulus, G′; loss modulus, G″; complex modulus, G*; phase angle, δ; and dynamic viscosity, η′. (From Leung, P. S. and Goddard, E. D., *Langmuir,* 7, 608, 1991. With copyright permission from the American Chemical Society.)

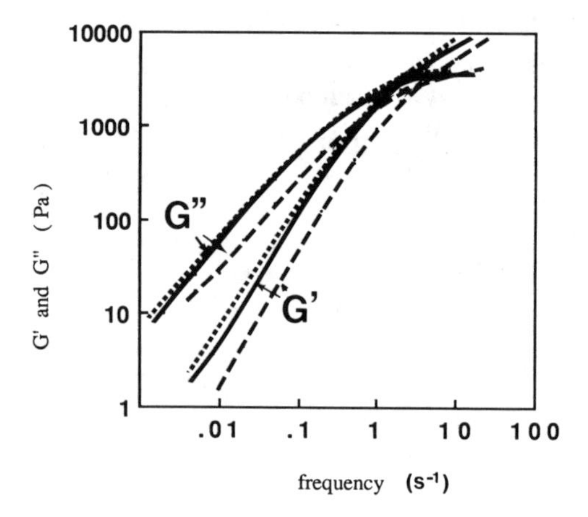

FIGURE 17. Storage (G′) and loss (G″) moduli as a function of oscillation frequency for two Hy-C_{14}TAB gels (full and dotted lines) and for a Hy solution of the corresponding concentration. (From Reference 71.)

Time-resolved fluorescence quenching studies of these gels give aggregation numbers, which are approximately the same as in the binary surfactant-water systems.[65] NMR spectra of the concentrated phase are quite similar to those of normal micellar solutions, thus indicating a close structural similarity. Increased linewidths are, however, found both in ^{1}H and ^{13}C spectra, indicating a faster relaxation due to the presence of the Hy polymer.[10,149] Recently, Wong et al.[10] performed a multifield ^{13}C relaxation study on the Hy-C_{14}TAB system, including also measurements of the differential line broadening, DLB, which is a T_2 type of measurement (i.e., a measure of the transverse relaxation). In order to fully decipher the surfactant dynamics in the presence of Hy, a "three-step" model had to be invoked, where the third step of motion (with a time-scale of 10^{-7} s) is due to the association of the micelle with the polyelectrolyte. Interestingly, the third step was not needed for samples with a high polyelectrolyte-to-surfactant ratio, and it was inferred that the surfactant micelles are rather evenly and densely covered with polyelectrolyte chains at high polyelectrolyte-to-surfactant ratios. In general, it was concluded that the order parameter, S, and the fast correlation time, τ_c^f, are quite similar in normal micelles and in polyelectrolyte-surfactant systems, pointing to a close structural similarity and a rather loose association between the polyelectrolyte chains and the surfactant micelles. This is in agreement with the small contribution from the bound micelles to the rheology of these systems (Figure 17), pointing to a loose and dynamic character of the binding. The mesh size of the polymer network was estimated according to de Gennes,[150] giving a value of about 5 nm. This is quite similar to the diameter of a C_{14}TAB micelle,[151] suggesting that the micelles fit well into the spaces between the Hy chains. We can thus picture the concentrated phase as an entangled polyelectrolyte solution with the surfactant micelles placed in the meshes of the network.

Many practically used nonionic polymers, like cellulose ethers and copolymers of ethylene oxide and propylene oxide, show a phase separation at elevated temperature in aqueous solution. As already mentioned, a strong increase in the cp results from the addition of ionic surfactants. The systems of nonionic polymer and ionic surfactant were considered to be particularly relevant for the attempt to produce novel structures with novel rheological properties since (physical) crosslinking effects can be modulated both by an ionic surfactant and by temperature. Thus, as these polymers become less polar at a higher temperature, they will offer better nuclei for surfactant self-assembly at a higher temperature. In the semidilute regime there should be a possibility for the surfactant micelles of crosslinking polymer chains.

Studies of these aspects have concentrated on samples of a nonionic cellulose ether, ethylhydroxyethyl cellulose (EHEC), with different polarities and thus, different cloud points. An aqueous solution of about 1% EHEC shows a moderate increase in viscosity compared to neat water; this decreases on increasing the temperature. In the presence of a small amount of surfactant (0.1 to 0.5%), the viscosity is relatively low and the solutions are easy flowing

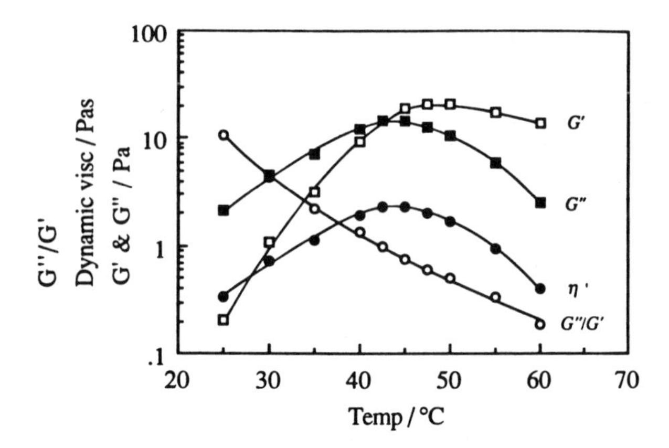

FIGURE 18. Viscoelastic properties vs. temperature at constant frequency (1.0 Hz) for 1.0 wt% EHEC and 0.36 wt% $C_{16}TAB$ in water. (From Carlsson, A., Karlström, G., and Lindman, B., *Colloids Surf.*, 147, 47, 1990. With permission.)

at low temperature. However, on increasing the temperature, a dramatic increase in viscosity is observed,[124,152] as illustrated in Figure 18. At low temperature the solutions are less viscous and even less elastic, and the rheological properties are dominated by loss effects. At higher temperatures, while the viscosity increases dramatically, the elastic effects become totally dominating. The system can be characterized as changing with temperature from a modestly viscous solution to a gel; the process is reversible and the gel melts on cooling. The molecular mechanism is apparently a crosslinking combined with surfactant self-assembly, this being facilitated at a higher temperature, where the polymer is more hydrophobic. The temperature for this gel formation is related to the cp of the polymer-water system and can be varied at will by changing the substitution.

Recently, different groups have characterized these gels by different techniques, such as (dynamic and static) light scattering, rheology, electrical conductivity, self-diffusion, and fluorescence quenching. Rehage[153] showed that on increasing the temperature, the elastic properties increase in importance and the frequency or time dependence becomes less pronounced, which is typical for dynamic or transient networks, which seem to be more stable at elevated temperatures. The results show the same characteristics as can, for example, be observed for entangled polymer solutions. No yield value was observed and relaxation occurs even at the higher temperatures, demonstrating that, as expected, there are no permanent crosslinking processes. Consistent results on the transient network structure was obtained in the light scattering work,[154] which also demonstrated a coil expansion with increasing temperature for the dilute solutions and structural changes in the network of a semidilute solution on approaching the sol-gel transition. All observations point to an enhanced binding of surfactant to the polymer at a higher temperature and this is also directly observed in CMC determinations by electrical conductivity,[155] and by determinations of the surfactant activity by surfactant-selective electrodes.[156] Counterion self-diffusion, conductivity, and fluorescence lifetime[155] all show that the micelle ionization degree increases with increasing temperature, indicating a more intimate polymer-surfactant association at a higher temperature. The fluorescence quenching work also demonstrates that the size of the polymer-bound surfactant aggregates decreases appreciably with increasing temperature; this was previously inferred indirectly from the slopes of cp curves.[47]

As pharmacologically active, water-soluble substances can be rather generally dissolved in these polymer-surfactant systems, it appeared that these systems, which have low viscosity at room temperature but spontaneously form a stiff gel at body temperature, can offer improvements in drug administration and drug delivery.[157] The gelling of the nonionic polymer-ionic surfactant system described above is largely independent of the presence of other species in the medium and requires very low concentrations of polymer and surfactant (98 to 99% water). Furthermore, the system is compatible with different ways of administration and has a long-term stability. The special EHEC quality used in this work has a cp in the range of 30 to 35°C. The gel-forming ability *in vitro* has been demonstrated in a number of aqueous solutions, provided the temperature is sufficiently high and an ionic surfactant is present in appropriate concentrations, which in the case of SDS means 0.05 to 0.15 wt%. The EHEC-ionic surfactant combination, forming a gel in simulated gastric juice, is a drinkable liquid at room temperature.

Such thermoreversible gels may have broad applicability as a "liquid carrier" for delivery of water-soluble drugs. The system has a viscosity that allows spraying, instilling, pouring, drinking, or spreading the dosage form into the intended biological cavity or part of the body. Upon administration the "liquid carrier" will adhere to the mucus or the biological membrane and form a high-viscosity or gel layer. It has been shown that it gels in gastric juice and that this gel is also retained in intestinal juice. This implies that on oral administration of a drug in the carrier a gel will form in the gastrointestinal tract giving a slow release of the active substance (and the surfactant) as well as an improved bioavailability. Once the gel is formed it is very resistant to salt and mechanical rupture.

VI. POLYMERS IN CONCENTRATED SURFACTANT SOLUTIONS. MICROEMULSIONS AND LIQUID CRYSTALS

Surfactant self-assembly at high surfactant concentrations shows a rich structural variation, governed by the spontaneous curvature of the surfactant aggregates and interaggregate interactions. Structures include rod and disc micelles, lamellar, hexagonal, and cubic liquid crystals, vesicles, and microemulsions. The incorporation of polymer molecules in these various structures as well as the effect of polymers on phase and aggregate stability are issues of broad fundamental, biological, and technical interest, but basic work in the field is very sparse. A broad and general discussion of key issues was attempted by de Gennes,[158] in particular on the conditions of incorporation of polymer molecules in surfactant films. A wide variety of behavior can occur, depending on the polymer. In particular, when mixing of the two solutes is favored, bilayers in a lamellar phase are predicted to be stiffened. When bilayers are exposed to polymer on one side only, strong bending is predicted. The theoretical arguments are compared with the observed effect of PEO on bilayer structure in a lamellar phase of surfactant and water observed by Kekicheff et al.[159]

Recent experimental work has included studies of the effect of polymer on the stability of rod micelles and vesicles and the incorporation of block copolymers and gelatin in W/O type microemulsions. Brackman and Engberts have demonstrated that long rod micelles may be transformed into small micelles on introduction of a polymer.[50,51] Tirrell and colleagues, in developing certain functional systems, have demonstrated that adsorption of polymer on a lecithin vesicle, phosphatidylcholine and poly(2-ethylacrylic acid), may induce transformation to small micelles.[160-164] This system provides an interesting approach to polymer-driven photoinduced release of vesicle contents. Wakita et al.[165] described formation of multilayered liposomes from C_{16}TAB and polyacrylate. The bromide counterions of the surfactant were first replaced by polyacrylate, which led to formation of a complex which was soluble in short-chain alcohols. Injection of the solution into water rendered a translucent liposome solution, which was stable for more than 48 hr.

If a polymer is added to a microemulsion, this is expected to strongly affect microstructure, stability, and dynamics. A nice system has been devised by Eicke's group[166,167] who, by a range of physicochemical techniques, study the addition of an ABA, poly(oxyethylene) (A)-poly(isoprene) (B), block copolymer to microemulsions of Aerosol OT-isooctane-water. It is found that for a range of conditions, the structure of nanometer-sized W/O droplets is retained. However, by the association of the A segments with the water cores of two different droplets, dimers of droplets may form. At higher polymer concentrations, larger aggregates are formed and, at very high copolymer concentrations, the formation of large aggregates and branching leads to a three-dimensional network and a gel.

The introduction of a gel-forming polymer into microemulsions has fascinated several groups during the past few years,[168-170] and all these groups chose the system of gelatin and Aerosol OT as a suitable model system. Physicochemical characterization has been undertaken by a very broad range of techniques, including different scattering methods, electrical conductivity, diffusion, rheology, Kerr effect, and optical rotation. Additionally, schematic phase diagrams have been worked out, showing, *inter alia,* the conditions for forming gels. In suggesting a structure of the gels formed (at relatively high gelatin contents), it is important to take into account both the conditions of gel formation in aqueous solutions and the geometrical constraints imposed

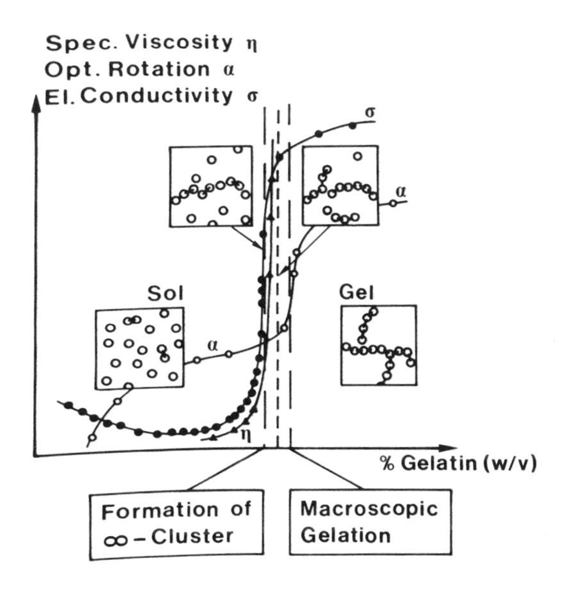

FIGURE 19. Schematic view of percolation in the gelatin — W/O microemulsion. (From Quellet, C., Eicke, H.-F., and Sager, W., *J. Phys. Chem.*, 95, 5642, 1991. With permission.)

by the surfactant concentration. Recent work from Eicke's group provides very significant insight.[168] As can be inferred from Figure 19, there is strong correlation among rheological properties and optical rotation and conductivity, showing that the formation of extended structures is coupled to conformational changes of the polymer. It is reasonable to assume that at low gelatin contents, the polymer is localized within the water droplets of the microemulsion, while at higher concentrations, there is an aggregation of droplets associated with an interdroplet helix formation. It is significant that the final step of gelation coincides with a steep rise of the helix fraction. The structure of these gels remains controversial and, for example, rather different exposure of the gelatin strands to the hydrocarbon solvent is found in different investigations. These differences can perhaps be accounted for by the rather different compositions which have been investigated. As Eicke has pointed out, it seems unlikely that the polymer molecules are much exposed to the nonpolar continuum, especially if enough water is available for solvating the polar groups.

VII. ADSORPTION FROM POLYMER-SURFACTANT SOLUTIONS

From a practical point of view, the very important problem of the interaction between an interface and a solution containing both a polymer and a surfactant has received much attention in applied studies. However, much more fundamental work by appropriate experimental techniques has yet to be made. Obviously, these systems are complex and many parameters are significant, and progress in the field presupposes a solid understanding of polymer-surfactant solutions, polymer-interface, and surfactant-interface interactions separately. Interactions between polymer and surfactant, either repulsive or attractive, both in the bulk solution and in an adsorbed layer, need to be discussed and we need to investigate polymer-surfactant complexes both at the interface and in the bulk. Information on the influence of one of the solutes on the adsorption of the other needs to be obtained. Electrostatic and hydrophobic interactions are typically the most important, and they can occur within any of the pairs interface-surfactant, polymer-surfactant, and interface-polymer. In most systems of interest, both polymer and surfactant will have a tendency to adsorb. For charged surfaces in the presence of polymer and surfactant, where one or both are charged, a complex pattern of electrostatic interactions emerges.

Polymer-surfactant systems at the air/water interface are one major field of study, and systems with various solid surfaces another. The significance of the former has mainly been assessed in terms of surface tension as a very convenient identification and measure of polymer-surfactant interactions (Chapter 4). Thus, the observation that the addition of an uncharged polymer to a surfactant solution raises the surface tension corresponds to a lowering of the surfactant activity due to (attractive) polymer-surfactant interactions. For very hydrophilic polymers, mainly the desorption of surfactant from the surface needs to be considered, but for many cases, both polymer molecules and polymer-surfactant complexes are surface-active as well and the analysis becomes more involved. A recent attempt to more fully analyze adsorption at the air/water interface from polymer-surfactant solutions is due to Chari and Houssain,[171] who (for PVP + SDS) complemented surface tension measurements with direct studies of component adsorption by a radiotracer technique. It is inferred that polymer as well as surfactant adsorbs at the surface and that PS complexes at the interface resemble those in the bulk.

Polymer-surfactant adsorption on solid surfaces is reviewed in Chapter 4. Two novel techniques, which allow a more detailed monitoring of adsorption, have recently been introduced. One is *in situ* ellipsometry, which allows a precise determination of the thickness of an adsorbed layer under equilibrium conditions as well as a function of time. The other technique is that of directly measuring the forces between two surfaces as a function of distance down to very small separations. The surface force technique has proved to be extremely useful for studies of adsorbed layers of a surfactant or a polymer alone.[20]

Claesson et al.[172] made a combined ellipsometry and surface force study of a system of a nonionic cellulose derivative, EHEC, and SDS interacting with hydrophobic surfaces. These were hydrophobized silica surfaces for the ellipsometry, and mica surfaces hydrophobized by a Langmuir-Blodgett technique for the surface force measurements.

Both EHEC and SDS adsorb separately on hydrophobic surfaces but the kinetics are very different, SDS adsorption taking place on a timescale of

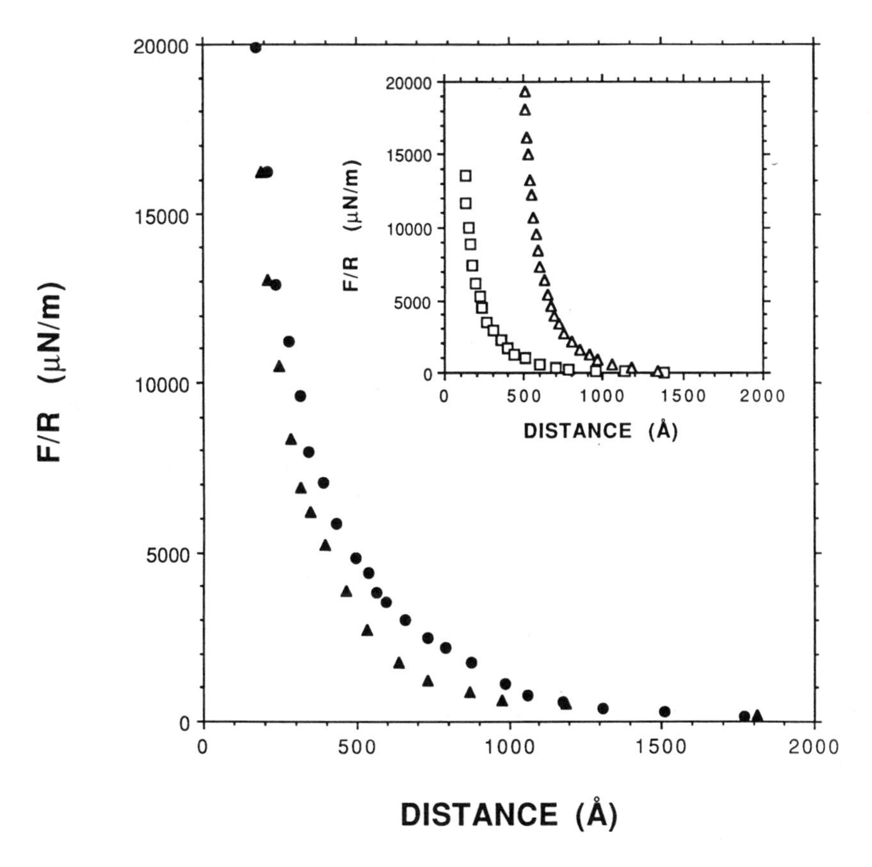

FIGURE 20. A comparison of the forces acting between hydrophobic surfaces across a solution containing 0.25 wt% EHEC and 4 m*M* SDS at 20°C (filled circles) and 35°C (filled triangles). The insert shows the forces measured without SDS at 20°C (open squares) and 37°C (open triangles). From Claesson et al., *Langmuir,* 7, 1441, 1991. With permission from the American Chemical Society.)

seconds while EHEC adsorption takes hours. EHEC adsorption on hydrophobic surfaces increases strongly with increasing temperature. On addition of SDS to a surface with preadsorbed polymer, the adsorbed amount decreases slowly to a lower level, but there is still significant polymer adsorption, and at the same time there is a decrease in the average adsorbed layer concentration.

As regards the forces between two EHEC-covered surfaces, they also show a peculiar temperature dependence with a much steeper force-distance profile at a higher temperature. However, as illustrated in Figure 20, this temperature dependence is eliminated by SDS. The observations of the effects of SDS of first reducing the adsorbed amount, but "swelling" the adsorbed layer, and second, of eliminating the strong temperature dependence on the adsorption and on the interaction forces, can be related to polymer-surfactant complexing in the bulk and at surfaces and competitive adsorption. The observation of surface complexes agrees with recent observations by Ma and Li[173,174] for quite different systems (SDS and PVP at TiO_2 and Fe_2O_3). These authors observed that the PS system stabilizes particle suspensions better at intermediate than at high surfactant concentrations, which they ascribe to a progressive change-over from surface to bulk PS complexes.

A surface force study of polyelectrolyte-surfactant systems was recently presented by Ananthapadmanabhan et al.,[175] who studied a cationic polymer (Polymer JR) interacting with an anionic surfactant (SDS) and negatively charged mica surfaces. These authors demonstrated major changes in the forces between the surfaces on introducing SDS, from being monotonically repulsive at both approach and separation to being attractive at certain distances on separation. This was attributed to the surfactant inducing hydrophobic interactions between the adsorbed layers.

On the basis of these examples, the potential of surface force measurements for the study of polymer-surfactant interactions is demonstrated and it should be a useful tool in the development of stabilizers for particle suspensions etc. A surface force study of a somewhat different type was earlier presented by Luckham and Klein,[176] who investigated the effect of added PEO on the interaction between mica surfaces onto which a nonionic surfactant was adsorbed. The observations indicate that the (rather high molecular weight) polymer replaces surfactant at the surfaces.

ACKNOWLEDGMENTS

We are grateful to Dr. Lennart Piculell for numerous enlightening discussions and suggestions and for reading the manuscript, and to Prof. Håkan Wennerström, Dr. Ilias Iliopoulos, and Dr. Anders Carlsson for discussions and help. The work was supported by NUTEK — Swedish Board for Industrial and Technical Development.

REFERENCES

1. **Robb, I. D.,** Polymer/surfactant interactions, in *Anionic Surfactants — Physical Chemistry of Surfactant Action,* Surfactant Sci. Ser., Vol. 11, Lucassen-Reynders, E., Ed., Marcel Dekker, New York, 1981, chap. 3.
2. **Goddard, E. D.,** Polymer-surfactant interaction, *Colloids Surf.,* 19, 255, 1986.
3. **Saito, S.,** Polymer-surfactant interactions, in *Nonionic Surfactants,* Schick, M. J., Ed., Marcel Dekker, New York, 1987, 881.
4. **Hayakawa, K. and Kwak, J. C. T.,** Interactions between polymers and cationic surfactants, in *Cationic Surfactants,* Rubingh, D. N. and Holland, P. M., Eds., Marcel Dekker, New York, 1991, 189.
5. **Stilbs, P.,** Fourier transform pulsed-gradient spin-echo studies of molecular diffusion, *Progr. Nucl. Magn. Reson. Spectrosc.,* 19, 1, 1987.

6. **Carlsson, A., Karlström, G., and Lindman, B.,** Characterization of the interaction between a nonionic polymer and a cationic surfactant by the Fourier transform NMR self-diffusion technique, *J. Phys. Chem.*, 93, 3673, 1989.
7. **Lindman, B., Söderman, O., and Wennerström, H.,** NMR studies of surfactant systems, in *Surfactant Solutions. New Methods of Investigation,* Surfactant Sci. Ser., Vol. 22, Zana, R., Ed., Marcel Dekker, New York, 1986, 295.
8. **Gao, Z., Wasylishen, R. E., and Kwak, J. C. T.,** Distribution equilibrium of poly(ethylene oxide) in sodium dodecyl sulfate micellar solutions: an NMR paramagnetic relaxation study, *J. Phys. Chem.*, 95, 462, 1991.
9. **Gao, Z., Wasylishen, R.-E., and Kwak, J. C. T.,** Carbon-13 NMR relaxation study of molecular dynamics and organization of sodium poly(styrenesulfonate) and dodecyltrimethylammonium bromide aggregates in aqueous solution, *J. Phys. Chem.*, 94, 773, 1990.
10. **Wong, T., Thalberg, K., Lindman, B., and Gracz, H.,** Surfactant ^{13}C relaxation and differential line broadening in a system of a polyanion and a cationic surfactant, *J. Phys. Chem.*, 95, 8850, 1991.
11. **Wennerström, H., Lindman, B., Söderman, O., Drakenberg, T., and Rosenholm, J.,** ^{13}C Magnetic relaxation in micellar solutions. Influence of aggregate motion on T_1, *J. Am. Chem. Soc.*, 101, 6860, 1979.
12. **Azzam, R. M. A. and Bashara, N. M.,** *Ellipsometry and Polarized Light,* North-Holland, Amsterdam, 1987.
13. **Luckham, P.,** Measurement of the interaction between adsorbed polymer layers: the steric effect, *Adv. Colloid Interface Sci.*, 34, 191, 1991.
14. **Zana, R.,** Luminescence probing methods, in *Surfactant Solutions. New Methods of Investigation,* Surfactant Sci. Ser., Vol. 22, Zana, R., Ed., Marcel Dekker, New York, 1986, 241.
15. **Zimm, B. H. and Bragg, J. K.,** Theory of the phase transition between helix and random coil in polypeptide chains, *J. Chem. Phys.*, 31, 526, 1959.
16. **Tanford, C.,** *The Hydrophobic Effect: Formation of Micelles and Biological Membranes,* 2nd ed., John Wiley & Sons, New York, 1980.
17. **Lindman, B. and Wennerström, H.,** Micelles. Amphiphile aggregation in aqueous solution, in *Top. Curr. Chem.*, 87, 1980, 1.
18. **Lindman, B. and Karlström, G.,** Polymer-surfactant systems, in *The Structure, Dynamics and Equilibrium Properties of Colloidal Systems,* Bloor, D. M. and Wyn-Jones, E., Eds., Kluwer Academic, Dordrecht, The Netherlands, 1990, 131.
19. **Karlström, G.,** A new model for upper and lower critical solution temperatures in poly(ethylene oxide) solutions, *J. Phys. Chem.*, 89, 4962, 1985.
20. **Israelachvili, J. N.,** *Intermolecular and Surface Forces,* Academic Press, London, 1985.
21. **Gunnarsson, G., Jönsson, B., and Wennerström, H.,** Surfactant association into micelles. An electrostatic approach, *J. Phys. Chem.*, 84, 3114, 1980.
22. **Flory, P. J.,** *Principles of Polymer Chemistry,* Cornell University Press, Ithaca, New York, 1953.
23. **Sjöberg, Å.,** The Mechanism of Phase-Separation in Aqueous Solutions of Poly(ethylene glycol) and Carbohydrates. A Theoretical and Experimental Investigation, Ph.D. thesis, University of Lund, Sweden, 1989.
24. **Chu, D. and Thomas, J. K.,** Effect of cationic surfactants on the conformational transition of poly(methacrylic acid), *J. Am. Chem. Soc.*, 108, 6270, 1986.
25. **Brackman, J.,** The Interaction between Water-Soluble Polymers and Surfactant Aggregates, Ph.D. thesis, University of Groningen, The Netherlands, 1990.
26. **Brackman, J. C., van Os, N. M., and Engberts, J. B. F. N.,** Polymer-nonionic micelle complexation. Formation of poly(propylene oxide)-complexed n-octyl thioglucoside micelles, *Langmuir,* 4, 1266, 1988.
27. **Schwuger, M. J.,** Mechanism of interaction between ionic surfactants and polyglycol ethers in water, *J. Colloid Interface Sci.*, 43, 491, 1973.

28. **François, J., Dayantis, J., and Sabbadin, J.,** Hydrodynamical behaviour of the poly(ethylene oxide)-sodium dodecylsulphate complex, *Eur. Polym. J.*, 21, 165, 1985.
29. **Lange, H.,** Wechselwirkung zwischen Natriumalkylsulfaten und Polyvinylpyrrolidon in wässrigen Lösungen, *Kolloid Z. Z. Polym.*, 243, 101, 1971.
30. **Shirahama, K.,** The binding equilibrium of sodium dodecyl sulfate to poly(ethylene oxide) in 0.1 *M* sodium chloride solution at 30°C, *Colloid Polym. Sci.*, 252, 978, 1974.
31. **Cabane, B. and Duplessix, R.,** Neutron scattering study of water-soluble polymers adsorbed on surfactant micelles, *Colloids Surf.*, 13, 19, 1985.
32. **Brackman, J. and Engberts, J. B. F. N.,** The effect of surfactant headgroup charge on polymer-micelle interaction, *J. Colloid Interface Sci.*, 132, 250, 1989.
33. **Chari, K. and Lenhart, W. C.,** Effect of polyvinylpyrrolidone on the self-assembly of model hydrocarbon amphiphiles, *J. Colloid Interface Sci.*, 137, 214, 1990.
34. **Gao, Z., Wasylishen, R. E., and Kwak, J. C. T.,** NMR studies in surfactant and polymer surfactant systems. Micelle formation of sodium ω-phenyldecanoate and interaction with poly(ethylene oxide), *J. Colloid Interface Sci.*, 137, 137, 1990.
35. **Saito, S. and Yukawa, M. J.,** Interactions of polymers and cationic surfactants with thiocyanate as counterions, *J. Colloid Interface Sci.*, 30, 211, 1969.
36. **Treiner, C. and Hguyen, D.,** Interactions of copper dodecyl sulfate aggregates with poly(ethylene oxide) or poly(vinylpyrrolidone), *J. Phys. Chem.*, 94, 2021, 1990.
37. **Shirahama, K., Himuro, A., and Takisawa, N.,** Binding of hexadecylammonium surfactants to water-soluble neutral polymers, *Colloid Polym. Sci.*, 96, 265, 1987.
38. **Shirahama, K., Oh-Ishi, M., and Takisawa, N.,** Interaction between cationic surfactants and poly(vinyl alcohol), *Colloids Surf.*, 40, 261, 1989.
39. **Perron, G., Francoeur, J., Desnoyers, J. E., and Kwak, J. C. T.,** Heat capacities and volumes in aqueous polymer and polymer-surfactant solutions, *Can. J. Chem.*, 65, 990, 1987.
40. **Hoffmann, H. and Huber, G.,** Electric birefringence and EMF-measurements on polymer/surfactant complexes, *Colloids Surf.*, 40, 181, 1989.
41. **Witte, F. M. and Engberts, J. B. F. N.,** An ESR spin probe study of micelle-polymer complexes. Poly(ethylene oxide)- and poly(propylene oxide)-complexed sodium dodecyl sulfate and cetyltrimethylammonium bromide micelles, *J. Org. Chem.*, 53, 3085, 1988.
42. **Witte, F. M. and Engberts, J. B. F. N.,** Micelle-polymer complexes: aggregation numbers, micellar rate effects and factors determining the complexation process, *Colloids Surf.*, 36, 417, 1989.
43. **Nagarajan, R.,** Thermodynamics of nonionic polymer-micelle association, *Colloids Surf.*, 13, 1, 1985.
44. **Nagarajan, R.,** Association of nonionic polymers with micelles, bilayers, and microemulsions, *J. Chem. Phys.*, 90, 1980, 1989.
45. **Ruckenstein, E., Huber, G., and Hoffmann, H.,** Surfactant aggregation in the presence of polymers, *Langmuir,* 3, 382, 1987.
46. **Mukerjee, P. and Mysels, K. J.,** Critical Micelle Concentrations of Aqueous Surfactant Systems, Natl. Stand. Ref. Data Ser., NSRDS-NBS 36, National Bureau of Standards, Washington, D.C., 1971.
47. **Karlström, G., Carlsson, A., and Lindman, B.,** Phase diagrams of nonionic polymer-water systems. Experimental and theoretical studies of the effects of surfactants and other cosolutes, *J. Phys. Chem.*, 94, 5005, 1990.
48. **Samii, A., Karlström, G., and Lindman, B.,** Phase behavior of a nonionic cellulose ether in nonaqueous solution, *Langmuir,* 7, 653, 1991.
49. **van Stam, J., Almgren, M., and Lindblad, C.,** Sodium dodecylsulfate-poly(ethyleneoxide) interactions studied by time-resolved fluorescence quenching, *Prog. Colloid Polym. Sci.*, 84, 13, 1991..
50. **Brackman, J. and Engberts, J. B. F. N.,** Polymer-induced breakdown of rodlike micelles. A striking transition of a non-Newtonian to a Newtonian fluid, *J. Am. Chem. Soc.*, 112, 872, 1990.

51. **Brackman, J. and Engberts, J. B. F. N.,** Effect of water-soluble polymers on rodlike and spherical micelles formed from 2-alkylmalonate salts, *Langmuir,* 7, 46, 1991.
52. **Winnik, F. M., Winnik, M. A., and Tazuke, S.,** Interaction of hydroxypropylcellulose with aqueous surfactants: fluorescence probe studies and a look a pyrene-labeled polymer, *J. Phys. Chem.,* 91, 594, 1987.
53. **Schild, H. G. and Tirrell, D. A.,** Sodium 2-(N-dodecylamino)naphthalene-6-sulfonate as a probe of polymer-surfactant interaction, *Langmuir,* 6, 1676, 1990.
54. **Bulpin, P. V., Cutler, A. N., and Lips, A.,** Cooperative binding of myristate to amylose, *Macromolecules,* 20, 44, 1987.
55. **Holmberg, C., Nilsson, S., Singh, S., and Sundelöf, L.-O.,** Hydrodynamic and thermodynamic aspects of the SDS-EHEC-water system, *J. Phys. Chem.,* 96, 871, 1992.
56. **Schild, H. G. and Tirrell, D. A.,** Interaction of poly(N-isopropylacrylamide) with sodium n-alkyl sulfates in aqueous solution, *Langmuir,* 7, 665, 1991.
57. **Ricka, J., Meewes, M., Nyffenegger, R., and Binkert, Th.,** Intermolecular and intramolecular solubilization — collapse and expansion of a polymer chain in surfactant solutions, *Phys. Rev. Lett.,* 65, 657, 1990.
58. **Winnik, F. M., Ringsdorf, H., and Venzmer, J.,** Interaction of surfactants with hydrophobically modified poly(N-isopropylacrylamides). II. Fluorescence label studies, *Langmuir,* 7, 912, 1991.
59. **Winnik, F. M., Ringsdorf, H., and Venzmer, J.,** Interactions of surfactants with hydrophobically modified poly(N-isopropylacrylamides). I. Fluorescence probe studies, *Langmuir,* 7, 905, 1991.
60. **Cabane, B.,** Structure of some polymer-detergent aggregates in water, *J. Phys. Chem.,* 81, 1639, 1977.
61. **Satake, I. and Yang, J. T.,** Interaction of sodium decyl sulfate with poly(L-ornithine) and poly(L-lysine) in aqueous solution, *Biopolymers,* 15, 2263, 1976.
62. **Thalberg, K. and Lindman, B.,** Interaction between hyaluronan and cationic surfactants, *J. Phys. Chem.,* 93, 1478, 1989.
63. **Binana-Limbelé, W. and Zana, R.,** Fluorescence probing of microdomains in aqueous solutions of polysoaps. I. Use of pyrene to study the conformational state of polysoaps and their comicellization with cationic surfactants, *Macromolecules,* 20, 1331, 1987.
64. **Satake, I., Takahashi, T., Hayakawa, K., Maeda, T., and Aoyagi, M.,** Effect of charge density on the cooperative binding isotherm of surfactant ion to polyelectrolyte, *Bull. Chem. Soc. Jpn.,* 63, 926, 1990.
65. **Thalberg, K., van Stam, J., Lindblad, C., Almgren, M., and Lindman, B.,** Time-resolved fluorescence and self-diffusion studies in systems of a cationic surfactant and an anionic polyelectrolyte, *J. Phys. Chem.,* 95, 8975, 1991.
66. **Shimizu, T., Seki, M., and Kwak, J. C. T.,** The binding of cationic surfactants by hydrophobic alternating copolymers of maleic acid, *Colloids Surf.,* 20, 289, 1986.
67. **Chandar, P., Somasundaran, P., and Turro, N. J.,** Fluorescence probe investigation of anionic polymer-cationic surfactant interactions, *Macromolecules,* 21, 950, 1988.
68. **Hayakawa, K., Santerre, J. P., and Kwak, J. C. T.,** Study of surfactant-polyelectrolyte interactions. Binding of dodecyl- and tetradecyltrimethylammonium bromide by some carboxylic polyelectrolytes, *Macromolecules,* 16, 1642, 1983.
69. **Hayakawa, K. and Kwak, J. C. T.,** Study of surfactant-polyelectrolyte interactions. II. Effect of multivalent counterions on the binding of dodecyltrimethylammonium ions by sodium dextran sulfate and sodium poly(styrenesulfonate) in aqueous solution, *J. Phys. Chem.,* 87, 506, 1983.
70. **Lindman, B., Puyal, M.-C., Kamenka, N., Rymdén, R., and Stilbs, P.,** Micelle formation of anionic and cationic surfactants from Fourier transform hydrogen-1 and lithium-7 nuclear magnetic resonance and tracer self-diffusion studies, *J. Phys. Chem.,* 88, 5048, 1984.
71. **Thalberg, K.,** Polyelectrolyte-Surfactant Interactions, Ph.D. thesis, University of Lund, Sweden, 1990.

72. **Thalberg, K., Lindman, B., and Bergfeldt, K.,** Phase behavior of systems of polyacrylate and cationic surfactant, *Langmuir,* 7, 2893, 1991.
73. **Abuin, E. B. and Scaiano, J. C.,** Exploratory study of the effect of polyelectrolyte-surfactant aggregates on photochemical behavior, *J. Am. Chem. Soc.*, 106, 6274, 1984.
74. **Almgren, M., Hansson, P., Mukhtar, E., and van Stam, J.,** Aggregation of alkyltrimethylammonium surfactants in poly(styrene sulfonate) solutions, submitted.
75. **Skerjanc, J., Kogej, K., and Vesnaver, G.,** Polyelectrolyte-surfactant interactions. Enthalpy of binding of dodecyl- and cetylpyridinium cations to poly(styrenesulfonate) anion, *J. Phys. Chem.*, 92, 6382, 1988.
76. **Skerjanc, J. and Kogej, K.,** Thermodynamic and transport properties of polyelectrolyte-surfactant complex solutions at various degrees of complexation, *J. Phys. Chem.*, 93, 7913, 1989.
77. **Gao, Z., Kwak, J. C. T., and Wasylishen, R.,** An NMR study of the binding between polyelectrolytes and surfactants, *J. Colloid Interface Sci.*, 126, 371, 1988.
78. **Gao, Z., Wasylishen, R., and Kwak, J. C. T.,** An NMR study of the solubilization of aromatic compounds in aggregates of poly(maleic acid-co-butyl vinyl ether) and dodecyltrimethylammonium bromide, *Macromolecules,* 22, 2544, 1989.
79. **Herslöf, Å., Sundelöf, L.-O., and Edsman, K.,** Hydrodynamic effects in the sodium hyaluronate/tetradecyltrimethylammonium bromide/sodium chloride/water system, *J. Phys. Chem.*, 96, 2345, 1992.
80. **Yang, J. T. and Kubota, S.,** Ordered conformation of poly(L-lysine) and its homologs in anionic surfactant solutions, in *Microdomains in Polymer Solutions,* Dubin, P., Ed., Plenum Press, New York, 1985, 311.
81. **Aoyagi, M., Maeda, T., Hayakawa, K., and Satake, I.,** The conformational changes of poly(L-ornithine) with organic counter ions in aqueous sodium 1-decanesulfonate solution, *Bull. Chem. Soc. Jpn.*, 64, 1378, 1991.
82. **Hayakawa, K., Fujita, M., Yokoi, S.-I., and Satake, I.,** Conformation of poly(L-lysine) and poly(L-ornithine) in α, ω-type surfactant solutions, *J. Bioactive Compatible Polym.*, 6, 36, 1991.
83. **Satake, I. and Hayakawa, K.,** The binding degree dependence of the coil-helix transition of poly(L-ornithine) in aqueous surfactant solution, *Chem. Lett. Chem. Soc. Jpn.*, 1051, 1990.
84. **Hayakawa, K., Fukutome, T., and Satake, I.,** Solubilization of water-insoluble dye by a cooperative binding system of surfactant and polyelectrolyte, *Langmuir,* 6, 1495, 1990.
85. **Dubin, P. L. and Oteri, R.,** Association of polyelectrolytes with oppositely charged mixed micelles, *J. Colloid Interface Sci.*, 95, 453, 1983.
86. **Dubin, P. L., Rigsbee, D. R., and McQuigg, D. W.,** Turbidimetric and dynamic light scattering studies of mixtures of cationic polymers and anionic mixed micelles, *J. Colloid Interface Sci.*, 105, 509, 1985.
87. **Dubin, P. L., Rigsbee, D. R., Gan, L.-M., and Fallon, M. A.,** Equilibrium binding of mixed micelles to oppositely charged polyelectrolytes, *Macromolecules,* 21, 2555, 1988.
88. **Dubin, P. L., Chew, C. H., and Gan, L. M.,** Complex formation between anionic polyelectrolytes and cationic/nonionic mixed micelles, *J. Colloid Interface Sci.*, 128, 566, 1989.
89. **Dubin, P. L., Thé, S. S., McQuigg, D. W., Chew, C. H., and Gan, L. M.,** Binding of polyelectrolytes to oppositely charged ionic micelles at critical micelle surface charge densities, *Langmuir,* 5, 89, 1989.
90. **Dubin, P. L., Vea, M. E. Y., Fallon, M. A., Thé, S. S., Rigsbee, D. R., and Gan, L. M.,** Higher order association in polyelectrolyte-micelle complexes, *Langmuir,* 6, 1422, 1990.
91. **Dubin, P. L., Thé, S. S., Gan, L. M., and Chew, C. H.,** Static light scattering of polyelectrolyte-micelle complexes, *Macromolecules,* 23, 2500, 1990.

92. **Yomota, C., Ito, Y., and Nakagaki, M.,** Interaction of cationic surfactant with arabate and chondroitin sulfate, *Chem. Pharm. Bull. Jpn.,* 35, 798, 1987.
93. **Tomalia, D. A., Baker, H., Dewald, J., Hall, M., Kallos, G., Martin, S., Roeck, J., Ryder, J., and Smith, P.,** Dendritic macromolecules: synthesis of starburst dendrimers, *Macromolecules,* 19, 2466, 1986.
94. **Caminati, G., Turro, N. J., and Tomalia, D. A.,** Photophysical investigation of starburst dendrimers and their interactions with anionic and cationic surfactants, *J. Am. Chem. Soc.,* 112, 8515, 1990.
95. **Binana-Limbelé, W. and Zana, R.,** Interactions between sodium dodecyl sulfate and polycarboxylates and polyethers. Effect of Ca^{2+} on these interactions, *Colloids Surf.,* 21, 483, 1986.
96. **Schwuger, M. and Lange, H.,** *Tenside,* 5, 257, 1968.
97. **Methemitis, C. and Morcellet, M.,** Interactions between partially hydrolyzed polyacrylamide and ionic surfactants, *Eur. Polym. J.,* 22, 619, 1986.
98. **Iliopoulos, I. and Audebert, R.,** Polymer complexes stabilized through hydrogen bonds: A semiquantitative theoretical model, *J. Polym. Sci. Part B,* 26, 2093, 1988.
99. **Winnik, F. M.,** Interaction of fluorescent dye labeled (hydroxypropyl)cellulose with nonionic surfactants, *Langmuir,* 6, 522, 1990.
100. **Iliopoulos, I., Wang, T. K., and Audebert, R.,** Viscometric evidence of interactions between hydrophobically modified poly(sodium acrylate) and sodium dodecyl sulfate, *Langmuir,* 7, 617, 1991.
101. **Wang, T. K., Iliopoulos, I., and Audebert, R.,** Synthesis, characterization and aqueous solution behavior of hydrophobically associating derivatives of poly(acrylic acid), in preparation.
102. **Wang, T. K., Iliopoulos, I., and Audebert, R.,** Solution behavior of hydrophobically associating poly(acrylic acid): effects of ionic strength and pH, in preparation.
103. **Wang, T. K., Iliopoulos, I., and Audebert, R.,** Aqueous solution behavior of hydrophobically modified poly(acrylic acid), *ACS Symp. Ser.,* 467, 218, 1991.
104. **Gelman, R. A.,** Hydrophobically modified hydroxyethylcellulose, in TAPPI Proceedings, International Dissolving Pulps Conference, 1987, 159.
105. **Tanaka, R., Meadows, J., Phillips, G. O., and Williams, P. O.,** Viscometric and spectroscopic studies on the solution behaviour of hydrophobically modified cellulosic polymers, *Carbohydr. Polym.,* 12, 443, 1990.
106. **Sivadasan, K. and Somasundaran, P.,** Polymer-surfactant interactions and the association behavior of hydrophobically modified hydroxyethylcellulose, *Colloids Surf.,* 49, 229, 1990.
107. **Peiffer, D. G.,** Hydrophobically associating polymers and their interactions with rod-like micelles, *Polymer,* 31, 2353, 1990.
108. **Peiffer, D. G.,** Solid state characterization of the structure of rod-like micelles and their mixtures with associating polymers, *Polymer,* 32, 134, 1991.
109. **Zugenmaier, P. and Aust, N.,** Rheological investigations on the interaction of cellulose derivatives with sodium dodecyl sulfate in aqueous solution, *Makromol. Chem. Rapid Commun.,* 11, 95, 1990.
110. **Winnik, F. M.,** Association of hydrophobic polymers in H_2O and D_2O: fluorescence studies with (hydroxypropyl)cellulose, *J. Phys. Chem.,* 93, 7452, 1989.
111. **Hu, Y.-Z., Zhao, C.-L., and Winnik, M. A.,** Fluorescence studies of the interaction of sodium dodecyl sulfate with hydrophobically modified poly(ethylene oxide), *Langmuir,* 6, 880, 1990.
112. **Tanaka, R., Meadows, J., Phillips, G. O., and Williams, P. A.,** Solution properties of hydrophobically modified hydroxyethyl cellulose, in preparation.
113. **Steiner, C. A. and Gelman, R. A.,** New cellulosic surfactants, in *Cellulosics Utilization: Research and Rewards in Cellulosics,* Inagaki, H. and Phillips, G. O., Eds., Elsevier, Barking, 1989, 132.

114. **Schild, H. G. and Tirrell, D. A.,** Microheterogeneous solutions of amphiphilic copolymers of N-isopropylacrylamide. An investigation via fluorescence methods, *Langmuir,* 7, 1319, 1991.
115. **Gilányi, T. and Wolfram, E.,** Interaction of ionic surfactants with polymers in aqueous solution, *Colloids Surf.,* 3, 181, 1981.
116. **Hall, D. G.,** Thermodynamics of ionic surfactant binding to macromolecules in solution, *J. Chem. Soc. Faraday Trans. 1,* 81, 885, 1985.
117. **Evans, D. F., Mitchell, D. J., and Ninham, B. W.,** Ion binding and dressed micelles, *J. Phys. Chem.,* 88, 6344, 1984.
118. **Balazs, A. C. and Hu, J. Y.,** Effects of surfactant concentration on polymer-surfactant interactions in dilute solutions: a computer model, *Langmuir,* 5, 1230, 1989.
119. **Balazs, A. C. and Hu, J. Y.,** A computer model for the effect of surfactants on the aggregation of associating polymers, *Langmuir,* 5, 1253, 1989.
120. **Hill, T. L.,** *An Introduction to Statistical Thermodynamics,* Dover, New York, 1960.
121. **Albertsson, P. Å.,** *Partition of Cell Particles and Macromolecules,* 3rd ed., John Wiley & Sons, New York, 1986.
122. **Cabane, B. and Duplessix, R.,** Organization of surfactant micelles adsorbed on a polymer molecule in water: a neutron scattering study, *J. Phys. (Paris),* 43, 1529, 1982.
123. **Cabane, B. and Duplessix, R.,** Decoration of semidilute polymer solutions with surfactant micelles, *J. Phys. (Paris),* 48, 651, 1987.
124. **Carlsson, A.,** Nonionic Cellulose Ethers, Interactions with Surfactants, Solubility and Other Aspects, Ph.D. thesis, University of Lund, Sweden, 1989.
125. **Carlsson, A., Karlström, G., and Lindman, B.,** Synergistic surfactant-electrolyte effect in polymer solutions, *Langmuir,* 2, 536, 1986.
126. **Lindman, B., Carlsson, A., Karlström, G., and Malmsten, M.,** Nonionic polymers and surfactants — some anomalies in temperature dependence and in interactions with ionic surfactants, *Adv. Colloid Interface Sci.,* 32, 183, 1990.
127. **Drummond, C. J., Albers, S., and Furlong, D. N.,** Polymer-surfactant interactions: (hydroxypropyl)cellulose with ionic and nonionic surfactants, *Colloids Surf.,* 62, 75, 1992.
128. **Samii, A. A., Lindman, B., and Karlström, G.,** Phase behaviour of some nonionic polymers in nonaqueous solvents, *Prog. Colloid Polym. Sci.,* 82, 280, 1990.
129. **Sjöberg, Å. and Karlström, G.,** Temperature dependence of the phase equilibria for the system poly(ethylene glycol)/dextran/water — a theoretical and experimental study, *Macromolecules,* 22, 1325, 1989.
130. **Karlström, G.,** A new model for upper and lower critical solution temperatures in poly(ethylene oxide) solutions, *J. Phys. Chem.,* 89, 4962, 1985.
131. **Zhang, K., Karlström, G., and Lindman, B.,** Phase behavior of a system of nonionic polymer, ionic surfactant and water, *Prog. Colloid Polymer Sci.,* 88, in press.
132. **Saito, S. V.,** Koagulation und Peptisation von Polyelektrolytlösung durch Detergent-Ionen. I, *Kolloid Z.,* 143, 66, 1955.
133. **Goddard, E. D. and Hannan, R. B.,** Cationic polymer/anionic surfactant interactions, *J. Colloid Interface Sci.,* 55, 73, 1976.
134. **Goddard, E. D. and Hannan, R. B.,** Polymer/surfactant interactions, *J. Am. Oil Chem. Soc.,* 54, 561, 1977.
135. **Thalberg, K., Lindman, B., and Karlström, G.,** Phase diagram of a system of cationic surfactant and anionic polyelectrolyte: tetradecyltrimethylammonium bromide-hyaluronan-water, *J. Phys. Chem.,* 94, 4289, 1990.
136. **Thalberg, K., Lindman, B., and Karlström, G.,** Phase diagram of systems of cationic surfactant and anionic polyelectrolyte: influence of surfactant chain length and polyelectrolyte molecular weight, *J. Phys. Chem.,* 95, 3370, 1991.
137. **Thalberg, K., Lindman, B., and Karlström, G.,** Phase diagram of a system of cationic surfactant and anionic polyelectrolyte: the effect of salt, *J. Phys. Chem.,* 95, 6004, 1991.

138. **Thalberg, K. and Lindman, B.,** Polyelectrolyte-ionic surfactant systems, in *Surfactants in Solution,* Mittal, K. and Shah, O. O., Eds., Plenum Press, New York, 11, 243, 1991.
139. **Ananthapadmanabhan, K. P., Leung, P. S., and Goddard, E. D.,** Fluorescence and solubilization studies of polymer-surfactant systems, *Colloids Surf.,* 13, 63, 1985.
140. **Bungenberg de Jong, H. G.,** Complex coacervation, in *Colloid Science,* Kruyt, H., Ed., Elsevier, Amsterdam, 1949, chap. 10.
141. **Frugier, D.,** Copolymeres Anioniques et Cationiques de Faible Densité de Charge: Synthese et Etude, Doctoral thesis, Université Pierre et Marie Curie, Paris, 1988.
142. **Piculell, L., Nilsson, S., Falck, L., and Tjerneld, F.,** Phase separation in aqueous mixtures of similarly charged polyelectrolytes, *Polym. Commun.,* 32(5), 158, 1991.
143. **Perrau, M. B., Iliopoulos, I., and Audebert, R.,** Phase separation of polyelectrolyte/nonionic polymer in aqueous solution: effects of salt and charge density, *Polymer,* 30, 2112, 1989.
144. **Zhang, K., Karlström, G., and Lindman, B.,** Phase behaviour of systems of a non-ionic surfactant and a non-ionic polymer in aqueous solution, *Colloids Surf.,* in press.
145. **Wormuth, K. R.,** Patterns of phase behavior in polymer and amphiphile mixtures, *Langmuir,* 7, 1622, 1991.
146. **Brackman, J. C.,** Sodium dodecyl sulfate induced enhancement of the viscosity and viscoelasticity of aqueous solutions of poly(ethylene oxide). A rheological study on polymer-micelle interaction, *Langmuir,* 7, 469, 1991.
147. **Goddard, E. D., Leung, P. S., and Padmanabhan, K.-P.-A.,** Novel gelling structures based on polymer/surfactant systems, *J. Soc. Cosmet. Chem.,* 42, 19, 1991.
148. **Leung, P. S. and Goddard, E. D.,** Gels from dilute polymer/surfactant solutions, *Langmuir,* 7, 608, 1991.
149. **Thalberg, K. and Lindman, B.,** Gel formation in aqueous systems of a polyanion and an oppositely charged surfactant, *Langmuir,* 7, 277, 1991.
150. **de Gennes, P. G.,** Scaling Concepts in Polymer Physics, Cornell University Press, Ithaca, NY, 1979.
151. **Berr, S. S.,** Solvent isotope effects on alkyltrimethylammonium bromide micelles as a function of alkyl chain length, *J. Phys. Chem.,* 91, 4760, 1987.
152. **Carlsson, A., Karlström, G., and Lindman, B.,** Thermal gelation of nonionic cellulose ethers and ionic surfactants in water, *Colloids Surf.,* 147, 47, 1990.
153. **Rehage, H.,** unpublished results.
154. **Nyström, B., Roots, J., Carlsson, A., and Lindman, B.,** Light scattering studies of the gelation process in aqueous system of a nonionic polymer and a cationic surfactant, *Polymer,* 33, 2875, 1992.
155. **Zana, R., Binana-Limbelé, W., Kamenka, N., and Lindman, B.,** Ethyl (hydroxyethyl) cellulose-cationic surfactant interactions: electrical conductivity, self-diffusion, and time-resolved fluorescence quenching investigations, *J. Phys. Chem.,* 96, 5461, 1992.
156. **Carlsson, A., Lindman, B., Watanabe, T., and Shirahama, K.,** Polymer-surfactant interactions. Binding of N-tetradecyl-pyridinium bromide to ethyl(hydroxyethyl)cellulose, *Langmuir,* 5, 1250, 1989.
157. **Lindman, B., Carlsson, A., Thalberg, K., and Bogentoft, C.,** Polymer-surfactant systems and formulation, with examples from the use for drug delivery, *L'actualitée chimique,* May-June, 1991, p. 181.
158. **de Gennes, P. G.,** Interaction between polymers and surfactants, *J. Phys. Chem.,* 94, 8407, 1990.
159. **Kekicheff, P., Cabane, P., and Rawiso, M.,** Macromolecules dissolved in a lamellar lyotropic mesophase, *J. Colloid Interface Sci.,* 102, 51, 1984.
160. **Seki, K. and Tirrell, D. A.,** pH-Dependent complexation of poly(acrylic acid) derivatives with phospholipid vesicle membranes, *Macromolecules,* 17, 1692, 1984.

161. **Borden, K. A., Eum, K. M., Langley, K. H., Tan, J. S., Tirrell, D. A., and Voycheck, C. L.,** pH-Dependent vesicle-to-micelle transition in an aqueous mixture of dipalmitoylphosphatidylcholine and a hydrophobic polyelectrolyte, *Macromolecules,* 21, 2649, 1988.
162. **Ferritto, M. S. and Tirrell, D. A.,** Photoregulation of the binding of a synthetic polyelectrolyte to phosphatidylcholine bilayer membranes, *Macromolecules,* 21, 3117, 1988.
163. **Tirrell, D. A., Takigawa, D. Y., and Seki, K.,** pH Sensitization of phospholipid vesicles via complexation with synthetic poly(carboxylic acid)s, in *Macromolecules as Drugs and as Carriers for Biologically Active Materials,* Annals of the New York Academy of Sciences, New York, 1999, 237.
164. **You, H. and Tirrell, D. A.,** Photoinduced, polyelectrolyte-driven release of contents of phosphatidylcholine bilayer vesicles, *J. Am. Chem. Soc.,* in press.
165. **Wakita, M., Edwards, K. A., Regen, S. L., Turner, D., and Gruner, S. M.,** Use of a polymeric counterion to induce bilayer formation from a single-chain surfactant, *J. Am. Chem. Soc.,* 110, 5221, 1988.
166. **Hilfiker, R., Eicke, H.-F., Steeb, C., and Hofmeier, U.,** Block-copolymer-induced structure formation in microemulsions, *J. Phys. Chem.,* 95, 1478, 1991.
167. **Struis, R. P. Q. J. and Eicke, H.-F.,** Polymers in complex fluids: dynamic and equilibrium properties of nanodroplet-ABA block copolymer structures, *J. Phys. Chem.,* 95, 5989, 1991.
168. **Quellet, C., Eicke, H.-F., and Sager, W.,** Formation of microemulsion-based gelatin gels, *J. Phys. Chem.,* 95, 5642, 1991.
169. **Petit, C., Zemb, Th., and Pileni, M. P.,** Structural study of microemulsion-based gels at the saturation point, *Langmuir,* 7, 223, 1991.
170. **Howe, A. M., Katsikides, A., Robinson, B. H., Chadwick, A. V., and Al-Mudaris, A.,** Structure and dynamics of microemulsion-based gels, *Prog. Colloid Polym. Sci.,* 76, 211, 1986.
171. **Chari, K. and Hossain, T. Z.,** Adsorption at the air/water interface from an aqueous solution of poly(vinylpyrrolidone) and sodium dodecyl sulfate, *J. Phys. Chem.,* 95, 3302, 1991.
172. **Claesson, P. M., Malmsten, M., and Lindman, B.,** Forces between hydrophobic surfaces coated with ethyl(hydroxyethyl)cellulose in the presence of an ionic surfactant, *Langmuir,* 7, 1441, 1991.
173. **Ma, C.,** Adsorption from mixed solutions of poly(vinyl pyrrolidone) and sodium dodecyl sulfate on titanium dioxide, *Colloids Surf.,* 16, 185, 1985.
174. **Ma, C. and Li, C.-L.,** Stability of dispersions of iron oxide in mixed solutions of polyvinylpyrrolidone and sodium alkyl sulfate, *Colloids Surf.,* 47, 117, 1990.
175. **Ananthapadmanabhan, K. P., Mao, G.-Z., Goddard, E. D., and Tirrell, M.,** Surforce force measurements on cationic polymer in the presence of anionic surfactants, *Colloids Surf.,* in press.
176. **Luckham, P. F. and Klein, J.,** Interactions between two surfaces with adsorbed nonionic surfactants in aqueous electrolyte and in a free polymer solution, *J. Colloid Interface Sci.,* 117, 149, 1987.

Chapter 6

HYDROPHOBE-MODIFIED POLYMERS

Ulrich P. Strauss

TABLE OF CONTENTS

0-8493-6784-0/93/$0.00 + $.50

I. INTRODUCTION

At the time this laboratory began investigating hydrophobe-modified polymers, there had developed considerable interest in the spontaneous formation and physical properties of micelles in surfactant solutions. Therefore it was decided in this laboratory to prefabricate micelles by chemically attaching surfactant molecules to polymer chains. The macromolecules created in this manner were called polysoaps.[1] It was expected that in this way micelles of chosen and stable structure could be formed, not susceptible to exchanges of participating surfactant molecules or to other whims of nature. Later in the progress of this investigation, as the essential role of the hydrophobic-hydrophilic balance in the functioning of natural macromolecules and membranes became recognized, the hydrophobe-modified polymers assumed an important new function as model systems in which relevant structural features could be systematically varied, thereby providing a means for enhancing insight into the underlying physical and chemical mechanisms of biological processes.

This review will deal chiefly with two kinds of hydrophobic polyelectrolytes investigated in this laboratory, namely cationic derivatives of polyvinylpyridine and anionic copolymers of maleic acid and alkyl vinyl ethers. While the properties of these systems may, to some limited extent, be due to effects caused by their specific chemical constituents, their behavior may generally be expected to be characteristic of the entire class of water-soluble polymers carrying hydrophobic groups.

II. POLY-2-VINYLPYRIDINE-DERIVED POLYSOAPS

The first polysoap investigated was prepared by quaternizing close to one third of the pyridine groups of poly-2-vinylpyridine with *n*-dodecyl bromide.[1] The reduced viscosity against concentration curves in water and ethanol exhibited the characteristic shapes of polyelectrolytes, but the reduced viscosity in water was only one third that obtained in ethanol, indicating exceptional compactness of the macromolecules in water. In contrast, ordinary polyelectrolytes, for example polyvinylbutylpyridinium bromide, exhibited a much higher reduced viscosity in water than in ethanol.[2] The compact conformation of the polysoap was ascribed to intramolecular micelle formation brought about by the hydrophobic forces between the dodecyl groups. This conclusion was confirmed by the observation, using vapor pressure measurements, that the polysoap solubilized isooctane in aqueous solution, just as micelle-forming surfactants do. The solubility of isooctane was proportional to the polysoap concentration, indicating that the solubilization was intramolecular and that no critical micelle concentration was necessary for solubilization to occur. The micelles needed for solubilization are present at any polysoap concentration, no matter how small. The efficiency of solubilization was 66 mg of isooctane per gram of polysoap. In contrast, the solubilization efficiency for isooctane by the related nonpolymeric surfactant, *n*-dodecylpyridinium bromide, was lower than that of the polysoap, on a weight-ratio basis, mainly because no solubilization occurs below the critical micelle concentration.[1] However, the difference becomes dramatic if one compares the solubilization efficiencies of prefabricated and natural micelles on a dodecyl group basis. For the polysoap the efficiency is 0.26 mol of isooctane per mole of dodecyl group, almost double the 0.15 mol of isooctane solubilized per mole of micellized dodecyl group by the surfactant.

The effect of the solubilized isooctane on the molecular dimensions of the polysoap molecules was monitored by viscosity measurements.[3] The viscosity of a given polysoap solution was found to decrease linearly with the isooctane concentration until the solubilization limit was reached, at which point the viscosity remained constant on further addition of isooctane. The viscosity method thus provided a convenient means for determining solubilization limits of hydrocarbons, especially useful for nonvolatile solubilizates for which the vapor pressure technique was inapplicable. By using this method, it was found that dodecane was less soluble than isooctane in the polysoap.[3]

Surprising effects were observed with benzene as the solubilizate, as shown in Figure 1, where the reduced viscosity is presented as a function of the benzene concentration for polysoap solutions of differing concentrations.[4] The curves are seen to pass through maxima which increase dramatically with increasing polysoap concentration. The solubilization limits, again determined at the point where the curves became horizontal, after correction for the solubility of benzene in water, were proportional to the polysoap concentration. Comparison with results obtained with *n*-octane, also shown in the figure, indicate that the solubility of benzene in the polysoap is about an order of magnitude larger than that of the aliphatic hydrocarbon.[4] With the insight provided by subsequent light scattering and conductivity data,[5] these results may be interpreted as follows: the benzene is not only solubilized in the hydrocarbon core of the micelles, but also at the micelle-water interface, where it causes clumping of ionic head groups, thus exposing hydrophobic regions to the solvent. The interactions between these exposed hydrophobic regions result in aggregation of polysoap molecules with increasing benzene solubilization. Initially, the aggregates are chain-like, arising from a small number of contact regions, which causes the viscosity increase. As the number of contact sites increases, the aggregates become more globular, with a consequent decrease in the viscosity. Subsequent studies with hexane, 1-hexene, cyclohexane, cyclohexene,[6] and diphenylmethane[7] showed that the viscosity maxima are characteristic of aromatic hydrocarbons, though not of aliphatic hydrocarbons even when they are unsaturated or cyclic.

Even more complex effects are seen with long-chain alcohols and similar polar-nonpolar molecules where both maxima and minima in the viscosity are observed. The topic is not pursued further here. However, those interested may refer to two papers in which the subject is discussed.[5,8]

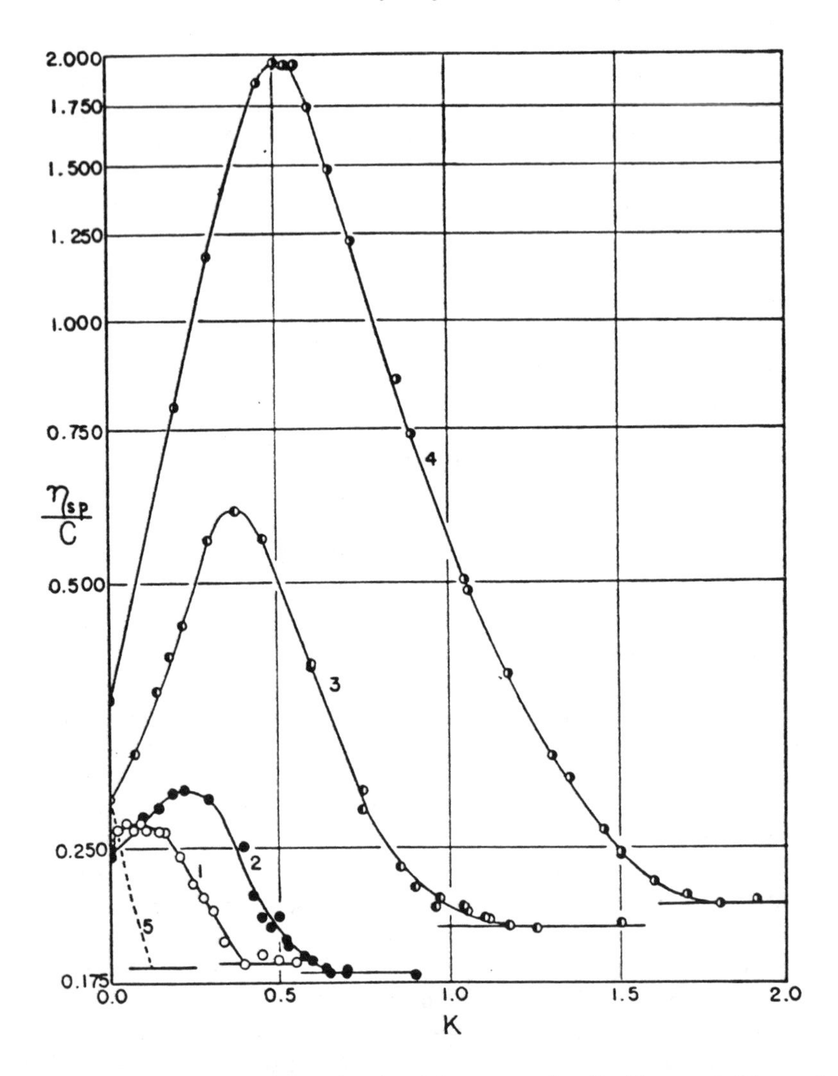

FIGURE 1. Reduced viscosity as a function of the concentration, K, of benzene (solid curves), and of *n*-octane (broken curve). Polysoap concentrations: (1) 0.5; (2) 1.0; (3) 2.0; (4) 3.0. All concentrations in g/100 ml. (From Layton, L. H. et al., *J. Polym. Sci.*, 9, 295, 1952. Reprinted by permission of John Wiley & Sons, Inc.)

III. TRANSITION FROM POLYELECTROLYTE TO POLYSOAP

While it was found impossible to quaternize more than one third of the pyridine groups of poly-2-vinylpyridine with dodecyl bromide, this limitation, presumably due to steric effects, was absent for poly-4-vinylpyridine. However, a polysoap prepared by close to complete quaternization of this polymer with *n*-dodecyl bromide turned out to be water insoluble.[9] Nevertheless, poly-

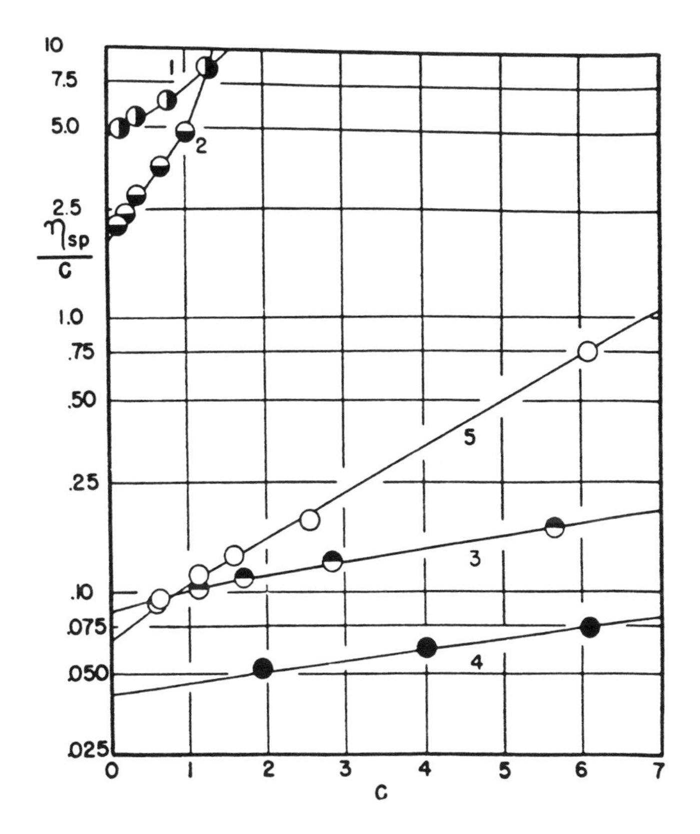

FIGURE 2. Effect of dodecyl group content on reduced viscosity of polyelectrolytes and polysoaps in 0.0226 *M* KBr: (1) PEB; (2) 6.7% polysoap; (3) 13.6% polysoap; (4) 28.5% polysoap; (5) 37.5% polysoap. (From Strauss, U. P. et al., *J. Phys. Chem.*, 60, 577, 1956. With permission of the American Chemical Society.)

4-vinylpyridine presented a convenient base for studying the transition from normal polyelectrolyte to polysoap by quaternizing part of the pyridine groups with dodecyl bromide and the remainder with ethyl bromide.[8] Figure 2 shows the effect of dodecyl group content on curves of the reduced viscosity against concentration for such poly-4-vinylpyridine derivatives obtained in 0.0226 *M* KBr at 25°C.[10] It is noteworthy that even with the ordinate given on a logarithmic scale, the curves fall into two distinct classes.

The first class contains curves 1 and 2, corresponding to an all ethyl bromide derivative (PEB) and a derivative in which 6.7% of the pyridine groups are quaternized with dodecyl bromide and most of the remainder with ethyl bromide (6.7% polysoap), respectively. These curves are characteristic of polyelectrolyte behavior. The intercepts, representing the intrinsic viscosities, are large, indicating extended random coil conformations due to intramolecular repulsion between the ionic groups. Curve 2 is seen to have a lower intercept, but a steeper slope than curve 1, indicating that replacing 6.7% of

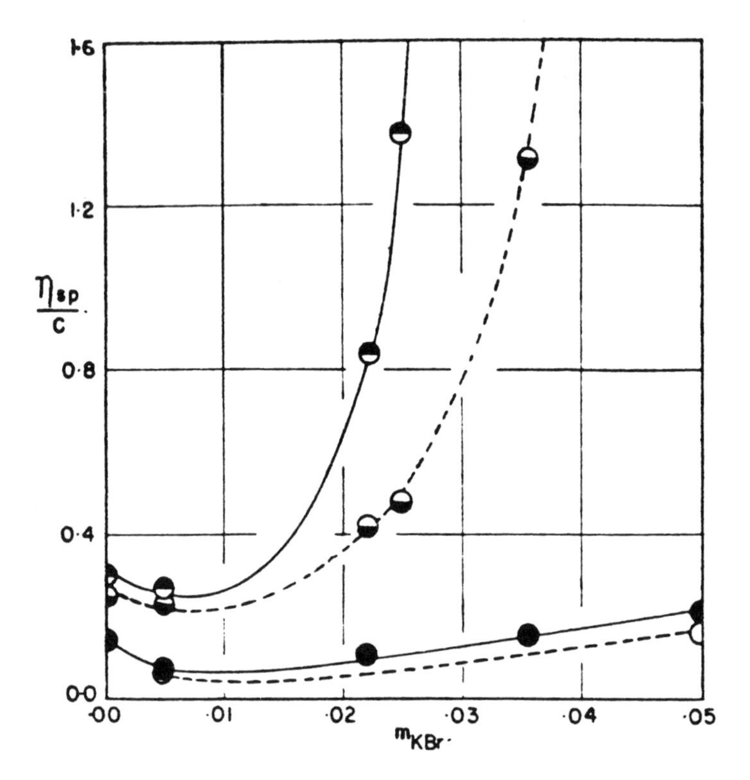

FIGURE 3. Effect of KBr on reduced viscosity of 37.9% polysoap. 1.00% polysoap solution: ●, 25°; ○, 45°. 6.00% polysoap solution: ◓, 25°; ◒, 45°. (From Strauss, U. P. et al., *J. Phys. Chem.*, 60, 577, 1956. With permission of the American Chemical Society.)

the ethyl by dodecyl groups has the effect of contracting the molecular dimensions but increasing the tendency towards intermolecular association.[10]

The second class containing curves 3, 4, and 5, representing 13.6, 28.5, and 37.9% polysoaps, respectively, is characterized by intercepts more than an order of magnitude smaller than those of the first class, indicating abnormally high compactness of the macromolecules. This behavior has been ascribed to intramolecular aggregation of the dodecyl groups caused by hydrophobic forces. The sharp drop in the intrinsic viscosity as the dodecyl content increases from 6.7 to 13.6% suggests the existence of a critical intramolecular micelle composition analogous to the critical micelle concentration characteristic of ordinary surfactants.[10]

It is noteworthy that both the intercepts and slopes of the curves pass through a minimum with increasing dodecyl content, indicating corresponding changes in molecular dimensions and interactions. The interactions of the 37.9% polysoap molecules are further enhanced with increasing polysoap and KBr concentrations, as shown in Figure 3, where the effect of temperature is also indicated.

The associating tendency of the 37.9% polysoap molecules is so strong that at 25°C their 6% solution becomes a clear thixotropic jelly as the KBr molality is raised to 0.035 (beyond the boundaries of Figure 3). Association is further confirmed by the slowness of viscosity changes after dilution or changes in temperature of dilute solutions of the 37.9% polysoap.[10] Such associating tendencies in similar hydrophobic water-soluble macromolecules has, in recent years, found many industrial applications, such as the development of emulsifying, dispersing, and thickening agents used for coatings, paints, varnishes, and in enhanced oil recovery.[11]

Solubilization of hydrocarbons depended strongly on the dodecyl content of the polyvinylpyridine addition products.[8] The aliphatic hydrocarbon, *n*-decane, was most strongly solubilized by the 37.9% polysoap. On a mole decane per mole of dodecyl group basis, the 28.5% polysoap solubilized only one third as much decane as did the 37.9% polysoap. Decane was not solubilized either by the 13.6% polysoap or by the derivatives with still smaller dodecyl group contents. However, the 13.6% polysoap did solubilize benzene, but, on the same mole per mole basis as indicated above, its solubilizing capacity was only one fourth that of the 37.9% polysoap, while on a gram benzene per gram polysoap basis it was one tenth.[8]

IV. SURFACE PHENOMENA OF POLYSOAPS

The surface activities of members of a similar set of poly-4-vinylpyridine derivatives, as determined by surface tension and interfacial tension measurements, were very small in the absence of simple electrolyte.[7] The finding supports the widely held opinion that micelles of ordinary surfactants have little, if any, surface activity. This property has been utilized in the emulsion polymerization of synthetic rubbers where monomer solubilization with minimal foam formation was desired. Addition of KBr produced surface activity for all derivatives containing dodecyl groups. In contrast, a bromopropylbenzene adduct of poly-2-vinylpyridine showed strong surface activity in aqueous solution even in the absence of simple electrolyte. This macromolecule produced high reduced viscosities in aqueous solution and did not solubilize hydrocarbons. The behavior observed indicates an extended conformation and the absence of micelle formation. Thus, the aromatic side chains are exposed to the solvent, resulting in the surface activity.[7]

A quaint phenomenon observed with most of the polysoaps described above was a formation of transient fibers when their solutions containing simple electrolyte were worked on in some manner, such as swirling, filtering, or pouring from one container to another.[10,12] The phenomenon was studied by means of "inversion tubes", which are glass tubes, about 1 ft long, filled three quarters with solution and sealed at both ends.[13] After the slow inversion of such a tube, the fibers were observed when the tube was returned to its vertical position.[14] The mechanism was explained as follows: during inversion a large air-solution interface, to which the surface-active polysoap molecules are attracted, is formed. After inversion the interface becomes very small and the adsorbed polysoap molecules are compressed before they can redissolve, resulting in transient fiber formation. Depending on conditions, fiber lifetimes ranging from a few seconds to more than 10 min have been observed in this manner.[14]

V. MALEIC ANHYDRIDE-ALKYL VINYL ETHER COPOLYMERS: CONFORMATIONAL TRANSITIONS

The second family of hydrophobic polyelectrolytes intensively investigated in this laboratory, the hydrolyzed copolymers of maleic anhydride and *n*-alkyl vinyl ethers, presented the opportunity for gaining additional insight into the behavior of polysoaps. In these copolymers the two monomers alternate regularly, so that the macromolecules may be considered as essentially homopolymers.

Since the acid groups are weak, one can vary their electrical charge over a wide range by titration with appropriate bases. Pronounced differences in behavior were observed with changes in the size of the alkyl group of the alkyl vinyl ether constituent. When the alkyl group contained one, two, or three carbon atoms, the polymers behaved as polyacids showing normal polyelectrolyte behavior.[15,16] With ten or more carbon atoms, the polymers behaved as typical polysoaps, similar to the polyvinylpyridine adducts discussed above, over the whole titration range.[17,18] However, when the alkyl group contained four to eight carbon atoms, the macromolecules underwent a transition from a compact hypercoiled conformation, typical of polysoaps, to an extended conformation, typical of normal polyelectrolytes, as their electrical charge was increased by the addition of base.[19] These conformational transitions have been observed by potentiometric[15] and calorimetric[16,19] titrations, by viscosimetry,[20] and by fluorescence spectroscopy with chemically attached optical probes.[21,22] Some of these findings will be presented in some detail below. It should be noted that the dicarboxylate groups of these copolymers show two distinct intrinsic pK values, differing by about three units, which under most conditions causes the deprotonation to proceed in two distinct successive steps. For this reason, it is convenient to define α, the degree of deprotonation, to cover the range from 0 to 2.[15]

The dansyl (1-dimethylaminonaphthalene-5-sulfonyl) group is well known for the sensitivity of its fluorescence spectrum to the polarity of its environment.[23,24] This probe was chemically attached to our copolymers by the reaction of dansyl (β-aminoethyl)amide with a small percentage (2 to 7%) of the copolymer anhydride groups.[25] The probe has an intense fluorescence maximum near 500 nm in nonpolar solvents. In water, the fluorescence intensity is very small and its maximum is red-shifted. Figure 4 shows the effect of increasing deprotonation of the dansylated butyl copolymer on the fluorescence intensity at 520 nm (F_{520}, in arbitrary units).[21] At low values of α, the fluorescence intensity is large, characteristic of a nonpolar environment.

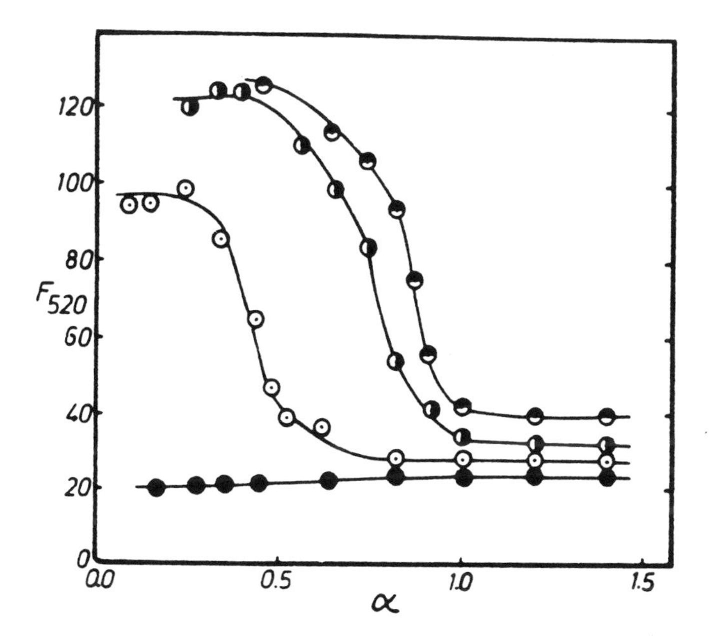

FIGURE 4. Dependence of F_{520}, the fluorescence emitted at 520 nm by dansylated copolymers, on α: (⊙, ◑, ◒), butyl copolymer in water, 0.2 *M* NaCl, and 0.5 *M* NaCl, respectively; (●) methyl copolymer in water. (Adapted from Strauss, U. P. and Vesnaver, G., *J. Phys. Chem.*, 79, 2426, 1975. With permission.)

Apparently, the dansyl groups are surrounded by hydrophobic microdomains which shield them from contact with the aqueous solvent. As α increases, the macromolecule goes through a conformational transition, caused by the repulsion between the ionic groups. As a result, the polymer molecules expand, the hydrophobic domains break up, and the dansyl groups become exposed to water, manifested by the low fluorescence intensity. As might be expected, the effect of added electrolyte is to reduce the ionic repulsions, thereby stabilizing the hypercoiled conformation, and shifting the transition to higher values of α. The graph also shows that the fluorescence of the methyl copolymer is small over the whole range of α, indicating that the dansyl groups are exposed to water and that this copolymer does not form hydrophobic microdomains.

An illustration of the effects of the conformational transition on the molecular dimensions of the butyl copolymer is shown in Figure 5.[20,26] Its intrinsic viscosity in 0.04 *M* NaCl at 30°C at low values of α is seen to be very small, revealing a rather compact conformation. As α increases, the intrinsic viscosity rises steeply with an overall 33-fold increase from α = 0.1 to α = 1.0. The intrinsic viscosity of the ethyl copolymer also rises over this region of α. However, its value at α = 0.1 is an order of magnitude larger than that of the butyl copolymer and over the range from α = 0.1 to α = 1.0 it shows an only 3.5-fold increase. This behavior of the ethyl copolymer (whose

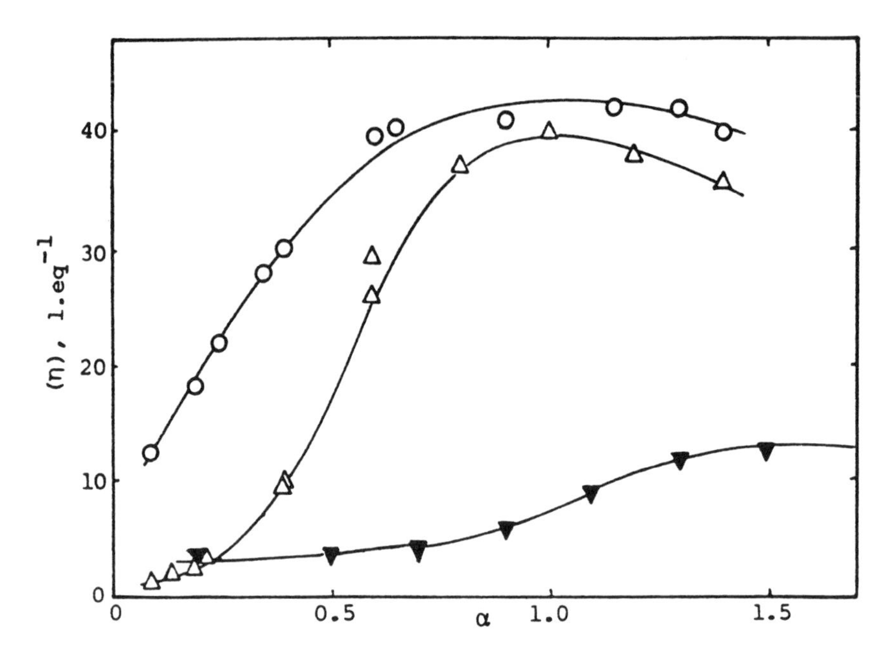

FIGURE 5. Intrinsic viscosity as a function of α in 0.04 *M* NaCl for ethyl (○), butyl (△), and hexyl (▼) copolymers. (From Strauss, U. P., in *Microdomains in Polymer Solutions,* Dubin, P., Ed., Plenum Press, New York, 1985, chap. 1. With permission.)

degree of polymerization is approximately 30% larger than that of the butyl copolymer) is characteristic of normal weak polyacids. The maxima in the curves at $\alpha = 1$ have been shown to be due to specific binding of sodium ion by the completely deprotonated dicarboxylate groups. The sodium ion contracts the molecular dimensions more strongly than does the hydrogen ion which it replaces. This effect is seen even more strongly with the hexyl copolymer whose degree of polymerization is about three times that of the butyl copolymer, but whose intrinsic viscosity remains small with increasing α. That this effect is caused by alkali metal ions but not by tetramethylammonium ion, which is not bound, is shown in Figure 6, where the reduced viscosity of the hexyl copolymer is seen to be almost constant beyond $\alpha = 1$ in 0.2 *M* LiCl and NaCl, but to rise sharply with increasing α when the supporting electrolyte is TMACl.[27]

Potentiometric titration results, shown in Figure 7, provide an excellent means for distinguishing between the different types of behavior produced by varying the alkyl group size in the copolymers.[20] For the region of α between 0 and 1, the curves for the methyl and ethyl copolymers coincide. It has been shown that these curves represent normal polyacid behavior. The curves for the butyl and hexyl copolymers show an initial rise which is strikingly steeper. This steepness is consistent with a compact conformation and the resulting higher negative charge density, which makes deprotonation more difficult. Eventually, these curves flatten out and overlap with the methyl

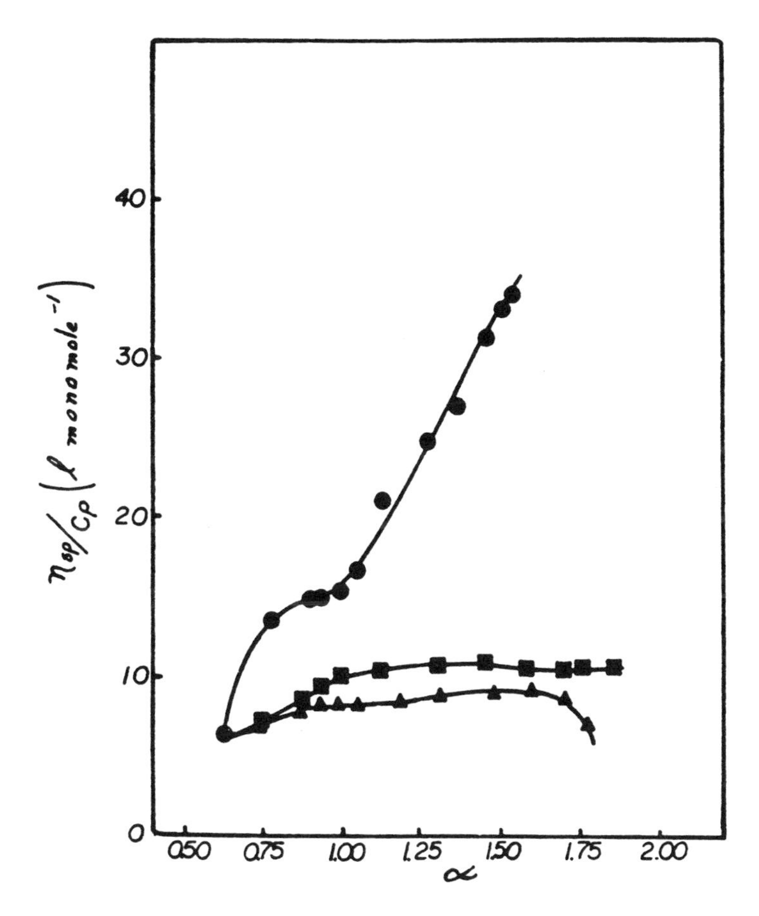

FIGURE 6. Reduced viscosity of 3.2×10^{-3} monomolar hexyl copolymer solution as a function of α in 0.2 *M* electrolyte solutions. (●) TMACl; (■) LiCl; (▲) NaCl. (From Lane, P. and Strauss, U. P., in *Copolymers, Polyblends and Composites,* Advances in Chem. Ser. 142, Platzer, N. A. J., Ed., American Chemical Society, Washington, D.C., 1975, 31. With permission.)

and ethyl copolymer curves. This flattening is indicative of a conformational transition from hypercoiled to random coil state, and is seen to occur at a higher value of α for the hexyl than for the butyl copolymer, as might be expected from the relative stabilizing powers of these copolymers for the hypercoiled state. The area between each of these curves and that of the methyl or ethyl copolymer is proportional to the free energy of stabilization for the uncharged compact state relative to the hypothetical uncharged extended state.[20] The remarkable similarity of these conformational transitions to the pH-induced denaturation of proteins is supported by the observation that urea, a well-known denaturant, diminishes the free energy of stabilization from 310 to 120 cal/mol of residue for the butyl copolymer, and from 1070 to 830 cal/mol of residue for the hexyl copolymer, as observed by potenti-

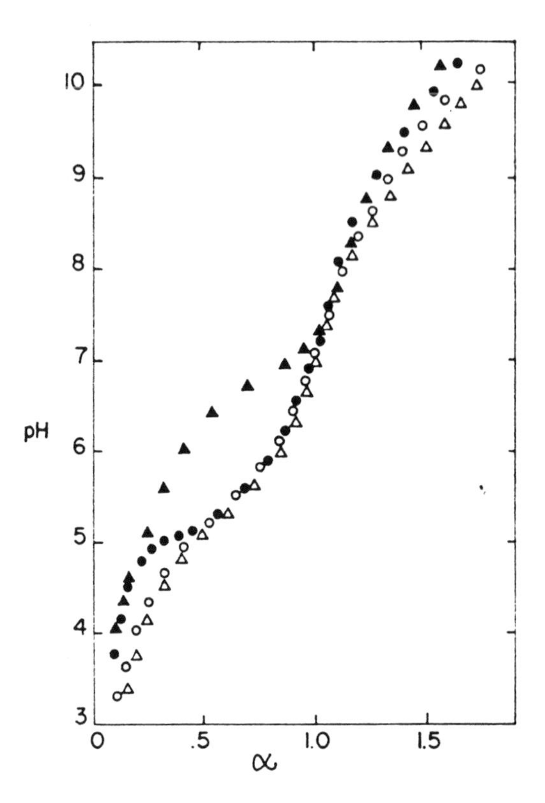

FIGURE 7. Potentiometric titrations in the absence of simple electrolyte for methyl (△); ethyl (○); butyl (●); and hexyl (▲) copolymers. (From Dubin, P. L. and Strauss, U. P., *J. Phys. Chem.*, 74, 2842, 1970. With permission of the American Chemical Society.)

ometric titrations.[28] Viscosity studies have shown that this destabilization of the hypercoiled state is due to an enhancement in the solvent affinities of the hydrophobic side chains as well as of the more polar backbone of the polyacids.[28]

With certain assumptions, one can calculate Θ_r, the fraction of residues in the random coil state by the equation

$$\Theta_r = (\alpha - \alpha_h)/\alpha_r - \alpha_h) \qquad (1)$$

where α_h and α_r are the values of α of the hypothetical hypercoiled and random coil states at the pH corresponding to the experimental value of α.[20] The dependence of α_h on the pH is obtained from the extrapolation of the experimental data for the butyl or hexyl copolymers from the hypercoiled into the transition region; the α_r values are assumed to be given by the experimental curve for the ethyl copolymer.[20] The progressions of transitions determined in this manner correspond quite closely to those observed by the fluorescence and viscosity techniques.[20,21]

A novel method for treating potentiometric titration data obtained with macromolecules is useful for elucidating the cooperativity of conformational transitions.[29,30] The method consists in treating the data as if they originated from a low molecular weight polyacid having N ionizable groups. In a certain sense, to be discussed below, this entity may be considered to be a subunit of the real polyacid, and its dissociation described by the overall reactions

$$AH_N \rightarrow AH_{N-i}^{-i} + iH^+ \tag{2}$$

In excess simple electrolyte, activity coefficients may be assumed to be constant, and overall dissociation constants may be defined by the equations

$$\beta_i = a_H^{+i}\,[AH_{N-i}^{-i}]/[AH_N] \tag{3}$$

which lead to the relations

$$\alpha = (\alpha_{max}/N)/\left(\sum_{i=0}^{N} i\beta_i h^i/\sum_{i=0}^{N} \beta_i h^i\right) \tag{4}$$

Here α_{max} is the value of α at complete deprotonation, chosen to equal the number of acidic groups per repeat unit, which is two for the polyacids treated here. The parameter h is the antilogarithm of the pH and represents the reciprocal of the hydrogen ion activity. Equation 4, after clearing of fractions, represents a set of linear equations in the β_is, one for each data point, which can be solved by conventional methods. The value of N has no intrinsic molecular significance, but it must be chosen large enough so that Equation 4 represents the experimental data within their precision limits. Equation 4 may thus be viewed as a practical empirical equation for representing potentiometric titration data of long-chain polyacids. With the currently attainable precision, the minimum number of parameters needed, i.e., the smallest allowable values of N, are four for polyacrylic acid and eight for the copolymer of maleic acid and butyl vinyl ether.[30]

However, the chief usefulness of this method arises from the fact that one can give the subunit a physical interpretation by defining it as any chain segment containing N acidic groups. The values of its dissociation constants, β_i, are affected by interactions with all ionic groups of the real polyacid molecule, located both within and outside the subunit. It has therefore been denoted by the term ''representative sample subunit'' or *RSSU*.[30] It can be shown that if a large number of *RSSU*s are selected independently and at random, the mole fraction of *RSSU*s with i dissociated hydrogen ions is given by the expression

$$x_i = \beta_i h^i/\sum_{i=0}^{N} \beta_i h^i \tag{5}$$

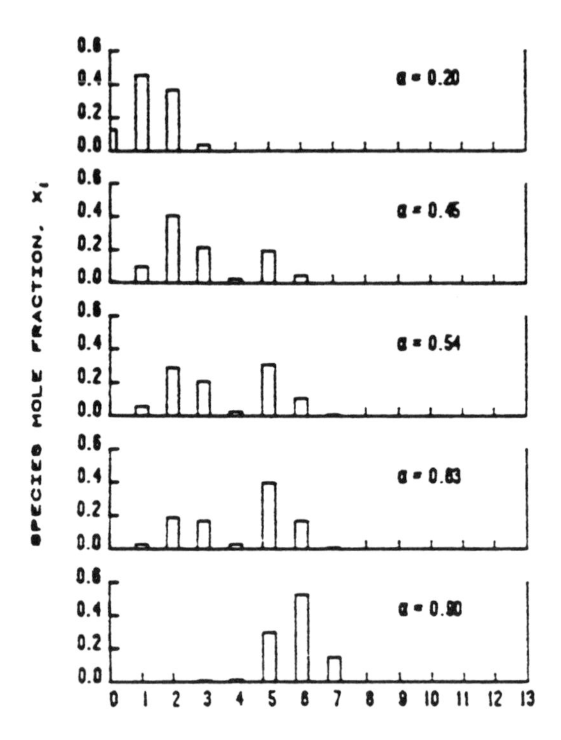

FIGURE 8. Progress of population distribution of *RSSU*s with i dissociated protons through conformational transition for N = 13. (From Strauss, U. P. et al., *J. Phys. Chem.*, 83, 2840, 1979. With permission of the American Chemical Society.)

The results of applying this method to potentiometric titration data obtained for a high molecular weight copolymer of maleic acid and butyl vinyl ether in 0.2 *M* LiCl is shown in Figure 8, where the distribution of mole fractions, x_i, is followed through the conformational transition. In the transition region, two population peaks, well separated by a deep minimum at i = 4 are clearly marked. One may assign the species with i <4 as belonging to the hypercoiled state, while the species with i >4 belong to the random coil state. The preservation of the bimodal distribution throughout the conformational transition shows convincingly that the transition is discontinuous and cooperative.[29]

The question now arises whether the polyacid molecule jumps as a whole from one state to the other or whether smaller portions of the molecule constitute the cooperative units. In the first case, the appropriate formula for the number of repeat units in a cooperative unit is[31]

$$n' = \frac{1}{\alpha_r - \alpha_h} \frac{d \log[\Theta_r/(1 - \Theta_r)]}{d\ pH} \tag{6}$$

For the butyl copolymer in 0.2 *M* LiCl the values of n′ so obtained vary from 4 to 15 over the range of the transition, a result inconsistent with the degree of polymerization of 2000.[32] It was therefore assumed that the cooperative units could be identified with micelles, the size of each micelle much smaller than the degree of polymerization. With the further assumptions that all the micelle had the same size, but that the random coil portions separating the micelles could be of any size, the following equation was derived for n, the number of repeat units per micelle:[32]

$$n = 1 + (n' - 1)/\Theta_r \qquad (7)$$

with n′ given by Equation 6. For the butyl copolymer, the value of n was found to be fairly constant, averaging 18 ± 2, over the range $\alpha = 0.35$ to 0.70, which corresponds to the major portion of the transition region. An improved method, developed subsequently, resulted in n = 19 for the butyl copolymer and in n = 13 for the pentyl copolymer.[33]

The presence of small intramolecular micelles in copolymers of this type was confirmed in this[34] as well as in other laboratories[35-38] using a recently developed luminescence quenching technique.[39] By this method, with the use of *tris*(2,2′-bipyridine)ruthenium(II) as the probe and 9-methylanthracene as the quencher, the average number of residues per micelle of the hydrolyzed copolymer of maleic anhydride and hexyl vinyl ether with DP = 1700 was found to be 24 ± 3, covering the transition range from $\Theta_r = 0$ to 0.465.[34] The rationale underlying this technique is that the probability for a fluorescent probe and a quencher molecule to be solubilized in the same micelle is inversely related to the concentration of micelles. It is significant that the micelle size was found to be constant with varying polymer concentration and with the extent of micellization.

VI. CONCLUDING REMARKS

Our work on hydrophobe-modified polymers has been built on by other laboratories, which have synthesized and studied polysoaps of different chemical structure,[40] including maleic anhydride copolymers of styrene[41] and of trimethylpentene.[42] These polymers show properties very similar to those of the systems described here. Because there appear to be no limits to their chemical design, hydrophobe-modified polymers offer challenging opportunities for a wealth of further exploration.

REFERENCES

1. **Strauss, U. P. and Jackson, E. G.,** Polysoaps. I. Viscosity and solubilization studies on an n-dodecyl bromide addition compound of poly-2-vinylpyridine, *J. Polym. Sci.*, 6, 649, 1951.
2. **Fuoss, R. M. and Strauss, U. P.,** Polyelectrolytes. II. Poly-4-vinylpyridonium chloride and poly-4-vinyl-N-n-butylpyridonium bromide, *J. Polym. Sci.*, 3, 246, 1948.
3. **Jackson, E. G. and Strauss, U. P.,** Polysoaps. II. Effect of added hydrocarbon on reduced viscosity of an n-dodecyl bromide addition compound of poly-2-vinylpyridine, *J. Polym. Sci.*, 7, 473, 1951.
4. **Layton, L. H., Jackson, E. G., and Strauss, U. P.,** Polysoaps. III. Influence of solubilized benzene on reduced viscosity of an n-dodecyl bromide addition compound of poly-2-vinylpyridine, *J. Polym. Sci.*, 9, 295, 1952.
5. **Strauss, U. P. and Slowata, S. S.,** The effect of solubilization on the equivalent conductance and reduced viscosity of a polysoap derived from poly-2-vinylpyridine, *J. Phys. Chem.*, 61, 411, 1957.
6. **Strauss, U. P. and Layton, L. H.,** A comparison of the effects of several solubilized C_6-hydrocarbons on the viscosity of a polysoap solution, *J. Phys. Chem.*, 57, 352, 1953.
7. **Jorgensen, H. E. and Strauss, U. P.,** Exploratory studies on the surface activity of polysoaps, *J. Phys. Chem.*, 65, 1873, 1961.
8. **Strauss, U. P. and Gershfeld, N. L.,** The transition from typical polyelectrolyte to polysoap. I. Viscosity and solubilization studies on copolymers of 4-vinyl-N-ethylpyridinium bromide and 4-vinyl-N-n-dodecylpyridinium bromide, *J. Phys. Chem.*, 58, 747, 1954.
9. **Strauss, U. P., Assony, S. J., Jackson, E. G., and Layton, L. H.,** Polysoaps. IV. Dissolution phenomena of a poly-4-vinylpyridine derivative, *J. Polym. Sci.*, 9, 509, 1952.
10. **Strauss, U. P., Gershfeld, N. L., and Crook, E. H.,** The transition from typical polyelectrolyte to polysoap. II. Viscosity studies of poly-4-vinylpyridine derivatives in aqueous KBr solutions, *J. Phys. Chem.*, 60, 577, 1956.
11. **Glass, J. E.,** Ed., *Polymers in Aqueous Media: Performance Through Association,* Advances in Chem. Ser. 223, American Chemical Society, Washington, D.C., 1989.
12. **Strauss, U. P. and Williams, B. L.,** The transition from typical polyelectrolyte to polysoap. III. Light scattering and viscosity studies of poly-4-vinylpyridine derivatives, *J. Phys. Chem.*, 65, 1390, 1961.
13. **Goldacre, R. J.,** Surface films, their collapse on compression, the shapes and sizes of cells and the origin of life, in *Surface Phenomena in Chemistry and Biology,* Danielli, J. F., Pankhurst, K. G. A., and Riddiford, A. C., Eds., Pergamon Press, New York, 1958, 278.
14. **Richlin, J. and Strauss, U. P.,** Polysoaps with surface properties resembling lipoproteins, *AChS Polym. Prepr.*, 27, 425, 1986.
15. **Dubin, P. and Strauss, U. P.,** Hydrophobic hypercoiling in copolymers of maleic acid and alkyl vinyl ethers, *J. Phys. Chem.*, 71, 2757, 1967.
16. **Martin, P. J., Morss, L. R., and Strauss, U. P.,** Calorimetric investigation of hydrolyzed copolymers of maleic anhydride with butyl and lower alkyl vinyl ethers, *J. Phys. Chem.*, 84, 577, 1980.
17. **Ito, K., Ono, H., and Yamashita, Y.,** Viscosity and solubilization studies on weak anionic polysoaps in water; effect of the counter-ion, *J. Colloid Sci.*, 19, 28, 1964.
18. **Varoqui, R. and Strauss, U. P.,** Comparison of electrical transport properties of anionic polyelectrolytes and polysoaps, *J. Phys. Chem.*, 72, 2507, 1968.
19. **Martin, P. J. and Strauss, U. P.,** Thermodynamics of hydrophobic polyacids, *Biophys. Chem.*, 11, 397, 1980.
20. **Dubin, P. L. and Strauss, U. P.,** Hydrophobic bonding in alternating copolymers of maleic acid and alkyl vinyl ethers, *J. Phys. Chem.*, 74, 2842, 1970.

21. **Strauss, U. P. and Vesnaver, G.,** Optical probes in polyelectrolyte studies. II. Fluorescence spectra of dansylated copolymers of maleic anhydride and alkyl vinyl ethers, *J. Phys. Chem.,* 79, 2426, 1975.
22. **Strauss, U. P. and Schlesinger, M. S.,** Effects of alkyl group size and counterion type on the behavior of copolymers of maleic anhydride and alkyl vinyl ethers. II. Fluorescence of dansylated copolymers, *J. Phys. Chem.,* 82, 1627, 1978.
23. **Weber, G.,** Polarization of the fluorescence of macromolecules. II. Fluorescent conjugates of ovalbumin and bovine serum albumin, *Biochem. J.,* 51, 155, 1952.
24. **Lagunoff, D. and Ottolenghi, P.,** Effect of pH on the fluorescence of dimethylaminonaphthalenesulfonate (DNS) and several derivatives, *Compt. Rend. Trav. Lab. Carlsberg,* 335, 63, 1966.
25. **Strauss, U. P. and Vesnaver, G.,** Optical probes in polyelectrolyte studies. I. Acid-base equilibria of dansylated copolymers of maleic anhydride and alkyl vinyl ethers, *J. Phys. Chem.,* 79, 1558, 1975.
26. **Strauss, U. P.,** Microdomains in hydrophobic polyacids, in *Microdomains in Polymer Solutions,* Dubin, P., Ed., Plenum Press, New York, 1985, chap. 1.
27. **Lane, P. and Strauss, U. P.,** Phase separation in solutions of maleic acid-alkyl vinyl ether copolymers, in *Copolymers, Polyblends and Composites,* Advances in Chem. Ser. 142, Platzer, N. A. J., Ed., American Chemical Society, Washington, D.C., 1975, 31.
28. **Dubin, P. and Strauss, U. P.,** Conformational transitions of hydrophobic polyacids in denaturant solutions. The effect of urea, *J. Phys. Chem.,* 77, 1427, 1973.
29. **Strauss, U. P., Barbieri, B. W., and Wong, G.,** Analysis of ionization equilibria of polyacids in terms of species population distributions. Examination of a ''two-state'' conformational transition, *J. Phys. Chem.,* 83, 2840, 1979.
30. **Strauss, U. P.,** Representation of polyacids by subunits in the analysis of ionization equilibria. Significance and thermodynamic applications, *Macromolecules,* 15, 1567, 1982.
31. **Ptytsyn, O. B. and Birshtein, T. M.,** Method of determining the relative stability of different conformational states of biological macromolecules, *Biopolymers,* 7, 435, 1969.
32. **Strauss, U. P. and Barbieri, B. W.,** Estimation of the cooperative unit size in conformational transitions of hydrophobic polyacids, *Macromolecules,* 15, 1347, 1982.
33. **Barbieri, B. W. and Strauss, U. P.,** Effect of alkyl group size on the cooperativity in conformational transitions of hydrophobic polyacids, *Macromolecules,* 18, 41, 1985.
34. **Hsu, J. L. and Strauss, U. P.,** Intramolecular micelles in a copolymer of maleic anhydride and hexyl vinyl ether: determination of aggregation number by luminescence quenching, *J. Phys. Chem.,* 91, 6238, 1987.
35. **Binana-Limbelé, W. and Zana, R.,** Fluorescence probing of microdomains in aqueous solutions of polysoaps. 1. Use of pyrene to study the conformational state of polysoaps and their comicellization with cationic surfactants, *Macromolecules,* 20, 1331, 1987.
36. **Binana-Limbelé, W. and Zana, R.,** Fluorescence probing of microdomains in aqueous solutions of polysoaps. II. Study of the size of the microdomains, *Macromolecules,* 23, 2731, 1990.
37. **Chu, D. Y. and Thomas, J. K.,** Photophysical and photochemical studies on a polymeric intramolecular micellar system, PA-$18K_2$, *Macromolecules,* 20, 2133, 1987.
38. **Chu, D. Y. and Thomas, J. K.,** Photophysical studies of hydrophobically modified polyelectrolytes, in *Polymers in Aqueous Media: Performance through Association,* Advances in Chem. Ser. 223, Glass, J. E., Ed., American Chemical Society, Washington, D.C., 1989, chap. 17.
39. **Turro, N. J. and Yekta, A.,** Luminescent probes for detergent solutions. A sample procedure for determination of the mean aggregation number of micelles, *J. Am. Chem. Soc.,* 100, 5951, 1978.
40. **Sinha, S. K., Medalia, A. I., and Harrington, D. P.,** Light scattering studies of polysoap solutions, *J. Am. Chem. Soc.,* 79, 281, 1957.

41. **Ohno, N. Nitta, K., Makino, S., and Sugai, S.,** Conformation transition of the copolymer of maleic acid and styrene in aqueous solution, *J. Polym. Sci. Polym. Phys. Ed.*, 11, 413, 1973.
42. **Barone, G., Di Virgilio, N., Elie, V., and Rizzo, E.,** Intramolecular and intermolecular aggregations in aqueous solutions of maleic anhydride copolymers, *J. Polym. Sci.*, Symp. No. 44, 1, 1974.

Chapter 7

PROTEINS IN SOLUTION AND AT INTERFACES

Eric Dickinson

TABLE OF CONTENTS

0-8493-6784-0/93/$0.00 + $.50

I. INTRODUCTION

Many biological systems of commercial or technological importance contain a mixture of proteins and low-molecular-weight surfactants. Proteins in solution contain a mixture of different types of chemical groups — nonpolar, polar, and electrically charged — and so it is not surprising that most small amphiphilic molecules will interact strongly with proteins. The chief driving force for the association of surfactant molecules into micelles and vesicles is the reduction in the hydrocarbon-water contact area of the alkyl chains.[1] This same driving force will favor association of surfactant alkyl chains with hydrophobic parts of protein molecules, while headgroups of ionic surfactants will tend to interact attractively with positively charged groups on proteins.

Both proteins and small-molecule surfactants have a strong tendency to adsorb at a wide range of hydrophilic and hydrophobic surfaces. In systems containing a mixture of protein + surfactant, competitive adsorption between the two species for the available interface will be influenced by the nature and strength of the protein-surfactant interaction.[2] Any complex formation between protein and surfactant in bulk solution reduces the amount of free surfactant available for competing with protein at the interface. In addition, any protein-surfactant binding may change the adsorption energy of the protein for the interface by affecting the net charge or the overall macromolecular hydrophobicity. There is a direct experimental link between studies of protein-surfactant binding and the competitive adsorption of proteins and surfactants, since many of the investigators, starting with the early work of Cockbain[3] and Knox and Parshall,[4] have used the measurement of surface tension to determine the presence and stoichiometry of protein-surfactant complexes in bulk solution.

In food emulsions such as salad dressing or ice cream, the distribution of proteins and low-molecular-weight emulsifiers between the oil-water interface and the bulk aqueous and oil phases is an important factor affecting the formation, rheology, and stability behavior.[5-8] The situation in real food colloids is complicated by the fact that there is invariably a mixture of several different individual proteins which are adsorbing competitively,[9] together with several low-molecular-weight surfactants, both oil and water soluble, which are competing for binding sites on the proteins and at the interface. In order to gain a more fundamental understanding, therefore, it is convenient in laboratory studies to consider model systems containing just a single pure protein and a single surfactant. It is with such systems that this chapter is mostly concerned.

II. STRUCTURE OF PROTEINS

A. GENERAL ASPECTS

Proteins are the complex unbranched polymers that play a crucial role in the structure and function of biological cells and organisms. The individual

L-α-amino acids are linked together head-to-tail by peptide bonds to form linear chains containing up to several hundred monomer units (residues). Protein molecular weights are typically in the range from 10^4 to 10^5 Da. Structures vary enormously depending on the sequence of amino acid residues, but they may be roughly classified into three main types: fibrous, globular, and disordered. Fibrous proteins are composed of polypeptide chains arranged to lie along a common linear axis. The functionality of fibrous proteins (collagen, keratin, etc.) is largely associated with their structural and mechanical properties in tissues such as bone, muscle, skin, and hair. In a globular protein, one or more polypeptide chains are folded compactly and uniquely together to form a three-dimensional structure with a roughly spherical shape and a complicated surface topology. Most globular proteins function as enzymes (trypsin, catalase, etc.), but some have other biological functions — as hormones (insulin), in transfer processes (hemoglobin), as a food supply (ovalbumin), and so on. Most fibrous and globular proteins become more disordered on heating. A few proteins have so little ordered structure in the native state that it is convenient to call them disordered. Notable examples are the milk proteins α_{s1}-casein and β-casein.

Proteins are amphoteric polyelectrolytes, being capable of holding both positive and negative charges. The ionizable groups have a wide range of pK values, and so the charge distribution is sensitive to pH. When the net charge in aqueous solution is zero, the system is said to be *isoelectric* (pH = pI). (The pH of zero charge in the absence of any ions except H_3O^+ and OH^- is called the *isoionic* point.) For many proteins the isoelectric point is close to pH 5, but some have much lower values (e.g., pepsin) and some much higher (e.g., lysozyme).

The physical properties of aqueous protein solutions strongly reflect the net charge on the macroions. To a first approximation, protein solubility in water increases with the proportion of polar and charged groups, and decreases with increasing molecular weight. Most proteins show a minimum in solubility at the isoelectric point where electrostatic interactions are minimal. The molecular size of a disordered protein is at its lowest near pI for the same reason. Addition of electrolyte attenuates the strength of intramolecular and intermolecular electrostatic interactions by shielding charged groups from one another. The properties of proteins in solution are therefore sensitive to both pH and ionic strength.

B. LEVELS OF STRUCTURE

To describe the structure of a particular protein, the absolute sequence of amino acid residues ($R_1R_2R_3$ . . . R_n) must first be specified. Table 1 lists the 20 normal amino acids together with their conventional three-letter abbreviations and the chemical formulae of the side chains. In the complete covalent structure, called the *primary structure,* there may be more than one polypeptide chain, as well as intrachain or interchain covalent crosslinks (disulfide bonds). The *secondary structure* is the local arrangement of the

TABLE 1
Properties of Amino Acid Residues in Proteins

Amino acid	Symbol	Side Chain	pK[a]	Hydrophobicity[b]
Alanine	Ala	$-CH_3$		3.1
Arginine	Arg	$-(CH_3)_2NHC(NH_2)_2^+$	12	3.1
Asparagine	Asn	$-CH_2CONH_2$		~0
Aspartic acid	Asp	$-CH_2CO_2^-$	4.4–4.6	
Cysteine	Cys	$-CH_2SH$	8.5–8.8	4.0
Glutamic acid	Glu	$-(CH_2)_2CO_2^-$	4.4–4.6	
Glutamine	Gln	$-(CH_2)_2CONH_2$		~0
Glycine	Gly	$-H$		0[c]
Histidine	His	$-CH_2(C_3N_2H_3)^+$	6.5–7.0	2.0
Isoleucine	Ile	$-CH(CH_3)CH_2CH_3$		12.3
Leucine	Leu	$-CH_2CH(CH_3)_2$		10.0
Lysine[d]	Lys	$-(CH_2)_4NH_3^+$	10.0–10.2	6.3
Methionine	Met	$-(CH_2)_2SCH_3$		5.4
Phenylalanine	Phe	$-CH_2(C_6H_5)$		11.0
Proline[e]	Pro	$-(CH_2)_3^-$		10.9
Serine	Ser	$-CH_2OH$		
Threonine	Thr	$-CH(CH_3)OH$		1.9
Tryptophan	Trp	$-CH_2(C_8NH_6)$		12.5
Tyrosine	Tyr	$-CH_2(C_6H_4)OH$	9.6–10.0	11.9
Valine	Val	$-CH(CH_3)_2$		7.1

[a] Values expected in a protein (see Reference 12, p. 49).
[b] Measured in kJ residue^{-1}; according to Tanford (Reference 17).
[c] Exactly zero by definition.
[d] In collagen, converted to δ-hydroxylysine by posttranslational enzymic modification.
[e] Secondary amino acid with cyclic five-membered ring structure; in collagen, converted to γ-hydroxyproline by posttranslational modification.

polypeptide backbone, i.e., the steric relationship of residues close to each other in the linear sequence. The two main elements of ordered secondary structure are the α-helix and the β-sheet. In the α-helix, the peptide oxygen of residue R_i forms a hydrogen bond with the peptide nitrogen of residue R_{i+4}. In the β-sheet, extended strands of backbone chain are held parallel to one another by crosslinking hydrogen bonds between peptide groups on adjacent strands. Sequences not in the α- or β-forms are known *faute de mieux* as ''random coils'', even though the local structure may be far from random. The complete three-dimensional structure of a single covalent species of the protein is called the *tertiary structure*. For water-soluble globular proteins, this arrangement involves most of the polar residues being located on the outer surface in direct contact with the solvent, and many of the nonpolar residues being buried in the interior. The highest level of protein structure is the *quaternary structure* resulting from the noncovalent association of independent tertiary structural units.

The various levels of protein structure are described in detail in standard textbooks.[10-13] What follows here is a brief overview of the key features of

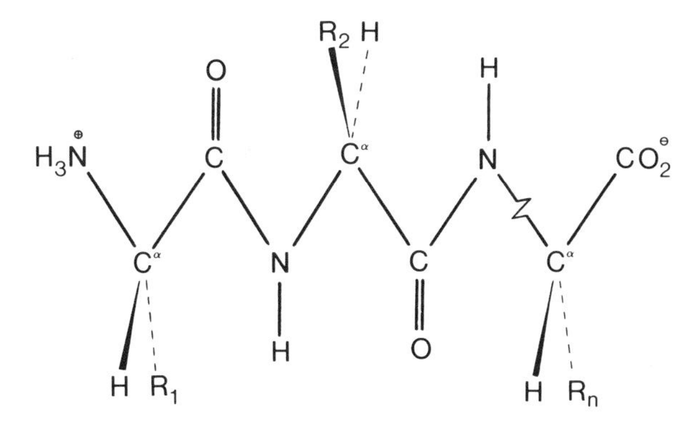

FIGURE 1. Structure of polypeptide chain showing peptide groups (CONH) and their associated C^{α} atoms with side chains R_i.

protein structure as a basis for approaching the topics of competitive adsorption and protein-surfactant interactions from a molecular viewpoint.

1. Primary Structure

The general structure of a short section of polypeptide is illustrated in Figure 1. The peptide linkage between two amino acid residues is a substituted amide bond. Compared with other biopolymers (notably polysaccharides), proteins are rather flexible polymers because many of the bonds in the side chains and backbone are rotationally permissive.[14] The main exception is the peptide bond itself, which is predominantly in the *trans*-configuration. The C–N bond has a partial double-bond character (about 40%) due to electron delocalization, and so the peptide group (CONH) and its adjoining α-carbon atom tend to remain in a common plane. Another consequence of delocalization is the lack of protonation of the NH group in the pH range 0 to 14.

A graph of net charge against pH for a protein in solution is called the titration curve. An idea of the pH range over which each side chain can normally be expected to titrate can be found from the pK values quoted in Table 1. A particular ionizable group goes from a state of being 1% ionized to a state of being 99% ionized over a range of approximately 4 pH units (i.e., pK $\pm$ 2). The actual pK value depends on a number of factors — the ionic strength and dielectric constant of the aqueous environment, the nature of adjacent side chains, and whether the protein is adsorbed at an interface. The uncharged asparagine and glutamine are easily hydrolyzed to Asp and Glu in the presence of acid or alkali. The aspartic acid and glutamic acid side chains are negatively charged at neutral pH. The residues with positively charged side chains at neutral pH are Lys, Arg, and His. The ε-amino and guanidino groups are responsible for the charges on lysine and arginine, respectively. The imidazole group of histidine may only be about 10% protonated at neutral pH.

The amino acids with nonpolar side chains are Ala, Ile, Leu, Met, Phe, Pro, Trp, and Val. These residues have a tendency to associate in water, and to adsorb at air-water and oil-water interfaces. Many scales of amino acid hydrophobicity have been proposed.[15] Probably the most widely accepted is that of Nozaki and Tanford[16] based on the free energy of transfer of 1 mol of amino acid from water into ethanol, relative to the same free energy change for glycine. Some values are listed in Table 1. We see that there is an increase in relative hydrophobicity with the length of the aliphatic side chain.

Several amino acid residues possess uncharged polar groups capable of forming hydrogen bonds with water and other hydrophilic side chains. The polarity of serine, threonine, and tyrosine is related to their hydroxyl groups, and the polarity of Asn and Gln to their amide groups. The polarity of cysteine is related to its thiol group, though this is often present in an oxidized state due to the free –SH groups of two cysteine residues combining to form a disulfide crosslink (cystine).

Amino acid compositions vary enormously from protein to protein, though certain amino acids are relatively rare (Met and Trp), while others are rather more common (Ala and Leu). Except in certain special cases (see below), the various amino acids are distributed apparently randomly along the polypeptide chain. There are, therefore, no simple general rules of protein structure, though the proportions of the different classes of residues can give a useful guide to certain aspects of structure and behavior. A high proportion of charged groups enhances electrostatic interactions, and this is important in relation to gelation, solubility, and surfactant binding. A high proportion ($\gtrsim$30%) of nonpolar residues generally implies good surface activity and emulsifying properties, and a tendency towards aggregation in aqueous solution. This aggregation may be reversed on cooling because hydrophobic interactions are weakened as the temperature is lowered. A high proportion of proline (or hydroxyproline) inhibits α-helix formation. Covalent bonds between pairs of oxidized cysteine residues have an important role in maintaining the structural integrity of globular proteins. It is noteworthy that the highly disordered hydrophobic protein β-casein contains a large proportion of proline and no cysteine. The disordered hydrophilic protein gelatin is rich in hydroxyproline.

2. Secondary Structure

To those familiar with synthetic polymers, the "random coil" might be thought to be the most obvious configuration for a protein in solution, favored as it is by the high conformational entropy and the extensive interaction with the solvent. In fact, however, the regularity of hydrogen bonding sites along the polypeptide chain generally leads to more energetically favored ordered structures — most notably the α-helix or the β-sheet.

The right-handed α-helix is a particularly stable structure. All side chains point outwards, and each peptide carbonyl in the helical structure serves as a hydrogen bond acceptor for the NH donor four residues away. The pitch of the helix is 0.54 nm, and there are 3.6 residues per turn. Proline is incompatible with the α-helix structure because it is a cyclic amino acid and cannot fit into the helix without bending the chain. Adoption of this structure may also be prevented by steric hindrance from bulky side chains (e.g., in polyisoleucine) or by strong electrostatic repulsion between side chains (e.g., in polyglutamic acid at high pH and low ionic strength).

In the β-sheet structure, the polypeptide chain is nearly fully extended. Individual strands associate side by side, forming hydrogen bonds between carbonyl and NH groups of the backbone. Proline has no NH group to participate in such bonding, and is stereochemically incompatible with the ordered structure. Whether a particular sequence of residues will tend to form a β-sheet or an α-helix, or no ordered structure at all, depends on the nature of the side chains. The classic case of a protein having an extended β-sheet structure is silk fibroin. In this protein long stretches of the chain have the six-residue unit Gly-Ser-Gly-Ala-Gly-Ala repeated many times. The sheets can pack closely together because all the Gly residues are on one side and all the Ala and Ser residues on the other.

A few other types of secondary structure also occur in proteins. In the reverse turn, the chain bends back on itself by forming a hydrogen bond between the carbonyl of residue R_i and the NH group of residue R_{i+3}. Since proline is incompatible with both α- and β-forms, it is not surprising that polyproline has its own special ordered structures. These are known as polyproline I, a right-handed helix containing *cis*-peptide bonds (3.3 residues per turn), and polyproline II, a left-handed helix containing *trans*-peptide bonds (3.0 residues per turn). The two structures can interconvert, and form II is the more stable in aqueous solution. The structural protein collagen is rich in proline (and hydroxyproline). Collagen has a triple-helix structure made by twisting together three polyproline II helices. Each polypeptide chain of collagen has Gly every third position, and these residues are located on the inside of the triple helix, with the bulky proline rings pointing outwards.

3. Tertiary Structure

The folding of the various types of secondary structure into the tertiary structure is mainly determined by the noncovalent interactions of the side chains. When two nonpolar groups come close together in an aqueous environment, some of the vicinal water molecules are released into bulk solution, leading to a large entropy increase associated with disruption of ordered water structure around the nonpolar groups. This hydrophobic effect[17] is the main thermodynamic driving force towards protein folding. In addition, tertiary structure is stabilized by ionic bonds or salt bridges arising from the interaction of charged residues; and some polar groups not directly involved in α- or β-structures are able to form hydrogen bonds with other polar side chains, e.g., –OH and =N–, or =NH and =O.

The interior of a globular protein is very closely packed. It is estimated[18] that about 75% of the interior space is occupied by atoms of size defined by their normal van der Waals radii. This contrasts with much lower values (45 to 60%) calculated for atoms or molecules in liquids. This tight packing of the globular protein interior is especially impressive in view of the fact that it must be compatible with the connectivity of the polypeptide chain and the necessity for all (unpaired) ionized groups to be exposed to the aqueous solvent. Atoms in the interior tend to exhibit substantially smaller thermal displacements than those near the surface. Atoms belonging to α-helices or β-sheets are less mobile than those in random coil configurations. Though many nonpolar residues are in the interior, many also are forced to reside at the protein surface. Ion pairs, when formed, tend to be located at the surface so as not to disrupt the aqueous hydration shell. Uncharged polar groups also reside preferentially at the surface; those that are buried are usually involved in hydrogen bonds to other polar side chains. In reality, of course, the distinction between the protein surface and its interior is rather blurred. The size and complexity of certain amino acids (e.g., Lys and Arg) make it difficult to classify them as just being either buried or exposed.[19] The overall folded structure of small proteins is usually roughly spherical in shape, but with a very irregular surface. Some proteins (e.g., insulin) are loosely folded, and the tertiary structure is held together by disulfide bonds.

C. PROTEINS IN SOLUTION

Much of the available information on protein structure has been derived from crystallographic studies. Three types of water molecules have been identified in protein crystals: bulk water, bound water, and buried water. The first is like ordinary liquid water with a rotational relaxation time of about 10^{-11} s. The so-called "bound water" has a longer relaxation time of approximately 10^{-9} s, and an altered freezing point due to strong hydrogen bonding interactions with the protein surface. Typically the amount of bound water is 0.4 ± 0.1 g g^{-1} which corresponds to 2 or 3 water molecules per residue on average. (A full monolayer of water molecules around the surface would come to about 1.0 g g^{-1}, which means that just 40% of the water surrounding a typical protein molecule in solution is "bound".) The third type of water molecule is buried tightly within the structure and is essentially immobile.

In many systems, proteins in solution have been shown to closely resemble proteins in crystals in terms of their spectroscopic properties. Nevertheless, it has been shown[20] by neutron magnetic resonance (NMR) (and other techniques that probe dynamical behavior) that there exists in aqueous solution a family of similar conformations, rather than one unique spatial arrangement as found in the crystal.

In terms of their thermodynamic behavior in solution, proteins are much more complex than synthetic polymers, owing to the diversity of chemical entities on the side chains. Standard statistical theories[21,22] are primarily devised to describe flexible chains composed of identical segments, where the relative segment-solvent interaction is quantified in terms of a single parameter (e.g., the Flory-Huggins χ-parameter) which measures solvent quality. For a protein, however, a solvent that is good for polar side chains will be poor for hydrophobic groups, and vice versa. This means that, for a protein solution that is close to being thermodynamically ideal, the concept of a theta solvent corresponding to an effective homopolymer composed of average protein segments is a considerable approximation to the real situation in which individual segments may experience a very nonideal local environment. In practice, the best solvents for disordered proteins are concentrated solutions of urea or guanidinium chloride.

Various physical and chemical treatments lead to the denaturation of proteins in solution. Denaturation is the uncoiling or unfolding of the tertiary and secondary structure without rupture of the covalent links of the primary structure. Though it may be accompanied by protein aggregation and loss of solubility, denaturation is generally a reversible process unless accompanied by oxidation of cysteine or disulfide interchange. The Gibbs free energy of stabilization of the native structure relative to the unfolded state is typically much less than 1 kT per amino acid residue, which means that for purely thermodynamic reasons the structure is susceptible to change by altering the temperature or the local solution environment. The folding/unfolding transition is a cooperative process, since the disruption of any significant portion of the folded structure leads to an unfolding of the rest. Most proteins are unfolded by guanidinium chloride in 6-*M* solution (approaching the solubility limit). The presence of the salt has the effect of greatly reducing the hydrophobic interaction between nonpolar residues by disrupting the ordered hydrogen-bonded structure of water. Proteins may also be denatured by heating, by changes in pH, or by addition of a solvent such as ethanol. Extremes of pH can induce unfolding because the native state has groups buried in the unionized form that can only ionize after unfolding. Most notable in this connection are the residues His and Tyr, whose presence can induce protein unfolding at low and high pH values, respectively.

III. PROTEIN-SURFACTANT INTERACTIONS

A. THERMODYNAMICS OF BINDING

The biological function of proteins depends crucially on their direct physical interaction with other molecules — nucleic acids, polysaccharides, lipids, and other proteins. It is not surprising, therefore, that proteins also interact with water-soluble surfactants, and indeed surfactants are widely used by biochemists in the solubilization, purification, and characterization of proteins (e.g., in sodium dodecyl sulfate [SDS] polyacrylamide gel electrophoresis, and in extraction of proteins from biological membranes[23]).

The binding of a ligand (L) to a single site on a protein (P) is defined by the equilibrium

$$P + L \rightleftharpoons PL \tag{1}$$

with an association constant

$$K = [PL]/[P][L] \qquad (2)$$

It is generally assumed that all species are present at concentrations low enough for thermodynamic ideality to apply. More realistic from the point of view of protein-surfactant interactions is the set of multiple equilibria:

$$PL_{i-1} + L \rightleftharpoons PL_i \qquad (1 \leq i \leq n) \qquad (3)$$

If the equilibrium constant K for each binding step is the same, then it follows that

$$K^n = [PL_n]/[P][L]^n \qquad (4)$$

The average number $\bar{n}$ of ligands bound per protein molecule is given by

$$\bar{n} = n(K[L])^n/\{1 + (K[L])^n\} \qquad (5)$$

Equation 5 refers to the situation in which the n binding sites are identical and independent. More realistically, however, ligand binding to one site may lead to an increase or decrease in the affinities of other sites — that is, there may be positive or negative cooperativity. This is allowed by empirically[24] by introducing a cooperativity coefficient ξ. The general expression for $\bar{n}$ then becomes

$$\bar{n} = n(K[L])^\xi/\{1 + (K[L])^\xi\} \qquad (6)$$

In positively cooperative binding ($\xi > 1$), the binding of ligand L_i enhances the binding of ligand L_{i+1}. Conversely, in negatively cooperative binding, the binding of one ligand weakens the binding of subsequent ligands. In a binding isotherm in which $\bar{n}$ is plotted vs. log [L], the steeper the plot the more cooperative is the binding.

With various globular proteins, experimental binding isotherms for ionic and nonionic surfactants are substantially different. This is illustrated by the data of Jones and co-workers[25-27] in Figure 2 for (a) SDS and (b) *n*-octyl β-glycoside (OBG) binding to lysozyme at pH 3.2 and pH 6.4, respectively. For SDS + lysozyme, the very steep initial part of the curve ($\xi \gg 1$) is typical of that for an ionic surfactant + a globular protein. It corresponds to specific cooperative binding of negatively charged SDS molecules onto cationic sites at the protein surface. The shift in the initial part of the binding curve to higher free SDS concentrations with increasing ionic strength is consistent with a weakening of the specific ionic binding as electrostatic interactions become screened by added electrolyte. At higher SDS concentrations, approaching the critical micelle concentration (CMC), further nonspecific binding occurs which is stronger at the higher ionic strength. In contrast to SDS + lysozyme, the system OBG + lysozyme exhibits only a nonspecific binding region as the CMC is approached.

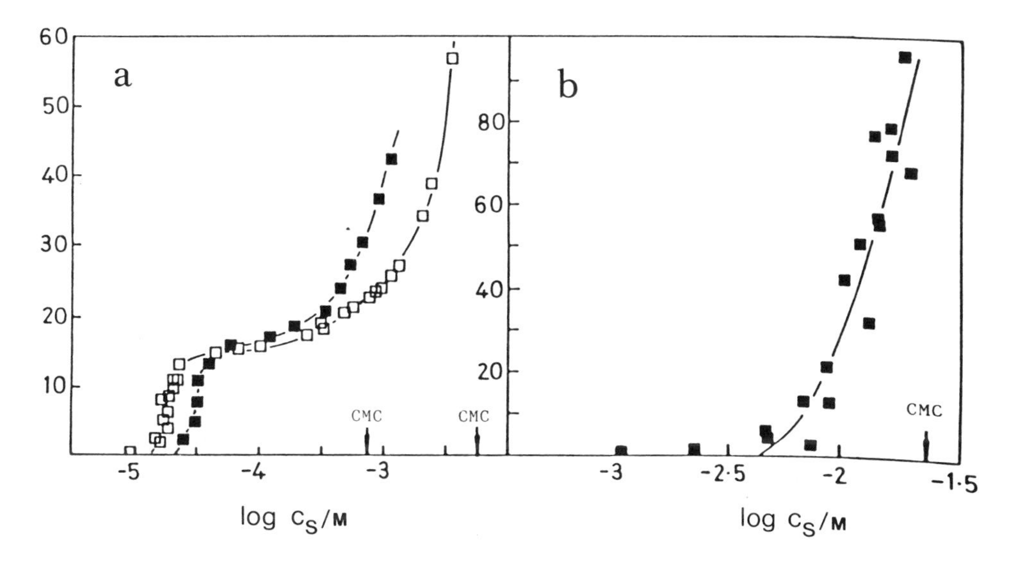

FIGURE 2. Binding isotherms of surfactants to lysozyme at 25°C. The average number $\bar{n}$ of ligands bound per protein molecule is plotted against the logarithm of the surfactant concentration. (a) Sodium dodecyl sulfate: (□) pH 3.2, ionic strength 0.012 *M*; (■) pH 3.2, ionic strength 0.212 *M*. (b) *n*-Octyl β-glycoside: pH 6.4, ionic strength 0.132 *M*. (From Jones, M. N. and Brass, A., in *Food Polymers, Gels, and Colloids*, Dickinson, E., Ed., Royal Society of Chemistry, Cambridge, 1991, 65. With permission.)

The experimental data in Figure 2 can be fitted to Equation 6 to obtain an intrinsic binding constant K. This, in turn, can be converted into an average standard Gibbs free energy per ligand bound:

$$\Delta G^{\circ}(\bar{n}) = -(RT/\bar{n}) \ln K \qquad (7)$$

For SDS + lysozyme in the specific binding region up to $\bar{n} = 18$, the analysis gives $\Delta G^{\circ}(\bar{n}) = -26$ kJ mol^{-1}. (A more refined analysis,[27] including an allowance for statistical factors, leads to a slightly less negative value of $\Delta G^{\circ}(\bar{n})$.) When the same analysis treatment is applied to the nonspecific binding of OBG up to $\bar{n} = 130$, the resulting free energy value is $\Delta G^{\circ}(\bar{n}) = -10$ kJ mol^{-1}. This means that the magnitude of $\Delta G^{\circ}(\bar{n})$ for the nonspecific (hydrophobic) interactions is less than one half of that for the specific (ionic) interactions.

B. MOLECULAR INTERACTIONS INVOLVED IN BINDING

1. Ionic Surfactants

The binding of charged surfactant molecules to protein can be envisaged as taking place in two distinct stages. Initially, the surfactant ligands bind to specific sites on the protein surface. With SDS, for instance, one expects an electrostatic interaction of the sulfate anion with the positively charged side chains of Lys, His, and Arg, together with a hydrophobic interaction between the dodecyl chain of the surfactant and nonpolar regions on the protein surface adjacent to the cationic sites. The involvement of hydrophobic interaction as well as electrostatic interaction in the specific binding is confirmed by computer simulation[27] and by the experimental observation[28] that binding affinity is dependent on the alkyl chain length. In contrast to the specific binding of SDS and sodium decyl sulfate (SDeS), the isotherm for sodium octyl sulfate (SOS) + lysozyme is characteristic of nonspecific cooperative binding at pH 3.2, indicating that, for specific binding to occur, the surfactant alkyl chain must be long enough to make hydrophobic contact with the protein surface. With sodium perfluorooctanoate (SPFO) + lysozyme, the even weaker interaction is attributed[29] both to the shortness of the surfactant tail and to the mutual antipathy of fluorocarbons and hydrocarbons.[30]

In the second stage of binding by ionic surfactants, the protein unfolds to expose its hydrophobic interior and hence further potential binding sites. In this nonspecific cooperative binding region, the main driving force is the hydrophobic interaction between the surfactant tails and the nonpolar residues of the unfolded protein. When the average number of ligands reaches 50 or more, and the sufactant concentration approaches the CMC, the structure of the protein-surfactant complex will closely resemble that of a surfactant micelle containing solubilized protein. Certainly, it appears from ^{13}C-NMR experiments that the alkyl chains are in a normal micelle-like environment at high binding levels ($\bar{n} > 80$). With lithium dodecyl sulfate (LDS) + bovine serum albumin (BSA), it is inferred from neutron scattering experiments[31] that, as LDS is continuously added, the globular proteins are gradually converted into random coil structures with strings of constant-size LDS micelles randomly decorating the polypeptide backbone (a "pearl necklace" structure). Many other studies of changes in protein conformation, including changes in α-helix and β-sheet structure, have been reported in the literature; a summary of these findings can be found elsewhere.[7]

2. Nonionic Surfactants

Nonionic surfactants exhibit nonspecific hydrophobic protein-surfactant interactions near the CMC, but specific binding does not occur except in certain special cases. The milk protein β-lactoglobulin forms a strong 1:1 complex with Tween 20 (polyoxyethylene [20] sorbitan monolaurate)[32] and probably also with other nonionic surfactants. BSA has a hydrophobic pocket on the protein surface which can accommodate 2 or 3 molecules of Triton X surfactants [$(CH_3)CCH_2(CH_3)_2CC_6H_4O(CH_2CH_2O)_nCH_2OH$].[33] The small endothermic enthalpy change indicates that the BSA-Triton X interaction is predominantly hydrophobic.[33]

As illustrated in Figure 2b, the system OBG + lysozyme shows only nonspecific binding as the surfactant concentration approaches the CMC. It is suggested[26] that OBG binds to hydrophobic regions on the native protein surface to form a prolate ellipsoidal complex. From microcalorimetry measurements, a value of $\Delta H°(\bar{n}) = 0.8$ kJ mol^{-1} is obtained for the standard enthalpy of binding per surfactant bound. We see that $\Delta H°(\bar{n})$ makes a rather small contribution to $\Delta G°(\bar{n})$ in comparison with the entropy term $T\Delta S°(\bar{n})$.

IV. ADSORPTION OF PROTEINS

A. THERMODYNAMIC AND STATISTICAL ASPECTS

Proteins adsorb spontaneously at a wide range of solid and liquid interfaces. A typical surface prior to protein adsorption is mainly hydrophobic with a number of charged groups relatively widely spaced. The predominant driving force for adsorption is the removal of nonpolar side chains from the unfavorable aqueous environment of the bulk solution. Molecular factors contributing to the Gibbs free energy of adsorption, roughly in order of decreasing importance, are as follows: (1) dehydration of hydrophobic regions of the surface; (2) changes in secondary and tertiary structure; (3) charge redistribution due to overlap of electric fields of protein and surface; (4) change in hydration of ions on transfer from bulk solution to adsorbed layer; (5) change in pK values of side chains; (6) van der Waals interaction between protein groups and surface. Adsorption is thermodynamically favored by factors (1) and (2), and opposed by factors (3) and (4). Factors (5) and (6) are of little quantitative significance. So, the major entropic driving force is displacement of ordered water molecules from the vicinity of the (hydrophobic) surface. An important secondary driving force is the unfolding of the (globular) protein on adsorption.

Due to their strong affinity for hydrophobic surfaces (e.g., oil-water interfaces), their unfolding to produce a ''denatured'' layer, and the fact that they cannot normally be desorbed by diluting with water, it is commonly stated that proteins adsorb irreversibly. The term ''irreversible'' is not really so appropriate, however, because adsorbed proteins can certainly be partially removed from the surface by changes in solvent conditions (pH, ionic strength, etc.) and they can be completely displaced by small-molecule surfactants. This type of behavior is also encountered with ordinary synthetic polymers, and so in many ways there is nothing special about the thermodynamics of protein adsorption. While the mean binding energy per residue may be small ($<$ kT), the energy per adsorbed molecule with many residues in contact with the surface may be so large that the Boltzmann factor is weighted overwhelmingly towards the adsorbed state. This is why desorption does not readily occur by dilution with water, though it may be induced by changing the solvent conditions to reduce the mean segment binding energy to a value close to zero.

An adsorbed protein film is a dense, thin macromolecular layer. Segments of a flexible adsorbed polymer lying in direct contact with the surface are called *trains:* the other segments in *loops* and *tails* protrude into the bulk phase forming an interfacial region that is much thicker than the width of the chain. Statistical distributions of segments of adsorbed homopolymers in terms

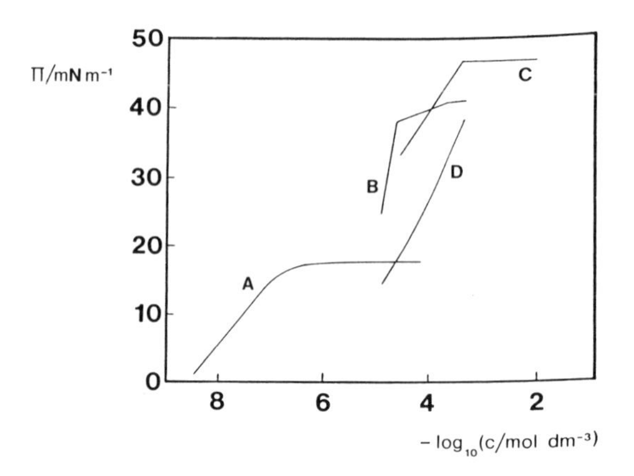

FIGURE 3. Comparison of adsorption behavior of lysozyme and three surfactants at the oil-water interface. Surface pressure Π is plotted against the logarithm of the bulk concentration c: (A) lysozyme; (B) sorbitan monooleate (Span 80); (C) octylphenoxyethoxyethanol; and (D) sodium dodecyl sulfate. (From Fisher, L. R. and Parker, N. S., in *Advances in Food Emulsions and Foams,* Dickinson, E. and Stainsby, G., Eds., Elsevier Applied Science, London, 1988, 53. With permission.)

of the proportions of trains, loops, and tails are known from statistical theories and numerical simulation.[35,36] At saturation coverage, only about one third of the available surface area is occupied by train segments. This low coverage of surface sites by flexible polymers arises from the configurational constraints of chain connectivity in the presence of other polymer chains. The many small gaps in the adsorbed layer structure are readily accessible to other small molecules (e.g., surfactants). While the train-loop-tail model is useful to a first approximation for describing the adsorbed state of a highly disordered protein such as β-casein, the model is inappropriate for globular proteins such as lysozyme or β-lactoglobulin, for which much of the secondary structure and some of the tertiary structure is retained on adsorption. It is more realistic to regard an adsorbed monolayer of globular protein molecules as a two-dimensional dense system of interacting deformable particles.[37,38]

At an oil-water interface, a protein like β-casein is a much more efficient adsorber than an ionic or nonionic surfactant, because the protein saturates the surface at a much lower bulk concentration. An effective monolayer saturation coverage is reached with the protein at a molar concentration some 10^3 or 10^4 times lower than for the surfactant. This greater affinity of proteins for the adsorbed state leads to a greater surface pressure than surfactants at low concentrations, but the better packing of the small amphiphiles in the vicinity of the Gibbs dividing plane leads to a larger surface pressure for surfactants than for proteins at high bulk concentrations. This is illustrated in Figure 3 by surface pressure results for lysozyme and three surfactants. Qualitatively similar behavior is found with other proteins and other surfactants. In particular, surfactants give a greater lowering of the surface free energy at high concentrations, which is why surfactants displace proteins from oil-water interfaces above a certain critical concentration.

When adsorbing at the surface of solid polystyrene latex particles, the disordered β-casein gives a high-affinity isotherm[39] like that found with linear synthetic polymers. With globular proteins, however, the behavior is more complex.[5,34] The plot of surface concentration against bulk concentration has a finite initial slope, like that for a small-molecule surfactant, even though the protein is not readily desorbed by solvent dilution. The explanation of this behavior is that the globular protein undergoes conformational change after initial adsorption involving only a few binding sites on the molecular surface. This unfolding increases the overall molecular binding energy, so making subsequent desorption statistically unlikely. It is observed that the desorption rate of globular proteins from solid surfaces depends on the residence time. On a short timescale, prior to conformational change, adsorption is reversible (low-affinity isotherm). On a long timescale, it is irreversible (high-affinity isotherm). The more prone a protein is to unfold, the less reversible is its adsorption.

The equilibrium structure of a protein film adsorbed from a concentrated solution may not be reached over the normal experimental timescale. The film may contain molecules in many different states of unfolding, as hydrophobic interaction between nonpolar residues, which stabilizes organization into α-helices and β-sheets in the native structure, is gradually replaced by hydrophobic interaction with nonpolar groups at the surface. Exchange between the different states of unfolding may be very slow due to the high density of close-range interactions and the slow diffusion in the adsorbed layer. At a fresh oil-water interface, for instance, the adsorbing proteins may not have had time to unfold properly before they are rapidly surrounded by others, producing a congested close-packed arrangement in which the scope for even modest configurational readjustment is severely hampered. Under these conditions, we may envisage the physical state of the adsorbed film to be more like that of a viscoelastic solid (a two-dimensional glass or gel) than that of a normal protein solution.

B. COMPETITIVE ADSORPTION OF PROTEINS AND SURFACTANTS

1. Theory

The problem of competitive adsorption has been tackled theoretically by two different types of statistical mechanical approach:[35] continuum models in which adsorbate molecules are represented as spheres of various sizes, and lattice-based models in which each polymer segment (or small molecule) occupies a single lattice site. Neither approach is ideally suited to the complexity of protein structure and interactions, but the latter type is probably a useful representation of competitive adsorption of a disordered protein (e.g., β-casein) with a simple noninteracting (nonionic) surfactant.

The lattice-based models are descendants of the statistical theories developed for describing the thermodynamics of polymer solutions.[21] Relative strengths of the various types of interactions (solvent-solvent, solvent-polymer, polymer-polymer, solvent-surface, polymer-surface) determine the equilibrium between bulk and surface phases. A useful concept[40] is the Flory-Huggins exchange parameter χ_s corresponding to the free energy change associated with formation of polymer-surface contacts at the expense of solvent-surface contacts. That is, if a surface site is occupied by either a solvent molecule or a polymer segment, $-\chi_s kT$ is the free energy difference between segment-surface and solvent-surface contacts. Theory predicts[41] that long flexible chains will not adsorb unless χ_s exceeds some critical value χ_s^c. The requirement for a finite χ_s arises because the decrease in free energy due to formation of polymer-surface contacts must outweigh the loss of conformational entropy when the polymer adsorbs. In a system containing adsorbed polymer, any change in conditions which cause χ_s to approach or become less than χ_s^c will lead to the polymer being partially or completely displaced from the interface. Such a change may be induced by adding surfactant to the system.

Simple expressions for the adsorption isotherm and the critical conditions for polymer displacement are obtained[42] if it is assumed that the surface can be approximated by a single plane of lattice sites. The justification for this one-layer approximation is that, close to the critical point ($\chi_s \approx \chi_s^c$), because the polymer surface excess is small, the segment density can be equated to its bulk value in all except the first (train) layer. Under these conditions, the adsorption isotherm for a mixed solution of polymer + surfactant is given by:

$$\theta = \theta_p + \theta_s = [(1 - \theta)^{fm}\phi_p K_p/(1 - \phi)^{fm}] + [1 - \theta)\phi_s K_s/(1 - \phi)] \quad (8)$$

In Equation 8, θ_p and θ_s are the fractions of surface sites occupied by polymer segments and surfactant molecules (one per site) respectively, ϕ_p and ϕ_s are the volume (site) fractions of polymer and surfactant, θ is the total solute surface fraction, $\phi = \phi_p + \phi_s$ is the total solute bulk volume fraction, m is the number of segments per polymer, f is the fraction of segments in trains, and K_p and K_s are Boltzmann factors related to the adsorption energies of polymer and surfactant:

$$K_p = \exp(-\Delta E_p/kT) \quad (9)$$

$$K_s = \exp(-\Delta E_s/kT) \quad (10)$$

The quantities ΔE_p and ΔE_s are the effective energy changes accompanying transfer of polymer and surfactant from bulk solution to the surface. In general, the values of ΔE_p and ΔE_s are dependent on the compositions of both the bulk solution and the surface layer. For the special case of athermal conditions, however, the quantities ΔE_p and ΔE_s can be treated as simple constants.

A theoretical displacement curve calculated from Equations 8 to 10 is given in Figure 4. Athermal solvent conditions are assumed for both polymer

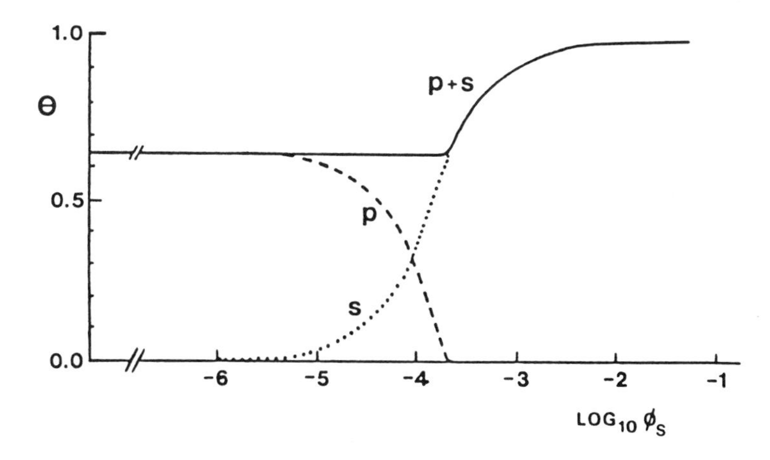

FIGURE 4. Competitive adsorption from simple statistical theory. The total solute surface fraction θ is plotted against the logarithm of the bulk surfactant volume fraction ϕ_s. Also shown are the separate contributions from polymer (p) and surfactant (s). (From Woskett, C. M., Ph.D. thesis, University of Leeds, Leeds, U.K., 1989.)

and surfactant with $\Delta E_p = -60$ kT (50 segments adsorbed, each with a binding energy of -1.2 kT) and $\Delta E_s = -12$ kT (10 methylene groups, each with a binding energy of about 1 kT). The deficiency of this simple theory is the overestimation of the polymer surface coverage θ_p compared with more complete theories. For the particular set of parameters used to calculate the adsorption isotherm in Figure 4, the critical surfactant volume fraction at which the polymer surface excess concentration just vanishes is $\phi_s^c \approx 2 \times 10^{-4}$.

Interactions per polymer and surfactant molecules can be allowed for in the above theory by introducing a Flory-Huggins polymer-surfactant interaction parameter χ_{ps}, although such a treatment is not appropriate for the strong specific and nonspecific interactions which occur in most protein + surfactant systems. Probably a more fruitful approach to modeling competitive adsorption in interacting systems of protein + surfactant is by computer simulation.[2,36,43] In a lattice-based Monte Carlo simulation of competitive adsorption of linear homopolymers (50 segments) + small displacer molecules (2 segments), the fractions of adsorbed segments in trains, loops, and tails have been computed as a function of displacer concentration, displacer adsorption energy, and the strength of the polymer-displacer interaction energy. The numerical results show[36] the crucial importance of displacer surface coverage in determining the extent of polymer displacement, which is also predicted by the simple lattice-based analytical theory.[41] More appropriate, perhaps, to systems containing globular proteins is a continuum model consisting of a mixture of large spheres (proteins) + small spheres (surfactants). In such a Monte Carlo model incorporating orientation-dependent surfactant-protein interactions, we find[43] that strong protein-surfactant binding leads to a tendency for individual protein/surfactant complexes to be repelled from

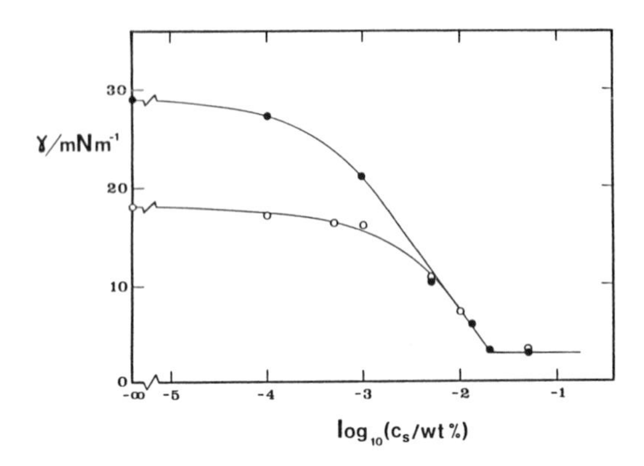

FIGURE 5. Interfacial tension γ of sodium caseinate + octaoxyethylene glycol dodecyl ether at the *n*-tetradecane-water interface (pH 7, 25°C) as a function of the logarithm of the surfactant concentration c_s: (●) no protein; (○) 0.1 wt% protein. (From Dickinson, E., Euston, S. R., and Woskett, C. M., *Prog. Colloid Polym. Sci.*, 82, 65, 1990. With permission.)

one another and also from the interface once it is covered with a surfactant monolayer. The nonbinding end of the surfactant points away from the protein/surfactant complex, and from the surfactant monolayer at the interface, providing an entropic barrier separating protein and interface. This is what we expect for a nonionic surfactant binding to protein via hydrophobic interaction — the binding makes the outside of the protein molecule more hydrophilic, and hence less surface-active at a hydrophobic surface than the native protein in the absence of surfactant.

2. Experiment

As an illustration of competitive adsorption, let us first consider a system which is most likely to resemble the simple theoretical treatment just described: a disordered protein (casein) + a nonionic surfactant. The chosen surfactant is octaoxyethylene glycol *n*-dodecyl ether ($C_{12}E_8$); it is water soluble with a CMC of 1.1×10^{-4} *M* and an estimated hydrophile lipophile balance (HLB) value of 13.1. The protein is in the form of sodium caseinate, which contains about 35% β-casein, together with roughly the same amount of α_{s1}-casein, and smaller amounts of other caseins. It has been demonstrated[44,45] that β-casein predominates at an oil-water interface in the presence of a mixture of α_{s1}-casein + β-casein. Surface pressures of β-casein and sodium caseinate are identical at the same bulk concentrations.[46]

Figure 5 shows the interfacial tension γ at the hydrocarbon-water interface as a function of surfactant concentration c_s for $C_{12}E_8$ alone and for $C_{12}E_8$ + 0.1 wt% sodium caseinate at pH 7 and 25°C.[2] At low surfactant concentrations, the tension is considerably lower with protein present, but the difference narrows as c_s increases, and it reaches zero at $c_s \approx 10^{-2}$ wt%. Beyond this (critical) surfactant concentration, one may infer that the protein is displaced from the primary interfacial layer since the $C_{12}E_8$ surface excess concentration

($\Gamma = 1.27$ mg m^{-2}), as calculated from dγ/d ln c_s (below the CMC) assuming the Gibbs adsorption equation, is the same in the presence or absence of protein. The value $c_s \approx 10^{-2}$ wt% corresponds to a surfactant-to-protein molar ratio R $\approx$ 5. It is of the same order as the theoretical critical surfactant concentration $\phi_s^c \approx 0.02$ vol% estimated from the simple analytic theory (see Figure 4). This suggests that the parameter values chosen (see legend of Figure 4) are reasonably sensible for modeling the competitive adsorption of a disordered protein + a nonionic (noninteracting) surfactant like $C_{12}E_8$.

In agreement with the simulation study,[2,36] one may infer from Figure 5 that protein segments predominate in the primary interfacial layer at a low surfactant concentration of $c_s = 10^{-4}$ wt%. However, while there is no protein displacement at this low surfactant concentration, the presence of adsorbed $C_{12}E_8$ in the gaps between the casein train segments has a substantial effect on the dynamics of the adsorbed protein layer. The film adsorbed from a 0.1 wt% solution of pure sodium caseinate develops a surface shear viscosity at the hydrocarbon-water interface of about 20 mN m^{-1} s after 24 hr, but it is found[2] that the presence of just 10^{-4} wt% surfactant prevents the formation of a film with any measurable surface viscosity (<0.2 mN m^{-1} s). The adsorbed surfactant appears to act as a sort of lubricant between the protein molecules enabling them to flow past one another more easily in the adsorbed layer, rather as a plasticizer at very low concentrations is able to modify the rheological properties of a bulk polymeric glass.

Qualitatively similar behavior to that described above is observed also in mixtures of nonionic surfactant + globular protein, e.g., Tween 20 + β-lactoglobulin. The globular protein adsorbs at the oil-water interface to give a highly viscoelastic film, but the film becomes highly mobile in the presence of Tween 20 at a concentration equivalent to a surfactant-to-protein molar ratio R $\approx$ 1.0.[47] The larger amount of surfactant required in this system to lower the protein film surface viscosity may be due to the fact that Tween 20 forms a specific 1:1 complex with β-lactoglobulin.

Addition of nonionic surfactant ($C_{12}E_8$, Tween 20, etc.) to an oil-in-water emulsion stabilized by protein (β-casein, β-lactoglobulin, etc.) leads to partial or complete displacement of protein from the oil-water interface depending on the bulk surfactant concentration. Figure 6 shows a plot of protein surface concentration against added surfactant concentration for a β-casein-stabilized hydrocarbon oil-in-water emulsion (20 wt% oil, 0.4 wt% protein, surface area 1.5 m^2 g^{-1}) with $C_{12}E_8$ added after emulsion formation.[48] We see that the protein is completely displaced from the oil-water interface at a surfactant-to-protein molar ratio of R = 18. Total displacement is achieved at an interfacial surfactant concentration of 1.1 mg m^{-2} (calculated assuming that all the surfactant present is adsorbed). This figure is close to the value $\Gamma = 1.27$ mg m^{-2} inferred from the tension data in Figure 5.

The strong specific binding of ionic surfactants to proteins makes their competitive adsorption behavior more complicated than that of the nonionics. Protein is again completely displaced from the interface at high bulk surfactant

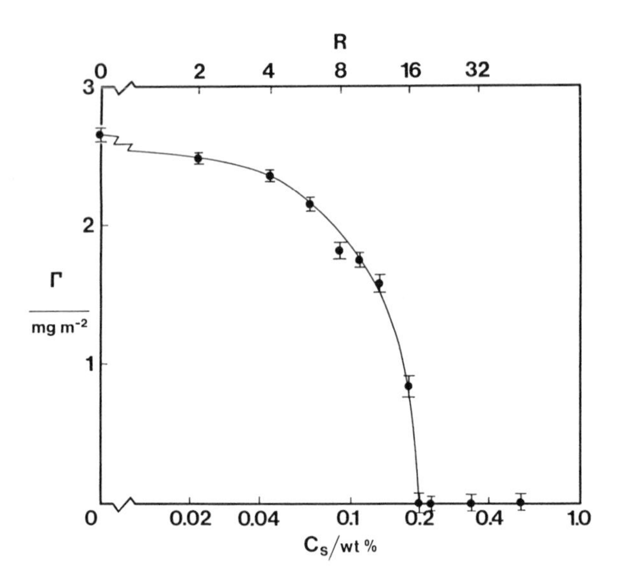

FIGURE 6. Displacement of β-casein from the oil-water interface by octaoxyethylene glycol dodecyl ether added after formation of *n*-hexadecane-in-water emulsion (0.4 wt% protein). Protein surface concentration Γ is plotted against surfactant concentration c_s and surfactant-to-protein molar ratio R.

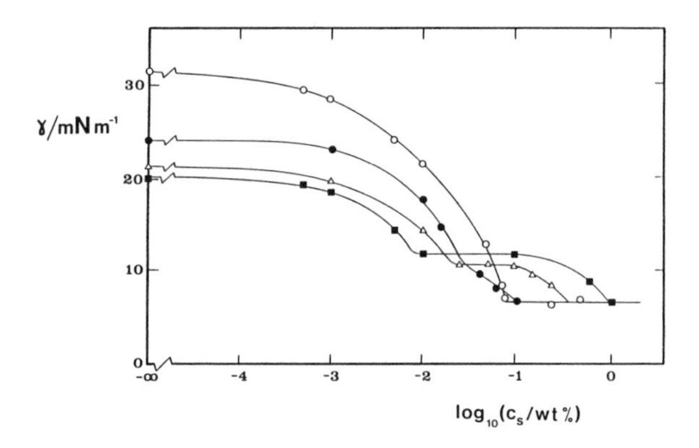

FIGURE 7. Interfacial tension γ of gelatin + sodium dodecyl sulfate at the *n*-tetradecane-water interface (pH 7, 25°C) as a function of the logarithm of the surfactant concentration c_s: (○) no protein; (●) 0.01 wt% protein; (△) 0.1 wt% protein; (■) 0.4 wt% protein. (From Dickinson, E., Euston, S. R., and Woskett, C. M., *Prog. Colloid Polym. Sci.*, 82, 65, 1990. With permission.)

concentrations, but at lower concentrations the interfacial composition and structure are strongly influenced by the protein-surfactant interactions.[49,50] Figure 7 shows interfacial tension data for SDS + gelatin (7.7×10^5 Da, pI 4.8) at three protein concentrations. At the highest SDS concentration (above the CMC), the tension for SDS + gelatin is essentially the same as that for SDS alone, but at lower concentrations the form of the plot of γ vs. $\log_{10} c_s$ is dependent on the bulk gelatin concentration.[2] At protein concentrations of 0.1 and 0.4 wt%, there is a distinct plateau in the plot extending over a considerable range of SDS concentration in the vicinity of the pure surfactant CMC. Presumably, in the plateau region, it is thermodynamically more favorable for surfactant molecules to bind cooperatively onto the protein than it is for them to displace hydrophobic protein train segments competitively from the interface. There is no discernible plateau region at 0.01 wt% protein, but there is still a crossover to tensions slightly above those for the pure surfactant at c_s values just above the CMC. The main effect of the protein-surfactant binding on the overall competitive adsorption behavior is to reduce the amount of SDS available to saturate the oil-water interface. So, in comparison to the nonionic surfactant, more of the ionic surfactant is required to displace protein from the interface, since some is used to bind to the protein and is therefore not available to act as a displacer.

REFERENCES

1. **Israelachvili, J. N.,** *Intermolecular and Surface Forces,* Academic Press, London, 1985.
2. **Dickinson, E., Euston, S. R., and Woskett, C. M.,** Competitive adsorption of food macromolecules and surfactants at the oil-water interface, *Prog. Colloid Polym. Sci.,* 82, 65, 1990.
3. **Cockbain, E. G.,** The interfacial activity and composition of bovine serum albumin + sodium dodecyl sulphate complexes, *Trans. Faraday Soc.,* 49, 104, 1953.
4. **Knox, W. J. and Parshall, T. O.,** The interaction of sodium dodecyl sulphate with gelatin, *J. Colloid Interface Sci.,* 33, 16, 1970.
5. **Dickinson, E. and Stainsby, G.,** *Colloids in Food,* Applied Science, London, 1982.
6. **Darling, D. F. and Birkett, R. J.,** Food colloids in practice, in *Food Emulsions and Foams,* Dickinson, E., Ed., Royal Society of Chemistry, London, 1987, 1.
7. **Dickinson, E. and Woskett, C. M.,** Competitive adsorption between proteins and small-molecule surfactants in food emulsions, in *Food Colloids,* Bee, R. D., Richmond, P., and Mingins, J., Eds., Royal Society of Chemistry, Cambridge, 1989, 74.
8. **Dickinson, E., Mauffret, A., Rolfe, S. E., and Woskett, C. M.,** Adsorption at interfaces in dairy systems, *J. Soc. Dairy Technol.,* 42, 18, 1989.
9. **Dickinson, E.,** Mixed proteinaceous emulsifiers: review of competitive protein adsorption and the relationship to food colloid stabilization, *Food Hydrocolloids,* 1, 3, 1986.
10. **Dickerson, R. E. and Geis, I.,** *The Structure and Action of Proteins,* Benjamin-Cummings, Melno Park, CA, 1969.
11. **Schulz, G. E. and Schirmer, R. H.,** *Principles of Protein Structure,* Springer-Verlag, New York, 1979.
12. **Cantor, C. R. and Schimmel, P. R.,** *Biophysical Chemistry. Part I. The Conformation of Biological Macromolecules,* W. H. Freeman, San Francisco, 1980.
13. **Creighton, T. E.,** *Proteins: Structures and Molecular Properties,* W. H. Freeman, New York, 1983.
14. **McCammon, J. A. and Harvey, S. C.,** *Dynamics of Proteins and Nucleic Acids,* Cambridge University Press, London, 1987.
15. **Nakai, S. and Li-Chan, E.,** *Hydrophobic Interactions in Food Systems,* CRC Press, Boca Raton, FL, 1988, 24.

16. **Nozaki, Y. and Tanford, C.,** The solubility of amino acids and two glycine peptides in aqueous ethanol and dioxane solutions, *J. Biol. Chem.*, 246, 2211, 1971.
17. **Tanford, C.,** *The Hydrophobic Effect: Formation of Micelles and Biological Membranes,* 2nd ed., John Wiley & Sons, New York, 1980.
18. **Richards, F. M.,** Areas, volumes, packing and protein structure, *Annu. Rev. Biophys. Bioeng.*, 6, 151, 1977.
19. **Manavalan, P. and Ponnuswarmy, P. K.,** A study of the preferred environment of amino acid residues in proteins, *Arch. Biochem. Biophys.*, 184, 476, 1977.
20. **Williams, R. J. P.,** The conformational properties of proteins in solution, *Biol. Rev.*, 54, 389, 1979.
21. **Flory, P. J.,** *Principles of Polymer Chemistry,* Cornell University Press, Ithaca, NY, 1953.
22. **de Gennes, P.-G.,** *Scaling Concepts in Polymer Physics,* Cornell University Press, Ithaca, NY, 1979.
23. **Houslay, M. D. and Stanley, K. K.,** *Dynamics of Biological Membranes,* John Wiley & Sons, Chichester, 1982, 30.
24. **Hill, A. V.,** The possible effects of the aggregation of the molecules of haemoglobin on its dissociation curves, *J. Physiol.*, 40, iv, 1910.
25. **Jones, M. N. and Manley, P.,** Binding of n-alkyl sulphates to lysozyme in aqueous solution, *J. Chem. Soc. Faraday Trans. 1,* 75, 1736, 1979.
26. **Cordoba, J., Reboiras, M. D., and Jones, M. N.,** Interaction of n-octyl-β-D-glucopyranoside with globular proteins in aqueous solution, *Int. J. Biol. Macromol.*, 10, 270, 1988.
27. **Jones, M. N. and Brass, A.,** Interactions between small amphiphilic molecules and proteins, in *Food Polymers, Gels and Colloids,* Dickinson, E., Ed., Royal Society of Chemistry, Cambridge, 1991, 65.
28. **Jones, M. N. and Manley, P.,** Interaction between lysozyme and n-alkyl sulphates in aqueous solution, *J. Chem. Soc. Faraday Trans. 1,* 76, 654, 1980.
29. **Fukushima, K., Sugihara, G., Murata, Y., and Tanaka, M.,** The effect of added NaCl on the binding of sodium perfluorooctanoate to lysozyme in aqueous solution, *Bull. Chem. Soc. Jpn.*, 55, 3113, 1982.
30. **Rowlinson, J. S.,** *Liquids and Liquid Mixtures,* 2nd ed., Butterworths, London, 1969, 155.
31. **Chen, S.-H. and Teixeira, J.,** Structure and fractal dimension of protein-detergent complexes, *Phys. Rev. Lett.*, 57, 2583, 1986.
32. **Coke, M., Wilde, P. J., Russell, E. J., and Clark, D. C.,** The influence of surface composition and molecular diffusion on the stability of foams formed from protein/surfactant mixtures, *J. Colloid Interface Sci.*, 138, 489, 1990.
33. **Sukow, W. W., Sandberg, H. E., Lewis, E. A., Eatough, D. J., and Hansen, L. D.,** Binding of the Triton X series of non-ionic surfactants to bovine serum albumin, *Biochemistry,* 19, 912, 1980.
34. **Norde, W.,** Proteins at Interfaces, Ph.D. thesis, Agricultural University, Wageningen, The Netherlands, 1976.
35. **Dickinson, E. and Lal, M.,** Statistical mechanics of physical adsorption in condensed phases, *Adv. Mol. Rel. Intern. Processes,* 17, 1, 1980.
36. **Dickinson, E. and Euston, S. R.,** Computer simulation of the competitive adsorption between polymers and small displacer molecules, *Mol. Phys.*, 68, 407, 1989.
37. **de Feijter, J. A. and Benjamins, J.,** Soft-particle model of compact macromolecules at interfaces, *J. Colloid Interface Sci.*, 90, 289, 1982.
38. **Dickinson, E. and Euston, S. R.,** Simulation of adsorption of deformable particles modelled as cyclic lattice chains: a simple statistical model of protein adsorption, *J. Chem. Soc. Faraday Trans.*, 86, 805, 1990.
39. **Dickinson, E., Robson, E. W., and Stainsby, G.,** Colloid stability of casein-coated polystyrene particles, *J. Chem. Soc. Faraday Trans. 1,* 79, 2937, 1983.

40. **Silberberg, A.,** Adsorption of flexible macromolecules. IV. Effect of solvent-solute interactions, solute concentration, and molecular weight, *J. Chem. Phys.,* 48, 2835, 1968.
41. **Scheutjens, J. M. H. M. and Fleer, G. J.,** Statistical theory of the adsorption of interacting chain molecules. I. Partition function, segment density distribution, and adsorption isotherms, *J. Phys. Chem.,* 83, 1619, 1979.
42. **Cohen Stuart, M. A., Fleer, G. J., and Scheutjens, J. M. H. M.,** Displacement of polymers. I. Theory. Segmental adsorption energy from polymer desorption in binary solvents, *J. Colloid Interface Sci.,* 97, 515, 1984.
43. **Dickinson, E.,** Monte Carlo model of competitive adsorption between interacting macromolecules and surfactants, *Mol. Phys.,* 65, 895, 1988.
44. **Dickinson, E., Rolfe, S. E., and Dalgleish, D. G.,** Competitive adsorption of α_{s1}-casein and β-casein in oil-in-water emulsions, *Food Hydrocolloids,* 2, 397, 1988.
45. **Dickinson, E.,** Competitive adsorption and protein-surfactant interactions in oil-in-water emulsions, *AChS Symp. Ser.,* 448, 114, 1991.
46. **Castle, J., Dickinson, E., Murray, B. S., and Stainsby, G.,** Mixed protein films adsorbed at the oil-water interface, *AChS Symp. Ser.,* 343, 118, 1987.
47. **Courthaudon, J.-L., Dickinson, E., Matsumura, Y., and Clark, D. C.,** Competitive adsorption of β-lactoglobulin + Tween 20 at the oil-water interface, *Colloids Surf.,* 56, 293, 1991.
48. **Courthaudon, J.-L., Dickinson, E., and Dalgleish, D. G.,** Competitive adsorption of β-casein and non-ionic surfactants in oil-in-water emulsions, *J. Colloid Interface Sci.,* 145, 390, 1991.
49. **Steinhardt, J. and Reynolds, J. A.,** *Multiple Equilibria in Proteins,* Academic Press, New York, 1969.
50. **Robb, I. D.,** Polymer-surfactant interactions, in *Anionic Surfactants: Physical Chemistry of Surfactant Action,* Lucassen-Reynders, E. H., Ed., Marcel Dekker, New York, 1981, 109.

Chapter 8

PROTEIN-SURFACTANT INTERACTIONS

K. P. Ananthapadmanabhan

TABLE OF CONTENTS

0-8493-6784-0/93/$0.00 + $.50

I. INTRODUCTION

Interaction of proteins with surfactants has been a subject of extensive study for over 50 years. As pointed out by Goddard,[1,2] the studies on protein-surfactant interactions have actually laid the groundwork for the current activities in the polymer-surfactant area. In this regard, proteins are essentially amphoteric polyelectrolytes with some hydrophobic groupings. However, unlike polyelectrolytes, proteins exhibit secondary and tertiary structures and this makes their interactions with surfactants much more complex. These interactions, often referred to as "surfactant binding", can lead to the unfolding of proteins and sometimes their denaturation.

Interactions of surfactants with proteins are of importance in a wide variety of industrial, biological, pharmaceutical, and cosmetic systems. Complex formation between proteins and sodium dodecyl sulfate (SDS) and certain properties of the complex are routinely made use of in the determination of the molecular weight of proteins by SDS polyacrylamide gel electrophoresis. Surfactant-induced skin irritation has been correlated with the interactions of the surfactant with stratum corneum proteins. Human hair and wool are two other proteinacious substrates which are frequently exposed to surfactants. Protein-surfactant combinations are found routinely in edible products. In industrial, pharmaceutical, and photographic applications gelatin enjoys a ubiquitous presence, often in combination with surfactants. Proteins and chemically modified proteins, because of their biodegradable nature, are potential substitutes for synthetic macromolecules and therefore are finding an increasingly important role in industry.

Regarding research activity, the pioneering efforts by such workers as Putnam, Lundgren, Neurath, Steinhardt, Tanford, Klotz, Strauss, Pankhurst, Reynolds, Jones, Chattoraj and many others laid the foundation for our current understanding of protein-surfactant interactions. A comprehensive review of the early work leading up to the late 1960s can be found in Reference 3. Since then, several reviews[4-11] have appeared on the recent developments in this field.

In his chapter on protein solutions (see Chapter 7), Dickinson has already provided a review of the chemistry of proteins in aqueous solution, together with a brief review of some aspects of protein-surfactant interactions. In this chapter, the interactions between proteins and surfactants, as studied by various classical and some modern spectroscopic and scattering techniques, will be reviewed. A section on surfactant binding to insoluble proteins such as stratum corneum and hair will also be included. Emphasis will be on the work done during the last 25 years, with particular attention paid to experimental techniques and structural information on protein-surfactant complexes. Several applications of protein-surfactant complexes will also be highlighted.

II. EVIDENCE FOR PROTEIN-SURFACTANT INTERACTIONS

The evidence for the existence of protein-surfactant complexes has come from indirect methods such as surface tension, rheology, and dye solubilization, and from direct measurements of binding by dialysis and gel filtration and the use of ion-selective electrodes. Some of these techniques also reveal the various characteristic properties of protein-surfactant complexes which may have a bearing on their use for various applications. In the sections to follow, the evidence for the existence, as well as some of the properties, of protein-surfactant complexes will be examined.

A. SURFACE TENSION

Boundary tension is a simple and useful method to study interactions between a surfactant and relatively nonsurface active component such as a protein or polymer. In fact, one of the early studies which provided evidence of interaction between bovine serum albumin (BSA) and SDS involved interfacial tension measurements.[12] Knox and Parshall[13] have reported the surface tension behavior of SDS in gelatin solutions and their results show significant interactions between the two components (see Figure 1). Interestingly, the latter behavior is similar to that of the polyethylene oxide (PEO)-SDS system reported by Jones[14] and it shows the existence of "T_1" and "T_2" transitions corresponding to the onset of interactions between the protein and the surfactant and the point of saturation of the protein. (See Chapter 4 for definitions of T_1 and T_2). Note that these interactions are at pH values above the isoelectric point (IEP) of the gelatin, where electrostatic factors will actually have opposed interaction. It is possible that hydrophobic interactions between the surfactant and the protein, and among various "bound" surfactant molecules, are responsible for the protein-surfactant complex formation observed at pH 7.0. Schwuger and Bartnik[5] have reported similar results for the binding of SDS to BSA which also show the presence of a distinct T_1 and T_2 concentration in the surface tension vs. log concentration plot. Evidently, surface tension, though an indirect technique, can be used to obtain quantitative information on the binding of surfactants to proteins.

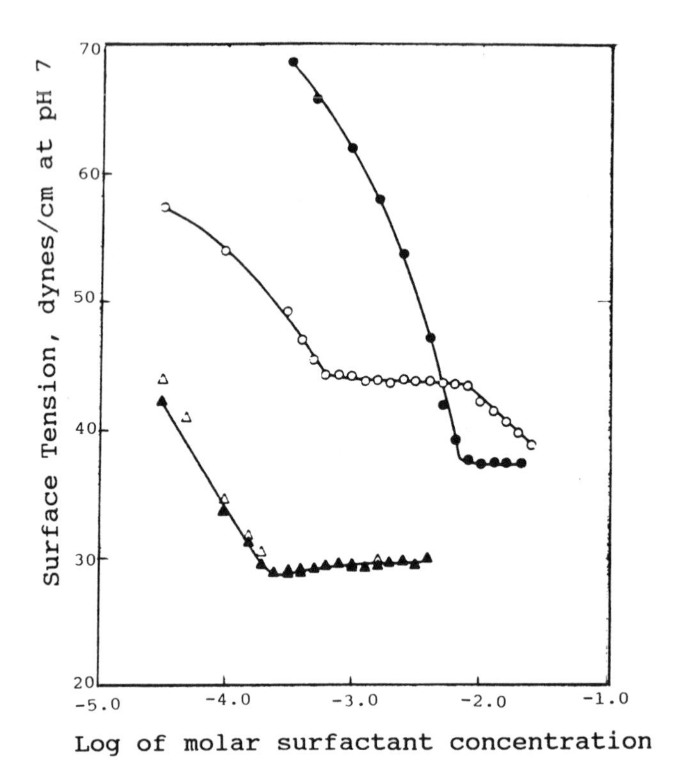

FIGURE 1. The effect of gelatin on the surface tension of SDS and Triton X-100: (●) SDS in distilled water; (○) SDS in 1% gelatin; (▲) Triton X-100; (△) Triton X-100 in 1% gelatin. (From Knox, W. L., Jr. and Parshall, T. O., *J. Colloid Interface Sci.*, 33, 16, 1970. With permission.)

It is interesting to see from the results in Figure 1 that a nonionic surfactant, Triton X-100 (polyoxyethylated octyl phenol), does not show any interaction with gelatin. This is possibly due to the low critical micelle concentration (CMC) of Triton X-100 which makes micelle formation in the bulk a more favorable process than interaction with gelatin. While low reactivity of nonionic surfactants towards polymers is well known, globular proteins such as BSA and lysozyme do show interaction with these surfactants. Nishikido et al.[15] have actually used surface tension data (see Figure 2) and the Gibbs equation (see Chapter 2) to determine the binding isotherm of hexaoxyethylene dodecyl ether (referred to as $C_{12}E_6$ in the figure) to BSA and lysozyme.

An analysis of the surface tension curves can, in principle, provide quantitative information on protein-surfactant interaction. Practically speaking, as pointed out in Chapter 4, complete interpretation of the data becomes more difficult when the protein itself is appreciably surface active or bears a net charge opposite to that of the surfactant.

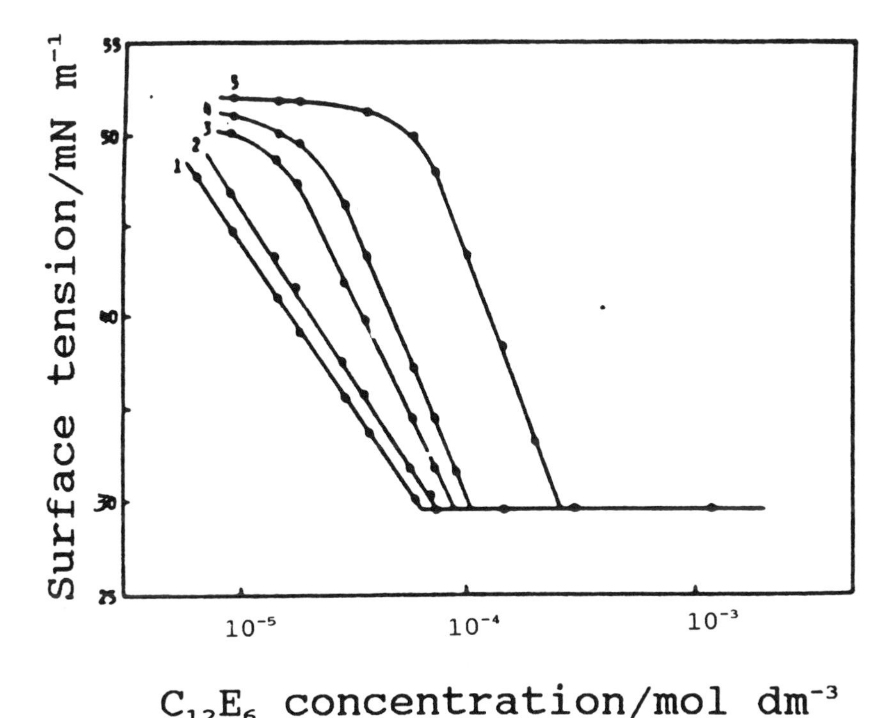

FIGURE 2. Plots of surface tension vs. logarithm of $C_{12}E_6$ concentration at various BSA concentrations for BSA-$C_{12}E_6$ systems at 298.15 K. BSA concentration: (1) $C_{12}E_6$ alone; (2) 1×10^{-6}; (3) 5×10^{-6}; (4) 1×10^{-5}; (5) 5×10^{-5} mol/dm^{-3}. (From Nishikido, N. et al., *Bull. Chem. Soc. Jpn.*, 55, 3085, 1982. With permission.)

B. VISCOSITY

As for simple polymers, interactions of proteins with ionic surfactants can result in significant changes in their rheological properties. The relative viscosity of alkali processed gelatin with SDS above the IEP (4.8) of gelatin is shown in Figure 3.[16] It is interesting that the onset of an increase in viscosity at various gelatin concentrations occurs at the same surfactant concentration. This is similar to the T_1 concept mentioned earlier. The viscosity exhibits a maximum followed by a minimum and then a sharp increase beyond the minimum. The position of the maximum and the minimum increases with gelatin concentration. The authors[16] of this work have attributed the increase in viscosity just above T_1 to the formation of micelle-like aggregates which essentially crosslink the gelatin molecules. Schwuger and Bartnik[5] have reported similar viscosity results for egg albumin-SDS systems. These authors attribute the initial increase in viscosity to the charging up of the protein molecule by binding of surfactants and the decrease above the maximum to the effect of counterions on the polymer coil. The subsequent increase at high

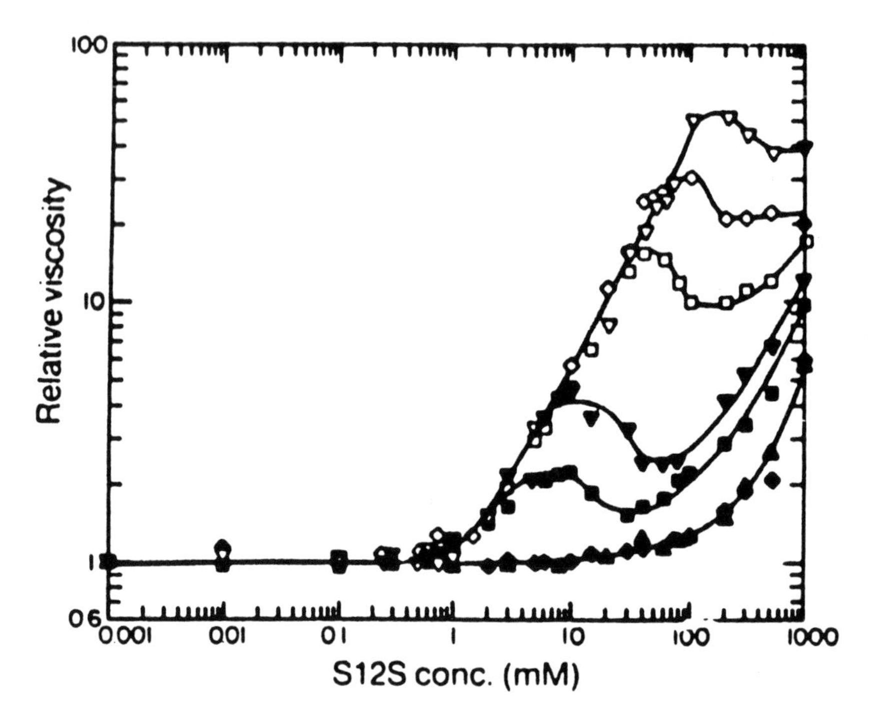

FIGURE 3. η_r vs. S12S (SDS) concentration. Effect of gelatin concentration: (▲) no gel; (◆) 0.1%; (■) 2.0%; (▼) 3.0%; (□) 5.0%; (◇) 7.0%, (▽) 10%. (From Greener, J. et al., *Macromolecules,* 20, 2490, 1987. With permission.)

surfactant concentration has been suggested to be due to the crosslinking of several aggregates by free surfactant micelles in the solution. Greener et al.[16] argue that the initial binding leading to an increase in viscosity is itself due to the formation of a micelle-type structure which crosslinks the gelatin molecules. These suggestions are similar to the interpretation of recently reported data for polyacrylic acid-tetradecyltrimethyl ammonium bromide systems in which the initial binding involves micelle-type aggregates.[17]

Some of the other conclusions from the work of Greener et al.[16] are the following: an increase in chain length of the alkyl sulfate lowers T_1; also, an increase in ionic strength reduces the viscosity enhancement at concentrations just above T_1. The latter indicates that the increase in viscosity, as mentioned earlier, is due to the charging of the protein by surfactant binding. The structure of protein-surfactant complexes in such systems will be discussed in a later section.

C. DYE SOLUBILIZATION AND STEADY-STATE FLUORESCENCE STUDIES

Information on the nature of protein-surfactant aggregates can be obtained from dye solubilization and fluorescence probe studies. Results obtained by Steinhardt et al.[18] for the solubilization of dimethylaminoazobenezene in oval-

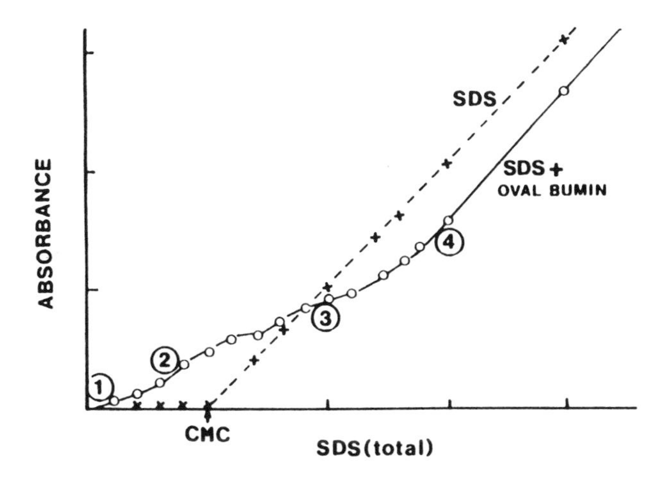

FIGURE 4. The solubilization (increase in absorbance) of dimethyl aminoazobenzene (probe) by ovalbumin-SDS complexes in pH 7.4 (0.033 ionic strength phosphate buffers) at 25°C. (From Chattoraj, D. K. and Birdi, K. S., *Adsorption and the Gibbs Surface Excess,* Plenum Press, New York, 1984, 379. With permission.)

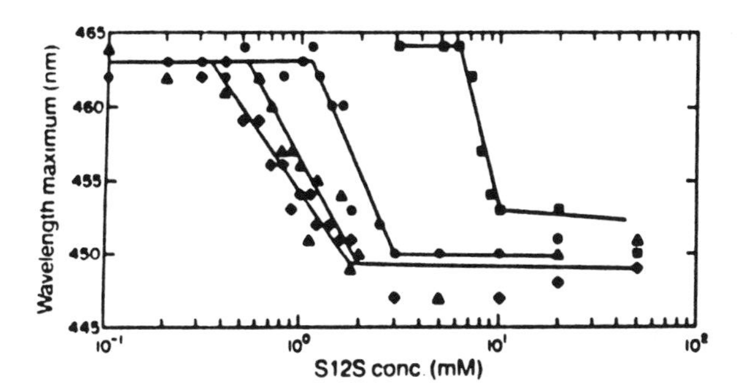

FIGURE 5. Fluorescence emission maximum for 1-pyrene carboxaldehyde vs. S12S (SDS) concentration. Effect of gelatin concentration: (■) 0%; (●) 0.5%; (▲) 1.0%, (◆) 5.0%. (From Greener, J. et al., *Macromolecules,* 20, 2490, 1987. With permission.)

bumin-SDS complexes are given in Figure 4. Solubilization of the dye at concentrations well below the CMC of SDS is a clear indication of the presence of hydrophobic aggregates in the protein-surfactant systems. It is interesting that the solubilization curve at very high SDS levels becomes parallel to that of SDS alone. The protein-surfactant curve crosses that for SDS alone and this is an indication of the smaller size of the hydrophobic aggregates, or the lower solubilization capacity, of the protein-surfactant complex vs. that of the conventional micelles.

Results obtained from fluorescence measurements of gelatin-SDS systems using pyrene carboxaldehyde as the probe are given in Figure 5. Note that the conditions used are essentially the same as those used in the rheological measurements given in Figure 3. Pyrene carboxaldehyde has a characteristic peak maximum (λ_{max}) at about 468 nm in water[19] and the position of the peak is related to the polarity/dielectric constant of the medium in which the probe is solubilized. Thus, the SDS concentration at which a reduction in the position of λ_{max} occurs corresponds to the onset of protein-surfactant aggregate formation in the system. Note that unlike the viscosity data, the fluorescence results show some dependence of gelatin concentration on the onset of interaction, i.e., of T_1, and this might be due to the higher sensitivity of the fluorescence probe technique over that of viscosity, especially in the initial stages of the protein-surfactant interaction. Interestingly, the results appear to indicate that the probe experiences a significantly lower polarity environment in protein-surfactant complexes than in conventional micelles. These are again consistent with the results obtained for the cationic polymer (Polymer JR)-SDS[20] and for the polyacrylic acid-tetradecyl trimethyl ammonium bromide systems.[17]

III. SURFACTANT BINDING STUDIES

Most studies in the past of binding of surfactants to proteins have used the dialysis technique.[3-8] Other techniques used to determine the binding isotherm include ultrafiltration/ultracentrifugation, potentiometry,[3] ion-selective electrode,[21] and surface tension.[15]

A. DIALYSIS MEASUREMENTS

The amount of surfactant bound to the protein is usually expressed in terms of moles (ν) of surfactant bound per mole of protein. In a typical dialysis experiment, a known concentration of the desired protein (c_p, volume V_i) in a dialysis bag is contacted with a surfactant solution of known concentration (c_{Ri}, volume V_o). The surfactant molecules diffuse through the dialysis bag and the system is allowed to attain equilibrium. At this point, the concentration of the surfactant outside the bag is determined by a suitable analytical technique such as titration, radiotracer measurements, surface tension, or absorbance. Knowing C_R, the final concentration of the surfactant, and taking into account the initial dilution of the surfactant by addition of the polymer solution, allows the binding to be calculated:

$$\nu = \frac{(C_{Ri} - C_R)}{C_p} \frac{(V_i + V_o)}{V_i} \tag{1}$$

where $c_{Ri} = c_{Ri'} [V_o/(V_o + V_i)]$. The technique bears a similarity to adsorption experiments onto solid particles, except that the protein is confined within the dialysis bag.

Some specific precautions needed in using the dialysis techniques are as follows. The presence of membranes can introduce Donnan inequalities and normally these effects are suppressed by conducting the experiments in high-salinity solutions. This, of course, introduces competition between the ligand/surfactant and the salt ions and these effects must be considered in interpreting the results. Another factor, not specific to this technique, is the practice of using buffers to maintain the pH of the protein solution. It is important to recognize that often these ions are not innocent bystanders.[3]

In dialysis experiments, it is essential to ensure that equilibrium is attained in the system during the time frame of the experiment. In concentrated protein solutions (>0.1%) this process can often take longer than 24 hr. Also, it is important to conduct control experiments to establish uptake, if any, of surfactant by the membrane itself and possible release of impurities from it.

B. ION-SELECTIVE ELECTRODES

Surfactant-sensitive electrodes offer a direct method of measuring the activity of the surfactant monomeric species in solution and, in fact, this technique has been used extensively to determine the amount of surfactant bound to proteins and polymers.[21-25] One of the early problems with these electrodes was their instability, especially in concentrated micellar solutions. Advances in the technology have led to electrodes which are stable in micellar solutions.[25] Proper precautions should, however, be taken to ensure that the protein and other buffer species do not adsorb on the electrode surface and cause problems.

C. OTHER TECHNIQUES

In principle, any UV, fluorescence, or other spectroscopic technique can be used to study binding, if the surfactant possesses the appropriate moiety and the signal for the monomer differs from those of micelles and protein surfactant complexes.

D. BINDING ISOTHERMS

High-affinity isotherms, typical of the binding of an anionic surfactant to a protein below its IEP, are presented in Figure 6 for the SDS-lysozyme system at pH 3.[26] The binding data are given in terms of the parameter ν. The isotherms exhibit an initial sharp increase followed by a region of slow increase and then a marked increase above a certain concentration. The initial sharp rise has been attributed to binding to high-affinity sites on the protein. These sites for anionic surfactants have been identified to be cationic amino acid residues of lysine, histidine, and arginine.[11] Lysozyme has 11 arginyl, 6 lysyl, and 1 histidyl residues.[26] Note that the first plateau corresponds to about 18 molecules of SDS bound to the protein (see Figure 6). Saturation of these high-energy sites results in the precipitation of the protein (not indicated in the figure) and this is followed by redissolution of the precipitate at higher surfactant concentration, corresponding to the second vertical increase in the isotherm. This region of the isotherm is usually referred to as

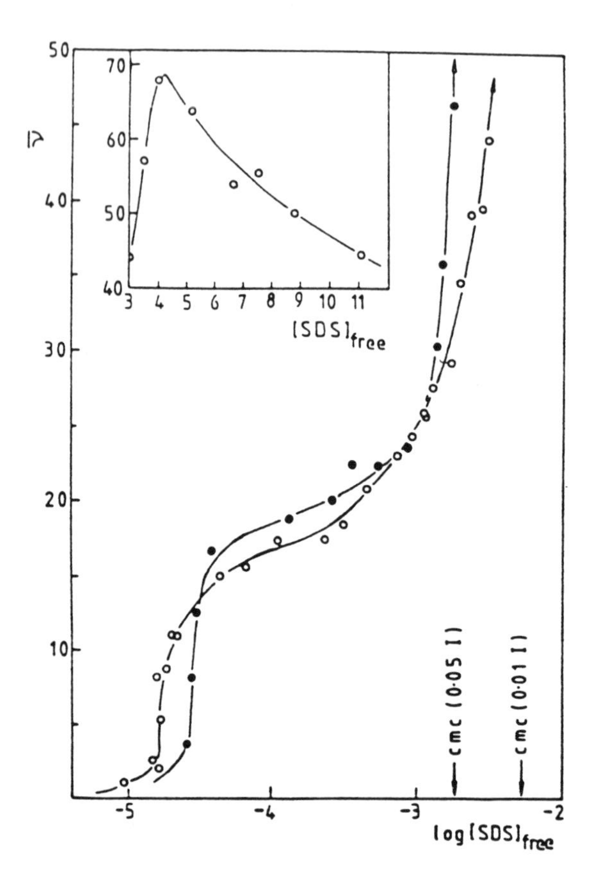

FIGURE 6. Binding isotherms for sodium dodecyl sulfate on interaction with lysozyme at pH 3.2, 25°C. (○) I = 0.0119; (●) I = 0.0519. Lysozyme concentration 0.125% (w/v). The inset shows the maximum in ν plotted as a function of $[SDS]_{free}$ at pH 3.2, I = 0.0119. (From Jones, M. N., *Food Polymers, Gels and Colloids,* Dickinson, E., Ed., Pub. Royal Soc. Chem. 1991. With permission.)

the cooperative binding region. Here unfolding of the protein occurs, resulting in the exposure of more sites and the binding of massive amounts of surfactant. It is important to note that saturation binding occurs at concentrations below the CMC of the surfactant.

The above shape of the isotherm is typical of the binding of ionic surfactants to proteins. Often the initial high-affinity binding occurs at very low surfactant concentrations and, therefore, in some protein systems the exact shape of the isotherm in this region can be missed. Precipitation of the protein by an anionic surfactant at pH values below the IEP of the protein at low surfactant concentrations is also very common. In fact, some of the early observations of precipitation of BSA with SDS and the redissolution of the precipitate were reported 50 years ago. The redissolution was attributed to the formation of a second surfactant layer with a reverse orientation on the already bound SDS molecules.[27] The excess negative charge on the protein upon formation of the second layer was thought to transform the macromolecule into a hydrophilic polyelectrolyte of opposite net charge to its initial one.

Saturation of all of the binding sites corresponds to about 1 to 2 g of surfactant per gram of the protein. The latter figure is for proteins in which all the disulfide bonds are reduced by chemical treatments using reagents such as mercaptoethanol. For such proteins without the disulfide bonds, Reynolds and Tanford[4,28] report that two distinct compositions of the complex exist; the first one corresponds to a stoichiometry of 0.4 g of surfactant per gram of protein (one surfactant molecule per seven amino acid residues) and the second to 1.4 g per gram (one surfactant molecules per two amino acid residues). Interestingly, this transition occurs for SDS in the range of pH 5.6 to 7.2 at about 6 to 7 $\times$ 10^{-4} mol/l at 25°C. Takagi et al.[29] measured the binding of SDS to several protein polypeptides and their results also show two phases, the first one with about 0.3 to 0.6 g of surfactant per gram of protein and the second one with 1.2 to 1.5 g per gram. Contrary to the results of Reynolds and Tanford,[4,28] Takagi et al.[29] reported some difference in the level of binding by different proteins at the first and the second phase.[29]

1. The Presence of Maxima in the Binding Isotherm

The insert in Figure 6 includes binding data for SDS levels well above the critical micelle concentration. As can be seen, the isotherm exhibits a maximum. Similar maxima were observed at other pH values.[30] Interestingly, for the lysozyme-SDS system such maxima at pH values 3.2 and 4.0 disappeared, and the one at a pH of 5.0 remained, when the disulfide bonds in the lysozyme were reduced by pretreatment of the enzyme with mercaptoethanol. Also, at higher NaCl concentrations, e.g., 0.15 mol/l, the isotherm showed a maximum followed by a minimum and another sharp increase. The presence of binding maxima has been attributed to the maximum in surfactant monomer activity reported for micellar solutions.[30] Koga et al.[31] have made similar observations and interpretations for the binding of *n*-methylammonium bromide on BSA.

It is relevant to mention that the presence of a maximum in the adsorption isotherm of surfactants on insoluble proteinaceous substrates such as wool (and indeed on mineral solids) has often been reported.[32-36] In addition to the presence of monomer activity maxima, explanations include the presence of impurities in the surfactant, exclusion of micelles from the interface because of electrostatic repulsion, size exclusion, incompatibility of hydration layers around the adsorbate and the adsorbent, and possible precipitation-redissolution of complexes/precipitates formed between the surfactant and other unavoidable species in the system.

Even though the binding maximum in protein-surfactant systems appears to be adequately accounted for in terms of the corresponding maximum in

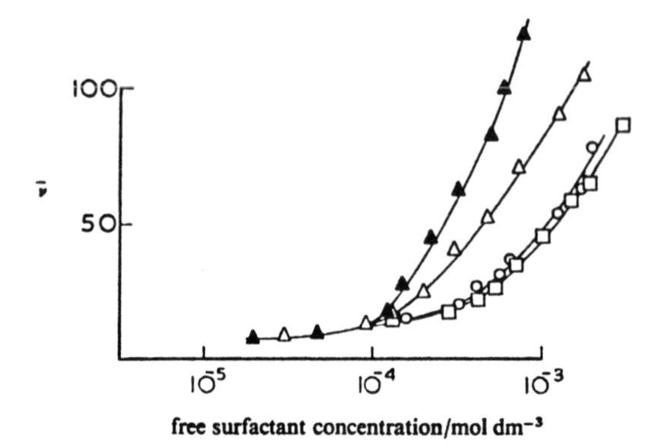

FIGURE 7. Surfactant binding to BSA in 10^{-2} mol dm^{-3} NaCl (open points) and in 10^{-1} mol dm^{-3} NaCl (filled points). (▲) SDS; (○) SDSo (sodium dodecyl sulfonate); and (□) OBS (octylbenzene sulfonate). (From Rendall, H. M., *J. Chem. Soc. Faraday Trans. 1*, 72, 481, 1976. With permission.)

the monomer activity,[30,31,37] the correlations have not been quantitative. For example, an auxiliary mechanism may be the exclusion of charged free micelles from the highly charged protein-surfactant complex. The presence of a minimum followed by an increase at higher surfactant concentrations in high-salinity solutions is in line with a reduction of repulsion, and therefore, of the exclusion.

In high-salinity solutions, the observed increase in surfactant uptake by lysozyme at SDS levels beyond the minimum[30] is possibly due to network formation resulting from the bridging of micelles by the proteins. The latter is in line with the observed increase in viscosity of these systems at high surfactant concentrations. The term "binding" may not, however, exactly represent the system's behavior under these conditions.

2. Effect of Variables

The binding of surfactants to proteins is influenced by a number of variables such as the nature and chain length of the surfactant, pH, ionic strength, temperature, and the nature of the protein. The effect of some of these variables is examined below.

a. Nature of the Hydrophilic Group

The binding of surfactants to proteins is dependent upon the nature of the hydrophilic group of the surfactant. In general, anionic surfactants interact much more strongly with proteins than do cationic and nonionic surfactants. Among anionics, alkyl sulfates appear to bind more strongly than alkylbenzene sulfonates and alkyl sulfonates.[21] The order of binding among equivalent-chain-length anionic surfactants can be represented as:[3,21,38] $SO_4^- > SO_3^- >$ benzene $SO_3^- > COO^-$. Typical binding isotherms for dodecyl sulfate, dodecyl sulfonate, and octyl benzene sulfonate are given in Figure 7. Unlike sulfates and sulfonates, the carboxylates exhibit pH-dependent charge characteristics. They achieve their fully ionized form only under alkaline pH conditions and since most proteins have their IEP values around pH 4 to 5, the condition of opposite charge of protein and surfactant, which is a primary factor in their interaction, cannot be achieved.

Note that the variation of the CMC of anionic surfactants with different polar groups, for a given alkyl chain length follows the order: carboxylates > sulfonates > sulfates (see Chapter 2). Evidently, the higher the tendency of the surfactant to form micelles, the stronger is its affinity to bind to proteins. This may be a reflection of the inherent surface activity of the surfactant as well as of similarities in the energetics involved in the two processes. This aspect will be discussed further in a section on the nature of protein-surfactant aggregates.

As mentioned earlier, cationic surfactants exhibit a lower tendency than anionic surfactants to bind to proteins. The shape of the isotherms in the two cases is, however, similar (see Figures 8a and b).[39,40] Nozaki et al.[39] found that at a pH of 5.6, tetradecyltrimethylammonium bromide (TTAB) initially adsorbed on four high-energy sites of BSA. This was followed by a sharp increase in adsorption corresponding to cooperative interactions at higher surfactant concentrations. Kaneshina et al.[40] reported the number of primary binding sites of BSA to be about 11 at a pH value of 6.8 (see Figure 8b). At higher concentrations, the cooperative binding of cationic surfactants can lead to denaturation.[40] The saturation binding of TTAB on BSA at pH 5.6 is about 0.6 g per gram of the protein.[39] Chemically reducing the sulfur bonds increases the binding to about 0.8 to 1.0 g per gram of BSA.[39]

Since the cooperative transition in the binding of SDS onto several proteins occurs at about the same concentration, Nozaki et al.[39] have argued that the specific features of amino acid sequence or other aspects of protein structure play a less critical role in the cooperative binding. Thus, the surfactant ions, while primarily attracted to the protein because of hydrophobic interactions, would tend to cluster around the side chains of opposite charge. On this basis, the binding of anionic surfactant would be favored because arginyl and lysyl side chains project further from the surface and contribute more $-CH_2-$ groups for incorporation in an aggregate than the glutamyl and aspartyl side chains which would be involved in the binding of cationic surfactants.

These considerations are consistent with the experimental observations that, in general, cooperative binding of anionic surfactants occurs at a much lower concentration than that of cationic surfactants (see Figures 7, 8a and b). In fact, the cooperative binding in the latter case occurs at concentrations close to the CMC. A consequence of this is the relatively low saturation binding of cationic surfactants as compared to anionic surfactants. It has been argued that the formation of micelles actually competes with the binding process, and, therefore, saturation binding in the case of cationic surfactants is micelle limited and complete saturation is never achieved. This is in line with the observation that the saturation binding of SDS to proteins is inde-

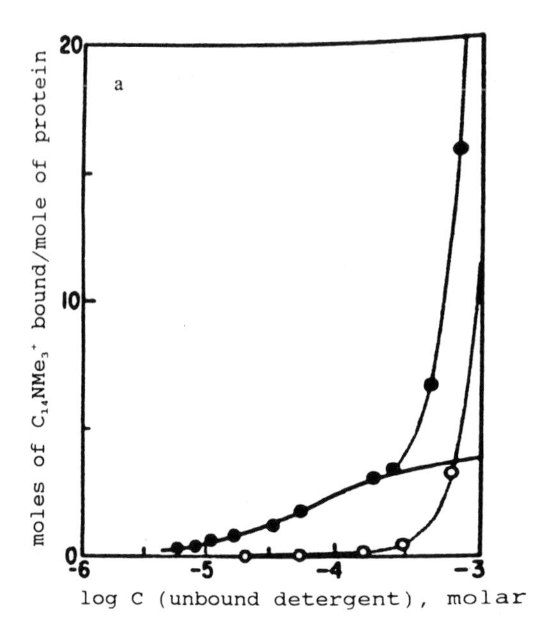

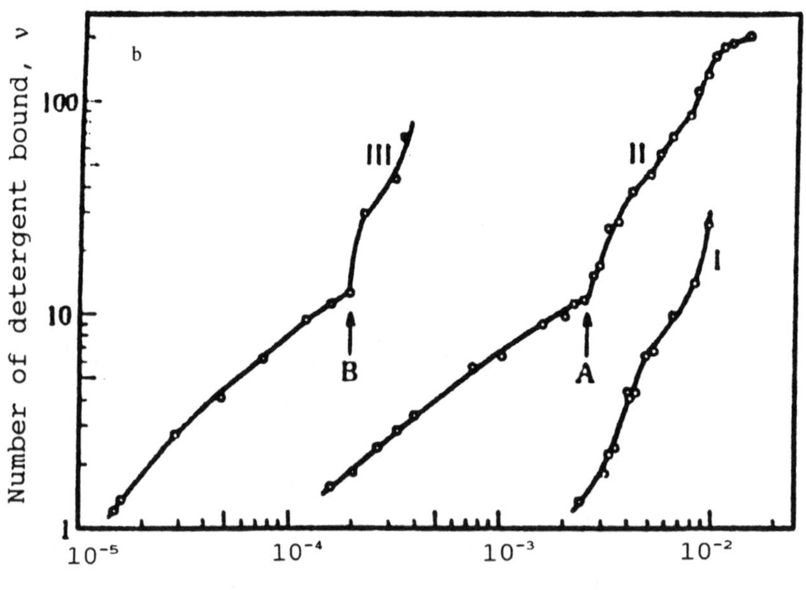

FIGURE 8. (a) Binding of $C_{14}NMe_3^+$ to ovalbumin (○) and serum albumin (●) at low detergent concentration in 0.015 *M* phosphate buffer, pH 5.6. The solid line is a theoretical curve for 4 identical noninteracting binding sites with an association constant of 1.5×10^4 l/mol. (From Nozaki, Y. et al., *J. Biol. Chem.*, 249(14), 4452, 1974. With permission.) (b) Binding isotherms for cationic detergents and BSA at 25°C and pH 6.8 (ionic strength 0.1). I, DeTAB; II, DTAB; and III, TTAB. (From Kaneshina, S. et al., *Bull. Chem. Soc. Jpn.*, 46, 2735, 1973. With permission.)

pendent of pH, but that of cationic surfactants increases with pH above the IEP of the protein.

In contrast to anionic and cationic surfactants, nonionic surfactants bind relatively weakly to proteins.[15,40] This observation shows the importance of ionic interactions in the binding of surfactants. Makino et al.[41] investigated the binding of Triton X-100 to BSA and showed that this protein has four principal binding sites. Unlike ionic surfactants, the nonionic surfactant does not exhibit a cooperative binding region and it does not unfold or denature the protein. This has been attributed to the low CMC of nonionic surfactants which makes micelle formation in the bulk solution a more favorable process than binding to the protein.

b. Surfactant Chain Length

The effect of surfactant chain length on the binding of cationic surfactants onto BSA can be seen in Figure 8b.[40] As expected, the onset of binding occurs at a lower surfactant concentration as the chain length increases. Also, the isotherm for the decyltrimethylammonium bromide does not exhibit a sharp transition from a high-affinity region to a cooperative region. Similar results for the binding of *n*-alkyl sulfates to lysozyme reveal the absence of a high-affinity region for the C_8 sulfate, but the presence of both a high-affinity and a cooperative region for C_{10} and C_{12} sulfates.[42]

It is clear that the higher the surface activity of the surfactant, the higher is the tendency of that surfactant to bind to proteins. A measure of this effect can be seen in Figure 9, in which the concentration, C_{50}, required to achieve the binding of 50 molecules of surfactant to a protein molecule (log C_{50}) is plotted against its CMC. As expected, log C_{50} increases directly with the CMC of the surfactant.[21]

c. Effect of Ionic Strength

The effect of salt on the binding of surfactants to proteins has been studied by several groups,[9,11,42-45] and it can be understood in terms of salt effects on various subprocesses which occur during binding. The main effects concern electrostatic interactions and cooperative hydrophobic interactions such as micellization/aggregation in solution and cluster formation at interfaces.

The effect of ionic strength on the binding of SDS to lysozyme can be seen in Figure 6.[11] The onset of surfactant binding, corresponding to the high-affinity region, increases with increase in ionic strength. This can be explained in terms of reduced electrostatic attraction for the surfactant monomer because of charge screening or of electrical double-layer compression effects. Interestingly, in the cooperative region, the isotherms reverse their positions. This cross-over of the isotherms is essentially due to the opposing effect of salt on electrostatic interactions vs. that on cooperative hydrophobic interactions. The latter, which is similar to the interactions in micellization of ionic surfactants (see Chapter 2), is enhanced by inorganic electrolytes. It is pertinent to mention that the effect of salt on adsorption/hemimicellization at the solid-liquid interface is similar to that on surfactant binding to proteins.[46,47]

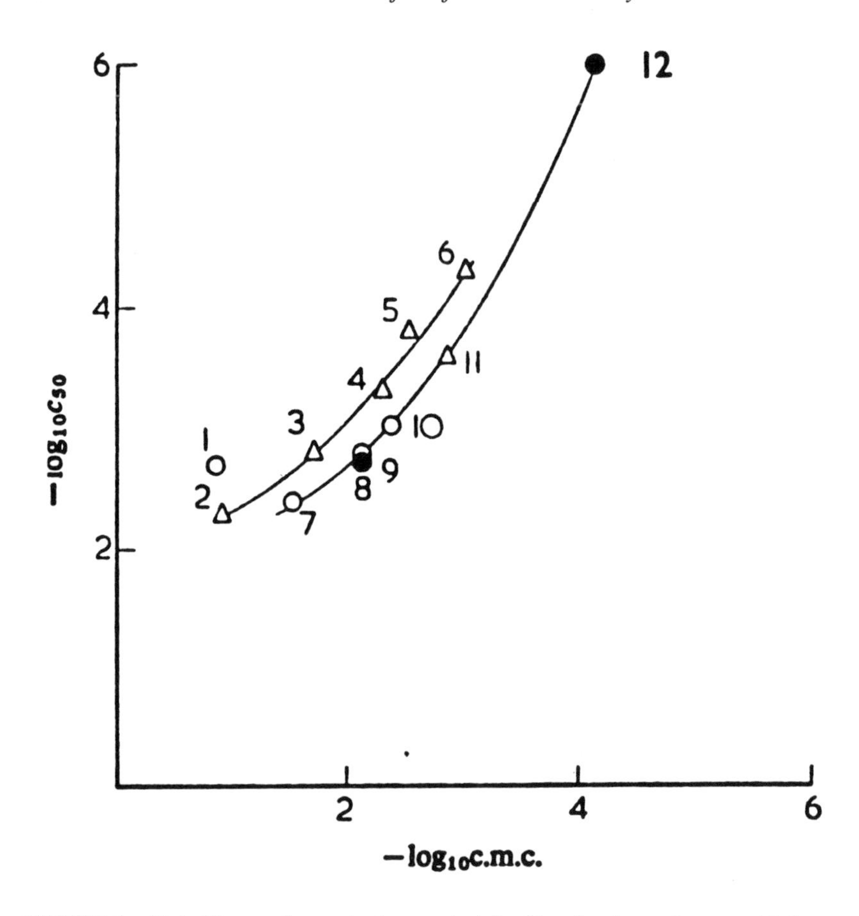

FIGURE 9. Plot of log c_{50} (concentration required for 50 molecules surfactant to bind to a single protein molecule) for surfactant binding to BSA against log CMC. (△, sulfate: 2, C_8[a]; 3, C_{10}[a]; 4, C_{12}[b]; 5, C_{12}[a]; 6, C_{14}[a]; 11, C_{12}[c]. ○, Sulfonate: 1, C_8[a]; 7, C_{10}[a]; 9, C_{12}[b]; 10, C_{12}[a]. ●, benzene sulfonate: 8, OBS (octyl benzene sulfonate); 12, DBS (dodecyl benzene sulfonate). [a]Data from Reference 3, Table 7–4, ionic strength = 0.033 mol/dm^{-3}; [b]data from Reference 21, ionic strength = 0.01 mol/dm^{-3}; [c]data from Reference 21, ionic strength 0.1 mol/dm^{-3}. (From Rendall, H. M., *J. Chem. Soc. Faraday Trans. 1*, 72, 481, 1976. With permission.)

In addition to its effect on the electrical double layer around the protein and on cooperative binding, salt can also influence the micellization process by reducing the CMC. Accordingly, addition of excess salt may render bulk micelle formation more favorable than binding to proteins. The term excess refers to levels which will lower CMC to concentrations lower than that required to initiate surfactant binding to proteins. Thus, the effect of salt on surfactant binding to proteins can be complex, depending upon the relative levels of salt and their effects on several phenomena occurring in the system. Results for the binding of myristyltrimethylammonium bromide onto DNA at different ionic strength levels, given in Figure 10, illustrates the complexity of salt effects on binding of this surfactant.[9]

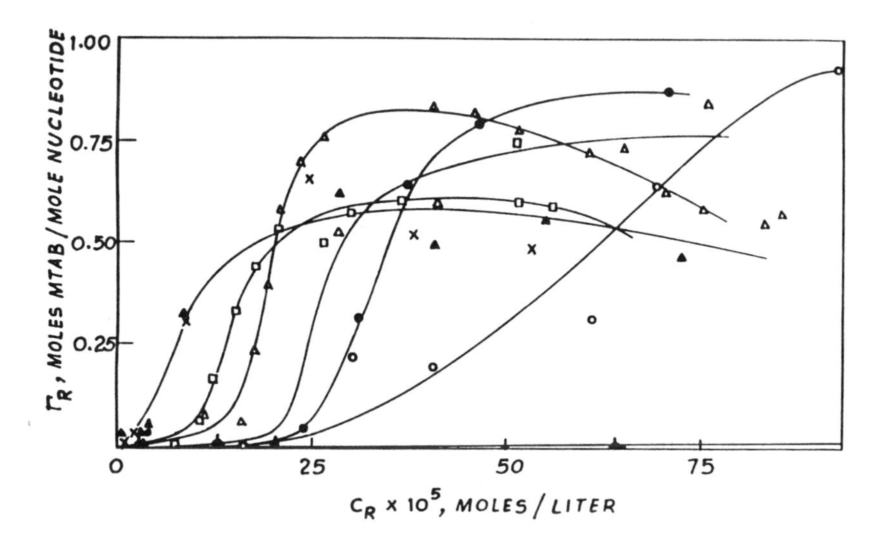

FIGURE 10. Γ_R^1 vs. c_R plot at pH 6 for the binding of MTAB (C_{14}), with DNA at 30°C. Ionic strengths (phosphate buffers): ▲, 0.0125; □, 0.025; △, 0.0625; ----, 0.125; ●, 0.25. Ionic strengths (NaCl solution): ○, 0.05; ▲, 2.0; × 0.0125 (phosphate buffer + denatured DNA). (From Chattoraj, D. K. and Birdi, K. S., *Adsorption and the Gibbs Surface Excess,* Plenum Press, New York, 1984. With permission.)

d. Effect of pH and Charge Density of the Protein

The pH of the solution, in relation to the isoelectric point of the protein, determines the effective charge of the protein and therefore its ability to attract cationic or anionic surfactants. Note that the isoelectric point of BSA is slightly less than pH 5.0.[48] IEP values for some of the other proteins are: pepsin $<$ 1, egg albumin 4.6, hemoglobin 6.8, and lysozyme 11.0. The binding of anionic surfactants onto the protein can, in general, be expected to occur at a higher free surfactant concentration as the pH is raised (see Figure 11).[49] For cationic surfactants, on the other hand, the isotherms would shift to higher surfactant concentrations with reduction in the solution pH.[50]

It is interesting that an anionic surfactant such as SDS adsorbs in significant amounts at pH values well above the IEP of the protein (see Figure 11). As previously noted, the shift in the onset of binding to higher surfactant concentration is the result of unfavorable electrostatic forces above the IEP. Yet the maximum binding of the surfactant under high pH conditions is similar to that observed at low pH conditions. This seems to suggest that the saturation binding for anionic surfactants is essentially controlled by cooperative hydrophobic interactions. For cationic surfactants, on the other hand, the saturation binding is limited by monomer activity because of competition between micellization and binding, and in this case, the pH has a significant effect on the saturation binding. For example, with increase in pH, electrostatic factors lower the concentration at which binding begins and in the process shift the competition in favor of binding over micelle formation.

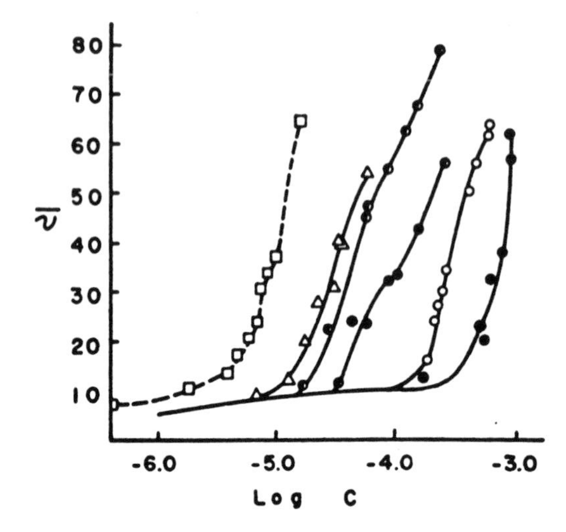

FIGURE 11. The effect of pH on the binding of dodecyl sulfate by BSA at 2°C and 0.033 ionic strength: (●) pH 6.8 (phosphate); (○) pH 5.6 (phosphate); (⊗) pH 4.8 (acetate); (◐) pH 4.1 (acetate); (△) pH 3.8 (acetate); (□) pH 3.8 (acetate), ionic strength 0.004. The lines drawn through the data are not theoretical curves. Data obtained at pH 6.2 and 7.5 are almost indistinguishable from the results at pH 6.8, but data at pH 8.3 are still further to the right. (From Reynolds, J. A. et al., *Biochemistry,* 9, 1232, 1970. With permission.)

An exception to the ''obvious'' effect of electrostatic factors can be encountered if the charge density of the substrate is high. For example, some of the recent work[24] on the binding of tetradecyltrimethylammonium bromide onto polyacrylic acid in 10^{-2} *M* NaBr solution showed that increasing the charge density of the polymer, by changing the solution pH, to ionization values above about 0.4 did not affect the position of the isotherm and, surprisingly, lowering the charge density to values below 0.4 actually lowered the onset of binding. These results are thought to involve the Manning[51] counterion condensation phenomenon according to which, above a certain critical charge density of the polymer, the effective charge density is reduced to a critical value by condensation of counterions onto the polymer backbone. The concentration of Na^+ ions in these systems is higher than that of the surfactant ions by almost 2 to 3 orders of magnitude. Therefore, the charge density of the polymer is essentially controlled by the Na^+ ions and is invariant above its critical charge density. This can account for the lack of charge density effects on the position of the surfactant binding isotherm at high ionization values. At low charge density values, in addition to electrostatic factors, conformational aspects of the polymer appear to play a role in determining the onset of binding and the shift in the isotherms to low surfactant concentrations.

e. Effect of Temperature

Temperature can influence several factors in a protein-surfactant system. An increase in temperature can affect the nature of the protein significantly and even lead to its denaturation. BSA, for example, undergoes a thermal transition above about 55 to 65°C,[52] depending upon the pH of the solution, and this in effect destroys its globular structure. Temperature can also affect cooperative hydrophobic interactions as well as electrostatic interactions.

The effect of temperature on the binding of *n*-alkyltrimethylammonium bromides onto BSA is illustrated in Figures 12 and 13.[53] A comparison of Figures 12 and 13 shows that an increase in temperature from 30 to 65°C increases the saturation binding of CTAB by well over an order of magnitude. Interestingly, some of the steps observed in the isotherm at 30°C are not seen at 65°C. These effects have been attributed to the thermal denaturation of the protein at 65°C and the consequent unfolding and exposure of more sites for binding. Even though at 30°C the isotherms do not show any trends with chain length, at the higher temperature an increase in the chain length increases the binding. The observed increase in the number of binding sites for myristyltrimethylammonium bromide is consistent with the results reported by Nozaki et al.[39]

Temperature dependence of binding of SDS to BSA shows a much more complex behavior than that of cationic surfactants.[9,43] The former isotherms obtained at a pH of 6.0 show a maximum followed by a minimum at 65°C. In this system, in the cooperative region, the temperature appears to decrease the extent of binding. Note that under the tested pH conditions, electrostatic factors will oppose binding and this may be partly responsible for the observed differences between the behavior of cationic surfactants vs. anionic surfactants. Schwuger and Bartnik[5] concluded from surface tension measurements that SDS binding to gelatin was reduced with increase in temperature.

The temperature-dependence data for the binding of surfactants to proteins can be used to obtain information on the enthalpy changes involved in the interaction. This aspect will be examined later.

f. Effect of Organic Additives

In general, additives which lower the CMC of an ionic surfactant will reduce its tendency to bind to proteins. An exception to this is salt addition which also lowers the CMC. Electrolyte addition does not always reduce surfactant binding to proteins and the reasons for this were discussed earlier. Addition of nonionic surfactants or small amounts of cationic or amphoteric surfactants will reduce the binding of anionic surfactants. These effects are applicable to both soluble and insoluble proteins; some of the specifics are discussed under the section on binding to insoluble proteins.

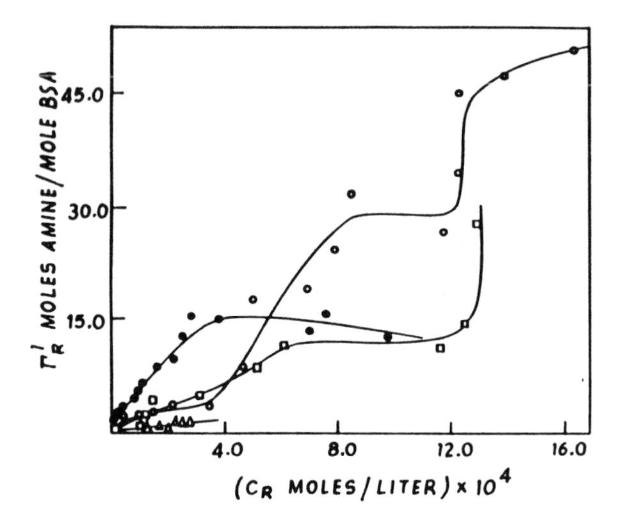

FIGURE 12. $\Gamma_R{}^1$ vs. c_R plot for binding of CTAB, MTAB and DTAB with BSA. Phosphate buffer pH 6.0, ionic strength, 0.125. ●, CTAB (30°C); ○, MTAB (30°C); □, DTAB (30°C); △, CTAB (15°C). (From Chattoraj, D. K. and Birdi, K. S., *Adsorption and Gibbs Surface Excess,* Plenum Press, New York, 1984. With permission.)

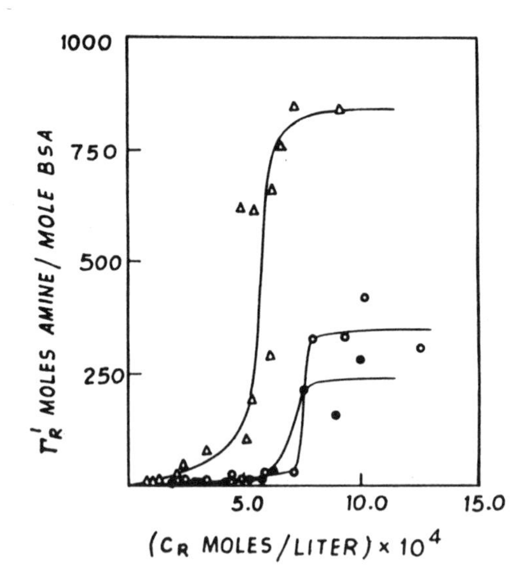

FIGURE 13. $\Gamma_R{}^1$ vs. c_R plot for binding of CTAB, MTAB, and DTAB with BSA at 65°C, phosphate buffer, pH 6.0, ionic strength = 0.125: (△) CTAB; (●) MTAB; (○) DTAB. (From Chattoraj, D. K. and Birdi, K. S., *Adsorption and the Gibbs Surface Excess,* Plenum Press, New York, 1984. With permission.)

3. Effect of Nature of the Substrate: Soluble vs. Insoluble Keratinous Proteins

As mentioned earlier, most proteins in their reduced state bind about 1.4 g of ionic surfactant per gram of the protein. This suggests that the binding to proteins is essentially nonspecific in nature. It must be pointed out that the binding observed in the initial stage does differ from protein to protein. In this section, however, inherent differences among proteins which are associated with differences in concentration as well as distribution of various amino acid residues and their structure will not be considered. Instead, some of the observed differences between water-soluble proteins and insoluble proteins such as zein and various forms of keratin such as wool and stratum corneum will be examined.

A knowledge of the binding behavior of surfactants to keratinous substrates is important for improving the understanding of surfactant-induced changes occurring on hair and skin. Zein, a corn protein, is important in this respect since studies have shown that the harshness of a surfactant towards skin can be correlated with its ability to solubilize zein.[5] Many studies have been carried out on the binding of surfactants to insoluble proteins,[5,32,54-63] but, unlike the case of soluble proteins, a full understanding of the interactions involved has not yet been established.

Zein has a finite, limited solubility in water. Addition of an anionic surfactant such as SDS increases its solubility markedly. Interestingly, the increase in solubility occurs at concentrations just above the CMC of the surfactant (see Figures 14a and b).[64] The exact mechanism by which zein is brought into solution has not been established. It is, however, known that nonionic surfactants and ethoxylated alkyl sulfates, which are relatively mild towards skin, solubilize significantly lower amounts of zein than does SDS.[5] These effects are similar to those observed in the case of soluble proteins, in the sense that nonionic and ethoxylated sulfates interact to a lesser extent with them.[10,13,59]

Insoluble proteins, in general, have a structure highly crosslinked with disulfide bonds. The extent of crosslinking usually determines properties such as solubility and swellability. The swelling nature of proteins and their resultant interactions with surfactant add another level of complexity in these cases. Note in this regard that both the skin and hair are highly swellable by water and surfactant systems.

The absence of significant crosslinking in water-soluble proteins allows them to expose a multitude of binding sites to surfactant molecules. Also, the secondary and tertiary structures supported by hydrogen bonding and hydrophobic-type interactions can be destroyed by surfactants. The latter effect is responsible for the unfolding of proteins even in the absence of disulfide reduction. The amount of SDS bound at saturation is an indication

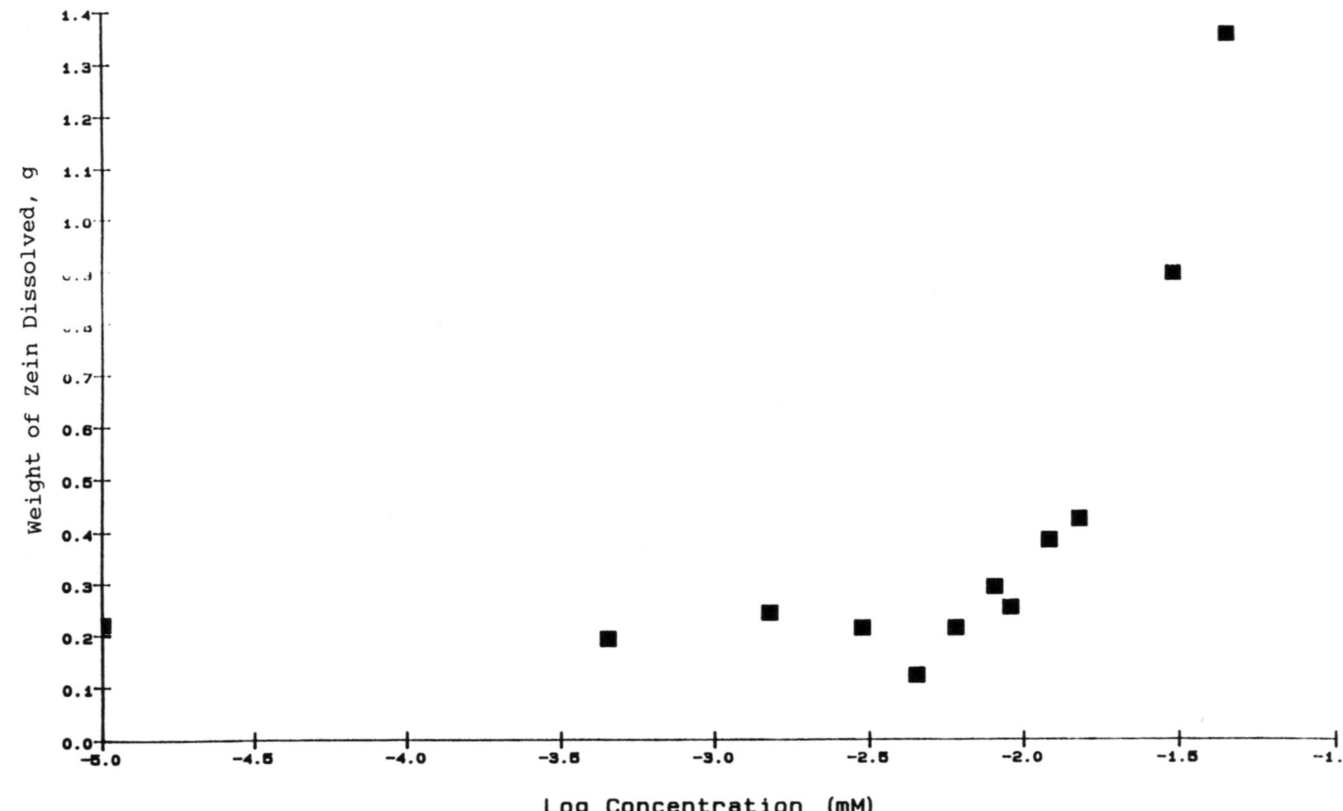

FIGURE 14a. (see the caption on page 341.)

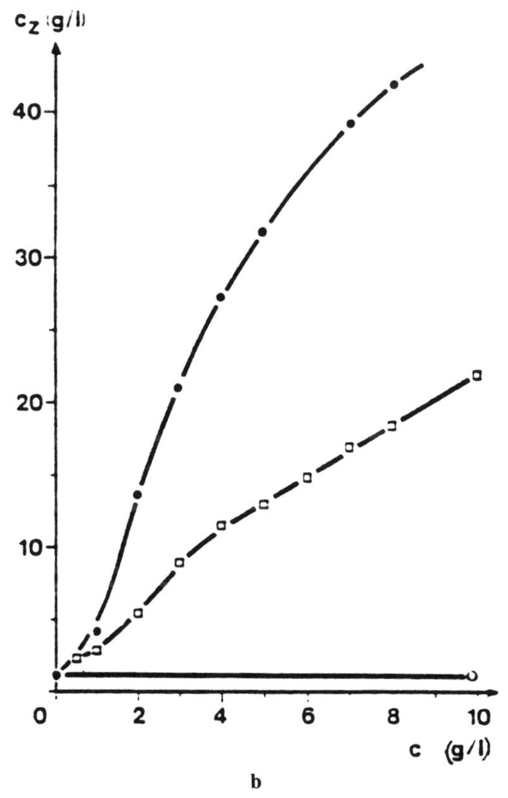

b

FIGURE 14. (a) Zein solubilization as a function of concentration of SDS. From Cho, S., unpublished results, Unilever, U. S., NJ, 1991. (b) Zein solubility, c_z, of saturated solutions as a function of surfactant concentration. Zein concentration, 50 g l^{-1}; mixing period, 2 h; temperature 40°C. (●) sodium dodecyl sulfate; (□) alkyl ether sulfate (2EO); (○) nonyl phenol (9EO). (From Schwuger, M. J. and Bartnik, F. G., in *Anionic Surfactants,* Surfactant Sci. Ser. Vol. 10, Gloxhuber, C., Ed., Marcel Dekker, New York, 1980, 1. With permission.)

of the extent of accessibility of the structure to penetrating molecules. For example, water-soluble proteins bind as much as 1.4 to 2 g of SDS per gram. In contrast to this, wool binds about 0.06 g of SDS per gram. Wool has a highly crosslinked structure and this explains why surfactants such as SDS cannot penetrate its matrix. Breaking the disulfide bonds increases the saturation binding markedly.

An important difference between the binding isotherms of soluble proteins and insoluble proteins is the continued binding of surfactant in the latter case at concentrations well above the CMC. As mentioned earlier, dissolution of the insoluble protein, zein, also occurs only above the CMC. It is a generally accepted notion that binding onto proteins, and indeed onto other substrates, involves only monomers, with micelles acting essentially as a reservoir of monomers. The exact reasons as to why saturation of all the sites does not occur below the CMC is not fully understood. It may involve both kinetic and thermodynamic factors. Breaking the hydrophobic and hydrogen bonding interactions which hold the protein structure together will require a certain energy and this may be derived from the continued binding of surfactant to sites which offer progressively increasing levels of access resistance. Building up an electrical charge will eventually supply the critical force/energy required to open up the protein structure. Breuer has discussed this aspect in terms of the energy required to overcome the elastic energy of the polypeptide network.[60]

The kinetic effect is likely to come from the slow diffusion of molecules into the inner areas of the folded protein. Faucher and Goddard[57] showed that the uptake of SDS by keratinous substrates, such as hair and stratum corneum, followed a linear dependence on the square root of time which is consistent with a diffusion-limited process (see Figure 15). Interestingly, their results showed that the slope of the uptake rate plot for this surfactant actually decreased above the CMC and this was interpreted as indicating that the monomer was the major diffusing species. Also, the rate of uptake for stratum corneum was significantly higher than that for hair. The latter effect may involve the presence of lipids in the stratum corneum, some of which, by getting solubilized in micellar solutions, could lead to faster diffusion. Furthermore, the keratin in stratum corneum, unlike hair, has very little disulfide linkages[63] and therefore it may be possible to ''open up'' the keratin in stratum corneum relatively easily.

As regards the question of kinetic vs. thermodynamic effects, the work by Griffith and Alexander[54,63] is noteworthy. Their results show that attaining equilibrium adsorption of SDS on wool at 35°C may take several weeks. The equilibration time diminishes at higher temperatures, estimates at 40 and 60°C being 5 and 3 days, respectively. Equilibrium adsorption of SDS onto wool at 40 and 60°C attains a plateau value at the CMC (see Figure 16). This clearly supports the notion that the binding is due to monomers and not micelles. The kinetic results reported by the above authors for hexadecyl sulfate show that the systems exhibit several local ''plateau values'' which could be considered as pseudo-equilibria (see Figure 17). These are surfactant concentration dependent and their unusual nature suggests that further work would be desirable. From an applications point of view, kinetic effects are important since the duration of contact of the substrate with the surfactant solutions may not be long enough to attain equilibrium.

The question as to whether or not all of the effects observed on the binding of surfactants to insoluble proteins above the CMC are due to kinetic effects is not clear at present. Another factor, as pointed out by Faucher and Goddard,[57] is the contribution to the total uptake which arises from swelling of the matrix. It is not clear whether the SDS uptake in these systems corresponds to bound SDS or bound SDS plus the SDS existing in the free solution inside the stratum corneum. In the case of zein solubilization, the

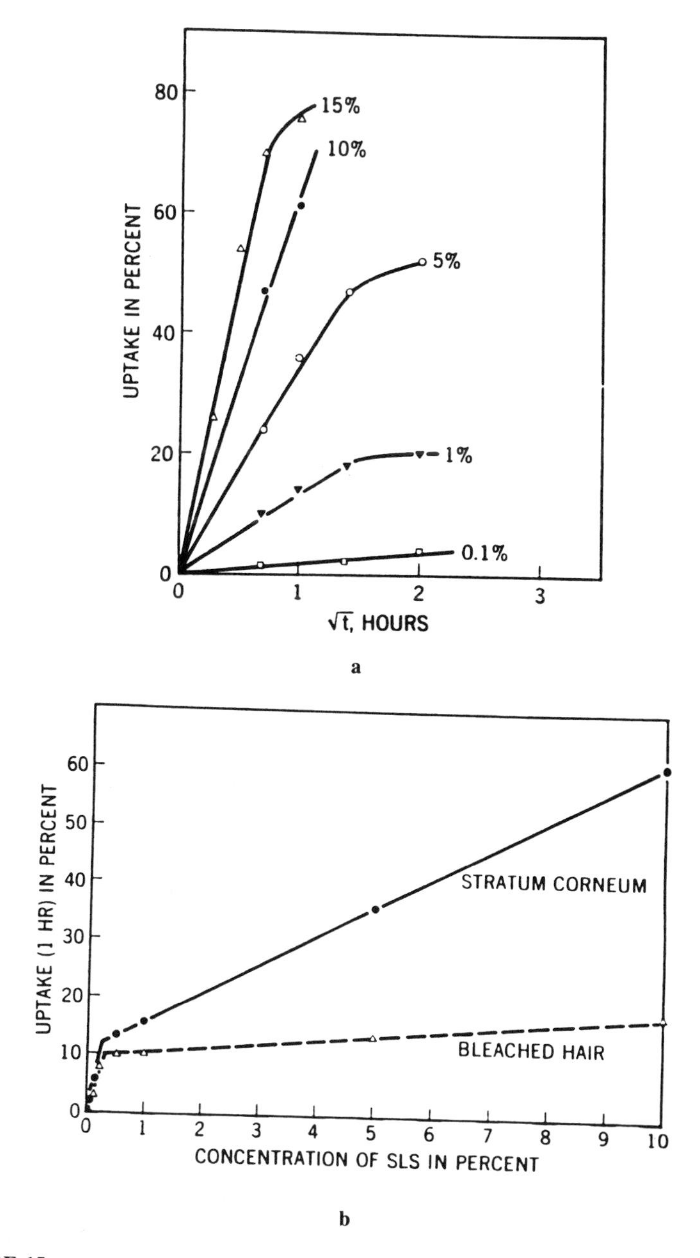

FIGURE 15. (a) Sorption of sodium lauryl sulfate by stratum corneum plotted as a function of uptake vs. square root of time % SLS indicated against each curve. (b) Uptake of sodium lauryl sulfate by stratum corneum and hair at the end of one hour of contact as a function of concentration of SLS. (From Faucher, J. A. and Goddard, E. D., *J. Soc. Cosmet. Chem.*, 29, 323, 1978. With permission.)

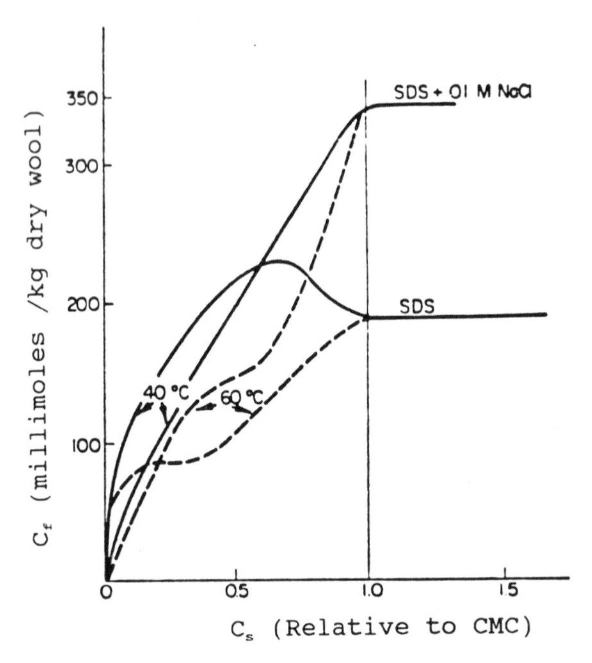

FIGURE 16. Isotherms for the adsorption of SDS by wool in the presence and absence of NaCl at 40 and 60°C as a function of reduced concentration (C/[CMC]). (From Griffith, J. C. and Alexander, A. E., *J. Colloid Interface Sci.*, 25, 311, 1967. With permission.)

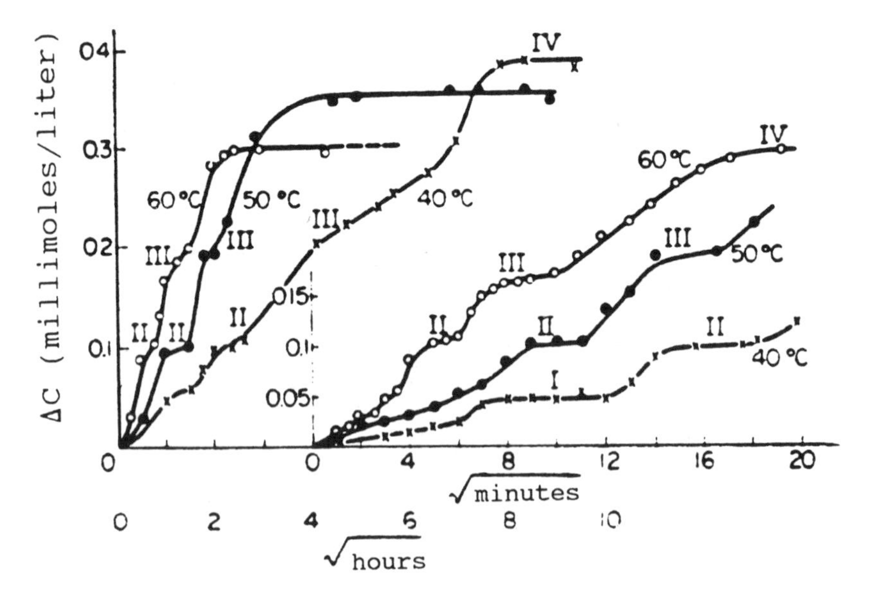

FIGURE 17. The rate of uptake of sodium hexadecyl sulfate by wool at 40°C (×), 50°C (●), and 60°C (○) from solution of initial concentration 1 m*M*. ΔC is the decrease in the concentration of solution. Inset shows the early data on an expanded scale. (From Griffith, J. C. and Alexander, A. E., *J. Colloid Interface Sci.*, 25, 317, 1967. With permission.)

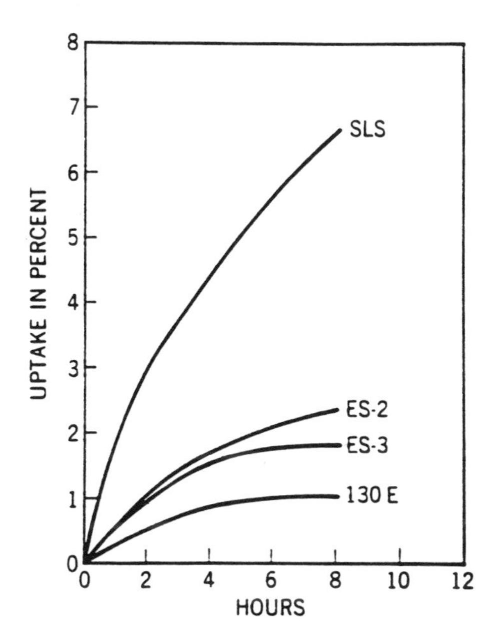

FIGURE 18. Sorption of lauryl ether sulfates by bleached hair from a 10% solution. Standapol ES-2, ES-3, and I30E are, respectively, 2, 3, and 12 mol ethoxylates of sodium lauryl sulfate. (From Faucher, J. A. and Goddard, E. D., *J. Soc. Cosmet. Chem.*, 29, 323, 1978. With permission.)

enhanced dissolution occurring in micellar solutions suggests the involvement of micelles in the process. It is clear from these discussions that further studies, hopefully that will employ techniques such as NMR, electron spin resonance (ESR), Raman, and fluorescence to probe the local environment of water and surfactant inside the protein network, are needed to shed more light on the mechanisms.

a. Effect of Variables and the Role of Additives

From a practical point of view, the mildness of surfactants towards skin and hair can be increased by reducing their binding to proteins. In general, factors which enhance micelle formation and reduce the affinity of the polar group for the protein would seem to be desirable.

As mentioned earlier, lauryl ether sulfates (ethoxylated alkyl sulfates) do not bind to keratin as avidly as does SDS[57] (see Figure 18). Also, as the number of EO groups in the molecule increases, the binding decreases. This effect has been attributed to the lower CMC of these ethoxylated materials and also to lower diffusion into the keratin structure because of the increase in their molecular size. It is significant that alkyl ether sulfates are considerably milder than SDS to the skin.

Additives can be used to reduce the binding of surfactant to proteins. For example, the uptake of SDS in the presence of Tergitol NP9, a nonyl phenol

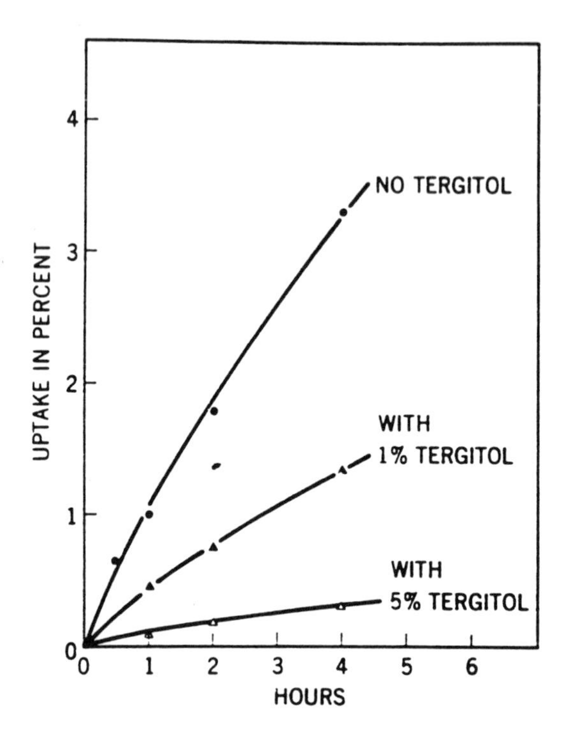

FIGURE 19. Effect of a nonionic surfactant on the sorption of sodium lauryl sulfate by bleached hair. (From Faucher, J. A. and Goddard, E. D., *J. Soc. Cosmet. Chem.*, 29, 323, 1978. With permission.)

ethoxylate, is significantly lower than that in its absence (see Figure 19). Addition of a nonionic surfactant results in the formation of mixed micelles, a lower CMC, and reduced activity of the anionic surfactant monomer. As stated above, addition of a small amount of a cationic or amphoteric surfactant can reduce the anionic surfactant CMC markedly[65,66] and the result will be a reduction in the binding of the surfactant.

Note that a commonly used method to reduce the CMC of an ionic surfactant is to add electrolyte. As discussed earlier, the effect of salt can be complex depending upon the level at which it is used. At low levels, the salt can increase the onset of surfactant binding to higher surfactant concentrations, but can enhance the binding in the cooperative region to such an extent that the overall binding may increase markedly (see Figures 6, 8, 16). If added in high enough levels to reduce the CMC to values lower than that required to initiate the binding to protein, then salt addition will promote micelle formation in solution over binding to proteins.

Other additives, such as water-soluble polymers, can be used to reduce the increase of monomer activity with total surfactant concentration, and so reduce the effective level of binding to proteins.[67,68] This is equivalent to providing the surfactant with a competing substrate for adsorption/binding.

IV. SURFACTANT-INDUCED DENATURATION OF PROTEINS

Ionic surfactants, in general, bind to proteins and initiate unfolding of the tertiary structure. Anionic surfactants such as SDS denature proteins more than do cationic surfactants. Nonionic surfactants such as the Tritons and Tergitols, on the other hand, do not alter the tertiary structure of the protein. The unfolding of the protein usually occurs in the region of the isotherm where a significant increase in surfactant binding by nonspecific cooperative interactions begins.

Methods which have been used to determine the degree of protein denaturation include the amount of sulfhydryl group liberated,[69,70] the change in optical activity determined by techniques such as CD and ORD,[71-75] and the degree of inhibition of invertase.[76] Yang, Satake, Shirahama and their respective co-workers have studied poly-amino acid-surfactant pairs extensively using these techniques and further reference to their work can be found in Chapter 4. Miyazawa et al.[77] have recently shown that a simple GPC (gel permeation chromatography) technique can be used to determine the degree of surfactant-induced denaturation of a protein. The applicability of the latter technique depends essentially on the protein-surfactant complex having a different elution peak from that of the protein and the surfactant.

SDS is considered to be a strong protein denaturant. In a series of dodecyl ethoxy $(EO)_n$ sulfates, Ohbu et al.[78] showed that the tendency to denature BSA decreased with increase in n and finally disappeared for homologs containing more than six EO groups. In general, denaturation by SDS can be reduced by addition of nonionic or amphoteric surfactants[77,79] and, as stated above, these effects are due essentially to the effect the additives have on binding of SDS or are a consequence of mixed micelle formation and reduced monomer activity.

The mechanism involved in the unfolding of proteins upon surfactant binding has been the subject of many investigations. The driving force for unfolding is thought to come from one or both of the following:[3,61] (1) electrostatic repulsion among charges of bound ligands which results in increased repulsion and eventual opening up of the structure; (2) penetration of hydrophobic tails of the surfactant into the apolar regions of the protein with resultant replacement of segment-segment hydrophobic interactions by interactions. A schematic of the sequence of events which lead to opening up of the structure is shown in Figure 20.

Note that all proteins are not denatured by anionic surfactants. For example, pepsin, papain, glucose oxidase, and bacterial catalase are not denatured by SDS.[80,81] Interestingly, these proteins do not bind significant amounts of SDS.[80]

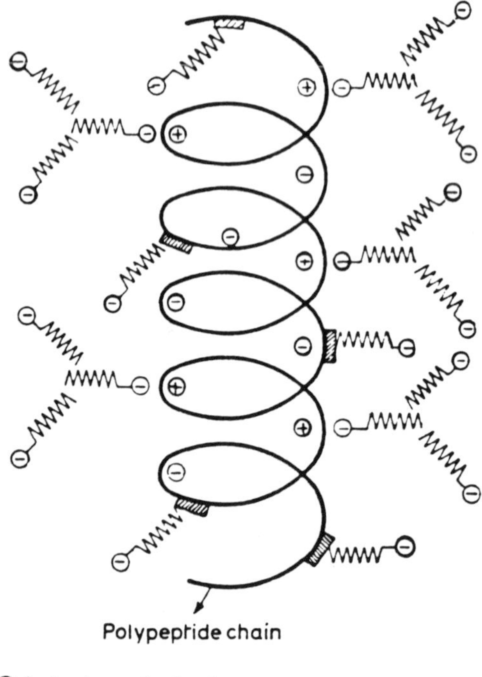

FIGURE 20. A diagrammatic representation of modes of binding of an anionic surfactant to a protein. (From Dominguez, J. G. et al., *J. Soc. Cosmet. Chem.*, 28, 165, 1977. With permission.)

V. STRUCTURE OF PROTEIN-SURFACTANT COMPLEXES

Information on the structure of protein-surfactant complexes has come from rheological,[16,82,83] spectroscopic,[84-87] electrophoretic,[88] binding,[4,28] and scattering studies.[89-93] In this section, the results of spectroscopic, scattering, and electrophoretic studies will be briefly reviewed.

Some of the results discussed earlier should be kept in mind while attempting to analyze the structure of protein-surfactant complexes. Even though the binding of an anionic surfactant such as SDS to water-soluble proteins differs with individual protein in the initial part of the binding isotherm, the saturation binding appears to be independent of the protein.[4,28] Furthermore, some of the results for the dependence on pH showed that the maximum binding, i.e., in the cooperative region of the isotherm, is relatively independent of pH.[49] These results suggest that the binding of SDS to proteins is nonspecific in nature. Dye solubilization and fluorescence results show that protein-surfactant complexes can actually solubilize water-insoluble dyes.[18] The latter behavior is commonly exhibited by surfactant micelles and this observation is an indication that protein-surfactant complexes possess micelle-like environments. This is further supported by the results of ESR[84] and NMR[85] measurements of the SDS-BSA system. Some of the general conclusions from these studies are (1) at low binding (ν <10 mole/mole of protein), both the surfactant headgroup and hydrophobic group are associated with the protein; (2) as ν values increase from 10 to about 50, the nature of the aggregate changes continuously until it becomes similar to that of micelles, as judged by the environment reported by the probe molecule; (3) as ν increases from 50 to about 180, the environment of SDS molecules remains essentially unchanged. Two other conclusions made from this study are that the mobility and environment of initially bound SDS did not change even after denaturation, and the secondary and tertiary structure also remain essentially intact even after denaturation. Oakes,[85] on the basis of NMR, ESR, and data from the ORD literature, has argued in favor of a micelle-like cluster on the protein. Other NMR studies[86,87] using sodium trifluorododecylsulfate and sodium *p*-butylphenol butane 1-sulfonate with BSA yielded chemical shifts in line with the formation of micelle-like aggregates in protein-surfactant complexes.

Reynolds and Tanford,[4,82] on the basis of their intrinsic viscosity studies of several protein surfactant complexes, suggested that the complex in the presence of excess SDS behaves similarly to rods whose length varies uniquely with the molecular weight of the protein. An assumption in the analysis was a compact structure of the aggregates. The simplest model proposed by the authors was a prolate ellipsoid. The semiminor axis of the ellipsoid was found to be constant for several protein-SDS pairs and equal to 18 Å. The major axis, on the other hand, was proportional to the molecular weight of the protein.

Weber and Osborn[94] studied the polyacrylamide gel electrophoresis of about 40 proteins in the presence of excess SDS and showed that the log (molecular weight) followed a linear dependence on mobility. This is the basis of SDS-PAGE electrophoresis for the determination of molecular weight of proteins. These results indicate that the charge per unit mass of protein-surfactant complexes is constant and that the hydrodynamic properties are a function of molecular length only. These conclusions are in accord with the model proposed by Reynolds and Tanford.[82]

Using the free-boundary electrophoresis technique, Shirahama et al.[88] showed that the mobility of protein-SDS complexes was independent of protein molecular weight. Interestingly, they also showed that the complex formed between SDS and a flexible synthetic polymer, PVP, exhibits electrophoretic mobilities independent of molecular weight and the values were essentially the same as that of protein-surfactant complexes. Moreover, the mobilities of protein-surfactant and polymer-surfactant complexes were comparable to that of SDS micelles. Based on these results, Shirahama et al.[88] proposed an alternative model of the complex according to which micelle-like aggregates formed along the protein or the polymer. This model, called the ''necklace model'', assumes a flexible chain which is free draining with respect to the solvent. The latter is the key difference between the rod-like model and the necklace model.

Wright et al.[95] and Rowe and Steinhardt[96] independently studied the electrical birefringence properties of protein-surfactant complexes. Wright et al. argued in favor of the rigid, prolate ellipsoid model for low-molecular-weight proteins and indicated that for high-molecular-weight proteins the necklace model and the prolate ellipsoid model become indistinguishable. Rowe et al., on the other hand, found that their results were inconsistent with a rigid prolate model, but did not find evidence either in support of or against the necklace model.

Mattice et al.[75] have pointed out that neither of the above models can account for the observed effects of SDS on the optical activity of the protein-surfactant complex. Also, some of the abnormal hydrodynamic behavior of SDS complexes with such proteins as lysozyme and collagen cannot be satisfactorily explained by the above models. To account for the differences in CD spectra of various protein-surfactant complexes, Mattice et al. assumed that arginyl, histidyl, and lysyl residues have an enhanced probability of propagating a helical segment in the presence of the detergent and developed a statistical model for complex formation. Interestingly, the model predictions showed that some of the complexes with helical content as high as 50% closely resembled a random coil rather than a rigid rod.

Recently, Lundahl et al.[97] proposed a theoretical model comprising a flexible capped cylindrical micelle around which hydrophilic segments of the protein chain are helically wound. The main binding force between the protein and the surfactant was suggested to be hydrogen bonding. A schematic representation[93] of the three types of structures proposed for protein-surfactant complexes is shown in Figure 21.

Some of the recent neutron scattering,[89-91] quasi-elastic light scattering,[92] and fluorescence/dynamic fluorescence results[98] appear to support the necklace model of protein-surfactant and polymer-surfactant complexes.[17] Guo et al.[90,91] and Tanner et al.[92] have conducted extensive small-angle neutron scattering (SANS) studies of BSA-dodecylsulfate and ovalbumin-dodecylsulfate systems to probe their size and microstructure. Absolute intensities were analyzed using a fractal model. Analysis of their results in the large Q range of the SANS distributions supported the idea that the surfactant molecules bound to the protein actually form micelle-like clusters. Analysis of the data in the small Q range, on the other hand, indicated that the arrangement of micelle-like clusters resembled a fractal packing of spheres. The protein-SDS complex was characterized essentially by four parameters extracted from the scattering experiments, namely, the average micelle size and aggregation number, the fractal dimension characterizing the conformation of the micellar chains, the correlation length giving the extent of unfolding of the polypeptide chains, and the number of micelle-like clusters in the complex.

(a) Necklace Model

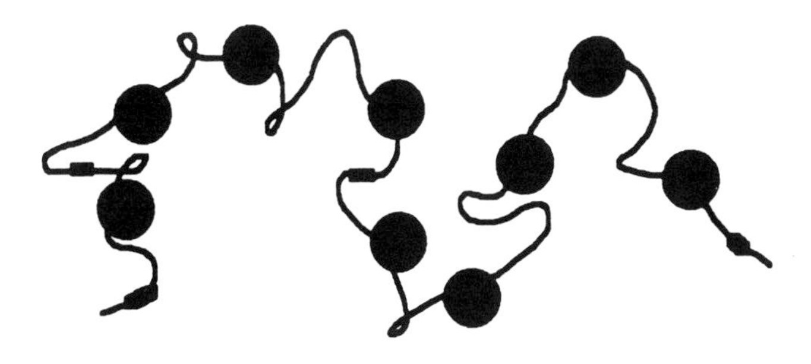

(b) Rod-like Model

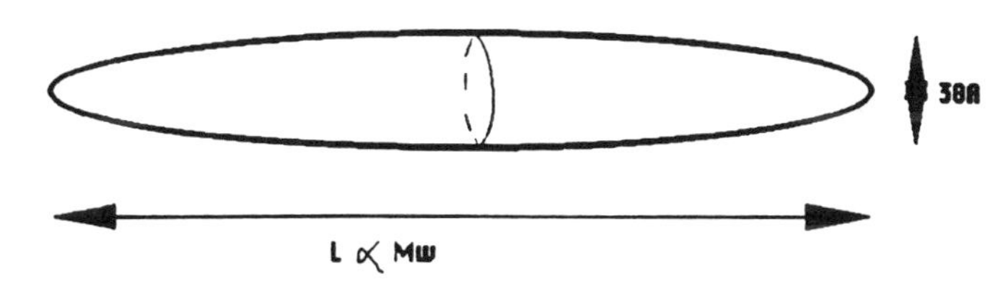

(c) Flexible Helix Model

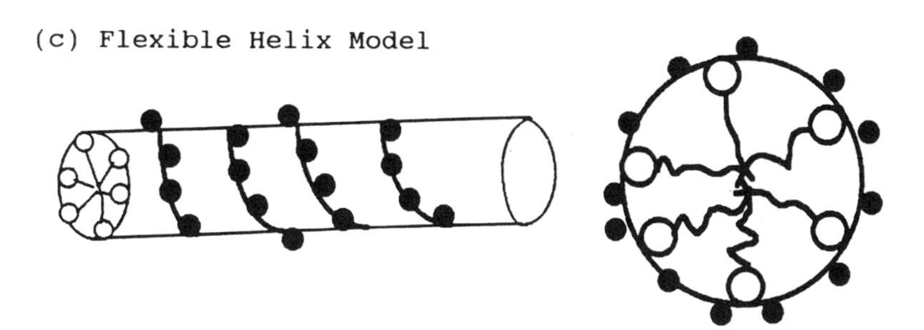

FIGURE 21. Schematic representation of the three types of structures proposed for the polymer-surfactant aggregates. (From Guo, X. H. and Chen, S. H., *Phys. Rev. Lett.*, 64(21), 2579, 1990. With permission.)

The "necklace and bead" structure of the protein-surfactant complex deduced from SANS results[90-92] is similar to those obtained by Cabane[99] for the PEO-SDS system and by Leung et al.[100] for a cationic cellulosic polymer-SDS system. This is consistent with Shirahama's observation that the protein-surfactant complexes and polymer-surfactant complexes behaved similarly in their free boundary electrophoresis studies.

The parameters used to fit the neutron-scattering data and the average size for a BSA-SDS system under two ionic strength conditions are reproduced from Reference 91 in Table 1. It appears that the size of the aggregate is smaller than that of regular micelles. Interestingly, a change in ionic strength

TABLE 1
Estimates of Various Parameters from the Neutron-Scattering Data

System 1: Ionic Strength = 0.6 *M*, 0.5 *M* NaCl, 0.1 *M* acetate buffer pD = 4.9

BSA/SDS	D	ξ	a	b	N	AGG
1/1	2.15	82	23	18	8	29
1/1.5	2.00	120	25	18	8	43
1/2	1.75	155	26	18	11	42
1/3	1.66	211	26	18	18	39

System 2: Ionic Strength = 0.2 *M*, 0.15 *M* NaCl, 0.02 *M* Na Phosphate, pD = 7.2

BSA/SDS	D	ξ	a	b	N	AGG
1/1	1.44	71	24.5	18	6	38
1/1.5	1.27	101	27	18.8	8	43
1/2	1.17	120	27	19	11	42
1/3	1.05	156	28	19	14	50

Note: Unit for ξ, a, and b is Å. D, fractal dimension; ξ, correlation length; a and b, semimajor and semiminor axes of an ellipsoid; N, number of micelle-like clusters in a protein-surfactant complex; AGG, mean aggregation number of micelle-like cluster.

From Guo, X. H., Zhao, N. M., Chen, S. H., and Teixeira, J., *Biopolymers,* 29, 335, 1990. With permission.

from 0.6 to 0.2 *M* did not alter the aggregate size markedly. Note that light-scattering studies show that an increase in the ionic strength from 0.2 to 0.6 *M* changes the aggregation number of regular micelles from about 100 to well over 200.[101]

Another technique which has become popular for determining the size of surfactant aggregates is the fluorescence decay method (see Chapter 9). Recently, this technique was used to determine the size of BSA-SDS aggregates[98] under conditions similar to those employed by Chen and Teixeira[89] and Guo et al.[90,91] in their SANS studies. Results of this study, tabulated in Table 2, show that the size of aggregates is smaller than that of regular micelles. The derived size of the complex, though somewhat larger than that obtained from SANS studies, is essentially in the same range.

The size of micelles formed in protein-surfactant complexes is similar to that found for TTAB micelles formed on polyacrylic acid in aqueous solution.[17] The latter measurements were obtained using the transient fluorescence

TABLE 2
Aggregation Number of SDS Micelles in Solution and on BSA as Determined by Pyrene Excimer Decay Measurements[90]

System	Aggregation number
1% SDS in water	66
1% SDS in buffer	108
1% BSA + 0.25% SDS in buffer	29
1% BSA + 0.5% SDS in buffer	49
1% BSA + 1.0% SDS in buffer	52

Note: Buffer: 0.15 *M* NaCl, 0.02 *M* Na phosphate, pH = 7.0.

measurements under conditions of excimer formation of pyrene. Static fluorescence measurements in this case showed that micelle-like aggregates on the polymer formed at binding fractions as low as 0.05 and the dynamic measurements showed that the size did not change markedly along the entire isotherm. These results are, therefore, consistent with the necklace model of the protein/polymer-surfactant complex.

A. SURFACTANT BINDING TO PROTEINS VS. POLYMERS

It is clear from the discussion above that many protein-surfactant and polymer-surfactant complexes have similar structures, namely a ''necklace'' structure with surfactant micellar beads decorating the polyelectrolyte chain. Typical examples are BSA-SDS and the polyacrylic acid-TTAB complexes. These general conclusions are valid under conditions of, or close to, saturation binding — in other words, in the cooperative binding region.

Typical protein-surfactant binding isotherms differ from the corresponding polymer-surfactant isotherms (compare Figure 7 with binding isotherms in Chapters 4 and 5). Binding isotherms with proteins often exhibit the initial so-called high-energy, binding region before the cooperative region. For polyelectrolytes such as polyacrylic acid or dextran sulfate, on the other hand, only a sharp rising region is observed. This difference stems from the fact that the polymers have available essentially all of their ''high-energy'' sites for binding. The strong electrostatic field raises the local concentration of the surfactant in the vicinity of the polymer chain to high enough values to induce cooperative interactions at very low binding. Thus, both the high-energy and the cooperative binding occur in the same concentration region. Proteins, on the other hand, have only a limited number of sites with charges opposite to those of surfactants. Furthermore, because of their secondary and tertiary structure, proteins have only some of the available sites exposed. For these reasons, the cooperative region occurs at higher concentrations than the initial high-energy region.

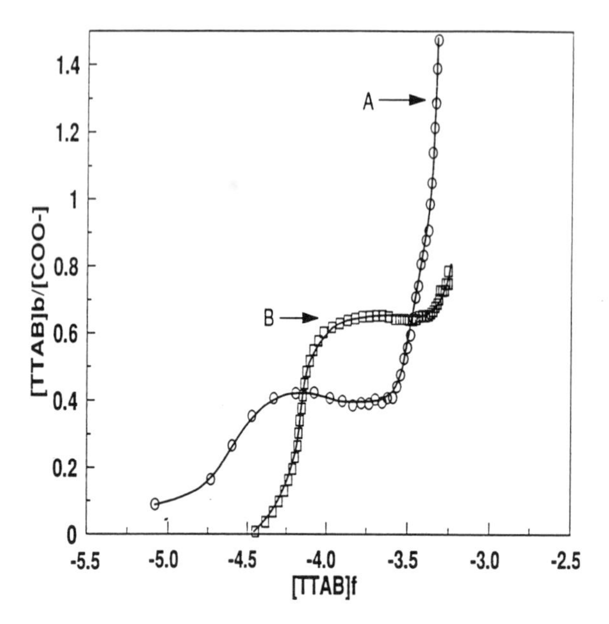

FIGURE 22. Effect of charge density on the binding of tetradecyltrimethyl ammonium bromide to polymethacrylic acid in 10^{-2} mol/l NaBr at 25°C. Curve A: I (Degree of ionization) = 0.25; Curve B: I = 1.0. Degree of ionization is changed by changing the solution pH. (From Kiefer, J. et al., submitted to *Langmuir,* 1992.)

Similar to proteins, polymers with a folded structure and limited number of charged groups may exhibit isotherms with distinct high-energy and co-operative regions. In fact, Kiefer et al.[24] have shown that the binding isotherms of TTAB onto polymethacrylic acid, a hydrophobic polymer, under low charge density conditions resemble those of protein-surfactant systems (see Figure 22). Under low charge density conditions, PMA exhibits a compact coil structure with hydrophobic and hydrogen bonding interactions. Therefore, surfactant binding during the initial stages involves only exposed sites, but with continued binding, the structure opens up in a way similar to that shown by protein.

It appears that it may be possible to tailor-make polymers with appropriate hydrophobic and amphoteric properties to exhibit complex "protein-like" structures which can be opened up in stages by surfactant or other ligand binding and, at the same time, develop interesting new technological applications.

VI. RELEVANCE AND APPLICATIONS OF PROTEIN-SURFACTANT INTERACTIONS

As mentioned earlier, the binding of surfactant to insoluble keratinous proteins, such as those in stratum corneum or hair, has relevance to personal care products. Surfactant-induced damage to stratum corneum and possible ways of prevention were discussed in an earlier section on surfactant binding to insoluble proteins. Indeed, methods to manipulate and control the binding of surfactants to proteins could be of interest in a wide variety of applications.

Complex formation between proteins and SDS has been used to determine the molecular weight of proteins by the so-called SDS-PAGE electrophoresis technique. The electrophoretic mobility of the protein-surfactant complex is proportional to the logarithm of molecular weight.[94] As mentioned earlier, Reynolds and Tanford[82] discussed the theoretical basis for this technique using their prolate ellipsoid model. Guo and Chen[93] have recently proposed a reptation model for the movement of SDS-protein complexes in polyacrylamide gels. This model assumes a flexible necklace structure. Essentially, the protein-SDS complex is assumed to move in a worm-like motion in a tube within the three-dimensional mesh of the gel. The tube is constituted of successive pores. Interestingly, this analysis leads to a simple expression for the mobility, μ, of the complex: $\mu = A/M$, where M is the molecular weight and A is constant for a given gel concentration and approximately inversely proportional to the gel concentration. According to this expression, the logarithm of molecular weight is proportional to the logarithm of mobility rather than being linearly dependent on mobility as indicated by previous analysis. Guo and Chen,[93] using the literature data for several protein-SDS complexes, have validated the above expression. This correlation with experimental data constitutes additional support for the flexible chain necklace model.

Similarities in the structure of protein-surfactant and polymer-surfactant complexes suggest that it may be possible to use the SDS-PAGE electrophoresis technique for the determination of molecular weights of water-soluble polymers, especially those with hydrophobic and amphoteric groups for which determination may be difficult by the conventional techniques.

Interactions of surfactants with proteins have been exploited in a number of protein separation and purification processes. One such process is the extraction of proteins into reverse micelle systems, formed with surfactant, oil, and water.[102-105] The extraction is normally carried out in a two-phase region in which a reverse micellar phase is in equilibrium with an aqueous phase. Hatton[105] has proposed a model for the separation process in which the proteins and micelles are considered to be in equilibrium with the protein-micelle complex. Some of the surfactants used for this purpose are Aerosol OT and trioctylmethylammonium chloride. This novel technique is providing an area of active current research. In such applications, it is important that the surfactant by itself does not denature the protein.

The ability of SDS to denature and solubilize serum proteins is made use of in the high-performance liquid chromatography (HPLC) determination of drugs in blood.[106,107] In conventional reverse-phase chromatography, the protein often precipitates at the head of the column, causing severe experimental problems. SDS, on the other hand, prevents precipitation as well as displacing drug bound to the protein.

Another protein separation method involves the use of the so-called "aqueous two-phase systems" in which either two incompatible polymers such as polyethylene glycol (PEG) and dextran, or a polymer-salt system, such as PEG and sodium sulfate, are used.[108-111] In the separation of alkaline protease in a PEG-salt system, the limiting process is the solubility of the protein in the PEG phase.[112] In this case, addition of SDS increased the solubility of the protein in the PEG phase without denaturing the protein.

The above few examples are given only to illustrate how the interaction of proteins with surfactants can be effectively used to manipulate the solution behavior of proteins to achieve desired end results and applications.

VII. THERMODYNAMICS OF BINDING

The binding of surfactants to proteins can be treated as a multiple-equilibrium phenomenon[3,11] which can be written in terms of the protein (P), the surfactant (R), and the complexes (PR_n) as follows:

$$P + R \rightleftharpoons PR_1 \tag{2}$$

$$PR_1 + R \rightleftharpoons PR_2 \tag{3}$$

$$PR_{n-1} + R \rightleftharpoons PR_n \tag{4}$$

where n is the maximum number of sites available on the protein. For such multiple equilibria, if the equilibrium constants K for all steps are identical, then,

$$K^n = \frac{[PR_n]}{[P][R]^n} \tag{5}$$

The number of surfactant molecules bound per protein molecule, ν, is given by:

$$\nu = \frac{n[PR_n]}{[P] + [PR_n]} = \frac{n(K[R])^n}{1 + (K[R])^n} \tag{6}$$

For the case where cooperative interactions may exist, Hill[113] has suggested the following modification:

$$\nu = \frac{n(K[R])^{n'}}{1 + (K[R])^{n'}} \tag{7}$$

If n' is equal to one there is no cooperativity. A value of n' less than one represents "anticooperativity" and n' greater than one corresponds to coop-

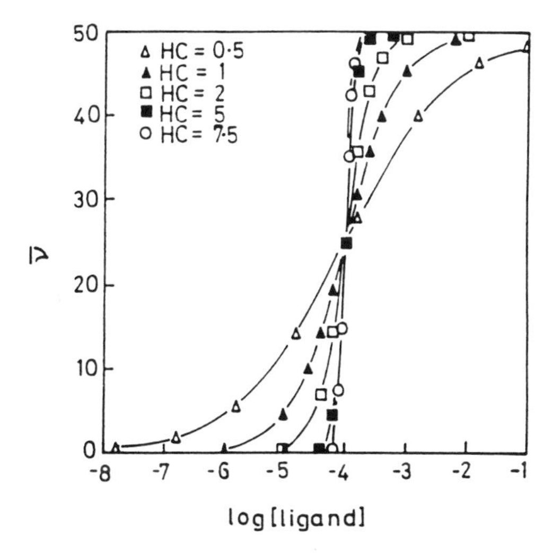

FIGURE 23. Binding isotherms (ν vs. log ligand concentration) calculated from the Hill equation for a protein with 50 binding sites (intrinsic binding constant 10^4) for a range of Hill coefficients (HC, same as n′ in Equation 7) from 0.5 to 7.5. (From Jones, M. N., in *Food Polymers, Gels and Colloids,* Dickinson, E., Ed., Pub. Royal Soc. Chem., 1991. With permission.)

erativity. For identical independent binding sites n′ = 1 and then Equation 7 can be written in the Scatchard[3,9,114] form as:

$$\nu/[R] = K(n - \nu) \tag{8}$$

and can be rearranged to give:

$$1/\nu = 1/n + 1/(nK)\{1/[R]\} \tag{9}$$

Thus, a plot of $1/\nu$ vs. $1/[R]$ can be used to determine the number of binding sites as well as the equilibrium constant, K. The Scatchard equation is identical in form to the Langmuir equation for the adsorption of ligands onto a solid surface.[9]

Hypothetical binding isotherms for a molecule with 50 binding sites and an intrinsic binding constant, K, equal to 10^4 for various values of n′, are shown in Figure 23.[11] These sigmoidal plots clearly show that with increase in cooperativity the binding isotherms become steeper. Scatchard plots derived from these plots are diagnostic of the type of cooperativity[11] (see Figure 24). Negative curvature and maxima are characteristic of negative and positive cooperativity, respectively.

Since the binding of surfactants to proteins involves formation of micelle-type aggregates along the backbone as described by the necklace structure, Scatchard-type analysis will not adequately account for the isotherm behavior,

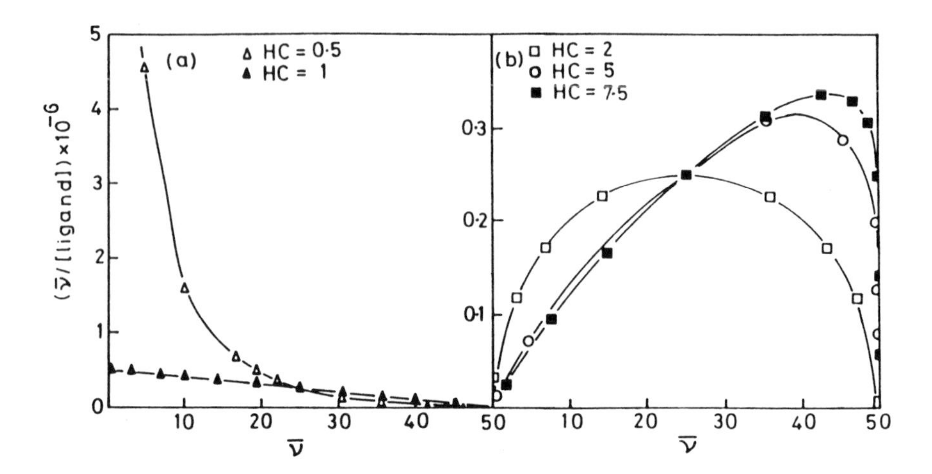

FIGURE 24. Scatchard plots ($\nu/[\text{ligand}]_{\text{free}}$ vs. ν for the isotherms in Figure 22 for a protein with 50 binding sites (intrinsic binding constant 10^4) for a range of Hill coefficients (HC, same as n' in Equation 7) from 0.5 to 7.5. (From Jones, M. N., in *Food Polymers, Gels and Colloids,* Dickinson, E., Ed., Pub. Royal Soc. Chem., 1991. With permission.)

especially in the cooperative binding region. Scatchard plots are, however, useful in determining the number of initial high-affinity binding sites. Other models, such as those based on site binding originally developed by Zimm and Bragg,[115] and later modified by Satake and Yang,[72] are also of limited applicability in their present form to describe necklace-type protein-surfactant structures. In this regard, the model developed by Nagarajan,[116] which accounts for the influence of polymers on the micellization process is noteworthy, but it needs to be modified to take into account the electrostatic interactions in the system. Some of these models are discussed in more detail in Chapters 4 and 5. One should mention that a thermodynamic model, such as the one developed by Hall,[37] without the need for any specific assumption on the structure of the aggregate itself, can be considered valid for the complexation process. Thus, the standard free energy change involved in the binding can be estimated without invoking any specific model for the process. Tanford[4] has proposed the following equation for a highly cooperative model:

$$\Delta G^{\circ} = RT \ln X_a$$

where X_a represents the mole fraction of the free ligand in solution. Use of this equation for the binding would be equivalent to treating the binding process as being similar to a micellization/phase separation process. Sen et al.[43] and Bull[117] have proposed other equations to determine the binding energy along the isotherm. Wyman[118] proposed the concept of a binding potential and described a procedure to estimate binding constants along the isotherm. Jones[11] has modified the latter to account for the changes in the entropic factors involved in the binding process. A comparison of free energy values obtained from different equations is difficult since they all involve different standard states.

Binding energy values estimated using the above models represent contributions from enthalpic and entropic factors and the latter two contributions can be separated out by either direct measurements of enthalpy changes[27,119] along the binding isotherm or by determining the isotherm at different temperatures and by using the standard equations.[9,43] Direct microcalorimetric measurements show that the binding of SDS to BSA[27] and lysozyme[119] is characterized by exothermic reactions in the initial part of the binding isotherm and the enthalpy increases[119] with increase of chain length and decrease in pH. At higher binding, endothermic contributions arising from the unfolding of protein also become important in the overall process.

Note that the estimated values of enthalpy and entropy are the net contributions from various subprocesses such as micellar dissociation for systems in which the initial concentration is above the CMC, effects arising from the binding, unfolding, changes in solvation, hydrophobic interactions, etc. Separating the overall contributions of enthalpy and entropy into various subprocesses is by no means easy. However, an understanding of the enthalpic and entropic contributions along the entire binding isotherm would be useful for developing an understanding of the binding mechanisms involved in the various regions.

VIII. CONCLUDING REMARKS

Interactions of surfactants with proteins have several similarities to interactions of surfactants with polymers. The amphoteric hydrophobic character of the proteins and their secondary and tertiary structure components, however, make them different from conventional polymer systems. Significant advances have been made in our understanding of the interactions of proteins with surfactants, together with the effect of such variables as surfactant chain length and structure, pH, ionic strength, temperature, and presence of additives. Methods to manipulate and control the binding of surfactants to proteins have essentially come from an understanding the effect of variables on binding.

Interesting differences exist in the binding properties of surfactants to soluble and insoluble proteins. Special attention should be paid to kinetic aspects when dealing with insoluble proteins. Keratin, the protein in stratum corneum and hair, binds anionic surfactants strongly and the extent of binding can be controlled by changing the surfactant structure and by additives. In general, factors which lower the CMC of the surfactant reduce surfactant binding to proteins.

The use of various scattering and probe techniques has resulted in a better understanding of the microstructure of protein-surfactant aggregates. While the exact structure of protein-surfactant complexes is still a subject of research, available results seem to support the string-of-beads structure of the complex, i.e., the formation of micelle-like clusters on the flexible protein chain under conditions of cooperative binding. The role of the inorganic counterion, for example of sodium in the case of SDS, in complex formation has yet to be addressed.

The mechanisms involved in the unfolding of proteins are understood only in a qualitative way and a need exists for the development of appropriate models which can quantitatively model such processes. The kinetics of formation and dissolution of the complex by techniques such as pressure jump, temperature jump, etc., may provide valuable information in this regard. Most studies on protein-surfactant interactions have been carried out using SDS as the surfactant. Interactions of other surfactants with different hydrophilic groups and different overall structures will be of practical importance to study. In short, several aspects of protein-surfactant interactions continue to provide challenging opportunities for researchers in this area.

REFERENCES

1. **Goddard, E. D.,** Polymer-surfactant interaction. I. Uncharged water soluble polymers and charged surfactants, *Colloids Surf.,* 19, 255, 1986.
2. **Goddard, E. D.,** Polymer-surfactant interaction. II. Polymer and surfactant of opposite charge, *Colloids Surf.,* 19, 301, 1986.
3. **Steinhardt, J. and Reynolds, J. A.,** *Multiple Equilibria in Proteins,* Academic Press, New York, 1969.
4. **Tanford, C.,** *The Hydrophobic Effect: Formation of Micelles and Biological Membranes,* 2nd ed., Wiley-Interscience, New York, 1980, chap. 14.
5. **Schwuger, M. J. and Bartnik, F. G.,** Interaction of anionic surfactants with proteins, enzymes and membranes, in *Anionic Surfactants,* Surfactant Sci. Ser. Vol. 10, Gloxhuber, C., Ed., Marcel Dekker, New York, 1980, 1.
6. **Makino, S.,** Interactions of proteins with amphipathic substances, in *Advances in Biophysics,* Vol. 12, Kotani, M., Ed., Japanese Science Society Press, Tokyo, and University Park Press, Baltimore, 1979, 131.
7. **Robb, I. D.,** Polymer-surfactant interactions, in *Anionic Surfactants, Physical Chemistry of Surfactant Action,* Surfactant Sci. Ser. Vol., Lucassen-Reyenders, Ed., Marcel Dekker, New York, 1981, 109.
8. **Jones, M. N.,** Physicochemical studies on the interactions between surfactants and globular proteins, *Commun. Jorn. Com. Esp. Deterg.,* 14, 117, 1983; CA 99 101122.
9. **Chattoraj, D. K. and Birdi, K. S.,** *Adsorption and the Gibbs Surface Excess,* Plenum Press, New York, 1984, 339.
10. **Saito,** Polymer-surfactant interactions, in *Nonionic Surfactants,* Surfactant Sci. Ser. Vol. 23, Schick, M. J., Ed., Marcel Dekker, New York, 1987, 881.
11. **Jones, M. N.,** Interactions between small amphipathic molecules and proteins, in *Food Polymers, Gels and Colloids,* Dickinson, E., Ed., Pub. Royal Soc. Chem., 1991; see also Surfactant interactions with biomembranes and proteins, *Chem. Soc. Rev.,* 21, 127, 1992.

12. **Cockbain, E. G.,** The interfacial activity and composition of bovine serum albumin + sodium dodecyl sulfate complexes, *Trans. Faraday Soc.,* 49, 104, 1953.
13. **Knox, W. L., Jr. and Parshall, T. O.,** The interaction of sodium dodecyl sulfate with gelatin, *J. Colloid Interface Sci.,* 33, 16, 1970.
14. **Jones, M. N.,** The interactions of SDS with polyethylene oxide, *J. Colloid Interface Sci.,* 23, 36, 1967.
15. **Nishikido, N., Takahara, T., Kobayashi, H., and Tanaka, M.,** Interaction between hydrophilic proteins and nonionic detergents by surface tension measurements, *Bull. Chem. Soc. Jpn.,* 55, 3085, 1982.
16. **Greener, J., Constestable, B. A., and Bale, M. D.,** Interaction of anionic surfactants with gelatin: viscosity effects, *Macromolecules,* 20, 2490, 1987.
17. **Kiefer, J., Somasundaran, P., and Ananthapadmanabhan, K. P.,** Size of tetradecyltrimethyl ammonium bromide aggregates on polyacrylic acid in solution by dynamic fluorescence, paper presented at the P & G UERP Symp. at the ACS meeting, New York, 1991, in press.
18. **Steinhardt, J., Scott, J. R., and Birdi, K. S.,** Differences in the solubilizing effectiveness of the sodium dodecyl sulfate complexes of various proteins, *Biochemistry,* 16, 718, 1977.
19. **Ananthapadmanabhan, K. P., Kuo, P. L., Goddard, E. D., and Turro, N. J.,** Fluorescence probes for CMC determination, *Langmuir,* 1, 352, 1985.
20. **Ananthapadmanabhan, K. P., Goddard, E. D., and Leung, P. S.,** Fluorescence studies of polymer-surfactant systems, *Colloid Surf.,* 13, 63, 1985.
21. **Rendall, H. M.,** Use of surfactant selective electrode in the measurement of the binding of anionic surfactants to bovine serum albumin, *J. Chem. Soc. Faraday Trans. 1,* 72, 481, 1976.
22. **Kresheck, G. C. and Constantinidis, I.,** Ion selective electrodes for octyl and decyl sulfate surfactants, *Anal. Chem.,* 56, 152, 1984.
23. **Hayakawa, K., Ayub, A. L., and Kwak, J. C. T.,** The application of surfactant selective electrodes to the study of surfactant adsorption in colloidal suspension, *Colloids Surf.,* 4, 389, 1982.
24. **Kiefer, J., Somasundaran, P., and Ananthapadmanabhan, K. P.,** Interaction of tetradecyltrimethyl ammonium bromide with polyacrylic acid and polymethacrylic acid — effect of charge density, submitted to *Langmuir,* 1992.
25. **Cutler, S. G., Meares, P., and Hall, D. J.,** Ion activities of sodium dodecyl sulfate solutions from electromotive measurements, *J. Chem. Soc. Faraday Trans. 1,* 74, 1758, 1978.
26. **Jones, M. N. and Manley, P.,** Thermodynamic studies on the interaction between lysozyme and sodium-n-dodecyl sulfate in aqueous solutions, in *Surfactant Solutions,* Vol. 2, Mittal, K. L. and Lindman, B., Eds., Plenum Press, New York, 1984, 1403.
27. **Goddard, E. D. and Pethica, B. A.,** On detergent-protein interactions, *J. Chem. Soc.,* 2659, 1953.
28. **Reynolds, J. A. and Tanford, C.,** Binding of dodecyl sulfate to proteins at high binding ratios. Possible implications for the state of proteins in biological membranes, *Proc. Natl. Acad. Sci. U.S.A.,* 66, 1002, 1970.
29. **Takagi, T., Tsujii, K., and Shirahama, K.,** Binding isotherms of SDS to protein polypeptides with special reference to SDS-polyacrylamide gel electrophoresis, *J. Biochem. (Tokyo),* 77, 939, 1975.
30. **Jones, M. N., Manley, P., and Midgley, P. J. W.,** Adsorption maxima in a protein surfactant solution, *J. Colloid Interface Sci.,* 82, 257, 1981.
31. **Koga, J., Chen, K. M., Yamazaki, Y., and Kuroki, N.,** Binding saturation of n-alkyl trimethylammonium bromide on bovine serum albumin, *J. Colloid Interface Sci.,* 91, 283, 1983.

32. **Sexsmith, F. H. and White, H. J., Jr.,** The absorption of cationic surfactants by cellulosic materials. I. The uptake of cation and anion by a variety of substrates, *J. Colloid Interface Sci.,* 14, 598, 1959; II. The effects of esterification of the carboxyl groups in the cellulosic substrates, *J. Colloid Interface Sci.,* 14, 619, 1959; III. A theoretical model for the absorption process and a discussion of maxima in absorption isotherms for surfactants, *J. Colloid Interface Sci.,* 14, 630, 1959.
33. **Mukerjee, P. and Anavil, A.,** Adsorption of ionic surfactants to porous glass: the exclusion of micelles and other solutes from adsorbed layers and the problem of adsorption maxima, in *Adsorption at Interfaces,* AChS Symposium Ser. No. 8, Mittal, K. L., Ed., The American Chemical Society, Washington, D.C., 1975, 109.
34. **Somasundaran, P., Celik, M. S., and Goyal, A.,** Precipitation and redissolution of sulfonates and their role in adsorption on minerals, in *Surface Phenomena in Enhanced Oil Recovery,* Shah, D. O., Ed., Plenum Press, New York, 1981, 641.
35. **Ananthapadmanabhan, K. P. and Somasundaran, P.,** Mechanism of adsorption maximum and hysteresis in sodium dodecylbenzene sulfonate/kaolinite system, *Colloids Surf.,* 7, 105, 1983.
36. **Trogus, F. J., Schechter, R. S., and Wade, W. H.,** A new interpretation of adsorption maxima and minima, *J. Colloid Interface Sci.,* 70, 293, 1979.
37. **Hall, D. J.,** Thermodynamics of ionic surfactant binding to macromolecules, *J. Chem. Soc. Faraday Trans. 1,* 81, 885, 1985.
38. **Reynolds, J. A., Herbert, S., and Steinhardt, J.,** The binding of some long chain fatty acid anions and alcohols by BSA, *Biochemistry,* 7, 1357, 1968.
39. **Nozaki, Y., Reynolds, J. A., and Tanford, C.,** The interaction of a cationic detergent with bovine serum albumin and other proteins, *J. Biol. Chem.,* 249(14), 4452, 1974.
40. **Kaneshina, S., Tanaka, M., and Kondo, T.,** Interaction of bovine serum albumin with detergent cations, *Bull. Chem. Soc. Jpn.,* 46, 2735, 1973.
41. **Makino, S., Reynolds, J. A., and Tanford, C.,** The binding of deoxycholate and Triton X-100 to proteins, *J. Biol. Chem.,* 248(14), 4926, 1973.
42. **Jones, M. L. and Manley, P.,** Binding of n-alkyl sulfates to lysozymes in aqueous solution, *J. Chem. Soc. Faraday Trans. 1,* 75, 1736, 1979.
43. **Sen, M., Mitra, S. P., and Chattoraj, D. K.,** Thermodynamics of binding of anionic detergents to bovine serum albumin, *Indian J. Biochem. Biophys.,* 17, 370, 1980.
44. **Chatterjee, R. and Chattoraj, D. K.,** Hydrophobic interactions of DNA with long chain amines, *Biopolymers,* 18, 147, 1979.
45. **Chattoraj, D. K., Bull, H. B., and Chaleley, R.,** Binding of histone to DNA, *Arch. Biochem. Biophys.,* 152, 778, 1972.
46. **Chandar, P.,** Effect of NaCl on the Adsorption of Sodium Dodecyl Sulfate onto Alumina, M.S. thesis, Columbia University, New York, 1983.
47. **Bitting, D. and Harwell, J. H.,** Effects of counterions on surfactant surface aggregates at the alumina/aqueous solution interface, *Langmuir,* 3, 500, 1987.
48. **Aoki, K. and Foster, J. F.,** Electrophoretic behavior of bovine plasma albumin at low pH, *J. Am. Chem. Soc.,* 79, 3385, 1957.
49. **Reynolds, J. A., Gallagher, J. P., and Steinhardt, J.,** Effect of pH on the binding of n-alkyl sulfates to bovine serum albumin, *Biochemistry,* 9, 1232, 1970.
50. **Subraminian, M., Sheshadri, B. S., and Venkatappa, M. P.,** Interaction of cationic detergents, cetyl- and dodecyl-trimethylammonium bromides with lysozymes, *J. Biochem.,* 95, 413, 1984.
51. **Manning, G. S.,** Limiting laws and counterion condensation in polyelectrolyte solutions, *J. Chem. Phys.,* 51, 294, 1969.
52. **Mitra, S. P. and Chattoraj, D. K.,** Thermal stability and excess free energy of hydration of BSA in the presence of neutral salts, *Indian J. Biochem. Biophys.,* 15, 239, 1979.
53. **Sen, M., Mitra, S. P., and Chattoraj, D. K.,** Thermodynamics of binding of cationic surfactants to bovine serum albumin, *Colloids Surf.,* 2, 259, 1981.

54. **Griffith, J. C. and Alexander, A. E.,** Equilibrium adsorption isotherms for wool detergent systems, *J. Colloid Interface Sci.*, 25, 311, 1967; see also *J. Colloid Interface Sci.*, 25, 317, 1967.
55. **Harrold, S. P. and Pethica, B. A.,** Thermodynamics of the adsorption of small molecules by proteins, *Trans. Faraday Soc.*, 54, 1876, 1958.
56. **Blank, I. H. and Gould, E.,** Penetration of anionic surfactants into skin, *J. Invest. Dermatol.*, 33, 327, 1959.
57. **Faucher, J. A. and Goddard, E. D.,** Interaction of keratinous substrates with sodium lauryl sulfate: sorption, *J. Soc. Cosmet. Chem.*, 29, 323, 1978.
58. **Kulkarni, R. D. and Goddard, E. D.,** Destruction of the electrophysiological potential of excised frog skin by surfactants, in *Advances in Bioelectrochemistry: Ions, Surfaces, Membranes,* Adv. Chem. Ser. No. 188, Blank, M., Ed., the American Chemical Society, Washington, D.C., 445.
59. **Imokawa, G. and Takeuchi, T.,** Surfactants and skin-roughness, *Cosmet. Toil,* 91, 32, 1976; Imokawa, G., Comparative study on the mechanism of irritation by sulfate and phosphate type anionic surfactants, *J. Soc. Cosmet. Chem.*, 31, 45, 1980.
60. **Breuer, M. M.,** The interaction between surfactants and keratinous tissues, *J. Soc. Cosmet. Chem.*, 30, 41, 1979.
61. **Dominguez, J. G., Parra, J. L., Infante, R. M., Pelejero, R. M., Balaguer, F., and Sastre, T.,** A new approach to the theory of adsorption and permeability of surfactants on keratinic proteins: specific behavior of certain hydrophobic chains, *J. Soc. Cosmet. Chem.*, 28, 165, 1977.
62. **Conrads, A., and Zahn, H.,** A study of the interaction of sodium dodecyl sulfate with the proteins of human heel stratum corneum, *Int. J. Cosmet. Sci.*, 9, 29, 1987.
63. **Baden, H. P. and Kuedar, J. C.,** *The Nail in Physiology, Biochemistry and Molecular Biology of the Skin,* Vol. I, Goldsmith, L. A., Ed., Oxford University Press, 1991, p. 697.
64. **Cho, S.,** personal communication, Unilever U. S., NJ, 1991.
65. **Rosen, M. J.,** *Surfactants and Interfacial Phenomena,* 2nd ed., John Wiley & Sons, New York, 1989.
66. **Rosen, M. J. and Zhu, B. Y.,** Synergism in binary mixtures of surfactants. III. Betaine-containing systems, *J. Colloid Interface Sci.*, 99(2), 427, 1984.
67. **Faucher, J. A., Goddard, E. D., and Kulkarni, R. D.,** Effect of polyoxyethylated materials on the interaction of surfactants with skin, *J. Am. Oil Chem. Soc.*, 56, 777, 1979.
68. **Faucher, J. A. and Goddard, E. D.,** Sorption of a cationic polymer, by stratum corneum, *J. Soc. Cosmet. Chem.*, 27, 543, 1976.
69. **Van Scott, E. J. and Lyon, J. B.,** A chemical measure of the effect of soaps and detergents on the skin, *J. Invest. Dermatol.*, 21, 199, 1953.
70. **Harrold, S. P.,** Denaturation of epidermal keratin by surface active agents, *J. Invest. Dermatol.*, 32, 581, 1959.
71. **Satake, I. and Yang, J. T.,** Effect of temperature and pH on the β helix transition of poly(L-lysine) in SDS, *Biopolymers,* 14, 1841, 1975.
72. **Satake, I. and Yang, J. T.,** Interaction of SDS with L-ornithine and poly L-lysine, *Biopolymers,* 15, 2263, 1976.
73. **Hayakawa, K., Fujita, M., Yokoi, S., and Satake, I.,** Conformation of poly(L-lysine) and poly(L-ornithine) in α,ω-type surfactant solutions, *J. Bioactive Compatible Polym.*, 6, 36, 1991.
74. **Imokawa, G., Sumara, K., and Katumi, M.,** Study on skin roughness caused by surfactants, *J. Am. Oil Chem. Soc.*, 52, 484, 1975.
75. **Mattice, W., Riser, J. M., and Clark, D. S.,** Conformational properties of the complexes formed by proteins and sodium dodecyl sulfate, *Biochemistry,* 15, 4264, 1976.

76. **Okamoto, K., Goda, K., and Kanda, Y.,** Evaluation of the skin-roughness caused by aqueous solutions of surfactants — relation between the inhibition of invertase activity by surfactant and the skin roughness, *Yakagaku,* 21, 151, 1972.
77. **Miyazawa, K., Ogawa, M., and Mitsui, T.,** The physico-chemical properties of protein denaturation potential of surfactant mixtures, *Int. J. Cosmet. Sci.,* 6, 33, 1984.
78. **Ohbu, K., Jona, N., Miyajima, M., and Kashiwa, I.,** Evaluation of the denaturing effect of surfactants on proteins as measured by circular dichroism, *Yakagaku,* 29, 866, 1980. CA: 94:26394x(1981).
79. **Ernst, R.,** Surface active betaines as protective agents against denaturation of an enzyme by alkyl sulfates, *J. Am. Oil. Chem. Soc.,* 57, 93, 1980.
80. **Nelson, C. A.,** The binding of detergents to proteins, *J. Biol. Chem.,* 246, 3895, 1971.
81. **Jones, M. N., Manley, P., Midgley, P. J. W., and Wilkinson, A. E.,** Dissolution of bovine and bacterial catalase by Na-n-dodecyl-sulfate, *Biopolymers,* 21, 1435, 1982.
82. **Reynolds, J. A. and Tanford, C.,** The gross conformation of protein-sodium dodecyl sulfate complexes, *J. Biol. Chem.,* 245, 5161, 1970.
83. **Rao, P. F. and Yakagi, T.,** Reassessment of the viscosity behavior of sodium dodecyl sulfate-protein polypeptide complexes, *J. Biochem.,* 106, 365, 1989.
84. **Laurie, O. and Oakes, J.,** Protein-surfactant interactions, Spin label study of interactions of BSA and SDS, *J. Chem. Soc. Faraday Trans. 1,* 1324, 1975.
85. **Oakes, J.,** Protein-surfactant interactions, NMR and binding isotherm studies of interactions between bovine serum albumin and sodium dodecyl sulfate, *J. Chem. Soc. Faraday Trans. 1,* 70, 2200, 1974.
86. **Smith, M. L. and Muller, N.,** Fluorine chemical shifts in complexes of sodium trifluoralkylsulfates with reduced proteins, *Biochem. Biophys. Res. Commun.,* 62(3), 723, 1975.
87. **Tsujii, K. and Takagi, T.,** Proton magnetic resonance studies of the binding of an anionic surfactant with a benzene ring to a protein polypeptide with special reference to SDS-polyacrylamide gel electrophoresis, *J. Biochem.,* 77, 511, 1975.
88. **Shirahama, K., Tsujii, K., and Takagi, T.,** Free boundary electrophoresis of sodium dodecyl sulfate-protein polypeptide complexes with special reference to SDS-polyacrylamide gel electrophoresis, *J. Biochem.,* 75, 309, 1974.
89. **Chen, S. H. and Teixeira, J.,** Structure and fractal dimension of protein-detergent complexes, *Phys. Rev. Lett.,* 57, 2583, 1986.
90. **Guo, X. H., Zhao, N. M., Chen, S. H., and Teixeira, J.,** Small-angle neutron scattering study of the structure of protein/detergent complexes, *Biopolymers,* 29, 335, 1990.
91. **Guo, X. H. and Chen, S. H.,** The structure and thermodynamics of protein-SDS complexes in solution and the mechanism of their transport in gel electrophoresis process, *Chem. Phys.,* 1990.
92. **Tanner, R. E., Herpigny, B., Chen, S. H., and Rha, C. K.,** Conformational change of protein sodium dodecyl sulfate complexes in solution: a study of dynamic light scattering, *J. Chem. Phys.,* 76, 3866, 1982.
93. **Guo, X. H. and Chen, S. H.,** Reptation mechanism in protein-sodium dodecylsulfate polyacrylamide-gel electrophoresis, *Phys. Rev. Lett.,* 64, 2579, 1990.
94. **Weber, K. and Osborn, M.,** The reliability of molecular weight determination by dodecyl sulfate-polyacrylamide gel electrophoresis, *J. Biol. Chem.,* 244, 4406, 1969.
95. **Wright, A. K., Thompson, M. R., and Miller, R. L.,** A study of protein-sodium dodecyl sulfate complexes by transient electric birefringence, *Biochemistry,* 14, 3224, 1975.
96. **Rowe, E. S. and Steinhardt, J.,** Electrooptical properties of reduced protein-sodium dodecyl sulfate complexes, *Biochemistry,* 15, 2579, 1976.
97. **Lundahl, P., Greijier, E., Sandberg, M., Cardell, S., and Eriksson, K. O.,** A model for ionic and hydrophobic interactions and hydrogen bonding in SDS-protein complexes, *Biochim. Biophys. Acta,* 873, 20, 1986.

98. **Lei, L. S., Ananthapadmanabhan, K. P., Turro, N. J., and Aronson, M.,** Static and transient fluorescence studies of BSA-SDS interactions, unpublished results.
99. **Cabane, B.,** Structure of some polymer-surfactant aggregates in water, *J. Phys. Chem.,* 81, 1639, 1977.
100. **Leung, P. S., Goddard, E. D., Han, C., and Glinka, C. J.,** *Colloids Surf.,* 13, 47, 1985.
101. **Zana, R.,** Luminescence probe techniques, in *Surfactant Solutions,* Surfactant Sci. Ser. Vol. 22, Zana, R., Ed., Marcel Dekker, New York, 1987, 241.
102. **Meier, P., Imre, E., Fleschar, M., and Luisi, P. L.,** in *Surfactants in Solution,* Mittal, K. L. and Lindman, B., Eds., Plenum Press, New York, 1984, 999.
103. **Dekker, M., Baltussen, J. W. A., Riet, K. V., Bijsterbosch, B. H., Laane, C., and Hilhorst, R.,** *Stud. Org. Chem. (Amsterdam),* 29, 285, 1987. CA: 108:203241a.
104. **Goklen, K. E. and Hatton, T. A.,** Liquid-liquid extraction of low molecular weight proteins by selective solubilization in reversed micelles, *Sep. Sci. Technol.,* 22, 831, 1987.
105. **Hatton, T. A.,** Reversed micellar extraction of proteins, in *Surfactant Based Separation Processes,* Surfactant Sci. Ser. Vol. 33, Scamehorn, J. F. and Harwell, J. H., Eds., Marcel Dekker, New York, 1989, 55.
106. **Wetlaufer, D. B.,** Surfactants in novel separation techniques, in *Surfactants in Emerging Technologies,* Rosen, M. J., Ed., Surfactant Sci. Ser. Vol. 26, Marcel Dekker, New York, 1987.
107. **DeLuccia, F. J., Arunyart, M., and Cline, L. J.,** Direct serum injection with micellar chromatography, for therapeutic drug monitoring, *Anal. Chem.,* 57, 1564, 1985.
108. **Albertsson, P.,** *Partition of Cell Particles and Macromolecules,* Wiley-Interscience, New York, 1971.
109. **Walter, H., Brooks, D. E., and Fisher, D.,** *Partitioning in Aqueous Two Phase Systems,* Academic Press, New York, 1985.
110. **Ananthapadmanabhan, K. P. and Goddard, E. D.,** Aqueous biphase formation in polyethylene oxide-inorganic salt systems, *Langmuir,* 3, 25, 1987.
111. **Ananthapadmanabhan, K. P. and Goddard, E. D.,** The relationship between clouding and aqueous biphase formation in polymer solutions, *Colloids Surf.,* 25, 393, 1987.
112. **Ananthapadmanabhan, K. P. and Goddard, E. D.,** Method for increasing the solubility of enzymes, U.S. Patent 4,738,925, 1988.
113. **Hill, A. V.,** *J. Physiol.,* 40, 40p, 1910.
114. **Scatchard, G.,** The attraction of proteins for small molecules and ions, *Ann. N.Y. Acad. Sci.,* 51, 660, 1949.
115. **Zimm, B. H. and Bragg, J. K.,** Theory of the one dimensional phase transition in polypeptide chains, *J. Chem. Phys.,* 13, 526, 1959.
116. **Nagarajan, R.,** Thermodynamics of nonionic polymer-micelle association, *Colloids Surf.,* 13, 1, 1985.
117. **Bull, H. B.,** Adsorption of bovine serum albumin on glass, *Biochim. Biophys. Acta,* 19, 464, 1956.
118. **Wyman, J.,** The binding potential, a neglected linkage concept, *J. Mol. Biol.,* 11, 631, 1965.
119. **Jones, M. N. and Manley, P.,** Interaction between lysozyme and n-alkyl sulfates in aqueous solution, *J. Chem. Soc. Faraday Trans. 1,* 76, 654, 1980.

Chapter 9

APPLICATIONS OF FLUORESCENCE SPECTROSCOPY TO THE STUDY OF POLYMER-SURFACTANT INTERACTIONS

Françoise M. Winnik

TABLE OF CONTENTS

0-8493-6784-0/93/$0.00 + $.50

I. INTRODUCTION

In Chapter 4, Goddard introduced the use of fluorescence probe experiments to investigate polymer-surfactant systems. Here the objective is to review experimental aspects of these techniques in more detail, to describe several new methods involving new probes or polymers covalently labeled with fluorescent substituents, to discuss the strengths of fluorescent probe and fluorescent labeling experiments, and to emphasize some of their inherent problems and limitations. The kind of information attainable is illustrated through examples taken from recent publications. For specific questions related to the spectroscopy or the data treatment, the reader is referred to general books on fluorescence spectroscopy[1,2] and to reviews on the applications of photophysical and photochemical techniques in the study of polymers[3-6] and surfactants.[7-11]

II. THE TOOLS AND TECHNIQUES

Two approaches are possible in applying fluorescence techniques to the study of polymer-surfactant interactions. In probe experiments one simply adds a fluorescent dye to the solution. This type of experiment is useful if the dye binds to the polymer or the polymer-surfactant aggregates. The difficulty with probe experiments is ascertaining the location of the probe in the system and to what extent a probe may disturb the system. Since only a very low level of fluorescent dye incorporation is needed, this is usually not a severe problem. Alternatively, one labels the polymer by attaching a dye covalently. Labeled-polymer experiments are often more informative because they report on phenomena from the aspect of the polymer. For polymers in organic solvents it is usually safe to assume that the label does not perturb the properties of the polymer. For water-soluble polymers the covalently bound dye may act as a hydrophobic substituent. In this situation it is important to test the consequences of varying the extent of labeling on the polymer in solution. Comparison of the results obtained by the two approaches for the same polymer-surfactant system also gives a good measure of the extent of the influence of the label on the properties of the polymer.

A. TECHNIQUES BASED ON THE USE OF FLUORESCENT PROBES

1. Characterization of the Surfactant-Polymer Aggregates

As described in previous chapters there is a vast array of experimental evidence establishing that microstructures formed in polymer-surfactant solutions have properties very different from those of the external aqueous solution. An organic dye added to polymer-surfactant solutions will reside preferentially in hydrophobic microdomains. Questions arise as to the location of the solute with respect to the aggregates. On the average a nonpolar dye may find itself (1) located "inside" the hydrophobic core; (2) adsorbed on the surface; or (3) located in the surface layer. An amphiphilic dye may be oriented with its polar portion in the surface layer and its nonpolar portion in the hydrophobic core. While average locations of probes in surfactant micelles are known with a reasonable degree of confidence, little information exists on the detailed location of probes in surfactant-polymer aggregates. Nevertheless, from the changes in the photophysical properties of a probe and the known behavior of the probe in various homogeneous solvents or surfactant micelles, it is possible to gain reliable new insight into the structure of the aggregates.

a. Fluorescence Emission Spectra

Various fluorescent dyes used in the study of polymer-surfactant aggregates are listed in Table 1. Pyrene is by far the most popular probe. It exhibits a medium-sensitive change in the vibrational fine structure of its emission spectrum. This characteristic of pyrene spectroscopy (the Ham effect) was exploited first by Kalyanasundaram and Thomas in a study of surfactant micelles[12] and has been applied by the groups of Turro[13] and Zana[14] to the study of polymer-surfactant aggregates. The effect is illustrated in Figure 1. The two spectra presented are typical of the emission of pyrene solubilized in either a hydrophilic environment or a hydrophobic environment. The spectra are reproduced from data by Chu and Thomas on the solution properties of poly(methacrylic acid) (PMA) and how these are affected by the presence of cationic surfactants.[15] The phenomena probed are related to the pH-induced conformational transition of PMA:[16] at low pH (<3) the polymer in its free acid form is collapsed into a tight hydrophobic coil. Above pH 4 the polymer coils tend to open. At a high pH, at which the carboxylic acid groups are almost completely ionized (pH >8), the PMA chains are stretched due to electrostatic repulsion among the anionic groups. At low pH, hydrophobic molecules are hosted within the polymer coils and as the polymer uncoils at

TABLE 1
Fluorescent Probes Used in the Study of Polymer-Surfactant Interactions

Probe	Measurement	Ref.
Pyrene (Py)	Fine structure (I_1/I_3), quenching	13, 14, 20, 27-30, 45, 54, 63
1-Pyrenecarboxaldehyde (PyCHO), acridine	Emission maximum shift	62
Anilinonaphthalene sulfonate (ANS), rhodamine 6G (R6G), proflavin	Fluorescence intensity, emission maximum shift	20-22, 49
Sodium 11-(3-hexyl-1-indolyl)-undecylsulfate	Emission maximum shift	13
4-(1-Pyrenyl)-butyrate	Excimer to monomer ratio	34
2-Methylanthracene	Depolarization, quenching	14, 27, 28
Rubidium bispyridyl chloride [$Ru(BiPy)_2^{3+}$]	Quenching	15, 27, 28
Dipyme	Fine structure ($[I_1/I_3]^{DP}$), excimer	17
Acridine	Emission maximum shift	19

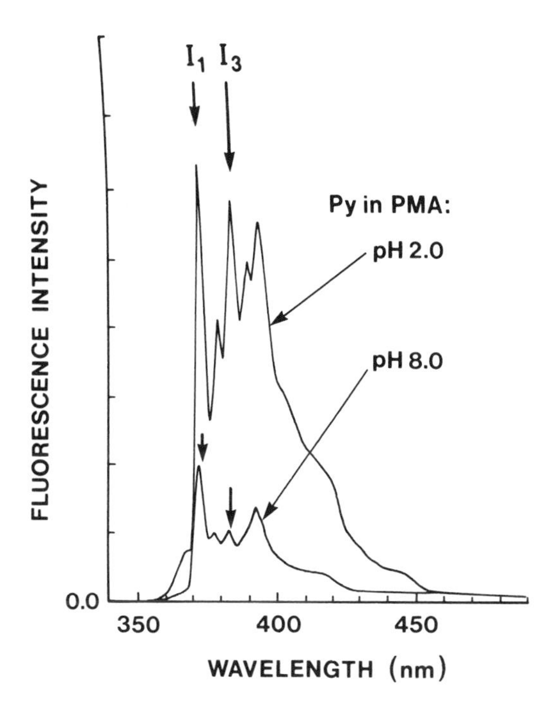

FIGURE 1. Fluorescence spectra of pyrene in aqueous solutions of poly(methacrylic acid) at pH 2 and 8. [Py] = 2×10^{-6} mol l^{-1}; [PMA] = 1 g l^{-1}; λ_{exc} = 340 nm. (Adapted from Kalyanarasundaram, K. and Thomas, J. K., *J. Am. Chem. Soc.*, 99, 2039, 1977. With permission.)

higher pH they are expelled into the aqueous phase. This change in environment is reported by the spectra of pyrene solubilized in PMA solutions. Note the key differences in the two spectra measured for pyrene in PMA at pH 2 (hydrophobic medium) and pH 8 (hydrophilic medium). First, the total fluorescence intensity is lower at pH 8.0 than at pH 2.0. This observation can be attributed to the fact that excited pyrenes in water have a shorter lifetime than those solubilized in hydrophobic media. Second, the relative intensities of the I_1 and I_3 bands (arrows) are affected. The I_1/I_3 ratio is low at pH 2.0, indicating that the probe experiences a hydrophobic environment; it is much larger at pH 8.0, pointing to a polar water-like solvent for the probe. Applications of the technique to polymer-surfactant systems are described in parts 1 and 2 of this chapter and in Chapter 6.

The photophysics of bis(1-pyrenylmethyl) ether (dipyme) provide a unique tool to study the interactions between hydrophobically modified polymers and surfactants.[17] Dipyme is a probe which provides information not only on the polarity but also on the microviscosity of an environment.[18] Dipyme, like the closely related 1,3-bis(1-pyrenyl)propane, forms an intramolecular excimer upon excitation (Figure 2). Excimer formation in both species depends

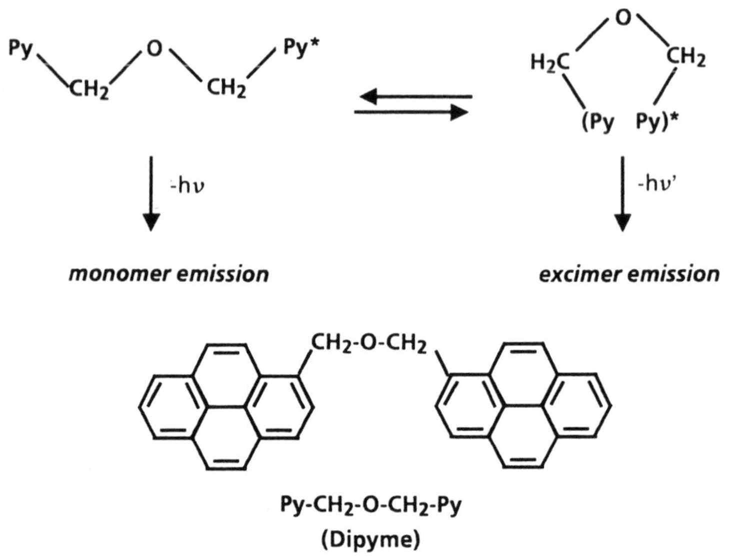

Py CH2 O CH2 Py* ⇄ H2C O CH2 (Py Py)*

-hν -hν'

monomer emission excimer emission

CH2-O-CH2

Py-CH2-O-CH2-Py
(Dipyme)

FIGURE 2. Representation of intramolecular excimer formation in dipyme bis(1-pyrenylmethyl) ether. (From Winnik, F. M. et al., *J. Phys. Chem.*, 95, 2583, 1991. With permission of the American Chemical Society.)

upon a change in conformation of the molecules. The rate of this motion is resisted by the local friction of the environment. As a consequence, the excimer-to-monomer intensity ratio, I_E/I_M, provides a measure of the "microviscosity". In addition, the vibrational fine structure in the dipyme monomer emission is, like that of pyrene itself, sensitive to the polarity of the probe environment. This effect is absent in 1,3-bis(1-pyrenyl)propane and most other 1-substituted pyrene derivatives. Representative emission spectra are shown in Figure 3 for dipyme in *n*-octyl β-D-thioglucopyranoside (OTG) micelles and in a solution of a copolymer of *N*-isopropylacrylamide and *n*-octadecylacrylamide, PNIPAM-C_{18}/100 (Figure 4). It is immediately apparent that the extent of excimer emission differs greatly for dipyme in the two environments. In OTG the ratio I_E/I_M is 0.88, while in PNIPAM-C_{18}/100 it is 0.11. The $[I_1/I_3]^{DP}$ values for the probe in the two solutions are similar, indicating that in both cases dipyme experiences a strongly hydrophobic environment. The lower I_E/I_M value measured for the probe in the solution containing polymer points to a much more rigid structure of the polymeric micelles, compared to surfactant micelles.

Other probes exhibit shifts in fluorescence maxima sensitive to the solvent medium. Pyrene-1-carboxaldehyde has been used in this context (see Part 3, this chapter). Other useful dyes include acridine,[19] anilino-naphthalenesulfonate (ANS),[20] rhodamine 6G (R6G), and proflavin.[21,22] For several of these dyes the spectral shifts are accompanied by significant changes in fluorescence

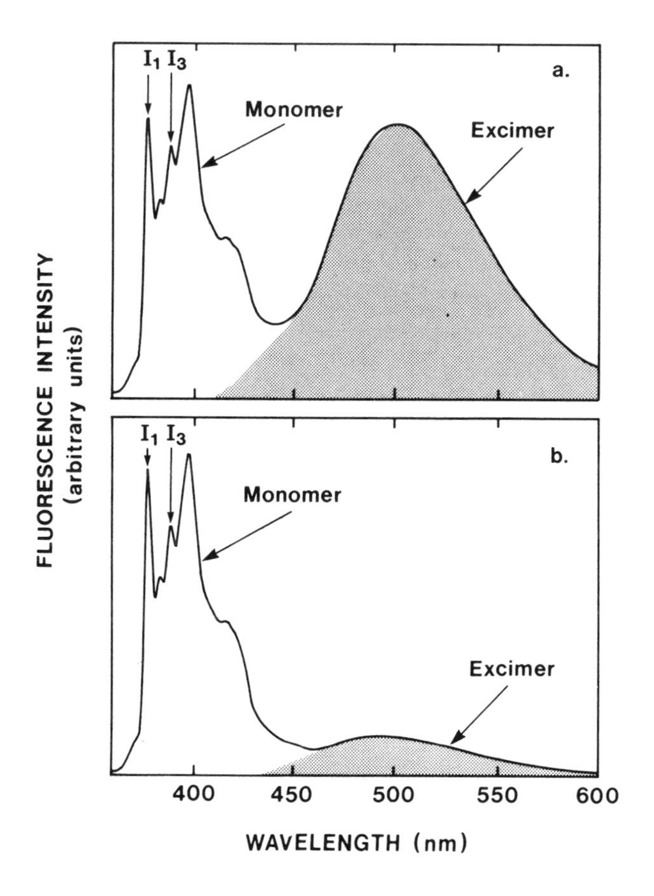

FIGURE 3. Fluorescence spectra of dipyme (a) in micelles of *n*-octyl β-D-thioglucopyranoside (1.67×10^{-2} mol l^{-1}) and (b) in an aqueous solution of PNIPAM-C_{18}/100 (1.87 g l^{-1}); 20°C, $\lambda_{exc} = 348$ nm; [dipyme] $= 2 \times 10^{-7}$ mol l^{-1}. (From Winnik, F. M. et al., *Langmuir,* 7, 905, 1991. With permission of the American Chemical Society.)

quantum yields (Φ_f) and fluorescence lifetimes (τ). Generally the quantum yields increase with decreasing solvent polarity. However it should be kept in mind that in addition to the polarity of the environment, many other factors may influence Φ_f. The presence of impurities, for example, may lead to fluorescence quenching which is difficult to distinguish from polarity effects on fluorescence intensity. Hence, caution should be exercised in the interpretation of experiments where one observes only changes in total emission intensity.

b. Fluorescence Depolarization

For a chromophore to absorb light a component of its transition moment must be parallel to the electric vector of the incident light. As a consequence, irradiating a sample of randomly oriented molecules with plane-polarized light

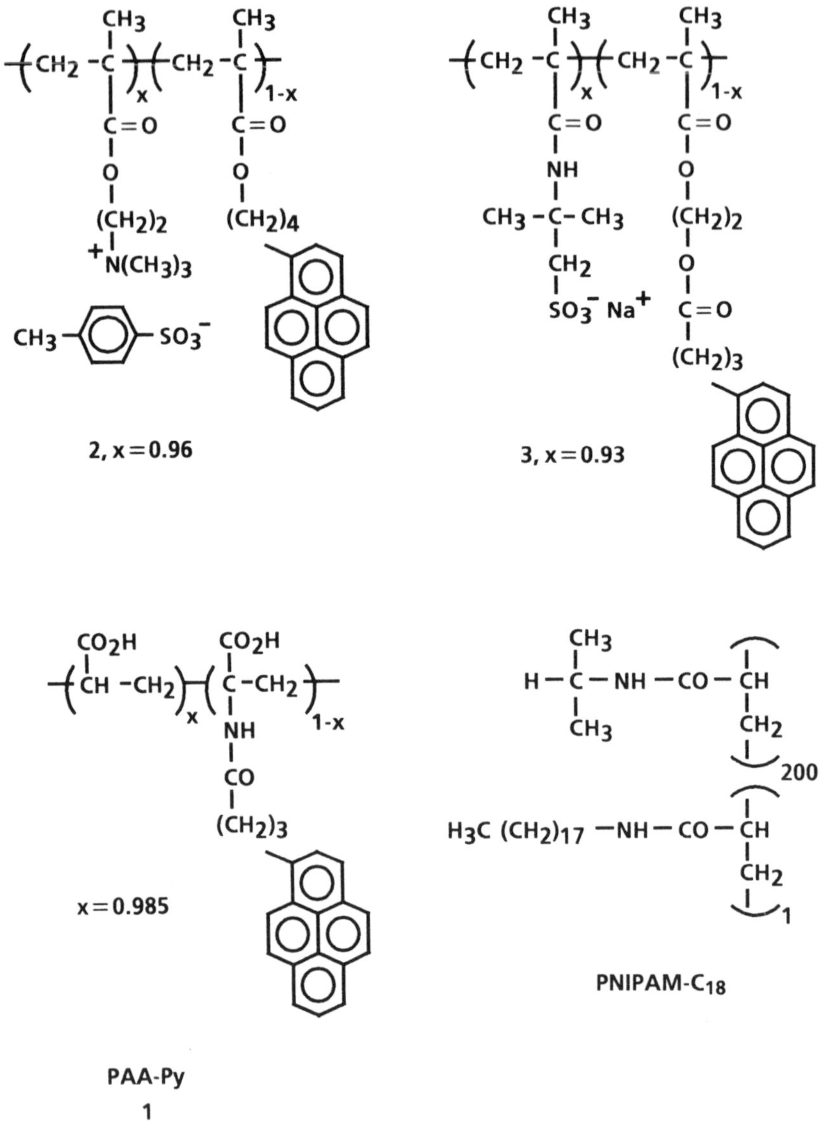

CH_3 CH_3
$-(CH_2-C)_x-(CH_2-C)_{1-x}-$
$C=O$ $C=O$
O O
$(CH_2)_2$ $(CH_2)_4$
$^+N(CH_3)_3$
$CH_3-C_6H_4-SO_3^-$

2, x=0.96

CH_3 CH_3
$-(CH_2-C)_x-(CH_2-C)_{1-x}-$
$C=O$ $C=O$
NH O
CH_3-C-CH_3 $(CH_2)_2$
CH_2 O
$SO_3^-\ Na^+$ $C=O$
$(CH_2)_3$

3, x=0.93

CO_2H CO_2H
$-(CH-CH_2)_x-(C-CH_2)_{1-x}-$
NH
CO
$(CH_2)_3$

x=0.985

PAA-Py

1

CH_3
$H-C-NH-CO-CH$
CH_3 CH_2
200
$H_3C\ (CH_2)_{17}-NH-CO-CH$
CH_2
1

PNIPAM-C_{18}

FIGURE 4. Chemical structures of polymers used in studies of polymer-surfactant interactions by fluorescence techniques.

results in an optical selection of which molecules are excited. If the motion of these molecules is slow compared to the timescale of fluorescence, the emitted light will also be polarized, and its degree of polarization will be related to the extent of motion of the molecule. In this way fluorescence depolarization can be used to determine the apparent microviscosity of a

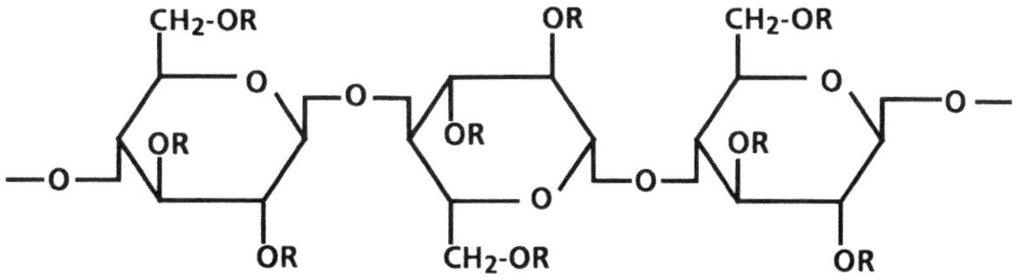

HPC $R = CH_2\text{-}CH(OH)\text{-}CH_3$

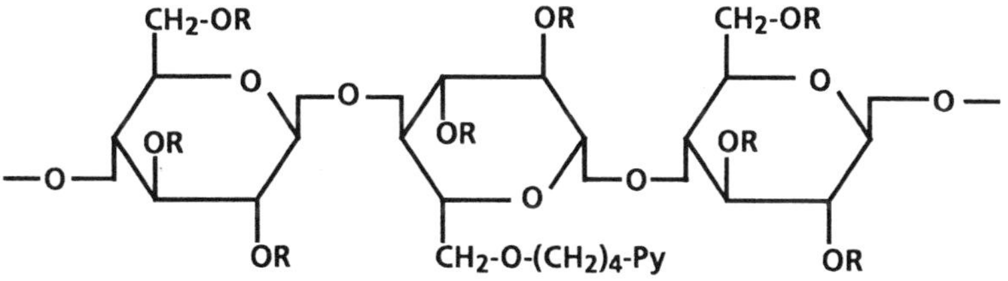

HPC-Py

FIGURE 4 (continued)

medium containing a suitable fluorescent probe. One can measure the time profile of polarization decay in a time-resolved experiment or one can measure the steady-state polarization with a fluorescence spectrometer equipped with polarizers.

In a steady-state experiment, one measures the fluorescence intensities detected through polarizers oriented parallel ($I_{\parallel}$) and perpendicular ($I_{\perp}$) to the plane of polarization of the excitation light beam. From these, one calculates either the steady-state polarization parameter P or, more meaningfully, the emission anisotropy, r (Equations 1 and 2).

$$P = \frac{I_{\parallel} - I_{\perp}}{I_{\parallel} + I_{\perp}} \tag{1}$$

$$r = \frac{I_{\parallel} - I_{\perp}}{I_{\parallel} + 2I_{\perp}} \tag{2}$$

To interpret these results one normally treats the probe as a spherical rotor, so that its motion can be described in terms of a single correlation time. This assumption is almost certainly not true in detail, but is still very useful for following changes in the mobility of the probe in different systems. In isotropic solvents, where the spherical rotor assumption is reasonably good, one often finds that the correlation time is related to η_o, the bulk solvent viscosity. Shinitzky et al.[23] were the first to apply these kinds of measurements to probes in microheterogeneous systems, such as surfactant micelles. The microviscosity, η, of the environment of a probe is estimated using Equation 3, where P_o and r_o are the limiting values of the polarization and emission anisotropy obtained in the absence of rotational motion, τ is the average lifetime of the fluorophore's excited state, k is the Boltzmann constant, T the absolute temperature, and V_o the effective molecular volume.

$$\frac{r_o}{r} = \frac{\frac{1}{P} - \frac{1}{3}}{\frac{1}{P_o} - \frac{1}{3}} = 1 + \frac{kT\tau}{\eta V_o} \tag{3}$$

Usually a calibration curve of r_o/r vs. $T\tau/\eta$ is constructed by employing a reference solvent series. By comparing the anisotropy of the chromophore in a given medium to that of the anisotropy of the same chromophore in a series of solvents of known viscosity, it is possible to assess the apparent microviscosity of the probe environment. Several assumptions are made in this treatment and care must be exercised in data interpretation.[24] The dye 2-methylanthracene has been used to study polymer/surfactant aggregates.[14] Other chromophores, such as perylene and diphenylhexatriene, which have proven to be extremely useful probes in the study of surfactant micelles, could be applied as well to polymer/surfactant systems.

2. Determination of Mean Aggregation Numbers ($\overline{N}$)

Fluorescence quenching methods have proven to be among the most reliable techniques for the measurement of the mean aggregation number, $\overline{N}$, of surfactant micelles. They are based on the quenching of a luminescent probe by a hydrophobic quencher. In these experiments one adds a quencher (Q) to a solution containing a known amount of surfactant and a very small amount of a fluorophore (F). Both F and Q are chosen to have a high affinity for the micelles and, in the data analysis, one assumes a Poisson distribution of F and Q among the micelles. In this way one obtains some micelles containing F but not Q, F and Q, and various levels of multiple occupancy. Quenching occurs only in micelles containing both F and Q.

Steady-state experiments — In a steady-state experiment one assumes instantaneous quenching of F* by Q so that the extent of decrease of fluorescence intensity provides a measure of the number of micelles containing both F and Q. Under these circumstances the ratio of luminescence intensities (I_o/I) in the absence of quencher to that in the presence of quencher can be related to $\overline{N}$ by the relation:

$$In \frac{I_o}{I} = \frac{[Q]}{[cluster]} = \frac{\overline{N}[Q]}{[surfactant] - [cac]} \quad (4)$$

where [Q] is the total quencher concentration, [cluster] is the concentration of polymer/surfactant aggregates, [surfactant] is the total surfactant concentration, and [cac] is the concentration of free surfactant molecules not incorporated into clusters or micelles. This methodology was developed originally by Turro and Yekta[25] for anionic micelles and has since been extended to cationic and nonionic micelles.[26] It works well if the following experimental criteria are met: (1) the micelles are not too large ($\overline{N}$ <100); (2) the fraction of both Q and F located in the aqueous phase is negligibly small; (3) F and Q do not form ground-state complexes; (4) Q is a very efficient quencher of F* (i.e., diffusion-controlled quenching); and (5) at the concentrations employed Q has a negligible absorbance at the wavelength used to excite F. Chromophore/quencher pairs used in the determination of the aggregation numbers of the surfactant/polymer aggregates include $Ru(bipy)_3^{2+}$ (F)/9-methylanthracene (Q),[27,28] pyrene (F)/benzophenone (Q),[29] 11-(3-hexyl-1-indolyl)-undecyl sulfate (F)/Co(II) (Q),[13] and pyrene (F)/dimethylbenzophenone (Q).[30] Note that the technique also permits one to determine the number of alkyl groups involved in the micellar structures formed in aqueous solutions of hydrophobically modified polymers, as described by Strauss in Chapter 6.

Fluorescence decay experiments — In time-resolved experiments one takes into account the idea that quenching of F* by Q occurs with a rate described by a quenching rate coefficient, k_q. Under these circumstances, the intensity decay profile is given by the following expression (Equation 5), where τ_o is the emission lifetime of the probe, and $\bar{n}$ is the mean occupation number, related to the mean aggregation number $\overline{N}$ by Equation 6. The underlying assumptions of the data treatment have been described in detail for the case of surfactant micelles.[8,31]

$$I(t) = I(0) \exp[-\frac{t}{\tau_o} + \bar{n}(e^{-k_q t} - 1)] \quad (5)$$

$$\bar{n} = \frac{[f]\overline{N}}{[cluster] - [cac]} \quad (6)$$

Quenching of excited pyrene by excimer formation (Equation 7) has been reported also as a method for the determination of micelle aggregation numbers. Excimers are formed in micelles containing two or more pyrenes. Although Equation 5 has been used in these experiments, the data analysis does not take into account excimer dissociation to reform excited pyrene. Therefore the $\overline{N}$ values obtained from this method will only be reliable under conditions (low temperature, high microviscosity) where excimer dissociation is relatively unimportant.

$$Py^* + Py \leftrightarrows [Py...Py]^* \rightarrow 2Py \qquad (7)$$

3. Limitations and Common Artifacts Associated with Fluorescence Probe Techniques

Some of the very features that make fluorescent dyes attractive as probes make them also a source of artifacts when applied to polymer/surfactant systems in water. One of the challenges is to establish the absence of such artifacts. One potential source of error is associated with sample preparation. The probes necessarily have a very low solubility in water. They are commonly introduced by injecting a small amount of a solution of the probe in a water-miscible organic solvent such as ethanol, acetone, tetrahydrofuran into the system. Under these conditions, microcrystals almost always form. These will interfere unpredictably in the fluorescence measurements. In the case of dipyme they are a source of severe error since they exhibit a strong excimer emission. Microcrystals disappear slowly as the probe dissolves molecularly in the hydrophobic domains. In this sense the artifact is a kinetic one, which can be monitored by following changes in emission spectra or preferably, especially in the case of dipyme,[32] in excitation spectra, as a function of equilibration times (hours to several days). If too much probe is added, the microdomains will become saturated. Excess probe in the form of microcrystals will remain in the aqueous phase. Separating this mixture is often impossible. It is important, therefore, to ascertain that there is no excess probe. With probes such as pyrene which have modest water solubility, it is usually best to prepare a (filtered) solution in water ([Py] $\sim 7 \times 10^{-7}$ mol l^{-1} at saturation). The polymer and surfactant are added to this solution and allowed to dissolve.

Another potential source of problems is related to the presence of oxygen, a powerful quencher of fluorescence. The solubility of oxygen is low in water compared to organic solvents, but it is higher in micellar solutions. Organic solutions can be degassed either by freeze-thaw cycles under high vacuum or by bubbling of nitrogen or argon. Neither technique works well with soapy aqueous solutions. As a consequence, measurements are usually carried out in untreated solutions. In some cases it is necessary to degas solutions, for example when one wants to distinguish quenching by oxygen from the effect of another quencher present in the system. An illustration of this situation can be found in a recent publication by Thalberg et al. in their study of the interactions between cationic surfactants (alkylated trimethylammonium bromides) with an anionic polyelectrolyte (hyaluronan).[30]

B. TECHNIQUES BASED ON THE USE OF FLUORESCENT LABELS

1. Synthesis of Labeled Polymers

Polymers may exhibit intrinsic fluorescence, as in the case of proteins or many water-insoluble polymers. Most water-soluble polymers of interest here have to be modified by incorporation of minute amounts of a fluorophore. Therefore, synthesis of a suitable ''polymer probe'' is the first step of this technique. There are three possible approaches to prepare such polymers: (1) a dye-substituted monomer is added as a comonomer to a polymerization mixture, to yield randomly labeled copolymers; (2) an unlabeled polymer is modified by reaction of a functional group of the polymer with a dye-substituted reagent, either at random along the chain or at specific sites; and (3) a specially synthesized ''reactive polymer'' is treated with an appropriate derivative of a fluorescent dye.

a. Copolymerization of a Dye-Substituted Monomer

Examples of such polymers used in studies of polymer-surfactant interactions include (1) polyelectrolytes, such as pyrene-labeled polyacrylic acid, (3, Figure 4) prepared by free-radical copolymerization in DMF of acrylic acid and 2-[4-(1-pyrene)butanoyl]aminopropenoic acid,[33] or the cationic pyrene-containing methacrylates (1,2),[34] and (2) neutral polymers such as pyrene-labeled hydrophobically modified poly-*N*-isopropylacrylamides.[35] In these syntheses the level of label incorporation is dictated by the initial monomer feeds and the reactivity ratios of the various comonomers.

b. Polymer Postmodification

By this technique commercial or natural polymers can be labeled, often without significant change in molecular weight or molecular weight distribution. In addition, it is sometimes possible to link the dyes to specific sites on the polymers, for example, to one or both endgroups. Chemical reactions employed to prepare labeled polymers by this route include:

- Ether formation — this reaction is applicable if the polymer to be labeled has free hydroxyl groups and if it is stable to strong bases. A polymer is treated with a strong base to generate alkoxides and subsequently reacted with a dye bearing a primary alkyl halide or tosylate substitutent. Since polyalkoxides tend to form gels, it is best to invert the normal order of adding the reagents in a Williamson ether synthesis. Instead of adding the alkyl halide or tosylate to the preformed alkoxide, the halide or tosylate is first dissolved in a solution of the polymer (for example in DMF or dimethyl sulfoxide [DMSO]). A strong base, such as sodium hydride, is then added to the mixture. The technique has been employed to label polysaccharides, such as hydroxypropyl cellulose (HPC).[36]
- Ester formation — the reaction has been employed to label olefin-maleic anhydride copolymers[37] and hydroxylated polymers bearing hydroxyl endgroups, such as poly(ethylene oxide).[38]
- Amide formation — this reaction has been useful to modify copolymers of maleic anhydride and alkyl vinyl ethers.[39]

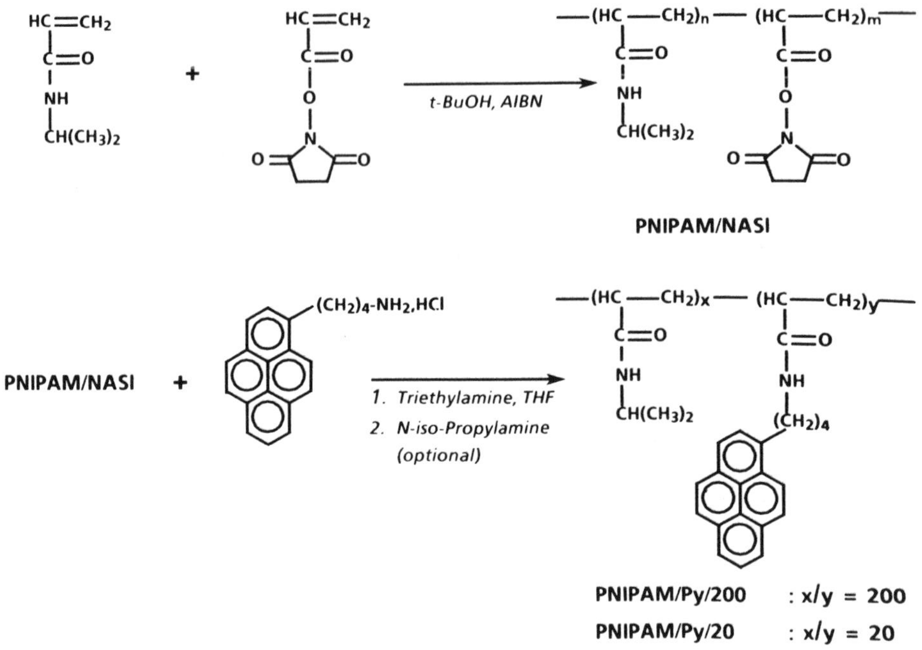

FIGURE 5. Synthetic scheme for the preparation of pyrene-labeled poly(*N*-isopropylacrylamide). (From Winnik, F. M., *Macromolecules,* 23, 233, 1990. With permission of the American Chemical Society.)

c. Labeling of "Reactive" Polymeric Intermediate

This approach is useful to label polymers having no functionality readily amenable to reacting with dye derivatives. The idea is to prepare a polymer carrying at random a small number of "reactive" functional groups. These are sites where fluorescent groups can be introduced. The reactive groups should be stable enough to allow storage of the polymer under normal conditions. The technique is illustrated here in the case of pyrene-labeled poly(*N*-isopropylacrylamide) (PNIPAM-Py, Figure 5).[40] This polymer was prepared by reaction of 4-(1-pyrenyl)butylamine with a random copolymer of *N*-isopropylacrylamide and *N*-(acryloxy)succinimide. By varying the initial ratio of amine to reactive polymer it was possible to synthesize PNIPAM-Py samples with different amounts of pyrene incorporation: after all the initial amine had been consumed, the polymer was treated with an excess isopropyl amine to convert residual unreacted *N*-oxysuccinimide groups to *N*-isopropylamide groups.

The main advantage of the method over the direct copolymerization of dye-substituted monomers (method a) is that it eliminates possible interferences of the chromophores during the polymerization. Aromatic derivatives often act as chain-transfer agents. Also, from a single reactive polymer it is possible to prepare polymers, identical with respect to their molecular characteristics, but labeled with different chromophores or with various amounts of the same chromophore.

2. Spectroscopic Tools

Most experiments on polymer-surfactant systems reported to date rely either on fluorescence depolarization or on pyrene excimer formation. Some other experiments examining changes in emission quantum yields of polymer-bound chromophores are described by Strauss in his review on hydrophobically modified polymers (Chapter 6). Since the techniques themselves have been described in the case of fluorescence probe experiments (*vide supra*), only properties specifically related to the use of labels are mentioned here.

Fluorescence depolarization — There has been an interesting report on the influence of surfactants on the depolarization of fluorescence of polymer-bound anthracene.[37] Three copolymers of α-olefins and maleic acid were labeled with 9-hydroxymethylanthracene (about 1 anthryl group per 390 repeat units). These were alternating copolymers (mol wt, about 20,000, Mn, about 7000) of maleic acid with 1-decene (PDMA), 1-tetradecene (PTMA), and 1-octadecene (POMA). At pH 8 in water they form micellar aggregates. The anthryl groups reside preferentially within the hydrophobic cores. Their anisotropy values (r) therefore give a measure of the local viscosity of these microdomains. Labels covalently attached to macromolecules usually do not satisfy the criteria of the Stokes-Einstein equation, which would permit calculation of the microviscosity of the probe environment from a knowledge of r_o, r, V_o, and τ (see Equation 3). It is still meaningful to follow trends in fluorescence anisotropy. The dependence of r on polymer composition: PDMA, 0.15; PTMA, 0.17; POMA, 0.19 indicates an increase in the local viscosity of the microdomains with increasing chain length of the hydrophobic comonomer. Addition of SDS to copolymer solutions results in each case in a decrease of the fluorescence anisotropy of the anthryl label. The decrease in r is not monotonic. It follows a discontinuous pattern, with a sharp breaking point at [SDS] about 8×10^{-4} mol l^{-1}, and a second transition at higher SDS concentration. The surfactant disrupts the microdomains, which become more fluid as a result of an expansion of the copolymer coils in the presence of surfactant molecules.

Pyrene excimer formation — The ratio of pyrene excimer to monomer emission intensities has been used to follow the interactions of surfactants with polyelectrolytes[34] and neutral polymers, such as PEO-Py,[41-44] HPC-Py,[45] and PNIPAM-Py.[46] In many of these systems excimers originate from preformed pyrene dimers or higher aggregates, rather than from the usual dynamic mechanism involving encounter of an excited pyrene with a pyrene in its ground state. The occurrence of pyrene aggregates in solution can be established from observations such as: (1) hypochromism and shifts in UV absorption spectra; (2) differences in the excitation spectra monitored for the monomer and excimer emissions; and (3) absence of a rising component in the time-dependent profile of the excimer emission, as measured in the nanosecond timescale. The fate of the chromophore aggregates as a function of added surfactant can be monitored easily. The changes one observes provide valuable information on polymer-surfactant interactions. Some of these experiments are described by Lindman and Thalberg in Chapter 5 (Table 2 and 3).

3. Limitations and Common Artifacts Associated with Fluorescence Label Experiments

Technical concerns — The most important requirement in experiments involving labeled polymers is to work with pure materials. Inadvertent presence of residual dye can lead to erroneous results. Repeated precipitation of the polymer is often successful in the separation of low-molecular-weight impurities from the polymer. This approach is not always effective. Other techniques such as dialysis may be more useful in some cases, and as a last resort one can use preparative size exclusion chromatography (SEC). It is helpful and often crucial to monitor the polymer purification by analytical SEC on a system equipped with refractive index and UV or fluorescence detectors in tandem. Characterization of the polymer is also important. In particular the level of dye incorporation should be determined, for example by comparison of UV absorption spectra of the polymer and a suitable model dye of known extinction coefficient. This is not always straightforward since the model compounds are often insoluble in water. Also, in the case of pyrene-labeled polymers the existence of aggregates should be assayed by measurements of absorption and excitation spectra. Another general problem with water-soluble polymers is that of hydrolysis of the dye-polymer linkage. Ester groups are notorious in this respect.[47] Amide or ether linkages are much more stable.

Properties of the polymers — All fluorescence experiments rely on the use of a dye. The question always arises about whether or not the dye perturbs the system under scrutiny. Probes are believed to be relatively unobtrusive at the low concentrations required for fluorescence measurements. This is often not the case for labels attached to water-soluble polymers. The chromophores are by nature hydrophobic and, thus, the labeled polymers may be viewed as "hydrophobically modified" aqueous polymers. As such, they may have solution properties quite different from those of the original polymer, as reported for example in studies of their lower critical solution temperature (LCST) behavior.[48] Little is known on the influence of a dye substituent on the polymer-surfactant interactions. For these experiments it is best to choose a low level of dye incorporation and when possible to compare results of labeled and unlabeled materials.

TABLE 2
Polymer-Surfactant Systems Studied by Fluorescence Probe Techniques: Neutral Polymers

Polymer		Surfactant	Probe	Ref.
Poly(ethylene oxide)	PEO	SDS	Py	13, 14, 28, 52
Poly(propylene oxide)	PPO	SDS, OTG	Py	66
Poly(*N*-vinylpyrrolidone)	PVP	SDS	Py	13, 14, 28
Hydroxypropyl cellulose	HPC	SDS, C_{16}TAC, OTG	Py, Dipyme	29, 45
Poly(*N*-ω-hydroxyalkylglutamines)		SDS	Acridine	19
Poly(*N*-isopropylacrylamide)	PNIPAM	SDS	Py	64
Hydrophobically modified	PNIPAM	SDS, C_{16}TAC	Py, Dipyme	65

TABLE 3
Polymer-Surfactant Systems Studied by Fluorescence Probe Techniques: Polyelectrolytes

Polymer		Surfactant	Probe	Ref.
Sodium polystyrene sulfonate	PSS	C_{12}TAB	Py, ANS	20
Sodium dextran sulfate	DxS	C_{12}TAB	R6G, proflavin	21
Sodium poly(vinyl sulfate)		C_{12}TAB	R6G	21
Poly(acrylic acid)	PAA	C_{10}TAB, C_{12}TAB	Py	57
Poly(methacrylic acid)	PMA	C_{10}TAB, C_{12}TAB	Py, R6G	15
Poly(maleic acid/alkylvinyl ether)		C_nTAB (n: 10, 12, 16)	Py, ANS	61
Poly(maleic acid/α-olefin)		C_{10}TAB, C_{12}TAB	ANS, Py	61
Hyaluronan		C_{10}TAB, C_{12}TAB	Py	30
Polymer JR 400		SDS	Py-CHO	62
Reten 220		SDS	Py-CHO	62

III. APPLICATIONS

A. POLYELECTROLYTE-SURFACTANT INTERACTIONS STUDIED BY IONIC FLUORESCENT PROBES

The experiments reported in Section II of this chapter were based on the use of hydrophobic fluorescent probes with extremely low solubility in water. A different approach has been taken by Hayakawa and co-workers.[22,49] They have promoted the use of water-soluble cationic dyes, such as rhodamine 6G (R6G) and 3,6-diaminoacridium to study the interactions of polyelectrolytes with surfactants. The solubility properties of these dyes complement those of the probes described earlier: they are very soluble in water but they tend to form dimers or higher aggregates in hydrophobic media,[50] and at elevated concentrations in water. The monomer and the dimer have distinct absorption and emission spectra. For example, the absorption spectrum of R6G in water at intermediate concentrations presents two bands with maxima at 527 nm (monomer, strong absorbance) and at 498 nm (dimer, weak absorbance).[51] The dye monomer fluoresces strongly (λ_{em} = 557 nm). This fluorescence is severely quenched in the dimers.

In aqueous solutions of polyelectrolytes such as dextran sulfate (DxS), poly(styrene sulfonic acid) (PSS), poly(acrylic acid) (PAA), or poly(methacrylic acid) (PMA), R6G exhibits a strong absorption centered at 498 nm (dimer form) and a weak fluorescence. Thus, the electrostatically driven binding of the dye to the polyions leads to an enhanced local concentration, promoting dimer and aggregate formation.

Upon addition of C_{12}TAB to solutions of the polyelectrolytes, the following spectroscopic changes occur: (1) in the absorption spectrum the dimer band disappears and a new band attributed to R6G monomer appears at 539 nm; and (2) in the fluorescence spectrum the emission maximum undergoes a red shift (572 nm) and an increase in intensity. This behavior is diagnostic of dimer dissociation, either through solubilization of the dye in a polar environment or displacement of the dye from the polymer, occurring as a result of competitive binding of the surfactant to the polyelectrolyte. Note that there are cases for which this kind of dye is unable to serve as a useful probe of the properties of polymer-surfactant aggregates. For example, in pectate solutions,[22] only minor changes occur in the absorption and emission spectra of R6G. Addition of surfactant leads to recovery of the free solution spectra, indicating only a disruption of the weak interaction of R6G with pectate upon formation of polymer-surfactant aggregates.

In the experiments with charged probes described above, the charge was borne by the chromophore itself. Another approach consists of using a neutral chromophore bearing a charged short side chain as a probe. The charged substituent serves as an anchor to the polyelectrolyte. This strategy is emplified in the study by Herkstroeter et al.[34] of the interactions between surfactants and either an anionic polyelectrolyte (sodium poly(γ-methacryloxypropylsulfonate) or a cationic polyelectrolyte, poly(β-methacryloxyethyltrimethylammonium) methanesulfonate. The ionic probes were a cationic pyrene derivative (3) and an anionic pyrene derivative, sodium 4-(1-pyrenyl)butyrate. At low concentrations in water both probes exhibit spectroscopic features characteristic of isolated pyrene chromophores, a strong well-resolved pyrene monomer emission with no contribution from pyrene excimer. In solutions of a polyelectrolyte of opposite charge, the probe emission undergoes dramatic changes. The overall spectral features become characteristic of preassociated chromophores, strong pyrene excimer emission and much weaker pyrene monomer emission. These effects are accompanied by changes in the excitation and absorption spectra, indicative of ground-state interactions between pyrene pairs. Binding of the cationic dyes to the polyelectrolytes takes place in part through electrostatic interactions between the charged substituents and the polyions. Hydrophobic forces between chromophores brought into close proximity along the polymer chain help to stabilize the dimeric species. Addition of the cationic surfactant $C_{12}TAB$ to this solution perturbs the dye-polymer interactions. The ground-state aggregates are broken up. The chromophores are solubilized in hydrophobic micellar microdomains formed along the polyelectrolytes.

B. NEUTRAL POLYMER/SURFACTANT INTERACTIONS: RECENT FLUORESCENCE EXPERIMENTS WITH PROBES AND LABELS

1. Aggregation Numbers ($\overline{N}$)

The size of the surfactant-polymer clusters and their structure have been the focal point of many studies. Undoubtedly the PEO-SDS system has been studied in the greatest detail. Recently Witte and Engberts[27] determined, via quenching measurements, the mean aggregation numbers of polymer/surfactant clusters in water and in the presence of salt for PEO as well as for PPO. Their results are compared here to those reported by Winnik and Winnik[29] for the HPC-SDS system. The polymers employed were poly(ethylene oxide) (PEO, mol wt 10,000), poly(propylene oxide) (PPO, mol wt 1000), and hydroxypropylcellulose (HPC, mol wt 100,000). The aggregation numbers were determined by Witte and Engberts from plots of $\ln I_o/I$ vs. [Q] (Equation 4) employing the $Ru(bipy)_3^{2+}$/9-methylanthracene fluorophore-quencher pair. The $\overline{N}$ values for HPC-SDS aggregates were obtained also by fluorescence quenching experiments, but with a different fluorophore-quencher pair (pyrene/benzophenone). In the PEO-SDS system the mean aggregation number ($\overline{N}$) are smaller than those of SDS micelles and increase with surfactant concentration (Figure 6). In the PPO-SDS system the measured $\overline{N}$ values are smaller than in the PEO-SDS system over the entire surfactant concentration range, but they follow the same trend. The results for the HPC-SDS system are quite different: the $\overline{N}$ values increase rapidly with SDS concentration and pass through a maximum (about 80) at an SDS concentration just above the

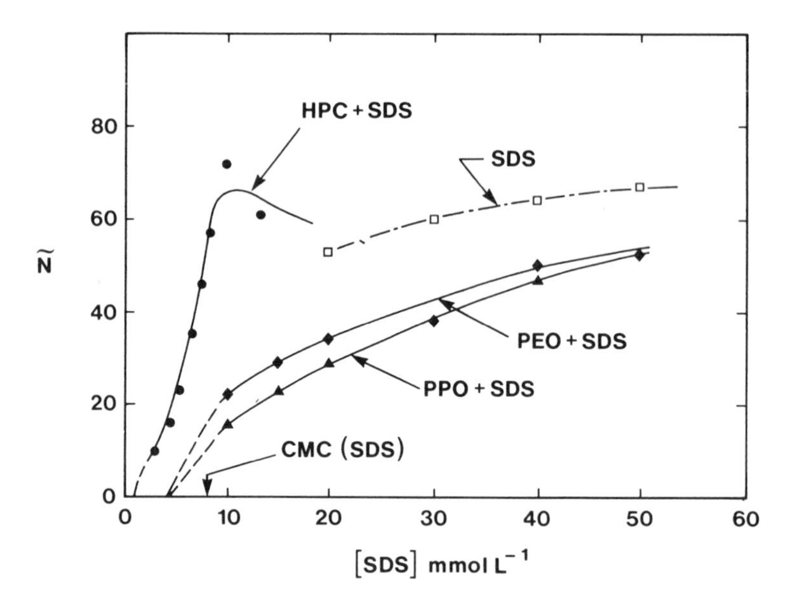

FIGURE 6. Plots of the aggregation number $\overline{N}$ of SDS-HPC, SDS-PEO, and SDS-PPO clusters, and of SDS micelles as a function of SDS concentration; [HPC] = 1 g l^{-1}, [PEO] = 3 g l^{-1}, [PPO] = 3 g l^{-1}.

critical micelle concentration (CMC) of SDS itself. In addition the $\overline{N}$ values depend on the ratio of SDS to HPC: they decrease as the amount of HPC increases relative to the SDS concentration (data not shown in Figure 6). Note that this behavior may be more general than predicted by most models of the polymer-surfactant interactions. In a recent study, van Stam and co-workers uncovered new details of the interactions between SDS and PEO.[52] Using time-resolved techniques, they monitored the quenching of fluorescence of pyrene by dimethylbenzophenone in various SDS-PEO systems. Aggregation numbers were determined at two temperatures (20 and 40°C) and two PEO concentrations (0.2 and 2%). From their data, van Stam et al.[52] conclude that, as in the case of the SDS-HPC system, the surfactant clusters along the PEO chain grow in size rather than in number with increasing SDS concentration.

As a final comment on the data in Figure 6, it should be kept in mind that $\overline{N}$ values determined at surfactant concentrations above its CMC represent an average over that of the polymer-surfactant mixed aggregates and that of the free micelles in the system. In order to be certain that $\overline{N}$ values refer only to the former, one must establish that sufficient polymer is present for the concentration of free micelles to be negligibly small.

2. Pyrene End-Labeled Poly(Ethylene Oxides)

Three publications have appeared recently on the interactions of surfactants with pyrene-labeled PEOs. Two of the polymers carry pyrene groups at

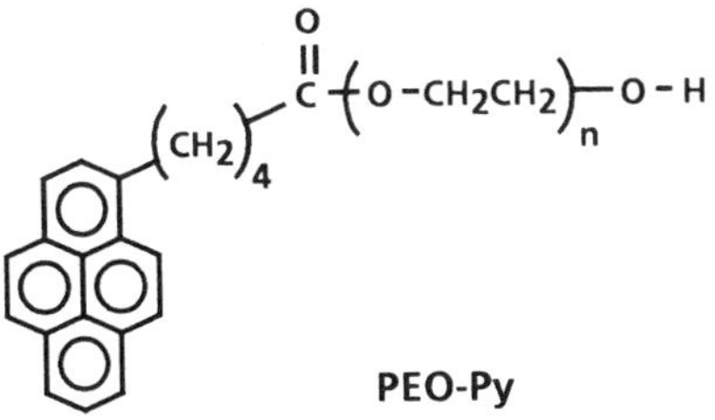

PEO-Py

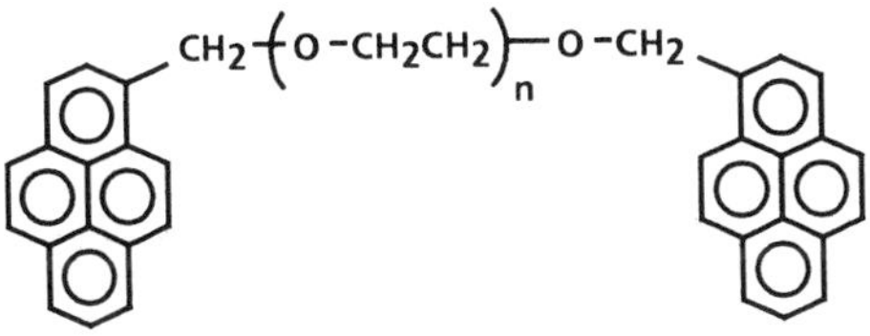

PEO-(O-CH_2-Py)$_2$

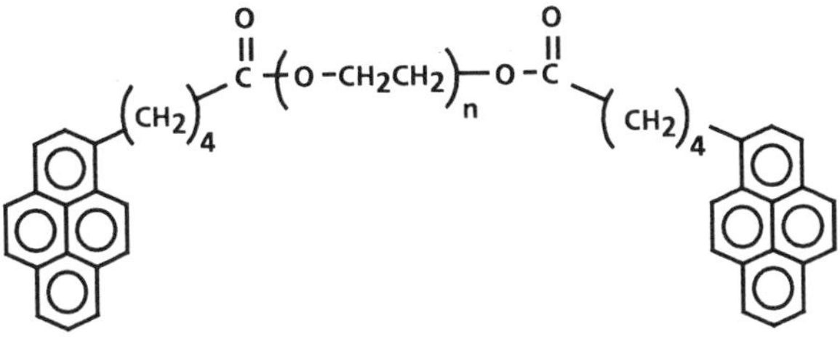

PEO-[O-CO-$(CH_2)_4$-Py]$_2$

FIGURE 7. Chemical structures of labeled poly(ethylene) oxides.

both ends via either an ether linkage or an ester linkage. The third polymer has been labeled (ester bond) at one end only (Figure 7). The singly labeled polymer was prepared from a sample of larger molecular weight (nominal 20,000) than the doubly labeled materials (8000 and 6000 to 7000, respectively).

The case of PEO labeled at one end only (PEO-Py)[44] — The fluorescence of PEO-Py in water (0.0125 to 0.2 wt%) is characterized by a strong pyrene monomer emission and a weak contribution from pyrene excimer. As SDS is added to a PEO-Py solution the excimer emission increases at the expense of monomer emission, the ratio I_E/I_M reaches a maximum value ([SDS] = 5×10^{-4} mol l^{-1}) and then decreases. At low SDS concentration ([SDS] $<1 \times 10^{-3}$ mol l^{-1}) the excimer formation is a fast process, with a rise time, τ_{rise}, <10 nsec. The process is slower at higher SDS concentration (τ_{rise} = 25 nsec, for [SDS] $>3.8 \times 10^{-3}$ mol l^{-1}). The most important observation in this study is that the interaction of the surfactant with the

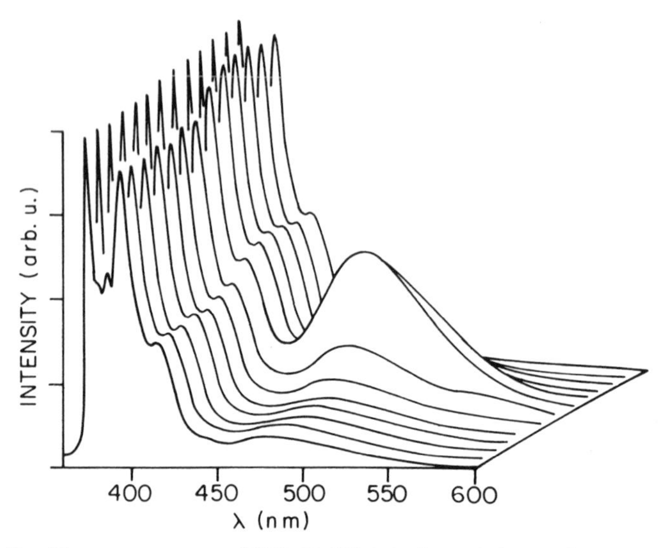

FIGURE 8. Fluorescence spectra of PEO-(O-CH_2-Py)$_2$ (1×10^{-6} mol l^{-1}) in the presence of various concentrations of SDS. The spectra shown correspond (front to rear) to [SDS] = 0, 1.0×10^{-6}, 5.1×10^{-5}, 1.0×10^{-4}, 2×10^{-4}, 4.0×10^{-4}, 6.1×10^{-4}, 8.1×10^{-4}, 1.0×10^{-3}, 2.8×10^{-3}, 3.5×10^{-3}, 1.0×10^{-2}, and 2.1×10^{-2} mol l^{-1}. (From Hu, Y.-Z. et al., *Langmuir,* 6, 880, 1990. With permission of the American Chemical Society.)

labeled polymer takes place at SDS concentrations considerably lower than those required ([SDS] $\sim 4 \times 10^{-3}$ mol l^{-1}) to produce micellization onto unlabeled PEO.[53] At low SDS concentration the polymer-surfactant aggregates comprise two or more PEO-Py chain ends, as evidenced by the increase in excimer emission intensity and the relatively short excimer rise time. At higher surfactant concentrations, the polymer chains are diluted in an increasing number of micellar structures of larger size; hence the observed decrease in excimer emission intensity and increase in excimer rise time.

The case of PEO labeled at both ends PEO-(O-CH_2-Py)$_2$[41] — In this study the ratio of pyrene excimer to monomer intensity of an aqueous solution of the polymer (1×10^{-6} mol l^{-1}, mol wt 8000, e.g., 8 ppm) was also monitored as a function of SDS concentration. The excimer makes only a modest contribution to the fluorescence of the polymer in water. Addition of SDS triggers a sharp increase in excimer emission which reaches a maximum intensity for [SDS] $\sim 8 \times 10^{-4}$ mol l^{-1}. Further addition of SDS to the solution results in a decrease in excimer emission intensity (Figure 8). A sharp change in the vibrational fine structure of the monomer emission is also observed at [SDS] 8×10^{-4} mol l^{-1}, indicating that the environment experienced by the chromophores undergoes a significant enhancement in hydrophobicity at that surfactant concentration. Both spectroscopic parameters give strong evidence for the formation of polymer-surfactant micellar structures with an apparent cac of 8×10^{-4} mol l^{-1}. Not only do the SDS-polymer interactions occur at surfactant concentrations well below the CMC of SDS or the cac of PEO/SDS concentrations, but also the surfactant molecules bind preferentially to the hydrophobic chain ends, at least at the onset of interactions. A model offered by the authors to describe the PEO-(O-CH_2-Py)$_2$/SDS system is discussed in a broader context in Chapter 6 (see Figure 6).

The most interesting aspect of this and the Lissi experiments[44] is the cooperative role of the pyrene end groups and the PEO chain in surfactant binding. In mixed aggregates of PEO and SDS, addition of a pyrene probe does not affect either the cac value or $\overline{N}$. When the pyrene is attached to the chain end, the cac value is lower, and there is some indication that $\overline{N}$ values are smaller.[54]

The case of PEO labeled at both ends PEO-[O-CO-(CH_2)$_4$-Py]$_2$[42,43] — A third study of the interactions between surfactant and dipyrene end-tagged PEO has been carried out recently by Maltesh and Somasundaram.[43] They monitored the effects of three surfactants on the spectroscopy of PEO-[O-CO-(CH_2)$_4$-Py]$_2$ in an attempt to distinguish the effects of the surfactant charge from those of their hydrophobic tails. Three surfactants with different headgroups were selected: sodium dodecyl sulfate (SDS), dodecyl trimethyl-ammonium chloride (DTAC), and *p*-nonylphenoxy polyethoxy alcohol (Triton X-100). Each was added to a solution of the polymer in water (20 ppm, mol wt 6000 to 7000). The addition of SDS to the solution triggered changes in I_E/I_M identical qualitatively to those described by Hu et al.[41] in the case of PEO-(O-CH_2-Py)$_2$ (Figure 8). The interesting new piece of information concerns the effects of the other surfactants. Both DTAC and Triton X-100 induce changes in the pyrene emission. The effects on I_E/I_M (Figure 9) are not as pronounced as those induced by SDS addition. Moreover, they occur at surfactant concentrations near their respective CMC (Triton X-100, 2×10^{-4} mol l^{-1}, DTAC, 2×10^{-2} mol l^{-1}). Note, however, that neither DTAC nor Triton X-100 interacts with PEO itself (see part 1, this chapter). This observation is reminiscent of a previous report on the interactions of nonionic surfactants with a hydrophobically modified hydroxypropyl cellulose.[55] In both instances, the unlabeled polymers do not interact with the surfactants, but the modified polymers somehow sense the presence of surfactant micelles.

C. MICRODOMAINS IN AQUEOUS POLYMER SOLUTIONS

Many facets of the fluorescence probe techniques and of their power in detecting hydrophobic microdomains in water have been reviewed to this point, with emphasis on solutions of either surfactants or polymer/surfactant mixtures. The technique has proven to be useful as well as to characterize microdomains created in solutions of copolymers bearing hydrophilic and hydrophobic groups. Such fluorescence probe experiments have been applied by Stauss in his thorough investigation of alternating copolymers of maleic anhydride and alkyl vinyl ether (see Chapter 6). Recent results from Binana-Limbelé and Zana[56] point to the importance of the length of the alkyl substituent of the vinyl ether comonomer. Pyrene fluorescence was used to probe the conformational state of copolymers where the alkyl group was *n*-butyl,

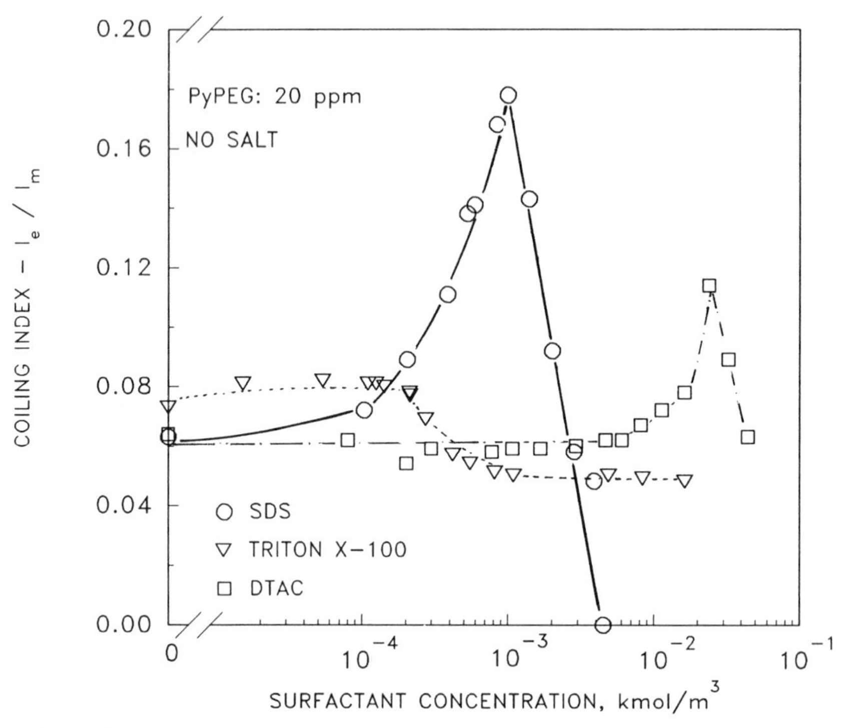

FIGURE 9. Interactions between pyrene end-labeled poly(ethylene oxide) PEO-[OCO-$(CH_2)_4$-Py] and SDS, DTAC, and Triton X-100. (From Maltesh, C. and Somasundaram, P., *Langmuir*, 8, 1926, 1992.)

n-decyl, and *n*-hexadecyl as a function of the neutralization degree of the carboxylic groups. Time-resolved fluorescence quenching experiments were employed to investigate the size of microdomains in solution of the hexadecyl- and decyl-substituted copolymers, with pyrene as the fluorophore and the dodecylpyridinium ion as the quencher.[56] Two conclusions of these experiments are noteworthy. First, in the case of the decyl-substituted copolymers the number of repeat units per microdomain decreases linearly upon increase of the degree of neutralization. Second, in the case of the hexadecyl-substituted copolymers, migration of the probe and of the quencher takes place on the fluorescence timescale, presumably between discrete microdomains created along a single polymeric chain.

A new class of hydrophobically modified maleic acid/ethyl vinyl copolymers has been described recently by McCormick and co-workers.[57] They modified the alternating copolymer structures by incorporation of 4-butylaniline in various amounts. The solution properties of these copolymers were investigated by the combined use of time-resolved fluorescence spectroscopy (pyrene probe) and fluorescence quenching with both hydrophilic (thallium salt) and hydrophobic (nitromethane) quenchers. Copolymers containing up to 50 mol% of butylaniline (relative to the total number of maleic acid groups) undergo a transition from a highly collapsed conformation at a low degree of ionization to an open hydrated conformation at a high degree of ionization, as expected for polysoaps. However, when the degree of incorporation of hydrophobic groups reaches the 70% level, the compact hydrophobic microdomains persist over the entire pH range.

A battery of fluorescence probe experiments was deployed in recent studies towards a better understanding of how the detailed molecular architecture of copolymers influences the structure of the hydrophobic microdomains existing in their aqueous solutions. Examples of polymers investigated include copolymers of acrylamide and styrene, either styrene-grafted polyacrylamides[58] or diblock copolymers,[59] and poly-*N*-isopropylacrylamides, carrying *n*-octadecyl groups grafted either at random along the polymer chains[35] or attached at one end of the chain.[60] In both cases the microdomains created by the grafted and the block copolymer were shown to differ in size, rigidity, and in their ability to screen hydrophobic probes from the aqueous environment.

IV. CONCLUSIONS

This section has examined certain experimental details of the application of fluorescence techniques to the study of polymer-surfactant interactions. Initially a distinction was drawn between probe experiments, where a fluorescent dye is added to a solution of the polymer plus surfactant, and labeling experiments, where the fluorescent dye is covalently attached to the polymer. Probe and labeling experiments provide different and often complementary information about these systems. Perhaps the most important types of information available from fluorescence experiments are the mean aggregation numbers of the micelle-like structures formed in the interactions, and measures of the mobility or internal fluidity of these structures. Labeling experiments offer a special advantage in that they allow one to follow changes in polymer-polymer association and in the global conformation of the polymer, as influenced by the presence of surfactants.

ACKNOWLEDGMENTS

I am grateful to Professor M. A. Winnik, University of Toronto, Canada, for sharing some of his latest results on the study of the PEO/Py/surfactant interactions and for his critical and helpful comments during the preparation of this manuscript. Thanks are due also to Professor P. Somasundaran (Columbia University, New York), for permission to cite part of his unpublished results on the PEO/Py/surfactant interactions. Finally, I thank Dr. E. D. Goddard for his suggestion to contribute a chapter to this book.

REFERENCES

1. **Whery, E. L., Ed.,** *Modern Fluorescence Spectroscopy,* Plenum Press, New York, 1976.

2. **Lakowicz, R.,** *Principles of Fluorescence Spectroscopy,* Plenum Press, New York, 1983.
3. **Winnik, M. A., Ed.,** *Photophysical and Photochemical Tools in Polymer Science,* D. Reidel Publishing, Dordrecht, 1986.
4. **Rabek, J. F.,** *Mechanisms of Photophysical Processes and Photochemical Reactions in Polymers,* John Wiley & Sons, Chichester, 1987.
5. **Honda, K., Ed.,** *Photochemical Processes in Organized Molecular Systems,* North-Holland-Elsevier, Amsterdam, 1991.
6. **Phillips, D. and Roberts, A. J., Eds.,** *Photophysics of Synthetic Polymers,* Science Review, Northwood, U.K., 1982.
7. **Kalyanasundaram, K.,** *Photochemistry in Microheterogeneous Systems,* Academic Press, Orlando, 1986.
8. **Zana, R., Ed.,** *Surfactant Solutions: New Methods of Investigation,* Marcel Dekker, New York, 1986.
9. **Ramamurthy, V., Ed.,** *Photochemistry in Organized and Constrained Media,* VCH Publishers, New York, 1991, chap. 2 and 3.
10. **Grieser, F. and Drummond, C. J.,** The physicochemical properties of self-assembled surfactant aggregates as determined by some molecular spectroscopic probe techniques, *J. Phys. Chem.,* 92, 5580, 1988.
11. **Malliaris, A.,** Fluorescence probing in aqueous micellar systems: an overview, *Intern. Rev. Phys. Chem.,* 7, 95, 1988.
12. **Kalyanasundaram, K. and Thomas, J. K.,** Environmental effects on vibronic band intensities in pyrene monomer fluorescence and their application in studies of micellar systems, *J. Am. Chem. Soc.,* 99, 2039, 1977.
13. **Turro, N. J., Baretz, B. H., and Kuo, P.-L.,** Photoluminescence probes for the investigation of the interactions between sodium dodecyl sulfate and water-soluble polymers, *Macromolecules,* 17, 1321, 1984.
14. **Zana, R., Lianos, P., and Lang, J.,** Fluorescence probe studies of the interactions between poly(oxyethylene) and surfactant micelles and microemulsion droplets in aqueous solutions, *J. Phys. Chem.,* 89, 41, 1985.
15. **Chu, D.-Y. and Thomas, J. K.,** Effect of cationic surfactants on the conformational transition of poly-(methacrylic acid), *J. Am. Chem. Soc.,* 108, 6270, 1986.
16. **Katchalsky, A. and Eisenberg, H.,** Molecular weight of polyacrylic acid and polymethacylic acid, *J. Polym. Sci.,* 6, 145, 1951.
17. **Winnik, F. M., Ringsdorf, H., and Venzmer, J.,** Interactions of surfactants with hydrophobically modified poly-(N-isopropylacrylamides). I. Fluorescence probe studies, *Langmuir,* 7, 905, 1991.
18. **Georgescauld, D., Desmasèz, J. P., Lapouyade, R., Babeau, A., Richard, H., and Winnik, M. A.,** Intramolecular excimer fluorescence: a new probe of phase transitions in synthetic phospholipid membranes, *Photochem. Photobiol.,* 31, 539, 1980; **Zachariasse, K. A., Vaz, W. L. C., Sotomayor, C., and Kühnle, W.,** Investigation of human erythrocyte ghost membranes with intramolecular excimer probes, *Biochim. Biophys. Acta,* 688, 323, 1982.
19. **Lin, T. H., Cherry, W. R., and Mattice, W. L.,** Acridine fluorescence as a probe of micelle formation by sodium dodecylsulfate in the presence of poly(N-(ω-hydroxyalkyl)-L-glutamines), *Polym. Commun.,* 27, 37, 1986.
20. **Abuin, E. B. and Scaiano, J. C.,** Exploratory study of the effect of polyelectrolyte-surfactant aggregates on photochemical behavior, *J. Am. Chem. Soc.,* 106, 6274, 1984.
21. **Birdi, K. S., Singh, H. N., and Dalsager, S. U.,** Interaction of ionic micelles with the hydrophobic fluorescent probe 1-anilino-8-naphthalenesulfonate, *J. Phys. Chem.,* 83, 2733, 1979.
22. **Hayakawa, K., Satake, I., Kwak, J. C. T., and Gao, Z.,** Spectroscopy of rhodamine 6G solubilized in complexes of anionic polymers with cationic surfactants, *Colloids Surf.,* 50, 309, 1990.

23. **Shinitzky, M., Diarnoux, A.-C., Gitler, C., and Weber, G.,** Microviscosity and order in the hydrocarbon region of micelles and membranes determined by fluorescent probes. I. Synthetic micelles, *Biochemistry,* 10, 2106, 1971.
24. See for example, Reference 11, Chapter 5.
25. **Turro, N. J. and Yekta, A.,** Luminescent probes for detergent solutions. A simple procedure for determination of the mean aggregation number of micelles, *J. Am. Chem. Soc.,* 100, 5951, 1978.
26. **Warr, G. G. and Grieser, F.,** Determination of micelle size and polydispersity by fluorescence quenching, *J. Chem. Soc. Faraday Trans. 1,* 82, 1813, 1986.
27. **Witte, F. M. and Engberts, J. B. F. N.,** Micelle-polymer complexes: aggregation numbers, micellar rate effects and factors determining the complexation process, *Colloids Surf.,* 36, 417, 1989.
28. **Lissi, E. A. and Abuin, E.,** Aggregation numbers of sodium dodecylsulfate micelles formed on poly(ethylene oxide) and poly(vinyl pyrrolidone) chains, *J. Colloid Interface Sci.,* 105, 1, 1985.
29. **Winnik, F. M. and Winnik, M. A.,** The interaction of sodium dodecylsulfate with (hydroxypropyl) cellulose, *Polym. J.,* 22, 482, 1990.
30. **Thalberg, K., van Stam, J., Lindblad, C., Almgren, M., and Lindman, B.,** Time resolved fluorescence and self-diffusion studies in systems of a cationic surfactant and an anionic polyelectrolyte, *J. Phys. Chem.,* 95, 8975, 1991.
31. **Atik, S. S., Nam, M., and Singer, L. A.,** Transient studies on intramicellar excimer formation. A useful probe of the host micelle, *Chem. Phys. Lett.,* 671, 75, 1979.
32. **Winnik, F. M., Winnik, M. A., Ringsdorf, H., and Venzmer, J.,** Bis(1-pyrenylmethyl) ether as an excimer-forming probe of hydrophobically modified poly(N-isopropylacrylamides) in water, *J. Phys. Chem.,* 95, 2583, 1991.
33. **Turro, N. J. and Arora, K. S.,** Pyrene as a photophysical probe for interactions of water-soluble polymers in dilute solutions, *Polymer,* 27, 783, 1986.
34. **Herkstroeter, W. G., Martic, P. A., Hartman, S. E., Williams, J. L. R., and Farid, S.,** Unique hydrophobic interactions of pyrene in aqueous solution as effected by polyelectrolytes and surfactants, *J. Polym. Sci. Polym. Chem. Ed.,* 21, 2473, 1983.
35. **Ringsdorf, H., Venzmer, J., and Winnik, F. M.,** Fluorescence studies of hydrophobically modified poly-(N-isopropylacrylamides), *Macromolecules,* 24, 1678, 1991.
36. **Winnik, F. M., Winnik, M. A., Tazuke, S., and Ober, C. K.,** Synthesis and characterization of pyrene-labeled (hydroxypropyl)cellulose and its fluorescence in solution, *Macromolecules,* 20, 38, 1987.
37. **McGlade, M. J. and Olufs, J. L.,** Adsorption of sodium dodecyl sulfate onto α-olefin/maleic acid copolymers in aqueous solutions, *Macromolecules,* 21, 2346, 1988.
38. **Oyama, H. T., Hemker, D. J., and Frank, C. W.,** Effect of the degree of ionization of poly(methacrylic acid) on the complex formed with pyrene end-labeled poly(ethylene glycol), *Macromolecules,* 22, 1255, 1989, and references therein.
39. **Strauss, U. P. and Vesnaver, G.,** Optical probes in polyelectrolyte studies. I. Acid-base equilibria of dansylated copolymers of maleic anhydride and alkyl vinyl ether, *J. Phys. Chem.,* 79, 1558, 1975.
40. **Winnik, F. M.,** Fluorescence studies of aqueous solutions of poly-N-(isopropylacrylamide) below and above their LCST, *Macromolecules,* 23, 233, 1990.
41. **Hu, Y.-Z., Zhao, C.-L., Winnik, M. A., and Sundararajan, P. R.,** Fluorescence studies of the interaction of sodium dodecyl sulfate with hydrophobically modified poly(ethylene oxide), *Langmuir,* 6, 880, 1990.
42. **Maltesh, C., Somasundaram, P., and Ramachandran, R.,** Conformational changes of poly(ethylene oxide) during association with sodium dodecylsulfate, *J. Appl. Pol. Sci. Appl. Pol. Symp.,* 45, 329, 1990.
43. **Maltesh, C. and Somasundaram, P.,** Effect of binding of cations to polyethylene glycol on its interactions with sodium dodecylsulfate, *Langmuir,* 8, 1926, 1992.

44. **Quina, F., Abuin, E., and Lissi, E.,** Effect of pyrene chain end labeling on the interaction of poly(ethylene oxide) with sodium dodecyl sulfate in aqueous solution, *Macromolecules,* 23, 5173, 1990.
45. **Winnik, F. M., Winnik, M. A., and Tazuke, S.,** Interaction of hydroxypropylcellulose with aqueous surfactants: fluorescence probe studies and a look at pyrene-labeled polymers, *J. Phys. Chem.*, 91, 594, 1987.
46. **Winnik, F. M., Ringsdorf, H., and Venzmer, J.,** Interaction of surfactants with hydrophobically-modified poly-(N-isopropylacrylamides). II. Fluorescence label studies, *Langmuir,* 7, 912, 1991.
47. See for example Reference 38.
48. **Winnik, F. M.,** Effect of temperature on aqueous solutions of pyrene-labeled (hydroxypropyl)cellulose, *Macromolecules,* 20, 2745, 1987.
49. **Hayakawa, K., Ohta, J., Maeda, T., Satake, I., and Kwak, J. C. T.,** Spectroscopic studies of dye solubilizates in micellelike complexes of surfactant with polyelectrolyte, *Langmuir,* 3, 377, 1987.
50. For a study of cationic probes in polyelectrolyte solutions, see Turro, N. J. and Pierola, I. F., *Macromolecules,* 16, 906, 1983.
51. **Selwyn, J. E. and Steinfield, I.,** Aggregation equilibria of xanthene dyes, *J. Phys. Chem.*, 76, 762, 1972.
52. **van Stam, J., Almgren, M., and Lindblad, C.,** Sodium dodecylsulfate-poly(ethyleneoxide) interactions studied by time-resolved fluorescence quenching, *Prog. Colloid Polym. Sci.*, 84, 13, 1991.
53. **Cabane, B.,** Structure of some polymer detergent aggregates in water, *J. Phys. Chem.*, 81, 1639, 1977.
54. **Hu, Y. Z. and Winnik, M. A.,** private communication.
55. **Winnik, F. M.,** Interaction of fluorescent dye labeled (hydroxypropyl)cellulose with nonionic surfactants, *Langmuir,* 6, 522, 1990.
56. **Binana-Limbelé, W. and Zana, R.,** Fluorescence probing of microdomains in aqueous solutions of polysoaps. II. Study of the size of the microdomains, *Macromolecules,* 23, 2731, 1990.
57. **McCormick, C. L., Hoyle, C. E., and Clark, M. D.,** Water-soluble copolymers: 26. Fluorescence probe studies of hydrophobically modified maleic acid-ethyl vinyl ether copolymers, *Polymer,* 33, 243, 1992.
58. **Dowling, K. C. and Thomas, J. K.,** Photophysical characterization of water-soluble styrene-grafted poly(acrylamide) copolymers, *Macromolecules,* 24, 2341, 1991.
59. **Dowling, K. C. and Thomas, J. K.,** A novel micellar synthesis and photophysical characterization of water-soluble acrylamide-styrene block copolymers, *Macromolecules,* 23, 1059, 1990.
60. **Winnik, F. M., Davidson, A. R., Hamer, G. K., and Kitano, H.,** Amphiphilic poly(N-isopropylacrylamides) prepared by using a lipophilic radical initiator: synthesis and solution properties in water, *Macromolecules,* 25, 1876, 1992.
61. **Binana-Limbelé, W. and Zana, R.,** Fluorescence probing of microdomains in aqueous solutions of polysoaps. I. Use of pyrene to study the conformational state of polysoaps and their comicellization with cationic surfactants, *Macromolecules,* 20, 1331, 1987.
62. **Ananthapadmanabhan, K. P., Leung, P. S., and Goddard, E. D.,** Fluorescence and solubilization studies of polymer-surfactant systems, *Colloids Surf.*, 13, 63, 1985.
63. **Binana-Limbelé, W. and Zana, R.,** Interactions between sodium dodecyl sulfate and polycarboxylates and polyethers. Effect of Ca^{2+} on these interactions, *Colloids and Surf.*, 21, 483, 1986.
64. **Schild, H. G. and Tirrell, D. A.,** Interaction of poly-(N-isopropylacrylamide) with sodium n-alkyl sulfates in aqueous solution, *Langmuir,* 7, 665, 1991.
65. **Brackman, J. C., van Os, N. M., and Engberts, J. B. F. N.,** Polymer-nonionic micelle complexation. Formation of poly(propylene oxide)-completed n-octyl thioglucoside micelles, *Langmuir,* 4, 1266, 1988.

Chapter 10

APPLICATIONS OF POLYMER-SURFACTANT SYSTEMS

E. D. Goddard

TABLE OF CONTENTS

0-8493-6784-0/93/$0.00 + $.50

I. INTRODUCTION

A great number of aqueous-based systems in nature and commerce contain one or more polymer and surfactant in the same solution or suspension. In these, any interaction between polymer and surfactant, as well as property changes conferred by any resulting "complex", can be of considerable importance. Included in this category are a multitude of biological systems, e.g., membranes (structure and functioning), carriers (lipid transport), and so on, and a variety of other systems including foods, pharmaceutics, cosmetics, detergents, and various chemical treating systems. It must be recognized that a knowledge of the interaction characteristics of the particular polymer(s) and surfactant(s) present will greatly aid the understanding and facilitate optimization of the properties of a system, although it is suspected that a great number of commercial products containing such mixtures are still assembled very much on an empirical, trial-and-error basis.

There are many changes in solution and surface properties which occur as a result of polymer/surfactant interactions, and these have been referred to in several places in the preceding chapters. In this chapter a number of the more important ones which appear to offer opportunities for exploitation are elaborated upon and are grouped under nine headings.

II. RHEOLOGY CONTROL; VISCOSITY ENHANCEMENT; GELS

It is known that polyelectrolytes tend to adopt a linear configuration in aqueous solution, unless the ionic strength of the solution is high. This configuration favors an increase in viscosity of the solution. When an un-ionized polymer binds an ionic surfactant, the polymer will acquire a charge and hence, on the basis of the "polyelectrolyte effect", an increase in viscosity would be anticipated. This area has been investigated by a number of authors, e.g., Jones[1] and Francois et al.,[2] who examined the effect of adding sodium dodecyl sulfate (SDS) to a series of polyethylene oxide (PEO) polymers. As discussed in Chapter 4, Part II, a sudden increase in viscosity occurs at a certain concentration (T_1) of surfactant, independent of polymer molecular weight, and this increase can be as high as fivefold, which is consistent with a polymer charging effect. Recently the viscosity behavior of aqueous solutions of a high-molecular-weight (5×10^6 Da) PEO specimen in the presence of SDS has been examined in detail by Brackman.[3] In addition to finding the anticipated enhancement of viscosity, considerable development of viscoelasticity, as manifested in the results of normal stress measurements, was encountered.

We recall that even larger increases in viscosity can be seen with certain oppositely charged polyelectrolyte/ionic surfactant pairs as illustrated in Figure 9 of Chapter 4, Part II. In this case involving the cationic cellulosic,

Polymer JR 400 (Union Carbide), relatively low levels of added SDS lead to substantial viscosity increases (as large as 200-fold), or even weak gel formation, in the immediate preprecipitation zone.[4] The most likely explanation is that ''super macromolecules'' are formed through association of the alkyl groups of surfactant ions bound to different polymer molecules. Such a structure would have pronounced shear thinning characteristics, thus providing opportunities for rheology control. It should be mentioned that thickening effects on addition of surfactant are dependent on the detailed molecular structure of the polyelectrolyte. Thus, the much more flexible polycation, Reten (Hercules), based on vinyl chemistry, did not show the viscosity increase displayed by the ''stiff backboned'' cationic cellulosic/SDS combination.

Combinations of SDS and the highest available molecular weight grade of Polymer JR (30M), where chain entanglements required for formation of a gel structuring network would be enhanced, were examined recently.[5,6] Strong gels were formed at concentrations of polymer of ~1% and about one tenth this amount of SDS. Rheological characterization showed that the complex modulus, G*, was dominated by the elastic component, G′; also, the ''zero frequency'' dynamic viscosity was very high (~10 kPa). G′ was found to increase with SDS concentration, up to 0.15%, and then to decrease. Similar trends were found with several other types of anionic surfactant (alkyl-ethoxysulfates, -benzenesulfonate, and -sulfosuccinate). It seems likely that gel formation in these systems results both from an increase in effective molecular weight of the polymer through crosslinking of bound clusters as well as from increased chain entanglement.

Gel formation in systems charged in the opposite sense, i.e., a polyanion plus a cationic surfactant, has recently been reported by Thalberg and Lindman[7] for hyaluronan and polyacrylate polymers mixed with members of the alkyltrimethylammonium bromide surfactant family. With reference once more to uncharged polymers, Carlsson et al.,[8] in an extension of their work on the phase diagrams of uncharged cellulosic polymer/charged surfactant combinations, demonstrated that the formation of high-viscosity systems, including gels, occurred under particular conditions (''windows'') of temperature and added concentration of surfactants (usually CTAB — the hexadecyl member of the above family) in the single-phase zone. Most of their work involved the polymeric ether ethylhydroxyethylcellulose.

Situations may arise in which it is desired to liquify a preformed gel. There are several possible approaches to achieve such thinning: (1) adding excess surfactant (or polymer) to move away from the optimum gelling ratio; (2) adding a competing surfactant or polymer; and (3) adding salt.[6] Brackman and Engberts[9] have reported an interesting case of viscosity *reduction* on adding a water-soluble polymer (polypropylene oxide or polyvinylmethyl ether) to the solution of a cationic surfactant (CTA^+X^-) in which the surfactant micelles are rod-like due to the presence of the strongly bound salicylate anion

(X^-). Competition between the polymer and the salicylate ion for the CTA^+ micelles is evidently responsible for the pronounced structural reorganization of the system which is de-gelled in the thinning process.

Interest in aqueous gel technology is traditional in the food and pharmaceutical industries, but it should be remarked that current patent literature now indicates an upsurge of interest in the cosmetics field[6] and to some extent in the detergents field as well. It is clear from the foregoing that the polymer/surfactant approach provides opportunities for gelling a variety of formulation types and for control of their rheology in general.

III. SOLUBILIZATION

Micellar surfactant solutions are well known for their ability to dissolve oil-soluble materials, e.g., dyes, hydrocarbons, esters, perfumes, and so on. To the extent that complex formation with a nonionized polymer can be regarded as a depression of the critical aggregation concentration of the surfactant (i.e., $T_1 <$ the critical micelle concentration [CMC]), enhanced solubilization by the complex can be anticipated. This effect has been confirmed using water-insoluble dyes[10-12] hydrocarbons,[13] and sparingly soluble fluorescers.[14] Much more pronounced effects have been found for polyelectrolyte/ionic surfactant pairs.[4] For the cationic cellulosic/SDS pair a solubilization region for the dye, Orange OT, occurs at very low concentration and the "main" solubilization zone is also widened (shifted to lower concentration) as compared to simple SDS solutions.[4] Solubilization at the very low concentration of SDS signifies clustering of surfactant around the positive charges of the polymer in the initial binding process. Likewise, the efficiency of a polymer/surfactant pair charged in the opposite sense, viz., a maleic anhydride/vinyl methyl ether copolymer (Gantrez S-95, GAF Corp.) and hexadecylpyridinium chloride, has been demonstrated for the solubilizing of chlorophenols.[15] The authors of this latter work point out the potential of this combination as an adjunct in a micellar-enhanced ultrafiltration process for clean-up of aqueous streams.

Since polymers and surfactants can associate in solution, it would not be surprising if they could influence *each other's* solubility as well as that of a third component. Perhaps the best-known case of this effect was reported by Isemura and Imanishi,[16] who showed that a polyvinylacetate polymer of very low solubility could be solubilized in solutions of SDS. This confirmed similar work carried out earlier by Sata and Saito.[17] Saito and Mizuta[18] showed that less strong, but still definite, solubilization of certain insoluble polymers could also be effected by certain cationic surfactants.

Solubilization processes of the above type allow the possibility of regenerating the original polymer in a different state, e.g., in the form of fibers, films, etc. Work related to this approach has been reported by Lundgren,[19] in which concentrated equi-weight mixtures of protein and anionic surfactant in water were extruded as continuous fiber into a magnesium sulfate coagulating bath prior to final drawing, washing, and drying. Another specific case of solubilization of polymers by surfactants is treated in the next section.

In an interacting nonionized polymer/ionic surfactant pair it is logical to expect that increased solubility could be manifested in the opposite sense, i.e., the polymer could increase the solubility of the surfactant since the monomer concentration required for aggregation of the surfactant is lowered in the presence of the polymer. As pointed out earlier (Chapter 4), such an effect has, in fact, been reported by Schwuger and Lange,[20] who showed that polyvinylpyrrolidone (PVP) can reduce the Krafft point of sodium hexadecyl sulfate by close to 10°C.[20]

Note: it is well known that many conditioning polymers are polycationic, and it has been pointed out that precipitation zones exist at certain ratios in combinations of such polyelectrolytes with anionic surfactants. In most cases, however, such precipitates can be solubilized in the presence of excess surfactant, or prevented by the copresence of a nonionic surfactant,[21-23] thus allowing the preparation of "acceptable" formulations in practice.

IV. ENHANCED WATER SOLUBILITY: CLOUD POINT ELEVATION

The ability of ionic surfactants to raise the cloud point of certain unionized water-soluble polymers has been referred to in Chapter 4. This subject has recently received new attention. Cloud point alteration has usually been considered to reflect a monotonic increase with increase in ionic surfactant concentration of the solution, but recent work has shown the first additions of ionic surfactant can sometimes lower the cloud point, especially when salt is present.[24-27] In fact, the phenomenon seems to be unusually sensitive to the presence of salt. Karlstrom et al.[24] have studied the phenomenon in detail and have presented phase relationships for a number of cellulosic polymer (especially ethylhydroxyethylcellulose)-ionic surfactant pairs. The actual cloud point vs. surfactant concentration plot is affected by a number of factors, viz., (1) the polymer itself; (2) the surfactant (structure and chain length); (3) the presence of salt and its concentration; and (4) the particular salt chosen. Specific ion effects, well known in salting out/in of polyether nonionic surfactants and polymers, turned out to be pronounced. In a qualitative way, it is proposed that the cloud point of the polymer is raised by association with an ionic surfactant because of electrical repulsion between the (now charged) polymer molecules. Evidently, this effect is very sensitive to, and can be offset by, electrical screening on adding salt.

V. SEPARATION AND PURIFICATION OF SOLUBLE POLYMERS

Early work by protein chemists, referred to in the chapters on proteins, had revealed that the properties of proteins in solution could be substantially changed by addition of ionic surfactants (see also the article by Dervichian[28]). In outline: (1) below their isoelectric point, proteins complex strongly with anionic surfactants, generally forming precipitates which can usually be solubilized on adding excess surfactant or salt. The analogous reactions above the isoelectric point take place with cationic surfactants; (2) even above (below) their isoelectric point, proteins will interact with anionic (cationic) surfactants, which interaction will alter their net charge and, hence, electrophoretic behavior.

These phenomena obviously provide means of separating proteins based on the addition of appropriate surfactants, at the appropriate pH, with control of electrolyte concentration, and so on, and will not be elaborated upon here (see, however, Chapter 8). Subsequent to the above work by protein chemists it was found that similar precipitation reactions occur between cationic surfactants and a variety of anionic polyelectrolytes, those investigated being generally anionic polysaccharides.[29-31] Scott,[30] for example, found for anionic polysaccharides with weak acidic groups (hyaluronic acid, alginic acid) that precipitation with CTAB could be inhibited as a result of deionization of their carboxyl groups. With carbohydrate monoester sulfates, such as dextran sulfate, no such inhibition was observed. He also reported that the "flocculation end-point" observed for this polyelectrolyte with CTAB corresponds to the formation of a stoichiometric salt.

The above dependence on pH of the precipitation of acidic polyelectrolytes with cationic surfactant clearly affords means of separating mixtures of such polymers based on their acid strength. In an interesting development, it was shown[30] that the method can be applied to the separation of neutral polysaccharides by working at sufficiently high pH to ionize their hydroxyl groups. A more convenient method, however, to develop polyanionic character and hence precipitability with added cationic surfactants, is to form the borate complex of the polysaccharide by simple addition of borate ion.[31]

Another fractionation method for polyanions came from the finding[32] that, like protein/oppositely charged surfactant complexes,[28] polyanion/cationic surfactant precipitates can be solubilized by addition of inorganic salt. Furthermore, solubility occurs at a critical electrolyte concentration (CEC) which was found to depend both on the surfactant itself and on the particular polyanion (its chemical structure, its anionic groups, and its molecular weight — with higher-molecular-weight analogs having higher CECs than lower analogs). This means, for example, that polyanions can be fractionated according to molecular weight, or if prefractionated, can have the molecular weight of their fractions determined from their CEC values and a calibration graph. By way of illustration, Laurent and Scott[32] applied this method to fractions of the polymers polyacrylate, chondroitin sulfate, and keratan sulfate.

A simple explanation of the CEC effect is based on electrical shielding by the added salt. A trend, supporting this explanation, is seen in the increase in critical adsorption concentration of dodecylpyridinium chloride with increase in added salt concentration, evident in the data of Malovikova et al.[33] for adsorption onto dextran sulfate. See Figure 7, Chapter 5. Several other examples of precipitating polycation/anionic surfactant systems, with an analysis of chain length effects, were given in this cited chapter. It is recalled for such systems that difficulties in resolubilization at high concentration of surfactant can be encountered if the charge density of the polycation is too high. Finally, we mention that Carlsson et al.[8] and Thalberg and co-workers[34-36] have recently published complete phase diagrams of polyanion/C_nTAB/water systems. The polyanions examined were hyaluronan[8,34,35] and polyacrylate.[36]

An interesting application of a polymer/surfactant approach for *purification* has been reported by Bennett,[37] who found that a natural discolorant in sugar was polyanionic in nature and could be removed by precipitation with a cationic surfactant (CTAB).

VI. REDUCTION OF MONOMER CONCENTRATION: MILDNESS OF SURFACTANTS

The fact that, in the presence of polymer, aggregates of surfactant can form at concentrations lower than the CMC, means that the maximum monomer concentration of surfactant is reduced. For nonionized polymer-ionic surfactant combinations the region involved would be in the T_1, T_2 concentration range (refer to Chapter 4, Part I). For polyelectrolyte/ionized surfactant pairs this effect would be maximal in the preprecipitation binding zone. Because of the strong bonding forces involved in this particular case, the reduction in monomer concentration would tend to be much higher than in the case of uncharged polymers.

Although the picture concerning the irritation to skin caused by exposure to surfactants, in particular anionic surfactants, is not completely clear, much evidence exists which suggests that lowered monomer concentration of the surfactant can correspond to lowered irritation.[38] Coupled with this, there is evidence[39,40] that addition of selected polymers to cosmetic formulations can reduce the irritation caused by the latter. The implication is that formulations of lowered irritation potential could be more reliably created if a knowledge of the binding characteristics of the particular polymer/surfactant combinations under consideration were established. On the other hand, a definite possibility always exists that any observed reduction of irritation occasioned by the presence of a polymer may involve more than one mechanism.[41]

VII. CONTROL OF DRUG DELIVERY

Alli and co-workers[42,43] have investigated the combination of hydroxypropylcellulose (HPC) and various anionic surfactants as a medium for release of selected drugs in tablet form. First of all, evidence of complex formation between the polymer and the surfactant SDS was obtained from solution viscosity measurements, and by differential scanning calorimetry (DSC) and mechanical extension tests on dry films: the DSC thermograms showed "new" peaks and the stress/strain curves showed extended elongation-to-break characteristics for the mixed species. Formation of the postulated complex was ascribed to a combination of ion-dipole attraction forces between the SDS and the high-molecular-weight HPC, reinforced by hydrophobic attraction among the bound SDS molecules.

The presence of SDS and other anionic surfactants was found, in most cases, to prolong the time of release of drug from a tablet containing HPC. The mechanism of prolongation is attributed to increased viscosity of a gel layer which forms on the surface of the tablet when it contacts an aqueous medium. The drugs tested were chloropheniramine maleate and sodium salicylate; the anionic surfactants, in addition to SDS, were sodium hexadecane sulfonate and sodium stearate, and the experiments were carried out at several pH values of the bathing aqueous solution.

The extensive current literature on novel drug delivery systems in fact reveals there is considerable interest in the use of gels as the delivery medium. To date, the classical gelling polymers have been the main types employed to provide the delayed diffusion characteristics associated with these media. Newer hydrophobe-modified polymers (see below) would also seem to have promise in this respect.

In the first section of this chapter, reference was made to the ability of a number of polymer/surfactant combinations to form water continuous gels. This field is now under active study and as scientific understanding increases, and gelling compositions are optimized, it is to be expected that "tailored" systems to meet specific drug delivery requirements will be in reach. An example, already given in Chapter 5, is of a composition which is liquid at room temperature but which gels at physiological temperature, affording interesting opportunities for new drug-dispensing forms.

VIII. SURFACE ACTIVITY; ADSORPTION; SURFACE CONDITIONING

An increase of surface tension of ionic surfactant solutions in the T_1, T_2 region on incorporation of a complexing nonionized polymer, coupled with direct evidence of binding of the surfactant by the polymer, clearly implies that the nonionized polymer reduces the surface activity of the surfactant by incorporating the latter in a weakly surface active complex. Recent direct measurements of the adsorption of radiolabeled SDS in the presence of PVP confirm this effect.[44] It is obvious that the properties of the mixed component interface will differ from those of the interface containing either surfactant or polymer alone, and in fact several references can be found in the literature

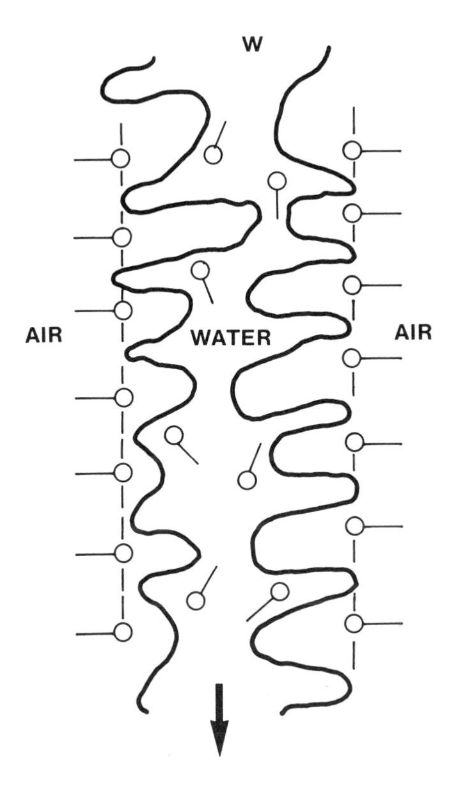

FIGURE 1. Schematic of a draining lamella in a foam prepared from a mixed solution of a polymer and a surfactant.

on the improvement of the foamability and foaming (or emulsifying) quality of ionic surfactants by the addition of nonionized polymers, such as polyvinylalcohol, PEO, modified starches, cellulosics, and so on. Figure 1 depicts a possible interaction structure in the draining foam lamella derived from a mixed polymer-surfactant system.

One drawback at the present time is that the combinations chosen in practice are largely empirical; in fact, in the author's opinion, a large fraction of optimized commercial formulations containing surfactants and polymers is still composed without any detailed knowledge of the interaction pattern of the components.

As regards solid surfaces, since both polymer and surfactant can adsorb on such surfaces, there has been much interest and considerable work done to determine what effect each has on the extent of adsorption of the other. Most of the work done has involved mineral (or latex) solids; see Chapters 4 and 5. Suffice it to say that positive and negative effects[45,46] have been found for both adsorbing components depending on conditions, i.e., the actual components themselves, addition sequence, the solid surface, the pH, and so on. As the adsorption energy per monomer unit of (especially) a nonionic polymer can be quite weak, it is not surprising that its adsorption overall can be affected by an added surfactant. The clear implication is that opportunities exist to modify the surface characteristics of chosen solids by appropriate choice of surfactants and polymers. Since the solid introduces a new phase (as in conditioning), the determination of altered, possibly improved, adsorption characteristics has in most cases to be done empirically, especially if other ingredients are present.

One recent example of systematic experimentation to establish a polymer/surfactant composition optimized for interfacial performance, comes from a study by Kilau and Voltz.[47] Their investigation was concerned with a solid surface conditioning problem involving the use of a complexing combination of a nonionic polymer and an anionic surfactant to achieve optimized wetting of the surface of coal. The polymers used were high-molecular-weight PEOs, namely POLYOX WSR-N-10 and POLYOX COAGULANT grades (Union Carbide) of molecular weight 0.1×10^6 and 5×10^6 Da, respectively. Synergistic wetting of hydrophobic coal surface was obtained with combinations of PEO and an anionic (sulfonated) surfactant, but not with nonionic surfactants. The fact that synergism was found with a hydrophobic coal, but not a hydrophilic coal, led the authors to postulate the model depicted in the figure below. Essentially, they argue that a "tails down-heads up" configuration of the adsorbed surfactant is required, and this can only be expected for a hydrophobic coal surface. Once adsorbed in this configuration, the surfactant molecules can readily interact with the polymer molecules, which can, in turn, stabilize the adsorbed array of surfactant molecules. In this way an interface is provided which is very hydrophilic and hence, easily wettable. Their depiction in Figure 2 is qualitatively similar to that in Figure 1 and to that of a polymer-stabilized micelle (Chapter 4, Part I, Figure 18), and the driving force for formation of the covered "hemimicelles" of this figure and of the regular polymer-wrapped micelle is evidently similar. Further evidence of adsorbed surfactant (SDS) on a solid (alumina) promoting the adsorption of a polymer (PEO) has been reported.[48] This work shows the polymer adsorbs in an extended form in the presence of the surfactant above its critical concentration for hemimicelle formation.

When the polymer is ionic and of opposite charge to that of the surfactant, strong synergism of interfacial activity can be encountered. Illustrations have been given in Chapter 4, Part II (Figures 1 to 3), and similar effects involving proteins and lipids were known to biochemists and membrane chemists at an early stage. (See References 28 and 49 and other articles in this compilation on lipoproteins). In general, the interfacial properties of polyion/oppositely charged surfactant systems can be expected to be a strong function of their mixing ratio, and it has been pointed out that in the Polymer JR/SDS system maximum activity, as measured by foaming (and also emulsification) exper-

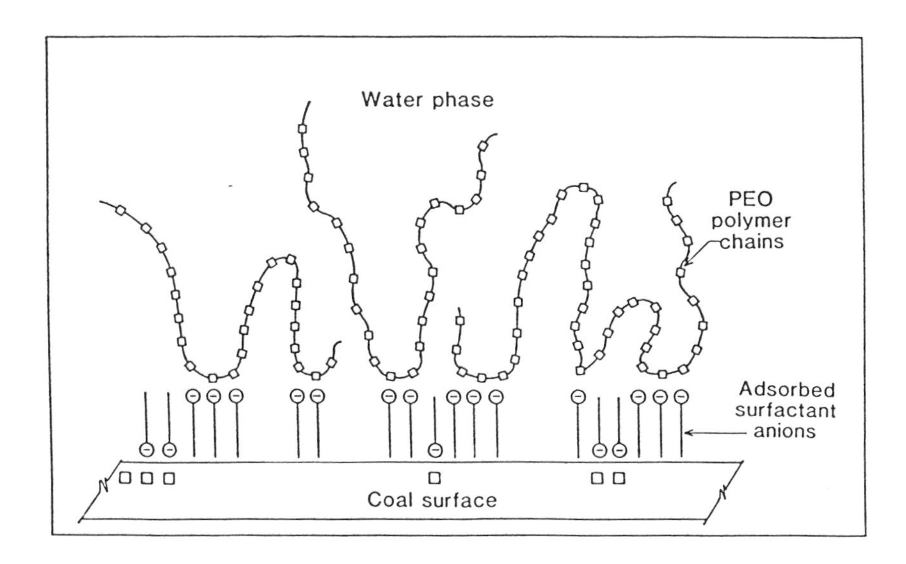

FIGURE 2. Synergistic wetting of coal by aqueous solutions of anionic surfactant and polyethylene oxide polymer. (From Kilau, H. W. and Voltz, J. I., *Colloids Surf.*, 57, 17, 1991. With permission.)

iments, is found near the stoichiometric 1:1 association ratio. Other examples of foam improvement by the employment of complexes have been reported.[50]

Combinations of proteins and/or natural gums and charged lipids are found in many commercial food products, where they contribute to the foaming, emulsifying, and texture of the aqueous composition. As stated previously, many, if not most, of these still appear to be assembled on an empirical basis.

As regards the use of polyion/ionic surfactant combination for the conditioning of solid surfaces, familiar illustrations come from the cosmetics industry in typical cremes, lotions, and shampoos. Since the keratin of skin and hair has a low isoelectric point and is therefore negatively charged, it is likely that an important step in the conditioning process in this case is the adsorption on keratin of the conditioning polycation and concomitant binding of anionic surfactant to this polyion. An effect similar to this was reported by Somasundaran and Lee,[45] who found a strong synergistic effect in the flotation of quartz by a mixture of a cationic polyacrylamide and sodium dodecanesulfonate (NaDDS). The "collector", NaDDS, by itself has no collection efficiency for the negatively charged quartz. Addition of as little as 10 ppm of the polymer to the solution allows close to 100% flotation efficiency. Presumably, by adsorbing, it acts as a collection "primer" and then provides positively charged binding sites for the surfactant. Other examples of polymer/surfactant combinations for use in flotation are provided in this article,[45] and we recall similar references in Chapter 4 of this book.

IX. POLYMERIC SURFACTANTS: ASSOCIATIVE THICKENING

As is evident in Chapter 6, there is at present a great renewal of interest in the physical chemistry of ''polymeric surfactants'' and there is also growing commercial recognition of their importance as so-called ''associative thickeners'' for use in latex and other formulated product systems. These materials are in effect conventional water-soluble polymers which have been modified by inclusion of hydrophobic moieties, in particular, alkyl groups.[51,52] They combine the properties of a surfactant/polymer mixture in one molecule and therefore display some of the properties of such mixtures. Thus, when dissolved in water they tend to self-associate by hydrophobic interaction, generating structures of high molecular weight once the ''polymer overlap concentration'', C*, is exceeded; hence, substantial increases in viscosity of their solutions result. Since the existence of apparently high-molecular-weight polymers involves physical association of the macromolecules, their solutions show substantial shear thinning — a factor of considerable importance, for example, in latex paints. Viscosity studies on long chain alkyl-substituted HEC (''HM-HEC'') solutions have been reported by Gelman and Barth,[53] Goodwin et al.,[54] and Landoll, who presents detailed information on the effect of polymer structure, length of the alkyl substituent, and so on.[55] Articles by McCormick et al.,[56] Schulz et al.,[57] and Peiffer[58] are examples of recent work on acrylamide-based polymeric surfactants; several papers on associative thickeners may be found in Reference 51.

Other consequences of the associative tendency of molecules with this type of structure are the development of solubilizing properties for water-insoluble materials, including dyes, and also a strong tendency to produce foams, especially foams of unusual stability: Goddard and Braun in fact showed that a hydrophobically modified cationic cellulosic polymer, QUATRISOFT LM 200 (Union Carbide), itself could form the basis of an aerosol mousse.[59]

It is recognized that polysoaps have great potential utility as emulsion and suspension stabilizers and also as surface conditioners. This arises from their combined functionality as surfactant and polymer. An example is the use of the copolymer of an alpha-olefin and maleic anhydride in cleaning/polishing compositions.[60] Reference 61 reports a fundamental study of the association characteristics of this type of polymer.

In very recent work, Li et al.[62] have shown that a similar copolymer (DAPRAL GE202 -AKZO), in which one of the carboxyl groups of maleic anhydride has been esterified with a polyglycol chain of 7 to 8 ethylene oxide units, and where the α-olefin comonomer introduces decyl pendent chains, shows interesting behavior as a flotation additive for coal fines. In general, its adsorption and flotation activity are greater at lower pH (3.5) at which electrical repulsion effects towards the coal surface are reduced. A most interesting observation is that the alkyl (decyl group) segments of adsorbed polymer molecules, which are "hidden" when the coal fines are in aqueous suspension, seem to be available for adsorption at the gas bubble interface to aid levitation. This bespeaks a considerable level of molecular flexibility of the adsorbed polymer. A loss of hydrophobicity of treated coal, which was observed if the concentration of the polymer was too high, is ascribed to the formation of hydrophilic micellar aggregates on the coal surface. These comprise "dangling" hydrophobic segments from adsorbed polymers and corresponding segments from polymer molecules still in solution which, together, form the core of the "surface" micelles.

Bassett[63] has prepared a qualitative depiction of possible structures involving association between hydrophobic polymeric species having two different molecular architectures (see Figure 3[63]). In one, the hydrophobic groups are terminal to the polymer chain; in the other, they are present along the chain in a "comb"-type configuration. As can be imagined, interesting effects on the properties, including the rheology, of solutions of polymeric surfactants can be expected on the addition of a conventional surfactant to their solution. Competitive "homo" and "hetero" association effects involving the two types of hydrophobic groups can occur, and evidence frequently points to maximum interaction when micelles of the simple surfactant are present (see Figure 4[63]). It is intuitively obvious that pronounced effects of concentration and mixing ratio of the two components will be encountered. When few surfactant micelles are present, they are likely to be "shared" by the hydrophobic groups of different polymer molecules and entities of effectively high molecular weight will result. This will confer a high viscosity of the solution. When a larger number of micelles are present, a single polymer molecule is more likely to wrap around individual micelles and the viscosity will tend to be much lower.

On defining "polymeric surfactants" as water-soluble polymers with alkyl group substituents of chain length $>C_{10}$, one actually finds relatively few published papers on the subject of mixtures of these polymers with conventional surfactants. We will refer briefly to a few such studies, starting with uncharged hydrophobically modified polymers. Sau and Landoll[64] have reported substantial viscosity-boosting effects on adding a nonionic surfactant to a dilute solution of hydrophobically modified hydroxyethylcellulose (HM-HEC). Furthermore, viscosity peaks were found by these authors on adding anionic surfactant to HM-HEC, and confirmed by Lindman and co-workers,[8] and also by Dualeh and Steiner,[65] who reported gel formation under certain conditions with added SDS.[66] Variable rheological effects for such systems are evident in the work of Sivadasan and Somasundaran,[67] who also report pronounced interaction between HM-HEC and the nonionic surfactant $C_{12}EO_8$ as seen by fluorescence techniques. One should note that the traditionally weak interaction between EO-based nonionic surfactants and polymers can be changed to a substantial interactive tendency when the polymer is a polymeric surfactant with long chain alkyl "anchoring" groups. Recently, Jen-

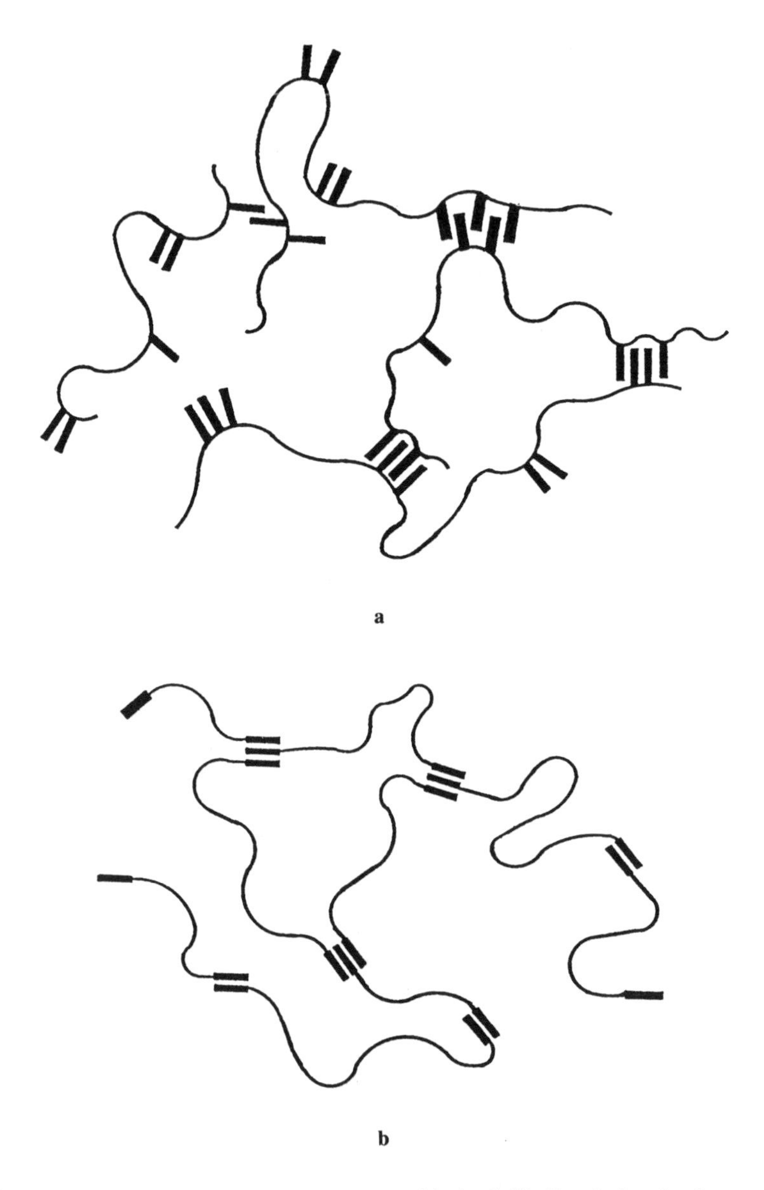

FIGURE 3. Depiction of association structures of hydrophobically substituted polymers: (a) end-substituted; (b) "comb".[63]

kins et al.[68] have shown that both PEO and polyacrylate-type water-soluble polymers, modified with undisclosed, "novel" hydrophobic groups, show especially strong interaction with certain nonionic surfactants, as revealed by pronounced increases in viscosity of the mixed solutions. The interaction is very sensitive to the number of EO groups in the surfactant, and the novel

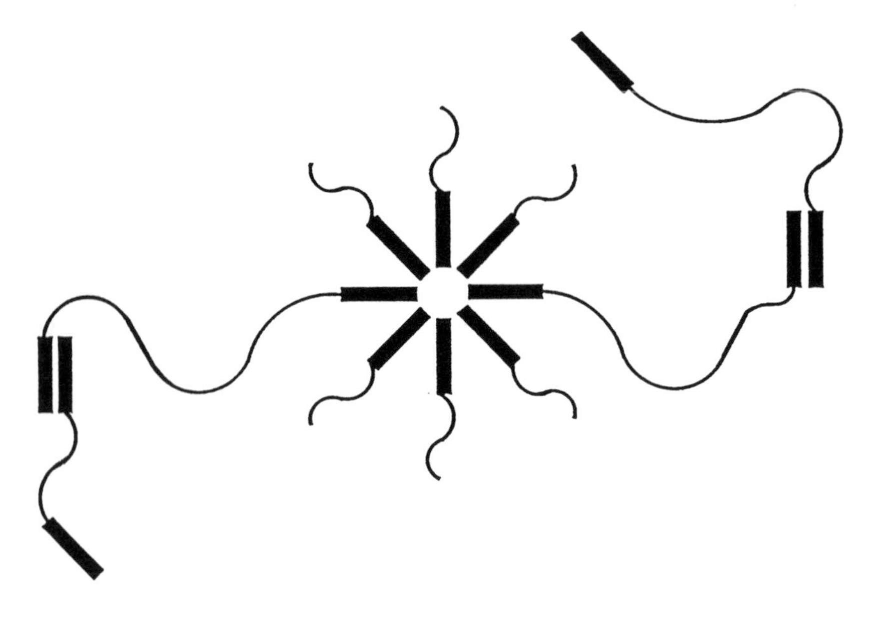

FIGURE 4. Depiction of an end-substituted polymer associating via a surfactant micellar bridge.[63]

hydrophobic groups lead to much more pronounced interaction effects than do corresponding hexadecyl or nonylphenyl groups.

Turning to oppositely charged pairs, with a hydrophobically(C_{12}) substituted cationic cellulosic polymer, Ananthapadmanabhan et al.[69] showed by pyrene fluorescence methods that association of SDS and polymer occurs at a much lower concentration than it does with a conventional cationic cellulosic, Polymer JR, that has a much higher degree of cationic substitution. This shows again the important effect on the association of the presence of hydrophobic groups in the polymer.

Another interesting development has been the demonstration[70] that the above combination can provide gelling compositions in the postprecipitation, as well as in the preprecipitation, surfactant concentration range. This signifies that the basic molecular structure of the gels attainable in the two concentration ranges is very different. Gels in the postprecipitation range are attainable with a variety of anionic surfactants, so providing a wide latitude of formulation opportunities.

A question of some importance, which has recently been answered, is whether or not a hydrophobically substituted charged polymer will interact with a surfactant of *like* charge: for example, will a polysoap interact with an anionic surfactant? McGlade et al.,[71] working with poly(1-decene-co-maleic acid) and poly (1-octadecene-co-maleic acid) polyelectrolytes, found clear evidence by fluorescence and surface tension measurements of interaction with SDS. In other words, interaction by hydrophobic association is

able to offset the effect of the adverse electrical field gradient between the two species. A second example comes from the study of Iliopoulos et al.[72] on hydrophobically modified poly(sodium acrylate). At certain ratios of added SDS (near its CMC) pronounced maxima, reflecting increments of three decades or more in viscosity, were encountered, again providing clear evidence of association.

A third example[70] involves the above-mentioned cationic polymer in mixture with cationic surfactants. In a homologous series of alkyltrimethylammonium bromides, precipitation reactions ("salting out") were encountered with the two lower-chain-length homologs (C_{10} and C_{12}) but not with the higher homologs (C_{14} and C_{16}). Solubilization encountered at high (> CMC) concentration in the former case reveals the anticipated interaction between the polymer and the surfactants (in micellar form in this case).

From the foregoing, one can confidently say that the potential of polymeric surfactants is considerable not only as thickeners but as specialty surfactants, specialty polymers, and surface conditioners as well. In light of the current level of research on this category of material, it is anticipated that many opportunities and applications will be forthcoming as new structures are developed and uses defined.

X. MISCELLANEOUS

A. POLYMER ACTIVATION BY DEIONIZATION

There are many illustrations in this book of the interaction of charged surfactants with oppositely charged polyelectrolytes, and there are several examples in this chapter of applications which depend on this combination. An implication is that, if the polyelectrolyte is of limited acid (or base) strength, the pH conditions have to be appropriate to achieve its ionization and develop a charge opposite to that of the surfactant in order to promote interaction. By the same token, no interaction is anticipated if ionization of the polymers leads to the formation of a polyelectrolyte with charges of the same sign as that of the surfactant[72,73,74] (unless the polyelectrolyte is hydrophobically modified — see above). An interesting finding has, however, been made recently of a polyelectrolyte's being activated for interaction by its *deionization*. Thus, Maltesh and Somasundaran[75] have developed evidence using fluorescence techniques that, when polyacrylic acid is ionized, it does not interact with SDS, but under acidic conditions it does. Two potential applications of this effect are the elevation of the viscosity of solutions of the unionized form of this polymer by addition of SDS,[76] and the enhancement of the flotation activity of SDS for hematite fines on addition of polyacrylic acid.[77] Opportunities thus exist for "on-off switching" of effects such as the above by simple alteration of the pH of the solutions, and the effect may involve not only the generation of the ionized form of the polymer but, indeed, sometimes its deionization. It has been found recently[78] that unionized polyacrylic acid also shows marked interaction with the cationic surfactant te-

tradecyltrimethylammonium bromide. This means that on the Breuer-Robb reactivity scale (see Chapter 4, Part I) it must be regarded as being quite hydrophobic.

B. CHEMOMECHANICS

A prototype polymer/surfactant combination which can generate mechanical motion has been described recently by Osada and co-workers.[79] The effect is based on the finding that a polyelectrolyte gel will shrink if exposed to an oppositely charged surfactant, owing to a reduction of osmotic pressure within the gel. The effect can be augmented if binding of the surfactant is assisted by application of an electrical field. In brief, in the prototype developed by Osada et al., a flat strip of hydrogel, made of weakly crosslinked poly(2-acrylamide-2-methyl propane) sulfonic acid is placed between two electrodes in an aqueous bath containing cetylpyridinium chloride. On applying a DC field surfactant molecules are bound preferentially onto one side of the gel, causing the strip to bow. The binding is reversible and when the field is removed the strain is relieved. This ''looping'' action — a type of motility — can be translated into sustained movement by turning the current on and off. The device is likened to a mechanical actuator or ''artificial muscle''. Related activity in this field is described in Reference 80.

REFERENCES

1. **Jones, M. N.,** The interaction of sodium dodecyl sulfate with polyethylene oxide, *J. Colloid Interface Sci.*, 23, 36, 1967.
2. **Francois, J., Dayantis, J., and Sabbadin, J.,** Hydrodynamical behavior of the poly(ethylene oxide)-sodium dodecylsulfate complex, *Eur. Polym. J.*, 21, 165, 1985.
3. **Brackman, J. C.,** Sodium dodecyl sulfate induced enhancement of the viscosity and viscoelasticity of aqueous solutions of poly(ethylene oxide). A rheological study on polymer-micelle interaction, *Langmuir,* 7, 469, 1991.
4. **Leung, P. S., Goddard, E. D., Han, C., and Glinka, C. J.,** A study of polycation-anionic surfactant systems, *Colloids Surf.*, 13, 47, 1985.
5. **Goddard, E. D. and Leung, P. S.,** Gels from dilute polymer/surfactant solutions, *Langmuir,* 7, 608, 1991.
6. **Goddard, E. D., Padmanabhan, K. P. A., and Leung, P. S.,** Novel gelling structures based on polymer/surfactant systems, *J. Soc. Cosmet. Chem.*, 42, 19, 1991.
7. **Thalberg, K. and Lindman, B.,** Thermal gelation of nonionic cellulose ethers and ionic surfactants in water, *Langmuir,* 7, 277, 1991.
8. **Carlsson, A., Karlstrom, G., and Lindman, B.,** Thermal gelation of nonionic cellulose ethers and ionic surfactants in water, *Colloid Surf.*, 47, 147, 1990.
9. **Brackman, J. C. and Engberts, J. B. F. N.,** Polymer-induced breakdown of rod like micelles. A striking transition of a non-Newtonian to a Newtonian fluid, *J. Am. Chem. Soc.*, 112, 872, 1990.
10. **Lange, H.,** Wechselwirkung zwischen Natriumalkylsulfaten und Polyvinylpyrolidon in wassrigen Losungen, *Kolloid Z. Z. Polym.*, 243, 101, 1971.

11. **Saito, S.,** Die Untersuching der Adsorptionskomplexe von Polymeren mit Netzmittionen, *Kolloid Z.*, 154, 19, 1957.
12. **Jones, M. N.,** Dye solubilization by a polymer-surfactant complex, *J. Colloid Interface Sci.*, 26, 532, 1968.
13. **Saito, S.,** Solubilization properties of polymer-surfactant complexes, *J. Colloid Interface Sci.*, 24, 227, 1967.
14. **Turro, N. J., Baretz, B. H., and Kuo, P. L.,** Photoluminescence probes for the investigation of interactions between sodium dodecylsulfate and water soluble polymers, *Macromolecules*, 17, 321, 1984.
15. **Lee, B.-H., Christian, S. D., Tucker, E. E., and Scamehorn, J. F.,** Effects of an anionic polyelectrolyte on the solubilization of mono- and dichlorophenols by aqueous solutions of N-hexadecylpyridinium chloride, *Langmuir*, 7, 1332, 1991.
16. **Isemura, T. and Imanishi, A.,** The dissolution of water-insoluble polymers in the surfactant solution. The polyelectrolyte-like behavior of the dissolved polymers, *J. Polym. Sci.*, 33, 337, 1958.
17. **Sata, N. and Saito, S.,** Solubilization of polyvinyl acetate, *Kolloid Z.*, 128, 154, 1952.
18. **Saito, S. and Mizuta, Y.,** Solubilization of polymers in aqueous cationic surfactant solution, *J. Colloid Interface Sci.*, 23, 604, 1967.
19. **Lundgren, H. P.,** Dispersion of proteins in aqueous detergents, U.S. Patent 2, 459, 708, 1949.
20. **Schwuger, M. J. and Lange, H.,** Interaction between surfactants and poly-(vinyl pyrrolidone) in aqueous solution, *Proc. 5th Int. Congr. Deterg.*, Vol. 2, Barcelona (1968), Ediciones Unidas S. A., Barcelona, 1969, 955.
21. **Goddard, E. D. and Hannan, R. B.,** Polymer/surfactant interactions, *J. Am. Oil Chem. Soc.*, 54, 561, 1977.
22. **Dubin, P. L. and Davis, D.,** Stoichiometry and coacervation of complexes formed between polyelectrolytes and micelles, *Colloids Surf.*, 13, 113, 1985.
23. **Dubin, P. L., Vea, M. E. Y., Fallon, M. A., The, S. S., Rigsbee, D. R., and Gan, L. M.,** Higher order association in poly-micelle complexes, *Langmuir*, 6, 1422, 1990.
24. **Karlstrom, G., Carlsson, A., and Lindman, B.,** Phase diagrams of nonionic polymer-water systems. Experimental and theoretical studies of the effects of surfactants and other cosolutes, *J. Phys. Chem.*, 95, 5005, 1990.
25. **Carlsson, A., Karlstrom, G., and Lindman, B.,** Synergistic surfactant-electrolyte effect in polymer solutions, *Langmuir*, 2, 536, 1986.
26. **Carlsson, A., Karlstrom, G., Lindman, B., and Stenberg, O.,** Interaction between (ethylhydroxyethyl)cellulose and sodium dodecyl sulfate in aqueous solution, *Colloid Polym. Sci.*, 226, 1031, 1988.
27. **Xie, F., Ma, C., and Gu, T.,** The effect of electrolytes on the cloud point of mixed solutions of polypropylene glycol and ionic surfactants, *Colloids Surf.*, 36, 39, 1989.
28. **Dervichian, D. G.,** Structural aspects of lipo-protein association, in *Lipo-Proteins, Discuss. Faraday Soc.*, No. 6, 1949, 7.
29. **Jones, A. S.,** Isolation of bacterial nucleic acids using cetyltrimethylammonium bromide, *Biochim. Biophys. Acta*, 10, 607, 1953.
30. **Scott, J. E.,** The reaction of long-chain quaternary ammonium salts with acidic polysaccharides, *Chem. Ind.*, 168, 1955.
31. **Palmstierna, H., Scott, J. E., and Gardell, S.,** The precipitation of neutral polysaccharides by cationic detergents, *Acta Chem. Scand.*, 11, 1792, 1957.
32. **Laurent, T. C. and Scott, J. E.,** Molecular weight fractionation of polyanions by cetylpyridinium chloride in salt solutions, *Nature*, 202, 661, 1964.
33. **Malovikova, A., Hayakawa, K., and Kwak, J. C. T.,** Surfactant-polyelectrolyte interactions. IV. Surfactant chain length dependence of the binding of alkylpyridinium cations to dextran sulfate, *J. Phys. Chem.*, 88, 1930, 1984.
34. **Thalberg, K. and Lindman, B.,** Interaction between hyaluronan and cationic surfactants, *J. Phys. Chem.*, 93, 1478, 1989.

35. **Thalberg, K., Lindman, B., and Karlstrom, G.,** Phase behavior of systems of cationic surfactant and anionic polyelectrolyte: influence of surfactant chain length and polyelectrolyte molecular weight, *J. Phys. Chem.,* 95, 3370, 1991.
36. **Thalberg, K., Lindman, B., and Bergfeldt, K.,** Phase behavior of systems of polyacrylate and cationic surfactants, *Langmuir,* 7, 2893, 1991.
37. **Bennett, M. C.,** A new industrial process for decolorising sugar, *Chem. Ind.,* 886, 1974.
38. **Rhein, L. D., Robbins, C. R., Fernee, K., and Cantore, R.,** Surfactant structure effects on swelling of isolated stratum corneum, *J. Soc. Cosmet. Chem.,* 37, 125, 1986.
39. **Prescott, F. J., Hahnel, E., and Day, D.,** Cosmetic PVP. I, *Drug and Cosmet. Ind.,* 93, 443, 1963.
40. **Ward, J. B. and Sperandio, G. J.,** Cosmetic applications of polyvinyl alcohol, *J. Soc. Cosmet. Chem.,* 15, 32, 1964.
41. **Faucher, J. A., Goddard, E. D., Hannan, R. B., and Kligman, A. M.,** Protection of the skin by a cationic cellulosic polymer, *Cosmet. Toiletries,* 92, 39, 1977.
42. **Alli, D.,** Interactions between Non-Ionic Cellulose Ethers and Anionic Surfactants and Their Influence on Drug Release, Ph.D. thesis, St. Johns University, New York, 1990.
43. **Alli, D., Bolton, S., and Gaylord, N. S.,** Hydroxypropylmethylcellulose-anionic surfactant interactions in aqueous systems, *J. Appl. Polym. Sci.,* 42, 947, 1991.
44. **Chari, K. and Hossain, T. Z.,** Adsorption at the air/water interface from an aqueous solution of poly(vinylpyrrolidones) and sodium dodecyl sulfate, *J. Phys. Chem.,* 95, 3302, 1991.
45. **Somasundaran, P. and Lee, L. T.,** Polymer-surfactant interaction in the flotation of quartz, *Sep. Sci. Technol.,* 16, 1475, 1981.
46. **Gebhardt, J. E. and Fuerstenau, D. W.,** The effect of preadsorbed polymers on adsorption of sodium dodecylsulfonate on hematite, in *Structure/Performance Relationships in Surfactants,* Rosen, M. J., Ed., Symposium Series, No. 253, American Chemical Society, Washington, D.C., 1984, 291.
47. **Kilau, H. W. and Voltz, J. I.,** Synergistic wetting of coal by aqueous solutions of anionic surfactant and polyethylene oxide polymer, *Colloids Surf.,* 57, 17, 1991.
48. **Ramachandran, R. and Somasundaran, P.,** Polymer-surfactant interaction, in *Flocculation Dewatering, Proc. Eng. Found. Conf.,* Moudgil, B. M. and Scheiner, B. J., Eds., Engineering Foundation, New York, 1988, 631.
49. **Matalon, R. and Schulman, J. H.,** Formation of lipo-protein monolayers, Part I and II, in *Lipo-Proteins, Discuss. Faraday Soc.,* No. 6, 1949, 21 and 27.
50. **Clark, K. P. and Falk, R. A.,** Polysaccharide/perfluoroalkyl complexes, Eur. Patent Appl. E. P. 311, 570, 12 Apr. 1989; U.S. Patent Appl. 107,434, 9 Oct. 1987.
51. **Glass, J. E.,** Ed., *Polymers in Aqueous Media,* Advances in Chemistry Ser. No. 223, American Chemical Society, Washington, D.C., 1989.
52. **Glancy, C. W. and Bassett, D. R.,** Effect of latex properties on the behavior of nonionic associative thickeners in paint, *Polym. Mater. Sci. Eng. Proc.,* 51, 348, 1984.
53. **Gelman, R. A. and Barth, H. G.,** Viscosity studies of hydrophobically modified hydroxy ethyl cellulose, in *Water Soluble Polymers,* Glass, J. E., Ed., Advances in Chemistry Ser. No. 213, American Chemical Society, Washington, D.C., 1986, 101.
54. **Goodwin, J. W., Hughes, R. W., Lam, C. K., Miles, J. A., and Warren, B. C. H.,** Viscosity studies of hydrophobically modified hydroxyethyl cellulose, in *Polymers in Aqueous Medica,* Advances in Chem. Ser. No. 223, Glass, J. E., Ed., American Chemical Society, Washington, D.C., 1989, 365.
55. **Landoll, L. M.,** Nonionic polymer surfactants, *J. Polym. Sci. Polym. Chem. Ed.,* 20, 443, 1982.
56. **McCormick, C. L., Nonaka, T., and Johnson, C. B.,** Water-soluble copolymers: 27. Synthesis and aqueous solution behavior of associative acrylamide/N-alkylacrylamide copolymers, *Polymer,* 29, 731, 1988.
57. **Schulz, D. N., Kaladas, J. J., Maurer, J. J., Bock, J., Pace, S. J., and Schulz, W. W.,** Copolymers of acrylamide and surfactant macromolecules: synthesis and solution properties, *Polymer,* 28, 210, 1987.

58. **Peiffer, D. G.,** Hydrophobically associating polymers and their interactions with rod-like micelles, *Polymer,* 31, 2353, 1990.
59. **Goddard, E. D. and Braun, D. B.,** A new surface active cationic cellulosic polymer, *Cosmet. Toiletries,* 100, 41, 1985.
60. **Sandvick, P. E.,** Cleaning and polishing compositions, U.S. Patent 4, 613, 646, 1986.
61. **Shih, L. B., Mauer, D. H. Verbrugge, C. J., Wu, C. F., Chang, S. L., and Chen, S. H.,** Small angle neutron scattering study of micellization of ionic copolymers in aqueous solutions — the effect of side-chain length and molecular weight, *Macromolecules,* 21, 3235, 1988.
62. **Li, C., Yu, X., and Somasundaran, P.,** Effect of a hydrophobically modified comb-like polymer on interfacial properties of coal, *Colloids Surf.,* 66, 39, 1992.
63. **Bassett, D. R.,** private communication, 1989.
64. **Sau, A. C. and Landoll, L. M.,** Synthesis and solution properties of hydrophobically modified (hydroxyethyl)cellulose, in *Polymers in Aqueous Media,* Advances in Chem. Ser. No. 223, Glass, J. E., Ed., American Chemical Society, Washington, D.C., 1989, 343.
65. **Dualeh, A. J. and Steiner, C. A.,** Hydrophobic microphase formation in surfactant solutions containing an amphiphilic graft copolymer, *Macromolecules,* 23, 251, 1990.
66. **Dualeh, A. J. and Steiner, C. A.,** Bulk and microscopic properties of surfactant-bridged hydrogels made from an amphiphilic graft copolymer, *Macromolecules,* 24, 112, 1991.
67. **Sivadasan, K. and Somasundaran, P.,** Polymer-surfactant interactions and the association behavior of hydrophobically modified hydroxyethylcellulose, *Colloids Surf.,* 49, 229, 1990.
68. **Jenkins, R. D., Sinha, B. R., and Bassett, D. R.,** Associative polymers with novel hydrophobe structures, *Polym. Mater. Sci. Eng.,* 65, 72, 1991.
69. **Ananthapadmanaban, K. P., Leung, P. S., and Goddard, E. D.,** Colloidal properties of surface-active cellulosic polymer, in *Polymer Association Structures,* El-Nokaly, M. A., Ed., AChS Symposium Ser. No. 384, American Chemical Society, Washington, D.C., 1989, 297.
70. **Goddard, E. D. and Leung, P. S.,** Studies of gel formation, phase behavior and surface tension in mixtures of a hydrophobically modified cationic cellulose polymer and surfactant, *Colloids Surf.,* 65, 211, 1992.
71. **McGlade, M. J., Randall, F. J., and Tcheurekdjian, N.,** Fluorescence probe studies of interaction between sodium dodecyl sulfate and anionic polyelectrolytes, *Macromolecules,* 20, 1782, 1987.
72. **Iliopoulos, I., Wang, T. K., and Audubert, R.,** Viscometric evidence of interactions between hydrophobically modified poly(sodium acrylate) and sodium dodecyl sulfate, *Langmuir,* 7, 617, 1991.
73. **Schwuger, M. J. and Lange, H.,** Über die Wechselwirkung zwischen Natriumcarboxymethylcellulose und Tensiden, *Tenside,* 5, 257, 1968.
74. **Binana-Limbele, W. and Zana, R.,** Interactions between sodium dodecyl sulfate and polycarboxylates and polyethers. Effect on Ca^{2+} on these interactions, *Colloids Surf.,* 21, 483, 1986.
75. **Maltesh, C. and Somasundaran, P.,** Evidence of complexation between poly(acrylic acid) and sodium dodecyl sulfate, *Colloids Surf.,* 69, 167, 1992.
76. **Eliassaf, J.,** Interaction of sodium dodecylsulfate with vinyl polymers in aqueous solutions, *J. Appl. Polym. Sci.,* 7, S3, 1963.
77. **Gebhardt, J. E. an Fuerstenau, D. W.,** Flotation behavior of hematite fines flocculated with polyacrylic acid, *Min. Met. Proc.,* 3, 164, 1986.
78. **Kiefer, J., Somasundaran, P. and Ananthapadmanabhan, K. P.,** Interaction of tetradecyltrimethylammonium bromide with polyacrylic acid and polymethacrylic acid, Effect of charge density, submitted to *Langmuir,* 1992.
79. **Osada, Y., Okuzaki, H., and Hori, H.,** A polymer gel with electrically driven motility, *Nature,* 355, 242, 1992.
80. **Kajiwara, K. and Ross-Murphy, S. B.,** Synthetic gels on the move, *Nature,* 355, 208, 1992.

INDEX

A

B

C

D

E

F

G

H

I

N

O

P

Q

T

U

V

W

X

Y

Z